MW01106209

Advances in Behavior Genetics

Series Editor
Yong-Kyu Kim

More information about this series at http://www.springer.com/series/10458

Jonathan C. Gewirtz • Yong-Kyu Kim
Editors

Animal Models of Behavior Genetics

Editors
Jonathan C. Gewirtz
Department of Psychology
University of Minnesota
Minneapolis, MN, USA

Yong-Kyu Kim
Janelia Research Campus
Howard Hughes Medical Institue
Ashburn, VA, USA

Advances in Behavior Genetics
ISBN 978-1-4939-3775-2 ISBN 978-1-4939-3777-6 (eBook)
DOI 10.1007/978-1-4939-3777-6

Library of Congress Control Number: 2016938080

Printed on acid-free paper

This Springer imprint is published by Springer Nature
The registered company is Springer Science+Business Media LLC New York

This book is dedicated to the memory of Irving Gottesman, who epitomized the endophenotypes of kindness and wisdom.

Preface

When the editors of this book were in graduate school in the early 1990s studying the biological basis of behavior in rats (JCG) and *Drosophila* (YKK), the availability of new approaches in molecular biology and genetics brought about a near-exponential growth in our capacity to explore the nervous system of invertebrate and vertebrate species. And it was around that time that a new form of an old word entered the lexicon of behavioral neuroscience: the noun, "phenotype," had transformed itself into a verb. Whereas "genotyping" means to identify the sequence of nucleotides in DNA, "phenotyping" came to refer to the task of identifying the structural, physiological, and behavioral consequences of genotypic variations. To these ends, new facilities were built with the express purpose of characterizing the behavior of genetically altered mice. These developments reflected recognition among geneticists and neurobiologists that behavior was an important endpoint of genotypic variation and even among some scientists that characterizing behavioral differences was a *sine qua non* for establishing the functional significance of a given genotype.

The expectation of many in the field was that this part of the scientific enterprise would be relatively straightforward. "Give the animal to the behavioral experts to run through their behavioral battery," ran the thinking, "and we will have our phenotype." In reality, however, discovery of the behavioral corollaries of genotypic variations frequently trailed the rapid pace of discovery of intracellular or morphological intermediaries within the CNS.

At least two problems stood in the way of the success of this enterprise. The first was that the process of discovery was all too often unidirectional, with genetics as the driver and behavior as a destination. An approach in which genetic, molecular, physiological, anatomical, and behavioral analyses are all mutually informative leads us in directions that are more interesting and better suited to the ultimate goal of understanding the bidirectional chains of causality between genes and behavior.

A second, not unrelated, obstacle was the way in which behavioral assays were seen as little more than diagnostic devices. "Phenotyping" was frequently reduced to a pattern of gains or losses of function among a set of standardized tests. The limitations of a "checklist" approach are obvious to anyone who has dedicated much time to understanding a single behavioral domain, such as those covered in this book. There is, for a start, a tendency to label behavioral profiles derived from these checklists with terminology taken straight from the realm of clinical psychopathology. This is despite the fact that the nosology of psychiatric disorders is plagued by issues, such as comorbidity among, and heterogeneity within, diagnostic categories (Helzer et al., 2009). If there can be so little agreement among clinicians as to how to diagnose psychiatric illnesses, what hope is there for the success of a similar approach in animal models?

The checklist approach also encourages the tendency to overdiagnose. The writer Jerome K. Jerome (1889) concluded after running through each letter of a medical encyclopedia that he suffered from every ailment listed except housemaid's knee. Now enter the behavioral experimenter, who runs each animal through an array of tests, each bearing an embarrassment of potential outcome measures. It is all too tempting under such circumstances for one to become a hypochondriac by proxy, taking a significant aberration in a single test as confirming a hypothesis that an animal of a given genotype suffers from an "anxiety-like" or some-other-psychiatric-disorder-like phenotype.

How can such pitfalls be avoided and the field be advanced? The purpose of this volume is to aid in answering these two questions. In common to all the chapters in this book is a focus on establishing reliable behavioral measures in animal models for identifying the functional consequences of genotypic variations. Some of the chapters highlight issues that are pertinent to the development of animal models in behavioral genetics in general. These include consideration of possible pleiotropic effects of genes on multiple behavioral domains, the sometimes-insidious influence of genetic background on mouse behavior, and the significance of the environment in which the animals are routinely housed while they await behavioral testing. Other chapters evaluate specific behavioral tests and how their results should be interpreted in light of differences observed in animals with different genetic characteristics. Finally, a number of the chapters apply this knowledge to summarize our best, current understanding of the genetics of specific domains of mentation and mental illness.

A compelling case has been made by our colleagues in human behavioral genetics that core variables undergirding mental illnesses should be viewed as falling on continua from normal and adaptive to abnormal and maladaptive. Of course, much of the usefulness of this approach depends on adequately characterizing the most important continua along which these diagnoses fall (see, e.g., Chap. 8 on the "endophenotype" concept). With this in mind, it may seem contradictory that we have divided the book into sections focusing on animal models of adaptive or "normal" behavior and maladaptive or "abnormal" behavior. This division reflects the fact that, in determining genetic influences on behavior, some models are relatively more focused on understanding normative behavior. One example of this approach is to study variations in naturally occurring behaviors across the genotypic spectrum of healthy, outbred rats (see Chap. 3). Other models are focused on phenomena at the dysfunctional ends of the behavioral spectrum, such as compulsive drug-taking behavior. Since the overarching goal of all of us is to understand the effects of genes on a given psychological or behavioral construct along the entire continuum, this division is recognized to be somewhat arbitrary. Nevertheless, we think it is a helpful organizational principle when setting out to describe, analyze, and summarize the field of animal behavioral genetics.

Part I of this volume covers animal models of behavioral genetics across a range of domains. Chapters 1 and 2 review our understanding of the genetics of the two closely associated domains of mating and aggressive behavior in *Drosophila*. It is appropriate that the book should begin with a look at fly behavior since this has been a formative and—for want of a better word—fruitful model species in the field of behavioral genetics for several decades. Above all, these chapters demonstrate the advantage of invertebrate models in being able to locate causal mechanisms—whereby genes affect neural substrates that affect behavior—with an extraordinary degree of specificity. Chapter 3 reviews the literature on the behavioral genetics of the trait of "impulsivity." It highlights translational research on genetic determinants of this trait in rodents and humans. In Chap. 4, the authors review the extensive literature on genetic influences on attachment and pair bonding, primarily from a developmental perspective. Much of the work reviewed utilizes the prairie vole, a microtine rodent that is especially suitable for study owing to its unusually monogamous lifestyle. Overall, therefore, the chapters of this section exemplify how, for modeling purposes, finding the most suitable animal model critically depends on matching the questions one is asking with the particular neurobiological and behavioral attributes of a given species.

Part II of this volume is concerned with animal models of cognition and cognitive decline. Chapter 5 assesses the relative merits of a broad array of approaches for modeling the genetics of intellectual disability. These range from studying behavior in different species of mammals to comparing morphological and molecular changes within homologous brain structures across more phylogenetically distant species. The usefulness of a given model for understanding clinical cases of intellectual disability is evaluated by the extent to which it fulfills a given set of criteria—criteria, indeed, that could be applied equally well to other domains of mental processes and mental illness. The following two chapters address animal models of highly genetically penetrant neurodegenerative diseases and Alzheimer's disease in particular. Chapter 6 provides an overview of the range of genes involved in producing different biomarkers of Alzheimer's disease and the relationship of these to cognitive deficits. Chapter 7 focuses on mouse models that have harbored mutations in ApoE genes, the family of genes perhaps most strongly implicated in the cognitive deficits associated with the later stages of this disease.

Part III dips into the genetics of neurodevelopmental and psychopathological conditions, as they have been characterized using animal models. The first two chapters in this section both address animal models of social interaction and how these can inform as to the genetics of autism spectrum disorders. Chapter 8 applies the "endophenotype" concept, which has been increasingly influential in the field of human behavioral genetics, to animal models of social interaction. Chapter 9 assesses how genetic variations affect key indicators of sociability in rodents, such as communication, imitation, and empathy. Importantly, this chapter highlights how genetic effects on expression of these behaviors can be moderated by the qualities of the environment in which the animals are housed.

Chapter 10 reviews the extensive and informative use of inbred and selective bred lines of rats to ascertain the nature of individual differences in vulnerability to compulsive drug taking. Chapter 11 establishes appropriate parameters for studying the phenomenon of "fear-potentiated startle" in mice as a means toward the clearer characterization of the genetics of fear and anxiety. As researchers in the fields of both drug addiction and affective disorders are aware, translating behavioral paradigms successfully from rats to mice is challenging. But overcoming the challenges can clearly be beneficial, given the greater accessibility of the mouse genome to experimental manipulation.

The final chapter considers future directions for animal models of behavioral genetics. Much of how the field is going to progress over the next 10 years will be consequent to advances in the technologies available for manipulating genes. As described in Chap. 12, these techniques are rapidly improving in terms of their spatial, temporal, and molecular specificity. If there is not one already, there ought to be an axiom in behavioral genetics much like Moore's law in the field of computing, which predicts the phenomenal rate of advancement of molecular genetic technologies. Chapter 12 concludes by reiterating our theme that instead of focusing on clinical syndromes, these new technologies should be harnessed to establish the genetic determinants of core factors that underlie different manifestations of normal and abnormal behavior.

The study of behavioral genetics in animal models has come a long way since the concept of "behavioral phenotyping" first emerged. It is to be hoped that the guiding principles embedded throughout this book, together with the burgeoning of new technologies, will help accelerate our accumulation of knowledge as to how specific nucleotide sequences influence adaptive and maladaptive behavior.

Minneapolis, MN, USA Jonathan C. Gewirtz
Ashburn, VA, USA Yong-Kyu Kim

References

Helzer, J. E., Kraemer, H. C., Krueger, R. F., Wittchen, H. U., Sirovatka, P. J., & Regier, D. A. (Eds.). (2009). *Dimensional approaches in diagnostic classification: Refining the research agenda for DSM-V*. American Psychiatric Publishing.

Jerome, J. K. (1889). *Three men in a boat (to say nothing of the dog)*. Bristol: J. W. Arrowsmith.

Acknowledgments

We are indebted to the following colleagues for reading earlier versions of the chapters of this book with such care and attention: Wyatt Anderson, David A. Blizard, Oliver Bosch, Patrice Bourgeois, Sietse F. de Boer, Lee Ehrman, Bill Falls, Shelley Flagel, Christian Grillon, Todd Gould, Elizabeth Hammock, Anthony Isles, Christoph Kellendonk, Terry Kosten, Tommy Patij, Mark Pisansky, Michael Potegal, Jacob Raber, John Ringo, Patrick Sullivan, Nathalie Thiriet, Brian C. Trainor, and Markus Wöhr. We thank Janice Stern, our editor at Springer, for her patience and encouragement throughout this project.

Contents

Contributors

Matthew A. Albrecht School of Psychology and Speech Pathology, Curtin University, Perth, WA, Australia

MPRC, Catonsville, MD, USA

Michèle Carlier CNRS UMR 7290 Psychologie Cognitive, Fédération de Recherche, Aix-Marseille Université, Marseille, France

Marilyn E. Carroll Department of Psychiatry, University of Minnesota, Minneapolis, MN, USA

Jeffrey W. Dalley Department of Psychology, University of Cambridge, Cambridge, UK

Department of Psychiatry, University of Cambridge, Cambridge, UK

Zoe R. Donaldson, Ph.D. Department of Molecular, Cellular and Developmental Biology and Department of Psychology and Neuroscience, University of Colorado, Boulder, CO, USA

Jonathan K. Foster School of Psychology and Speech Pathology, Curtin University, Perth, WA, Australia

Neurosciences Unit, Health Department of WA, Perth, WA, Australia

Marc V. Fuccillo Department of Neuroscience, Perelman School of Medicine, University of Pennsylvania, Philadelphia, PA, USA

Jonathan C. Gewirtz Department of Psychology, University of Minnesota, Minneapolis, MN, USA

Irving I. Gottesman Department of Psychology, University of Minnesota, Minneapolis, MN, USA

Nathan A. Holtz, Ph.D. Department of Psychiatry and Behavioral Sciences, University of Washington, Seattle, WA, USA

Christopher Janus, Ph.D. Department of Neuroscience, Center for Translational Research in Neurodegenerative Disease, University of Florida, Gainesville, FL, USA

Bianca Jupp Department of Psychology, University of Cambridge, Cambridge, UK

Yong-Kyu Kim Janelia Research Campus, Howard Hughes Medical Institute, Ashburn, VA, USA

Garet P. Lahvis Department of Behavioral Neuroscience, Oregon Health & Sciences University, Portland, OR, USA

Marc T. Pisansky Graduate Program in Neuroscience, University of Minnesota—Twin Cities, Minneapolis, MN, USA

Anna K. Radke Department of Psychology, Miami University, Oxford, OH, USA

Patrick E. Rothwell Department of Neuroscience, University of Minnesota, Minneapolis, MN, USA

Pierre L. Roubertoux INSERM UMR S910, Génétique Médicale et Génomique Fonctionnelle, Aix-Marseille Université, Marseille, France

Hans Welzl, Ph.D. Division of Neuroanatomy and Behavior, Institute of Anatomy, University of Zurich, Zurich, Switzerland

Daisuke Yamamoto Department of Developmental Biology and Neuroscience, Tohoku University Graduate School of Life Sciences, Sendai, Miyagi, Japan

Larry J. Young, Ph.D. Department of Psychiatry and Behavioral Sciences, Center for Translational Social Neuroscience, Yerkes National Primate Research Center, Emory University, Atlanta, GA, USA

Part I
Animal Models of Normative Behavior

Chapter 1
Male Fruit Fly's Courtship and Its Double Control by the *Fruitless* and *Doublesex* Genes

Daisuke Yamamoto

Introduction

Since Morgan (1910), *Drosophila melanogaster* has been one of the most favored animal models in genetics, a field that has experienced, by now, several rounds of renovations and revolutions in techniques and conceptual frameworks. Its small genome size (compiled in large chromosomes, 1–3, and a very small chromosome 4) with a minimum redundancy in functional genes (total *ca.* 15,000 genes) makes genetic analysis easier and simpler (Adams et al. 2000), offering opportunities to explore the molecular and cellular bases of complex higher biological functions, including the behavior and underlying neural mechanisms. The existence of numerous mutants and other genetic rearrangements together with balancer chromosomes that suppress recombination between the homologous chromosomes is the firm basis for the classic genetic approach in *Drosophila*. The germ line transformation achieved by introducing an engineered DNA construct into the fly genome is now a routine task, and several companies currently perform this service on a commercial basis. New analytical technologies specialized for *Drosophila* genetics come up every year, making this organism even more attractive for a wide range of biological research.

Drosophila displays a vast spectrum of behaviors, some of which are instinctive and others of which are learned. One of the most successful genetic analyses of the instinctive behavior in *Drosophila* stems from isolation of the circadian rhythm mutants *period*^{S} (*per*^{S}), *per*^{l} and *per*^{0} (Konopka and Benzer 1971), which stimulated an impressive expansion of circadian clock research and led to the

D. Yamamoto (✉)
Department of Developmental Biology and Neuroscience, Tohoku University Graduate School of Life Sciences, 2-1-1 Katahira, Aoba-ku, Sendai, Miyagi 980-8577, Japan
e-mail: daichan@m.tohoku.ac.jp

J.C. Gewirtz, Y.-K. Kim (eds.), *Animal Models of Behavior Genetics*, Advances in Behavior Genetics, DOI 10.1007/978-1-4939-3777-6_1

discovery of the conserved molecular mechanism for biological timing. The genetic studies on learning in *Drosophila* were initiated by the isolation of a mutant in a cAMP phosphodiesterase-coding locus, *dunce* (*dnc*; Dudai et al. 1976), which demonstrated the involvement of the cAMP-CREB pathway in learning and memory, and reinforced the hypothesis that cAMP-mediated synaptic enhancement is the key for classical conditioning, a notion derived from physiological studies in *Aplysia* (Kandel 1976). These two divergent streams of the genetic basis of behavior in *Drosophila* have further enhanced the status of this organism as a model for the study of complex functions relevant to those in higher organisms, including humans, and prompted us to apply this neurogenetic approach in deciphering the neural basis for other behaviors.

The progress in studies on mating behavior is particularly intriguing; the principle of decision-making, the prime integral function of the brain, is now emerging from the neurogenetic study of mating behavior in *Drosophila* (Yamamoto and Koganezawa 2013). In this chapter, such cutting edge research results will be reviewed, together with a historical overview of critical experiments that led to the recent explosion of neurogenetic researches on *Drosophila* mating behavior.

Current Research

The Field of Mating Research Began with Fruitless

In 1963, Gill reported in a scientific meeting a male-sterile mutant he named *fruity* (*fru*; later, the name was changed to *fruitless* by Hall and the original allele isolated by Gill was designated *fru*^{1}), the males of which vigorously courted both males and females but copulated with neither (Gill 1963). Gill wanted to isolate mutants with defects in gametogenesis, so for him, *fru*^{1} was just a side product and consequently this mutant was not further characterized for an additional 15 years, when Hall finally published the first paper on *fru*^{1} (Hall 1978). Hall (1978) quantified the male preference of *fru*^{1} mutant males in courtship and observed that, when placed together, these males court each other by forming a suitors' chain, now known as the "courtship chain" (Fig. 1.1). He also noted that even wild-type males court *fru*^{1} mutant males although they usually do not court other males (Hall 1978). This indicates that *fru*^{1} males have two distinct phenotypes: courting other males (the courting male phenotype) and being courted by other males (the courted-by-male phenotype). Gailey and Hall (1989) mapped cytologically the courted-by-male phenotype to 90C and the courting male phenotype to 91B, at which two break points of an inversion associated with the *fru* mutation are located, respectively. Therefore, the courted-by-male phenotype and the courting male phenotype are induced by mutations at two independent loci. The courting male phenotype mapped at 91B corresponds to the *fru* locus in

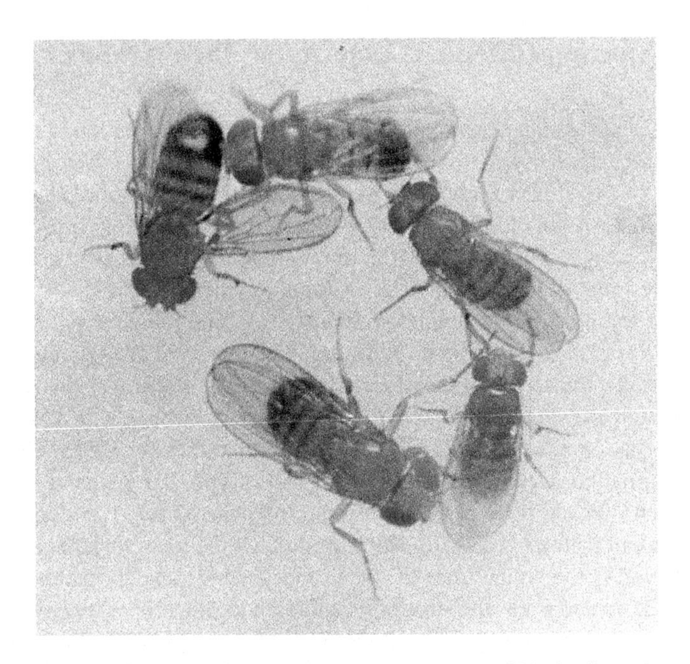

Fig. 1.1 Male-to-male courtship in *fru*^{sat} mutant males (courtesy of Morito Ogawa)

the current definition while the other locus associated with the courted-by-male phenotype remains uncharacterized. The courted-by-male phenotype might reflect a change in the pheromone profile of *fru*^{1} mutant males (Tompkins et al. 1980), in contrast to the courting male phenotype, which results from impaired neural functions (see below).

A Muscle Tells About the Sex of the Brain

In addition to the behavioral phenotype just described, a developmental defect was found in *fru* mutant males by Gailey et al. (1991), i.e., the absence of a male-specific muscle, which was designated as the muscle of Lawrence (MOL) after the name of its discoverer, Peter Lawrence. The MOL is a large pair of dorsal longitudinal muscles present in the fifth abdominal segment (A5) of adult *melanogaster* males (Lawrence and Johnston 1984, 1986). Lawrence and Johnston (1984) have demonstrated that the MOL forms in sexually mosaic flies regardless of the sex of the muscle cell itself provided that the innervating motor nerve is male. This suggests that the loss of MOL in *fru* mutant males could be due to an altered sexual fate of the innervating motoneurons rather than any defects in the muscle itself.

An RNA Splicing Difference Determines the Fly's Sex

Taylor (1992) carried out a thorough analysis of the MOL formation in flies that are mutant for *Sexlethal (Sxl), transformer (tra), transformer2 (tra2),* or *doublesex (dsx),* four known genes composing the *Drosophila* sex determination cascade (Marín and Baker 1998). Unlike wild type flies, female (XX) flies with mutations in the *Sxl, tra,* or *tra2* locus had the MOL, in contrast to *dsx* mutants in which, as in wild type flies, the MOL forms in males and not females (Taylor 1992). *Sxl* is positioned at the top rung of the sex determination cascade, followed by *tra* and *tra2*, then finally *dsx* (Marín and Baker 1998). Sxl, Tra, and Tra2 are splicing regulators whereas Dsx is a transcription factor (Marín and Baker 1998). Tra binds to the *dsx* mRNA precursor to splice it in a sex-specific manner, resulting in an mRNA coding either male-type DsxM or female-type DsxF (Marín and Baker 1998). The fact that the MOL formation depends on *Sxl* and *tra/tra2* and not *dsx* suggests that the sex determination cascade bifurcates downstream of *tra/tra2* into *dsx*-dependent and *dsx*-independent pathways. Taylor (1992) postulated the involvement of a gene she called *ambisex* as the counterpart of *dsx* in the *dsx*-independent pathway. Indeed, Taylor (1992) treated *fru* as the strongest candidate for the role of *ambisex*, and thus placed *fru* in the output path of the sex determination cascade (Taylor et al. 1994).

Molecular Identification of the *fru* Gene

Our group (Ito et al. 1996) cloned the *fru* gene by chromosomal walking initiated from the P-element insertion which was responsible for homosexual courtship and loss of the MOL in males of *satori*, an allele of *fru* that we isolated in a previous study (Yamamoto et al. 1991). The Baker group (Ryner et al. 1996) performed a genomic library screen for Tra-binding sequences on the assumption that *fru* is a target of Tra as is *dsx*. The *fru* gene thus cloned spans over 150 kb on the genome, carrying at least four promoters, one of which yields a sexually dimorphic set of transcripts expressed selectively in the nervous system (Ito et al. 1996; Ryner et al. 1996). The male-type transcripts encode a series of proteins of the BTB-zinc finger family collectively called FruM (M stands for "male-specific"), which are putative transcription factors (Ito et al. 1996; Ryner et al. 1996), whereas the female-type transcripts yield no protein products (Lee et al. 2000; Usui-Aoki et al. 2000). The FruM proteins are composed of at least five isoforms with distinct C-terminal sequences, FruAM–FruEM (Usui-Aoki et al. 2000; Song et al. 2002; Note that the terminology for isoforms is different between the two papers). When the Fru proteins are artificially expressed in motoneurons in females which otherwise lack them, these females gain the male-specific MOL (Billeter et al. 2006; Nojima et al. 2010; Usui-Aoki et al. 2000). This result supports the notion that the decisive factor for the MOL formation is the neuronal maleness, which is primarily determined by Fru.

Demir and Dickson (2005) engineered a fly strain, *fru*^{M}, in which the *fru* mRNA precursor is always spliced in the male pattern even in females (or the other way around, *fru*^{F}). The *fru*^{M} females court other females displaying the stereotypic male-type courtship ritual, which wild-type females never show (Demir and Dickson 2005). These authors claimed that Fru expression in neurons is necessary and sufficient for the generation of male-typical courtship behavior (Demir and Dickson 2005). Practically, the same conclusion was drawn by the Baker group (Manoli et al. 2005), who demonstrated that, using independently generated fly lines expressing Fru in females, such females exhibit male-type courtship behavior.

Some *fru*-Expressing Neurons are Sexually Dimorphic

We (Kimura et al. 2005; 2008) attempted to identify *fru*-expressing neurons by labeling them in small clones as generated by the Mosaic Analysis with a Repressible Cell Marker (MARCM) technique (Lee and Luo 1999), visualizing the entire structure of single cells, i.e., from the soma to the tip of every neurite. We identified approximately 50 *fru*-expressing neuronal clusters in this attempt (Kimura et al. 2008). One of these clusters, mAL (the acronym stands for "medially located just above the antennal lobe"; Lee et al. 2000) revealed conspicuous sex differences in three ways (Fig. 1.2): (1) the number of cells composing the cluster is 5 in females and 30 in males; (2) the neurite contralateral to the soma terminates in the suboesophageal ganglion with Y-shaped endings in females while it forms terminals with a horse tail appearance in males; (3) the male mAL cluster possesses the ipsilateral neurites extending to the

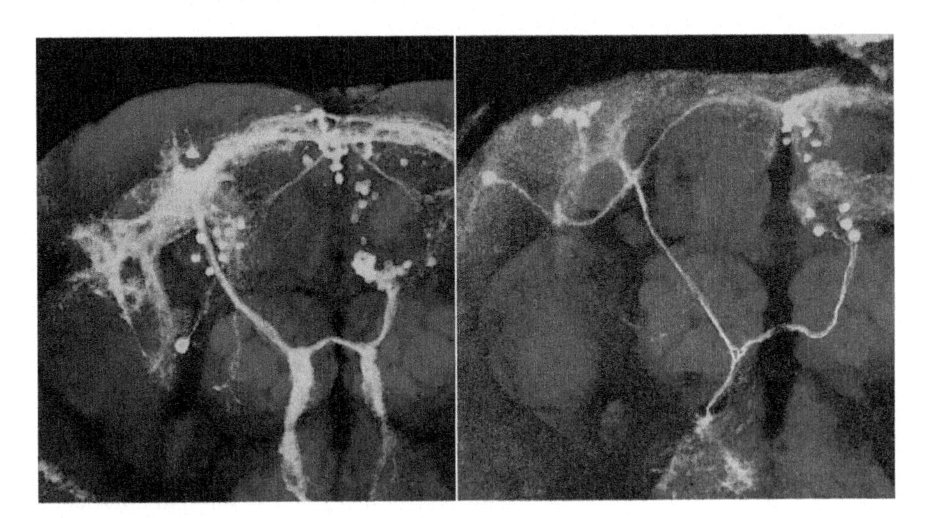

Fig. 1.2 Sexual dimorphism of mAL interneurons (modified from Kimura et al. 2005). *Left-hand side*: mAL neurons in the male brain. *Right-hand side*: mAL neurons in the female brain

suboesophageal ganglion while the female mAL cluster has no corresponding neurites (Kimura et al. 2005). Importantly, *fru* null mutant males have mAL neurons that are female-type in all three parameters (Kimura et al. 2005). This indicates that the Fru protein is required in the mAL neurons for masculinization.

The Mechanisms by Which Fru Generates Sex Differences in Neurons

The question arises as to how the Fru protein produces sexual dimorphism in mAL neurons. The difference in cell number results from either "hyper" proliferation of the neuroblast in males or female-biased cell death of mAL neurons. The mAL clones homozygous for a deficiency, *H99*, which deletes three major cell death genes, *hid*, *grim,* and *reaper*, contained excess cells (up to 29) in females, in contrast to males, in which the cluster is composed of 30 cells as is the wild-type cluster (Kimura et al. 2005). This result demonstrates that the difference in the number of cells contained in the mAL cluster is accounted for by female-specific cell death. The Fru protein existing only in male neurons (Lee et al. 2000; Usui-Aoki et al. 2000) protects those cells susceptible to the cell death mechanism in males.

The *H99* homozygous mAL cluster in females developed the ipsilateral neurites that are absent from wild-type females. This observation appears to be consistent with the idea that the neurons fated to develop into male-type are selectively eliminated by cell death in females. However, protection against cell death is not the sole mechanism whereby the Fru protein exerts its masculinizing action on mAL neurons. This is because the excess cells generated by blocking cell death in females retained the female-typical Y-shaped terminal structure of the contralateral neurites in the suboesophageal ganglion (Kimura et al. 2005). This means that the male-typical shaping of the contralateral neurites requires Fru functions in those mAL neurons that escaped cell death. This proposition was further supported by the finding that knockdown of *Hunchback (Hb)* transforms the contralateral neurite structure from male-type to female-type without affecting the number of mAL neurons or the structure of ipsilateral neurites extending to the suboesophageal ganglion (Goto et al. 2011). A genetic screen for *fru* modifiers (described in a later section) had yielded this well-known gap gene encoding a zinc finger protein with a function to specify the anterior segment of the embryo (Goto et al. 2011). These observations collectively indicate that three sex-specific characteristics of mAL neurons are all governed by Fru, but each through a different mechanism.

Ito et al. (2012) showed that Fru protein binds more than 100 specific sites on chromosomes, to which the transcriptional cofactor Bonus (Bon) is recruited. Bon further recruits either of two chromatin regulators, Histone deacetylase 1 (HDAC1) or Heterochromatin protein 1 (HP1), suggesting that Fru acts through chromatin modification. Phenotypic analysis at the single cell resolution revealed that HDAC1 supports the masculinizing action of Fru, whereas HP1 counteracts the masculinization. Notably, *fru* hypomorphic mutants carry mAL cluster in which neurons with the male-type structure and those with the female-type structure are mixed and no neuron exhibits a structure that is intermediate of the female-type and male-type. Reducing the dose of HDAC1 increases the number of female-type neurons whereas decreasing the dose of HP1a increases the number of male-type cells. Thus, the Fru action in neuronal masculinization is an all-or-none process at the single cell level (Yamamoto and Koganezawa 2013). No Fru target genes have been firmly identified except for several candidate genes deduced from analyses by the DamID (Neville et al. 2014), SELEX (Dalton et al. 2013), or ChIP (Vernes 2014) method.

Some Neurons are Present Only in the Female or Male Brain

In addition to sexually dimorphic clusters such as mAL, there are neurons that exist only in either sex. One such example is the P1 cluster, which is male specific (Kimura et al. 2008). The P1 cluster is composed of 20 interneurons per hemisphere, having cell bodies in the lateral protocerebrum (lpr) extending transmidline fibers with extensive ramification in the bilateral lpr. The P1 cluster formed ectopically in the female brain if it was homozygous for the deficiency *H99* (Kimura et al. 2008), reminiscent of extra mAL neurons with ipsilateral neurites in the female brain (Kimura et al. 2005). Contrary to our expectations, the P1 cluster was retained in *fru* mutant males that were deficient in the Fru protein, the condition under which mAL neurons were feminized ("male-type" mAL neurons were eliminated by cell death). Subsequent genetic analysis revealed that the expression of DsxF (the female-specific product of *dsx*) actively eliminates P1 from the female brain, making the P1 cluster male-specific (Kimura et al. 2008). In fact, P1 neurons express both Fru and Dsx, unlike the majority of *fru*-expressing neurons, which do not express *dsx* (Kimura et al. 2008). Conversely, many *dsx*-expressing neurons do not express *fru* (Kimura et al. 2008; Lee et al. 2002). In *fru* mutant males, the P1 contralateral neurites terminate in an incorrect position, suggesting that Fru is directly or indirectly required for path finding of neurites while it is dispensable for the protection against cell death in this neuronal cluster (Kimura et al. 2008).

Do Flies with Sexually Mosaic Brains Behave as Female or Male?

A gynandromorph is an individual composed of female tissues and male tissues. Such sexually mosaic individuals are generated at a relatively high rate in particular *Drosophila* strains; they carry an unstable X chromosome known as the Ring-X chromosome which is often lost during mitosis, producing XO cells in an embryo otherwise composed of XX cells. The cells with two X chromosomes (XX) develop as female whereas those with a single X chromosome (XO and XY) develop as male, so the loss of the Ring-X chromosome from an XX embryo results in the generation of a sexually mosaic individual, a gynandromorph. Gynandromorphism thus provides an opportunity to localize the tissue which determines the sex type of a particular behavior because the tissue border to separate the sexes runs at random (more exactly quasi-random), and therefore, each mosaic fly has a unique sexual composition of cells, allowing it to correlate the sex of a particular tissue with the sex-type of the behavior concerned. Sturtevant (1915), one of the founders of *Drosophila* genetics, described sexual behavior of sexually mosaic flies a century ago, and systematic mosaic analysis of the mating behavior of *Drosophila* was first done by Hotta and Benzer around 1970 (Hotta and Benzer 1972). In their first attempt, visible markers on the body surface were used to deduce the sex of nearby internal tissues, with which the sex-type of the mating behavior was correlated; the recessive marker *yellow (y)* changes the body color from brown to yellow when XR is lost, making it possible to distinguish the male cuticle (XO) from the female cuticle (XX). They found that the majority of mosaic flies that displayed male-type courtship behavior had a male cuticle on at least one side of their head. This implies that some structure close to the head (i.e., the brain) produces the male courtship behavior. In the syncytial blastoderm stage, nuclei synchronously divide deep in the embryo, then after the 13th division, move to the surface and are separated individually by membranes to form the cellular blastoderm. The two-dimensional position in this stage is one of the factors that specifies the fate of individual cells. As sibling cells tend to locate close to each other in the cellular blastoderm, it is inferred that two cells distantly located at this stage have more chance to be separated by the mosaic border at the adult stage than a closely located pair of cells do. Therefore, the distance between any two cells in the cellular blastoderm must be deduced from the probability of these cells being separated by the mosaic border at the adult stage. By triangulation with these probability values, one can construct a map of primordia for adult organs on the virtual blastoderm, the blastodermal fate map. Hotta and Benzer (1970) produced a comprehensive fate map, in which internal tissues including the brain were implemented using neural phenotypes of some mutants as functional markers (Ikeda and Kaplan 1970a, b). When male-type courtship behavior was plotted on this map, the focus of this behavior (the site showing positive correlation with the occurrence of this behavior) was found to coincide with the brain (Hotta and Benzer 1972). In addition, they found that the mosaic flies in which the brain was unilaterally (or indeed bilaterally) male performed male-type courtship, whereas the flies with a brain that was bilaterally female never displayed the male-type courtship behavior (Hotta and Benzer 1972). This indicates that the male brain is "domineering" over the female brain (or the female brain is "submissive" to the male brain) in producing the male-type courtship behavior.

Which Part of the Brain Needs to be Male for the Fly to Behave as a Male?

Hall (1979) determined the focus for male-courtship behavior by directly scoring the sex of internal tissues on serial sections of mosaic flies. To distinguish the sex of internal tissues, Hall (1979) used an enzyme, Acid phosphatase-1 (Asph-1), which produces dark brown deposits on tissues when treated appropriately. Hall (1979) used a translocation of the *Asph-1* gene from the 3rd chromosome to the X chromosome, introduced into the *Asph-1* null mutant. Under these conditions, the translocated X-linked *Asph-1* is the sole functional copy of *Asph-1*. He used the mutation *paternal loss* (*pal*) to induce stochastic loss of the paternally derived X chromosome to produce XO (male) tissues in an otherwise XX (female) fly. By scoring serially sectioned brains of 180 mosaic flies, Hall (1979) concluded that a single focus located near the mushroom body in the dorsal posterior brain must be male in at least one side of the brain for the fly to show early steps of male courtship behavior, including tapping, chasing, and unilateral wing extension and vibration. Tompkins and Hall (1983) conducted a similar mosaic analysis for female sexual receptivity, identifying a single focus for this in the dorsal anterior brain, which must be bilaterally female for the fly to be receptive to courting males.

In 1993, Brand and Perrimon (1993) invented the GAL4-UAS system, in which an arbitrarily chosen gene fused to the UAS sequence can be expressed only in the cells that express GAL4 from an independent transgene. Ferveur and Greenspan (1998) applied this system to sexually transform parts of the brain from male to female by expressing a feminizing gene *tra*, confirming the critical role of the posterior-dorsal brain for male courtship behavior.

Greenspan and his colleagues (Broughton et al. 2004) further attempted to narrow down the candidate regions in the brain for regulating male courtship; they mapped the brain regions that facilitate or inhibit male courtship behavior by forcibly activating or inactivating portions of the brain using the GAL4-UAS system. The tool used here to inactivate neural activities was *shibire*^{ts1} (*shi*^{ts1}), which is a temperature-dependent dominant negative mutation in the *shi* locus encoding Dynamin, a protein required for endocytosis (Kitamoto 2002). *shi*^{ts1} mutants become paralyzed when the temperature is raised over 29 °C because the synaptic vesicles dry up due to a failure in endocytosis-mediated recycling, and thus synaptic transmission is blocked; the paralysis is then reversed if the temperature is decreased to 25 °C (Ikeda et al. 1976). The tool used for enhancing neural activities was *ether-á-gogo*$^{\Delta 932}$ (*eag*$^{\Delta 932}$), a dominant negative mutation in the gene coding for a potassium channel. The Eag potassium channel is involved in hyperpolarizing the neural membrane (Wu et al. 1983) and, in *eag*$^{\Delta 932}$ mutants, defects in this process depolarize the membrane making neurons easier to excite. Greenspan and his colleagues (Broughton et al. 2004) expressed these genes in different regions of the brain using the GAL4-UAS system in male flies, which were then tested for courtship behavior. In this effort, they found a posterior region of the lateral protocerebrum to be a neural center to initiate male courtship and a region flanking anteriorly to this region was found to repress the initiation (Broughton et al. 2004).

Which Cells in the Male's Brain Decide to Court?

These studies demonstrated that the courtship-initiating function is confined to particular areas in the brain, but which neurons are involved? We used MARCM to approach this problem (Lee and Luo 1999), which allowed us to manipulate and label a small population of neurons in the brain. In the first series of experiments, we used the tra^1 mutation to achieve female-to-male transformation of a population of neurons that were simultaneously labeled by a fluorescent marker mCD8::GFP as driven by fru^{NP21}, an enhancer trap strain with GAL4 expression in *fru*-positive cells (Kimura et al. 2005). These flies were chromosomally female yet had masculinized cells that are a clone for tra^1 (we call them "MARCM females"). Although wild-type females never display male-type courtship behavior, some of the MARCM females chase other females, even extending and vibrating one of the wings as do the male flies in courtship (Kimura et al. 2008). Such male-type courtship behavior was observed in 16 out of 205 MARCM females (Kimura et al. 2008). After examining the courtship behavior, the brains were dissected out for histochemistry to determine which *fru*-expressing neurons in the brain were mCD8::GFP-positive and thus masculinized by the action of tra^1. If there exists any cluster that must be male for the fly to show male-type courtship behavior, that cluster must be mCD8::GFP-positive in all 16 flies that displayed male-type courtship behavior. In contrast, the same cluster should be mCD8::GFP-negative for the remnant flies that did not display male-type behavior. Our analysis revealed that all but one of the scored clusters (approximately 50 *fru*-expressing clusters were scored) did not show any correlation between the cellular sex of the neuron and the behavioral sex of the fly (Kimura et al. 2008). The only exception was P1, a male-specific *fru*-expressing cluster. In 81.3 % of MARCM females that displayed male-type courtship behavior (13 out of 16 flies), P1 was mCD8::GFP-positive, i.e., its cellular sex was male (Kimura et al. 2008). Because P1 is nonexistent in wild-type females, "P1-masculinization" means the ectopic formation of P1 in the female brain. This indicates that when a fly has the P1 cluster in its brain, it is highly likely that the fly will display male-type courtship behavior. However, the correlation was not perfect, and therefore, P1 is not absolutely necessary for inducing male-type courtship behavior (e.g., three MARCM flies without P1 displayed male-type courtship behavior; Kimura et al. 2008). P1 was also not sufficient for inducing male-type courtship behavior, because 12 out of 189 MARCM flies that did not display male-type courtship behavior had P1 in their brain (Kimura et al. 2008). Nevertheless, we consider that P1 is the prime decision-maker of courtship initiation in the male brain for its strong (though not absolute) effect on this process. Note also the fact that fru^M females display male-type courtship behavior (Demir and Dickson 2005), even though they are females and thus lack P1.

The P1 cluster is composed of 20 interneurons, each of which extends a trans-midline neurite connecting the left and right lateral protocerebrum, forming a fiber tract within the superior medial protocerebrum (the SMP tract; Ruta et al. 2010).

Forced Activation of P1 Triggers Courtship Behavior in a Solitary Male

The roles of the P1 cluster in male courtship behavior have been evaluated by experiments in which these neurons were directly activated in the absence of any courtship target. *fru*-expressing neurons were activated through a warmth-sensitive channel, dTrpA1, which opens when the ambient temperature is raised to, for example, 29 °C (Hamada et al. 2008) using the GAL4-UAS system. When all *fru*-expressing neurons were activated simultaneously in this way, the male flies placed alone in a chamber suddenly commenced the courtship actions, which included tapping, wing extension and vibration, licking and attempted copulation (Kohatsu et al. 2011). A recent report from the Baker laboratory showed that direct activation of all *dsx*-expressing neurons had qualitatively similar effects (Pan et al. 2011). To reduce the number of neurons to be activated, we produced MARCM clones expressing dTrpA1 (Kohatsu et al. 2011). For the same purpose, the Dickson group developed a binary expression system (Yu et al. 2010), in which only a subpopulation of *fru*-expressing neurons delineated by overlapping expression of an arbitrary chosen GAL4 line can be activated by dTrpA1 (von Philipsborn et al. 2011). Both groups identified two types of brain interneurons as the prime activators of male courtship behavior, the P1 cluster and a descending interneuron cluster, P2b (terminology by Kimura et al. 2008) or a pIP10 neuron (terminology by Yu et al. 2010). Although the pIP10 neuron has not been firmly established to be a member of the P2b cluster, they both have somata in the posterior brain and extend descending axons that terminate in pro-, meso-, and meta-thoracic neuromeres in the ventral ganglia (Kohatsu et al. 2011; von Philipsborn et al. 2011), where the presumptive motor center for courtship song generation is located (von Schilcher and Hall 1979). In an experiment to induce male-type courtship behavior in females by *tra*^{1}-mediated masculinization of MARCM clones (Kimura et al. 2008), P1 but not P2b (pIP10) was identified, whereas these two clusters were found to induce male-type courtship behavior in the experiment with dTrpA1-mediated direct activation (Kohatsu et al. 2011; von Philipsborn et al. 2011). This might suggest that P1 is involved in the decision-making for male courtship behavior while P2b (pIP10) has no such ability and instead plays a role in conveying the decision to the motor center. This "epistatic" relationship between P1 and P2b (pIP10) clusters is consistent with the observation that the arborizations of these neurons are positioned close to each other in the lateral protocerebrum (Kohatsu et al. 2011; von Philipsborn et al. 2011).

P1 Neurons Are Activated Upon Female Contact

While the observation that direct activation of P1 and/or P2b induces courtship behavior in solitary males is intriguing, one might argue that dTrpA1-induced behavior is likely to be executed by a mechanism different from that for normally occurring behavior under such radically different conditions. If the P1 cluster is indeed the prime decision-maker for courtship, it must be activated prior to the execution of behavior by a natural releaser, i.e., a female. To determine whether this is the case, we invented a system in which neuronal activities can be monitored from the immobilized brain (head) of a male fly, which is nevertheless able to perform courtship behavior (Kohatsu et al. 2011). In this tethered male preparation, a metal wire holds a male fly at its dorsal thorax, while a Styrofoam ball is given to him to hold with his legs, so that he rotates the ball by walking on it (Fig. 1.3). By opening a window on the head cuticle, the brain is exposed for optical recordings of neural activities from the fly "on the move." Using this system, we first demonstrated that the tethered male fly begins to chase a moving female presented in front of him, often displaying the unilateral wing vibration typical of courtship, provided that the male touches the stimulus female at least one time with his foreleg, where chemosensory (gustatory) cells exist (Kohatsu et al. 2011). Contact by a stimulus male did not induce persistent chasing behavior in the tethered male. These observations indicate that the perception of chemical cues by gustatory sensory cells on the tarsal segment of forelegs is critical for the initiation of male courtship behavior, at least under the tethered conditions. In addition, a visual target is required for maintaining the courtship pursuit initiated by contacting a female. Although male flies need to contact a female to initiate courtship under the tethered conditions, freely moving males may initiate courtship if appropriately stimulated visually in the presence (Bath et al. 2014; Pan et al. 2012) or absence (Agrawal et al. 2014) of simultaneous artificial subthreshold depolarization of P1 neurons. Even under the tethered conditions, a moving dummy, such as an array of light spots moving horizontally on the computer display, can induce male courtship if subthreshold depolarization is imposed on the P1 cluster (more exactly, the *dsx*-expressing pC1 cluster inclusive of the *fru*-expressing P1 cluster) via Channelrhodopsin activation (Kohatsu and Yamamoto 2015). This result suggests that artificial depolarization of P1 can be a substitute for the excitatory pheromone input that is normally produced by contacting a female, and that no female-like features are required in the visual target in order to maintain male courtship once the courtship has started. Thus, this study established a virtual reality paradigm for courtship behavior, in which two natural courtship inducers, the contact pheromone and the sight of a female, were

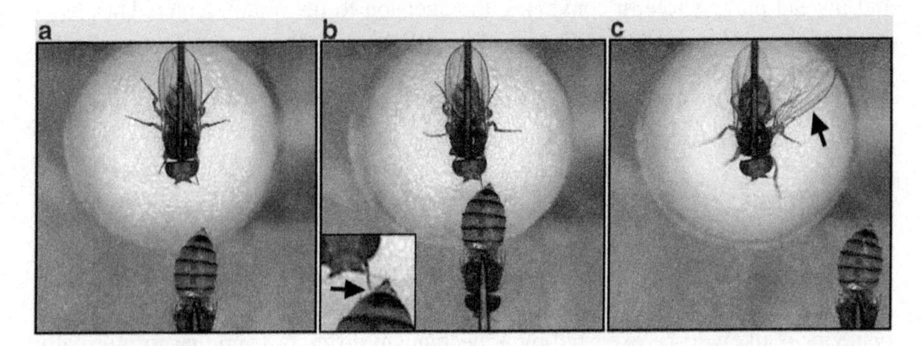

Fig. 1.3 A tethered male courting a female on a treadmill (reproduced from Kohatsu et al. 2011)

successfully replaced with two artificial stimuli, neuronal depolarization via Channelrhodopsin-activation and light spots on a screen. In the same study, Kohatsu and Yamamoto (2015) identified another neural cluster, pC2l, whose stimulation via Channelrhodopsin can induce courtship in a male fly, if he is simultaneously stimulated by a moving visual target. pC2l is a *dsx*-positive and *fru*-negative neural cluster (K.-i. Kimura, unpublished observation). Activation of pC2l results in target chasing accompanied by wing extension and vibration for courtship song generation as does activation of pC1 (P1), along with two late components of courtship behavior, licking and attempted copulation, which are not typically induced by activation of pC1 (P1) (Kohatsu and Yamamoto 2015). The wing extension and vibration induced by pC2l stimulation are confined to the contralateral side (Kohatsu and Yamamoto 2015), in contrast to the case of pC1 (P1) stimulation, which induces alternative bilateral wing movements (Kohatsu et al. 2011). It is tempting to speculate that, when driven by pC1 (P1), pC2l combines the early motor program and late motor program to complete an entire program for courtship actions, while coordinating wing motions that need to be strictly unilateral.

To measure neuronal activities in the brain, a Ca^{2+}-sensitive fluorescent protein was expressed as driven by fru^{NP21}, a GAL4 enhancer trap line for the *fru* locus. The use of a tethered male under a fluorescent microscope allowed us to monitor Ca^{2+} activity changes associated with female contact. Tarsal contact by a female abdomen immediately induced a transient rise of Ca^{2+} activity in the lateral protocerebrum but not in the mushroom body or optic tubercle in the brain of the tethered male (Kohatsu et al. 2011). The Ca^{2+} response of the lateral protocerebrum to male contact was significantly smaller than that induced by female contact. Interestingly, contact by a glass rod coated with hexane extract of cuticles, but not an uncoated glass rod, produced a Ca^{2+} response, further supporting the idea that chemical components rather than mechanical stimulation activate neurons in the lateral protocerebrum of the tethered male when contacted by a female abdomen (Kohatsu et al. 2011).

The lateral protocerebrum is densely innervated by P1 and other *fru*-expressing neurons. To address whether P1 neurons contribute to the Ca^{2+} transients recorded from the lateral protocerebrum in response to female contact, the Ca^{2+} indicator Yellow cameleon was expressed in MARCM clones in the brain of tethered males, which were subjected first to Ca^{2+} imaging for neuronal activities upon female contact and then to histochemistry for the identification of the neurons from which recordings were made. The Ca^{2+} transients were recorded in response to female contact from 6 males with Yellow cameleon-expressing MARCM clones in the lateral protocerebrum and in all these cases, it was the P1 cluster that yielded the Ca^{2+} responses recorded (Kohatsu et al. 2011). This result demonstrates that P1 excites prior to the execution of courtship in response to female contact. A more recent imaging study with GCaMP as a Ca^{2+} indicator in the virtual reality paradigm demonstrated that pC1 (P1) neurons exhibit periodical Ca^{2+} rises in phase with every contralateral turn of the courtship pursuit by a male fly (Kohatsu and Yamamoto 2015). This result suggests that transient depolarization of P1 neurons is obligatory for triggering each courtship bout.

The Motor Center for Courtship Song Generation

The Dickson group also identified a neuronal cluster in the thoracic neuromere with the ability to induce courtship actions by forced activation (von Philipsborn et al. 2011). This cluster, expressing both *fru* and *dsx* (Rideout et al. 2007) was suggested to be a component of the courtship song pattern generator in the motor center (von Philipsborn et al. 2011). In fact, Clyne and Miesenböck (2008) have shown that decapitated (brainless) males generate courtship songs when all *fru*-expressing neurons in the ventral ganglia are forcibly excited by the transgenically expressed ATP-activated channel. They reported that even females generate sounds similar to courtship songs when their *fru* (transcript)-expressing neurons are activated in a similar way although the song pattern was distorted. dTrpA1-mediated activation of *fru*-expressing neurons also induced distorted courtship songs in females (Pan et al. 2011) although higher temperature (stronger stimulation) was required to induce the courtship-like actions than in males (Kohatsu et al. 2011). These observations imply that the core portion of the motor pattern generator for courtship songs is not entirely male specific, yet the production of proper song patterns requires tuning by higher order neurons, which are likely male specific.

The courtship song of *Drosophila* is generally composed of two components, the pulse song and hum song (Riabinina et al. 2011; von Schilcher 1976). A male fly changes the ratio of the pulse and sine components in his song depending on how closely he places himself to the target female at every moment of the courtship pursuit (Coen et al. 2014; Trott et al. 2012). In *melanogaster*, the pulse song is a series of tone spikes tonically generated with an inter-pulse interval (IPI) having a mean of ~35 msec, while the hum song is a sine wave oscillation with a frequency of about 160 Hz (von Schilcher 1976). The claim that the IPI fluctuates rhythmically and this *per*-dependent, species-specific rhythm is important for species recognition by females (Kyriacou and Hall 1980, 1984, 1986; Wheeler et al. 1991; Yu et al. 1987) was recently shown to be based on an error in data analysis (Stern 2014). The songs of different species differ with respect to these parameters and even lack either or both songs (Riabinina et al. 2011; Spieth 1952).

Several genes have been implicated in species differences in the song pattern by quantitative trait loci (QTL) analysis, including *fru* and *croaker* (Gleason and Ritchie 2004). A more recent study has challenged the idea that the *fru* gene has evolved to generate divergent male courtship in *Drosophila* (Cande et al. 2014): *melanogaster* males carrying heterospecific genomic *fru* retain the courtship repertoire of *melanogaster*. On the other hand, fru^1 mutant males generate a pulse song with altered IPI (Villela et al. 1997), and males harboring stronger loss-of-function *fru* alleles do not produce any song (Villela et al. 1997). Another study showed that loss of FruBM completely blocks the sine song generation, whereas loss of FruBM and FruEM prolongs the IPI without blocking the generation of pulse song (Neville et al. 2014). In light of these findings, it is conceivable that *fru*-expressing neurons can be substrates for diversified male courtship behavior in *Drosophila* species, even though the *fru* gene per se has not evolved for species-specific differences in courtship behavior.

The defects in courtship behavior observed in *fru* mutants likely result from the combinatorial effects of *fru* mutations on the higher brain center and thoracic motor center. When synaptic outputs from P1 neurons were bilaterally blocked by shi^{ts} expression, the males treated in this manner failed to court females (Kimura et al. 2008), indicating that the song inhibition originated in the brain. As mentioned above, the Dickson group (Yu et al. 2010) employed an intersectional genetics strategy to cross a fly line bearing flippase activity in *fru*-expressing neurons with an arbitrary GAL4-enhancer trap line, yielding offspring in which UAS-driven transgenes are activated only in cells that are doubly positive for *fru* and GAL4. By choosing adequate GAL4 lines, transgene expression can be restricted to a few defined *fru*-expressing neurons. Using this strategy, the P1 cluster was shown to consist of brain neurons that are important for pulse song generation (see above): the blocking of synaptic outputs from some of the P1 cluster neurons with *TNT* (*tetanus neurotoxin light chain*) significantly reduced the number of pulses the males produce to court virgin females while their forced activation by dTrpA1 resulted in pulse song generation in a solitary male (von Philipsborn et al. 2011). In this series of experiments, they identified three thoracic clusters, dPR1, vPR6, and vMS11, as those with the ability to produce pulse songs without the involvement of descending signals from the brain (von Philipsborn et al. 2011). These *fru*-expressing clusters are thus likely to be among the elements composing the motor pattern generator for courtship songs in the thoracic ganglia. Rubinstein et al. (2010) examined the effect of thoracic-specific knockdown of *fru* by expressing *fruIR* via *teashirt-GAL4*. Pulse songs recorded from these males showed the expanded pulse- width yet had normal IPIs (Rubinstein et al. 2010), implying that the temporal organization of pulse trains and pulse shape are controlled by separate mechanisms.

Dsx is Required for Sine Song Production

Males double heterozygous for *dsx* and *fru* were less active in courting females than males singly heterozygous for either locus (Shirangi et al. 2006). A similar synergy with *fru* and *dsx* in male courtship has been reported for a mutation in *retained* (*retn*), a gene encoding an ARID-box transcription factor (Shirangi et al. 2006). While the roles of *retn* in song production remain obscure, the importance of *dsx* in this process has been well established: males harboring a *dsx* loss-of-function mutation do not produce sine songs and generate pulse songs that are indistinguishable from those recorded from wild-type males (Villella and Hall 1996). The Goodwin group engineered a *dsx* knock-in allele, dsx^{Gal4} (here referred to as dsx^{Gal4+} to make it distinguishable from another dsx^{Gal4} generated by the Baker group), which expresses functional *dsx* mRNAs in addition to the Gal4 transcript under the control of the endogenous *dsx* promoter (Rideout et al. 2010). The generation of both pulse and sine songs was completely inhibited by expressing *UAS-TNT* as driven by dsx^{Gal4+} and was relieved if *elav-Gal80* was additionally expressed to repress Gal4 activity in all neurons. TNT expressed via dsx^{Gal4+} inhibited both the sine and pulse songs presumably because it blocked synaptic transmission in *dsx*-expressing brain neurons with the command function, i.e., P1, which expresses *dsx* in addition to *fru* (Kimura et al. 2008). As the P1 cluster forms even in *dsx* mutant males, these males sing. However, *dsx* mutant males are unable to generate sine songs, which presumably require intact *dsx*-expressing cells in the thoracic motor circuit or other peripheral tissues. Interestingly, Shirangi et al. (2013) identified a male-enlarged muscle, named hg1, which is dedicated to the generation of the sine song but plays no role in pulse song production. The male-type *dsx* product DsxM promotes hg1 muscle growth and the female-type *dsx* product DsxF suppresses it, whereas *fru* function does not influence the hg1 size (Shirangi et al. 2013). Another muscle, ps1, is specifically required for maintaining the pulse carrier frequency and pulse amplitude of the pulse song, but plays no role in sine song production. Ps1 is sexually monomorphic, yet innervated by *fru*-positive motoneurons (Shirangi et al. 2013). These findings shed light on the importance of modular modification of peripheral structures (the neuromuscular system in the above example) in the diversification of song patterns among species.

The Baker group (Robinett et al. 2010) created the other types of *dsx* knock-in alleles, $dsx^{Gal4(1)}$ and $dsx^{Gal4(\Delta2)}$, which are slightly different from each other in construct design yet are both null for *dsx* functions (here these alleles are collectively called dsx^{Gal4-}). When dsx^{Gal4-}-expressing neurons were forcibly activated with dTrpA1 via $dsx^{Gal4(\Delta2)}$ a solitary male sang both the pulse and sine songs (Pan et al. 2011), in keeping with the above result that the synaptic block from all *dsx*-expressing cells by TNT inhibited the generation of both pulse and sine songs (Rideout et al. 2010). When $dsx^{Gal4(1)}$ was used instead of $dsx^{Gal4(\Delta2)}$, only sine songs were produced by dTrpA1-mediated activation (Pan et al. 2011). The reason why $dsx^{Gal4(1)}$ and $dsx^{Gal4(\Delta2)}$ behave differently is elusive. An intriguing possibility is that $dsx^{Gal4(1)}$ may not reproduce *dsx* expression in the command element in the brain, and, therefore, dTrpA1 could generate courtship songs only through its action on the motor center in the thorax, in which *dsx*-expressing neurons might contribute solely to the production of sine songs as implied by the sine-song-less phenotype of loss-of-function *dsx* mutants. Interestingly, even when the male fly was excited forcibly with dTrpA1 via $dsx^{Gal4(\Delta2)}$, he generated not only sine songs but also pulse songs provided that a target female was present (Pan et al. 2011). This observation is compatible with the idea that the $dsx^{al4(\Delta2)}$ males generate pulse songs if the brain command system is adequately stimulated by natural sensory cues.

Sensory Control of Courtship Behavior

Although sensory inputs play submissive roles in the motor pattern generation for courtship, they serve a primary role in the initiation of courtship. Gustation, olfaction, vision, audition, and tactile sensation all participate in successful mating, and their relative importance changes according to the context and condition (Krstic et al. 2009).

Given the above findings, it is probably safe to say that gustation generally plays the decisive role in this process, as the major sex pheromones in *Drosophila* are cuticular hydrocarbons (CHs) that are typically in solid or liquid phases (Jallon 1984). In *D. melanogaster*, the contents of some CHs are sexually dimorphic, e.g., 7-tricosene (7-T) is enriched in males whereas 7,11-Heptacosadiene (7,11-HD) is nearly female specific (Billeter et al. 2009; Ferveur 1997; Jallon 1984; Yew et al. 2009). These and other CHs, many of which remain to be identified, have a strong impact on mating success. In this review, I will not go into further details of CHs and their synthesis, but those who are interested may consult the other articles listed above. Instead, I will here focus on the perception of CH sex pheromones.

Gustatory Receptors for Nonvolatile Pheromone Sensation

Among the *ca*70 gustatory receptors (GRs), GR68a was the first that was suggested to function as a receptor in tarsal chemosensory hairs to detect a CH sex pheromone during tapping of the female abdomen with forelegs by a courting male. Inactivation of GR68a-expressing cells with TNT impaired male-to-female courtship without affecting male-to-male courtship, which could be observed even in wild-type males at a low level (Bray and Amrein 2003). However, subsequent studies by Ejima and Griffith (2008) demonstrated that GR68a plays more prevalent roles in audition, since *Gr68a*-inactivated males did not show worse courtship achievement than wild-type males if their performance was evaluated under conditions in which phonetic cues derived from a target female were withheld. The fact that axons of *Gr68a*-expressing sensory neurons terminate in the antennal mechanosensory and motor center AMMC (Kamikouchi et al. 2006) in the suboesophageal ganglion supports the notion that *Gr68a* plays a role in audition (Ejima and Griffith 2008; Koganezawa et al. 2010).

Gr39a and *Gr32a* are close relatives of *Gr68a* (Gardiner et al. 2008). Loss-of-function mutations in *Gr39a* shortened courtship bouts in males, suggesting that *Gr39a* plays a role in sustaining courtship (Watanabe et al. 2011). On the other hand, knocking out *Gr32a* was found to increase male-to-male courtship (Miyamoto and Amrein 2008) when the target male was decapitated (Wang et al. 2011). In addition, *Gr32a* knockout eliminates male-to-male aggression, which requires 7-tricosene to occur (Wang et al. 2011). Inactivation of *Gr32a*-expressing tarsal sensory neurons also disturbed the male mating posture (Koganezawa et al. 2010). Wild-type males extend and vibrate one of two wings during courtship (unilateral extension) while the wing of the other side remains strictly in the resting position. In contrast, the males with inactivated *Gr32a*-expressing neurons or *Gr32a*-knockout males are unable to hold the wing of the "wrong side" in the resting position and, instead, extend both wings to a different degree (simultaneous extension). This change in the male posture results in the distortion of species-specific song patterns: tone pulses change from monocyclic to polycyclic (Koganezawa et al. 2010). Wang et al. (2011) reported a failure in replicating the observation by Koganezawa et al. (2010) that *Gr32a* knockout led to simultaneous wing extension during courtship. This failure is likely due to the fluorescent illumination used by Wang et al. (2011); for unequivocal demonstration of the gustatory component in the regulation of wing usage, the observation must be done under dim red light conditions to eliminate visual contributions that compensate for the deprived gustatory information (Koganezawa et al. 2010).

In the tarsus, *Gr32a*-expressing sensory neurons are housed in several chemosensory hairs, packed together with supporting cells that express the odorant-binding proteins OBP57d and OBP57e (Koganezawa et al. 2010). Odorant-binding proteins are obligatory elements for the reception of volatile pheromones although they are often dispensable for other odorants by olfactory receptors (Laughlin et al. 2008). It is therefore plausible that OBP57d/e cooperates with GR32a, which functions as a gustatory receptor for nonvolatile pheromones. Notably, knockout of *Obp57d* but not *Obp57e* phenocopied the *Gr32a* mutants to cause simultaneous wing extension during courtship (Koganezawa et al. 2010). Another protein secreted from support cells in tarsal sensory hairs is CheB42a, which appears to attenuate male sexual activities (Park et al. 2006). Although GR32a is dispensable for normal taste, it is coexpressed with several of the other receptor proteins involved in bitter taste perception (Isono and Morita 2010; Weiss et al. 2011). Importantly, the Ferveur group has demonstrated with single unit recordings from bitter cells in the labellum that 7-tricosene activates their firing, which exhibits cross adaptation when treated with caffeine, a bitter tastant (Lacaille et al. 2007). *Drosophila* gustatory receptor proteins seem to form a functional receptor after hetero-oligomerization, and an essential component common to possibly all "bitter" receptors is GR33a, knockout of which has been shown to enhance male-to-male courtship (Moon et al. 2009). Considering all these results together, it is tempting to hypothesize that GR32a is a receptor for 7-tricosene, which has at least three distinct functions, inhibiting homosexual courtship, promoting aggression, and shaping the male courtship posture.

Ligand-Activated Channels as Mediators of Pheromonal Communication

The *pickpoket* (*ppk*) family of genes encode a group of ligand-activated Degenerin/epithelial Na$^+$ (DEC/ENaC) channels. The *D. melanogaster* genome harbors 30 *ppk* genes, but many of them have been poorly characterized. Among these, the *ppk23*, *ppk25, and ppk29* genes have been shown to contribute to pheromone perception (Lin et al. 2005; Lu et al. 2012; Starostina et al. 2012; Thistle et al. 2012; Toda et al. 2012). The products of these three genes appear to form a complex to operate as a functional channel (Liu et al. 2012). *ppk23, 25,* and *29*-expressing sensory neurons, some of which coexpress *fru*, are distributed widely on the legs and wings (Thistle et al. 2012), and their axon terminals in the thoracic ganglion exhibit sexual dimorphism: they cross the midline only in males (Mellert et al. 2010; Possidente and Murphey 1989). Thistle et al. (2012) reported that a pair of sensory neurons both expressing Ppk23 and Ppk29 are contained in a single sensory hair in male and female flies, and one cell called the female sensing cell preferentially responds to female pheromones such as 7,11-HD and 7,11-nonacosadiene, whereas the other, called the male sensing cell, responds to male pheromones such as 7-T, 7-pentacosene and *cis*-vaccenylacetate (cVA). It remains to be seen whether *ppk23, ppk25, and ppk29* are directly and specifically involved in sexual behavior because at least one of these (the *ppk23* gene) has been identified as the gene which reduces aggression when mutated (Edwards et al. 2009). Yet other candidates for the gustatory pheromone receptors are IR52c and IR52d, which belong to the family of IR (Ionotropic Receptor) proteins that are structurally related to ionotropic glutamate receptors (Koh et al. 2014). Mutational loss of Ir52c and Ir52d impairs male courtship performance, and Ir52c-expressing sensory neurons exhibit elevated activities only when the males are housed with females (Koh et al. 2014), based on monitoring by the Ca^{2+}-sensitive activity reporter cassette, the Nuclear Factor of activated T cells (NFAT; Masuyama et al. 2012). IR52c or IR52d-expressing cells are housed in leg sensory hairs, particularly those associated with the male-specific group of bristles known as the "sex comb" (Koh et al. 2014), which is brought into contact with female genitalia during copulation attempts. The ligands for IR52c and IR52d remain to be determined, however.

Circuitry for Gustatory Pheromone Processing

It is largely unknown how primary gustatory inputs are processed to regulate motor outputs, particularly during courtship behavior. Koganezawa et al. (2010) documented that the axon terminals of *Gr32a*-expressing sensory neurons were closely apposed to the contralateral dendritic process of mAL interneurons in males, whereas no obvious contact between the two was detected in females due to the sex difference in the mAL structure. Remarkably, inactivation of an mAL MARCM clone with dominant negative *shi* led to simultaneous wing extension similar to that induced by the inactivation of *Gr32a*-expressing sensory neurons (Koganezawa et al. 2010). It is envisaged that laterally asymmetric sensory inputs from the left and right legs provoke excitation of mAL interneurons on both the left and right sides, and the asymmetry in the strength of evoked activities in mALs is then intensified by reciprocal lateral inhibition between the mAL pair of the two sides (Koganezawa et al. 2010), sharpening the laterality in wing usage during courtship. The mAL interneurons are indeed GABAergic (Koganezawa et al. 2010). This study shows that the sexually dimorphic neurite structure allows gender-specific processing of otherwise non-sex-specific sensory inputs to shape sex-specific behavior.

Progress in an effort to map all neurons in the standardized brain offered the opportunity to predict likely synaptic contacts among the neurons (e.g., Chiang et al. 2011). For *fru*-expressing neurons in particular, a wiring diagram of possible connections has been proposed (Yu et al. 2010) based on an intersectional genetic strategy to label a few defined neurons (Yu et al. 2010) and MARCM-aided single-neuron labeling (Cachero et al. 2010). These databases point to the possibility that mAL neurons are presynaptic to P1 neurons. If this were the case, mAL interneurons could function as the sensory filter to extract the courtship-specific key stimulus, which would gate firing of P1 neurons that release the innate program for courtship. *Gr32a* has also been implicated in premating reproductive isolation: *Gr32a*-knockout induces a high level of interspecific courtship in *melanogaster* males (Fan et al. 2013). DT6, an *fru*-expressing interneuron cluster, similarly induces a high level of interspecific courtship in *melanogaster* males if these neurons are prevented from being masculinized by *fru* knockdown (Fan et al. 2013). The *Gr32a* sensory neuron–*fru*-expressing mAL interneuron pathway likely conveys the information on species differences in the hydrocarbon profile to DT6 interneurons because the proposed circuit diagram (Yu et al. 2010) predicts a synaptic connection between mAL and DT6.

The Olfactory System Regulating Courtship Behavior

Kondoh et al. (2003) discovered conspicuous sexual dimorphisms in the volume of two out of 51 glomeruli in the antennal lobe, globular neuropil compartments of the primary olfactory center; the DA1 and VA1v (=VA1lm/l; Jefferis et al. 2007) glomeruli were 62 % and 33 % larger in males than in females, respectively. Kondoh et al. (2003) presumed that these two glomeruli would process olfactory pheromone information, since their dorsolateral location in the antennal lobe corresponds to that of the macroglomerular complex in moths and cockroaches, the well-established pheromone-specific primary olfactory center (Schneiderman et al. 1982; Watanabe et al. 2010). Subsequently, Stockinger et al. (2005) found that *fru*-expressing olfactory receptor neurons (ORN) selectively innervate these two and neighboring VL2a glomeruli. In insects, each ORN typically expresses a variable ligand-binding receptor subunit and a constant subunit, Olfactory receptor 83b (OR83b, also known as Olfactory receptor co-receptor: Orco) (Vosshall 2008), and they appear to form a functional receptor that forms an ion channel (Sato et al. 2008). Among 62 olfactory receptor genes in the *D. melanogaster* genome, OR67d and Or65a are specifically responsive to cVA (Ha and Smith 2006; van der Goes van Naters and Carlson 2007), a substance originally found in the male accessory gland (Butterworth 1969) and subsequently shown to reduce the sexual attractiveness of fertilized females when transferred, together with sperm from a male to a female during copulation (Jallon et al. 1981; Tompkins 1984). cVA is not just a sex pheromone. It also stimulates aggregation in both sexes (Bartelt et al. 1985) and modulates aggression in males (Wang and Anderson 2010). In responding to cVA, the *Or67d*-expressing neurons require two additional proteins; a transmembrane protein called sensory neuron membrane protein (SNMP; Benton et al. 2007; Jin et al. 2008) and the Odorant binding protein LUSH (Laughlin et al. 2008). ORNs expressing the same variable subunit project to the identical glomerulus. One of two cVA-responsive ORN groups, *Or67d*-expressing ORNs, projects to DA1 (Kurtovic et al. 2007), a male-enlarged glomerulus in *melanogaster*, and the other group of cVA-responsive ORNs, *Or65a*-expressing ORNs, project to DL3 (Lebreton et al. 2014), a glomerulus which is male-enlarged in some Hawaiian *Drosophila* species but not in *melanogaster* (Kondoh et al. 2003). Yet another male-enlarged glomerulus in *melanogaster*, VA1v, is innervated by ORNs-expressing OR47b (Wang et al. 2011), which responds to unidentified odor compounds shared by both sexes (van der Goes van Naters and Carlson 2007). Although the VA1v projection neurons have been implicated in increasing sexual receptivity in virgin females (Sakurai et al. 2013), their roles in males remain obscure. Or67d mediates an acute effect of cVA to elicit aggression in males (Wang and Anderson 2010), whereas Or65a mediates a chronic effect of cVA to suppress male courtship (and thus contributes to a reduction in the sex-appeal of mated females; Ejima and Griffith 2008) and aggression (Liu et al. 2011). Recent work has shown that the aggregation-inducing effect of cVA in females varies depending on their mating status: females mated with a starved male carry less cVA in their bodies and are attracted to cVA, whereas females mated with a fed male carry more cVA and are not attracted to cVA (Lebreton et al. 2014). Lebreton et al. (2014) further suggested that the Or67d-dependent attraction to cVA is inhibited by Or65a-mediated afferents in the antennal lobe, which attenuate the activities of DA1 projection neurons.

Kohl et al. (2013) employed patch-clamp recordings from cVA-responsive *fru*-expressing central neurons and revealed that the DA1 projection neuron–aSP-f neuron synaptic connection is present in males but not in females, where DA1 projection neurons synapse on an alternative target, aSP-g neurons: cVA-responses can be recorded from aSP-f but not aSP-g in males, whereas the opposite is true in females. The aSP-f neurons that are *tra*^{1} mutant (thus masculinized)-MARCM clones in the female brain generate excitatory postsynaptic potentials (EPSPs) upon the antennal stimulation with cVA, whereas similar *tra*^{1} mutant aSPg in the female brain becomes unresponsive to cVA (Kohl et al. 2013). Thus, sensory input provoked by cVA is processed in a sex-specific manner, which relies on *tra* (and potentially *fru*)-dependent sexually dimorphic synaptic connections between DA1 projection axons and aSP-f (in males) or aSP-g (in females). This finding points to the importance of cell-autonomous sex determination in each neuron for the formation of functional circuitry underlying sex-specific behaviors.

Inhibitory Pheromone Inputs Modulate the Decision to Court

As described above, cVA is regarded as an inhibitory pheromone that attenuates courtship vigor in males toward fertilized females. Indeed, coating with cVA significantly diminished the attractiveness of virgin females for males. In accord with this, cVA dramatically reduced the Ca^{2+} response of *fru*-expressing neurons in the lateral protocerebrum of a tethered male when stimulated by the virgin female contact (Kohatsu et al. 2011). This observation suggests that inputs from the cVA processing circuitry to P1 neurons play a critical role in decision-making by the latter.

Auditory Input Affects Male Courtship

Courtship song plays an important role in recognizing a conspecific mating partner. In *D. melanogaster*, only males sing, yet the song affects behavior of not only females but also males: in virgin females, it increases sexual receptivity whereas, in males, it promotes locomotion which often accompanies unilateral wing vibration for singing (von Schilcher 1976). Courtship song is perceived by the sensory cells housed in the Johnston's organ (JO), which is located in the second segment of antenna. Axons of JO neurons (JONs) project to specific zones within the AMMC and those sensitive to courtship song terminate mainly in zones A and B (Yorozu et al. 2009). The secondary neurons that receive synaptic input from JONs extract low frequency components in the primary afferent signals, conveying temporal features of courtship song by their slow potential changes without spiking (Tootoonian et al. 2012). Two specific classes of neurons in the AMMC, aPN1 and aLN(al), impair male mating success when artificially silenced; aPN1 projection neurons send ascending axons to the bilateral "wedge" region and aLN(al) represents local interneurons that are GABAergic (Vaughan et al. 2014). GCaMP imaging revealed that aPN1 dendrites are best tuned to the frequency range within which IPIs of pulse song vary (Vaughan et al. 2014). It remains to be explored how aPN1-mediated song information contributes to the decision-making by P1 neurons.

Experience Modulates Innate Courtship Behavior

Male flies change their mating behavior by experience. For example, males who have experienced rejection by a unreceptive fertilized female display diminished courtship when he subsequently encounters another female even if the second female is potentially receptive to mate (Siegel and Hall 1979). This courtship suppression has been interpreted as counterconditioning (Zawistowski 1988), in which the male associates the attractive pheromonal cue with the punishment of unsuccessful mating with the first female (training), reducing courtship activity toward the second female (test). More recent work proposes an alternative interpretation that the male becomes sensitized to cVA at the training so that he becomes less active to court a fertilized female who carries cVA (Keleman et al. 2012). Another form of plasticity in male mating behavior is manifested by prolongation of the copulation duration in the presence of rival males (Kim et al. 2013). For this plasticity in copulation duration to occur, peptidergic signaling involving pigment-dispersing factor (PDF) and neuropeptideF (NPF) among several clock neurons, s-Lv, LNd, and s-LNv in the brain needs to be functional (Kim et al. 2013). Some of these neurons are sexually dimorphic and even fru-positive (Kim et al. 2013). It remains to be examined how these clock neurons modulate functions of the central courtship circuitry.

Chaining in courtship is one of the hallmarks of *fru* mutant males (Hall 1978). Expression of this behavior in *fru* mutant males was recently shown to be experience dependent (Pan and Baker 2014). Deprivation of social interaction by isolating *fru* mutant males on the day of eclosion, significantly suppressed male-to-male courtship chaining when tested in a group several days later (Pan and Baker 2014). Acquisition of chaining propensity is abrogated when *dsx* and *fru* double-positive neurons are silenced throughout development (Pan and Baker 2014). The male-specific DsxM is indispensable for developing chaining behavior in *fru* mutant males (Pan and Baker 2014). Viewing sibs after adult emergence is prerequisite for the acquisition of chaining propensity whereas expressing chaining behavior requires fly odor, which is dispensable for acquiring this property (Pan and Baker 2014). Kohatsu and Yamamoto (2015) demonstrated, with the virtual reality paradigm, that *fru* mutant males exhibit courtship pursuit in response to moving light spot without prior tapping of a target fly or without Channelrhodopsin-mediated depolarization of pC1 (P1) neurons, either of which is essential for inducing courtship pursuit in wild-type males. Importantly, courtship pursuit of a moving light spot by *fru* mutant males is greatly attenuated when the flies are isolated at eclosion (Kohatsu and Yamamoto 2015), suggesting that group-housed experience causes visual hypersensitivity, resulting in vision-induced courtship pursuit in response to inappropriate targets including moving light spots and other males. Indeed, *dsx*-positive neurons in the lateral protocerebrum, that include pC1 (P1), exhibit a Ca^{2+} rise in response to moving light spots in group-housed *fru* mutant males but not in single-housed *fru* mutant males or wild-type males of either housing conditions (Kohatsu and Yamamoto 2015). This result raises the intriguing possibility that pC1(P1) neurons are the site where innate courtship behavior is modified by social experience. In contrast, Inagaki et al. (2013) reported that courtship activities in wild-type males decrease when they are group housed, and this decline in courtship activities is correlated with reduced excitability of P1 neurons as detected by the reduced Ca^{2+} rise in response to Channelrhodopsin-mediated neural stimulation. It remains to be determined why group-housing-modified P1 neuron responsiveness is more excitable in *fru* mutant males and less excitable in wild-type males. The discovery of experience-dependence of male–male courtship in fru mutants paves the way for genetic analysis of molecular mechanisms for gene–environment interactions in the determination of sexual orientation.

Conclusion

By analyzing *fru*-expressing neurons, the core part of the neural circuitry for controlling male courtship behavior is now being unveiled (Fig. 1.4). Although *fru* has been proposed to be a switch gene that can specify an entire innate behavior (Demir and Dickson 2005), in a sense comparable to the selector or master gene that triggers the development of a complete anatomical structure, thorough analysis of the *fru* function has disclosed the existence of more redundancy and flexibility in the neuronal architecture than initially thought, inviting an alternative hypothesis for the organizational principle of courtship circuitry. Efforts to describe all neuronal connections are being made to this end. In parallel with the development of a realistic understanding of *fru*'s role, the importance of the other sex determination factor *dsx* in the courtship circuitry was rediscovered. Why, then, are *fru* and *dsx* needed to shape the courtship circuitry? One possibility is that dual control by two genes makes the system more robust, and the sexual differentiation of the brain less vulnerable to environmental stress. Another possibility that merits consideration is

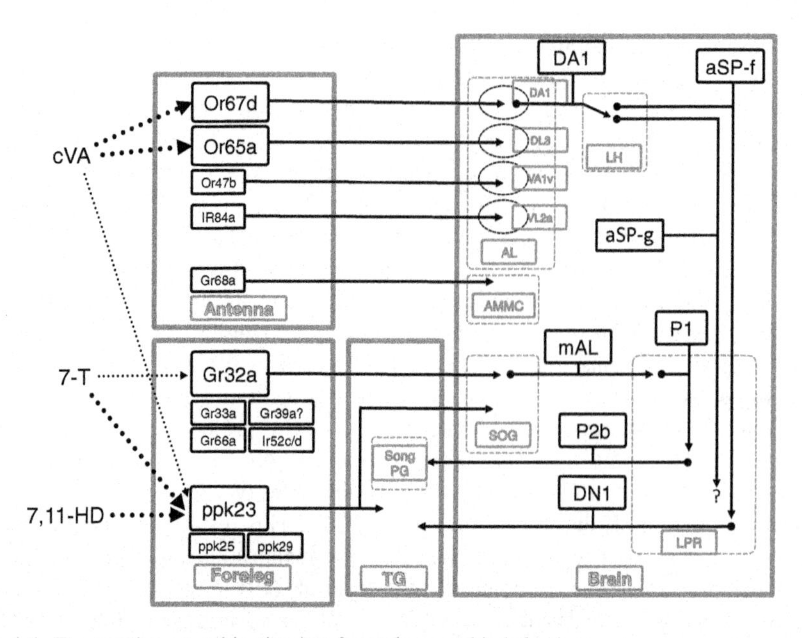

Fig. 1.4 The putative courtship circuitry for male courtship behavior

that dual control might confer more flexibility rather than robustness on the courtship system. Radical changes in *dsx* expression and function are risky, as this gene is the key for the sexual differentiation of many reproductive functions. Even when we just look at neural expression, *dsx*-positive cells are highly enriched in the abdominal ganglion, the most "visceral" part of the nervous system. In contrast, *fru*, at least in terms of its products governed by the distal promoter, has more freedom to change because it has no vital functions. Recent studies by the Baker group show that the forced activation of *fru*-expressing neurons by dTrpA1 induced, in *melanogaster* males, heterospecific courtship, which was not seen when *dsx*-expressing neurons were activated in a similar way (Pan et al. 2011). This might mean that evolutionary changes in the courtship system, including those that could lead to speciation, are driven by non-vital *fru*, whereas conserved reproductive functions are retained by *dsx*. Regardless of whether this interpretation is correct or not, *fru* and *dsx* together establish the sex differences in the nervous system which themselves are shaped through sexual selection upon courtship behavior, their own product. The courtship circuit will therefore remain a topical subject in biology for the coming decade.

Acknowledgements I thank Masa Koganezawa for critical reading of the manuscript, and Mayura Suyama and Hiromi Sato for secretarial assistance. Our studies were supported by Grants-in-Aid for Scientific Research from the Japanese Government Ministry of Education, Culture, Sports, Science and Technology (26113702, 26114502, 24113502 and 23220007) to D.Y., the Japan-France Bilateral Joint Research Project Grant from Japan Society for the Promotion of Science to D.Y. and the Life Science Grant from the Takeda Science Foundation to D. Y.

References

Adams, M. D., Celniker, S. E., Holt, R. A., Evans, C. A., Gocayne, J. D., Amanatides, P. G., et al. (2000). The genome sequence of *Drosophila melanogaster. Science, 287*, 2185–2195.

Agrawal, S., Safarik, S., & Dickinson, M. (2014). The relative roles of vision and chemosensation in mate recognition of *Drosophila melanogaster. Journal of Experimental Biology, 217*, 2796–2805.

Bartelt, R. J., Schaner, A. M., & Jackson, L. L. (1985). *cis*-Vaccenyl acetate as an aggregation pheromone in *Drosophila melanogaster. Journal of Chemical Ecology, 11*, 1747–1756.

Bath, D. E., Stowers, J. R., Hormann, D., Poehlmann, A., Dickson, B. J., & Straw, A. D. (2014). FlyMAD: Rapid thermogenetic control of neuronal activity in freely walking *Drosophila. Nature Methods, 11*, 756–762.

Benton, R., Vannice, K. S., & Vosshall, L. B. (2007). An essential role for a CD36-related receptor in pheromone detection in *Drosophila. Nature, 450*, 289–293.

Billeter, J. C., Atallah, J., Krupp, J. J., Millar, J. G., & Levine, J. D. (2009). Specialized cells tag sexual and species identity in *Drosophila melanogaster. Nature, 461*, 987–991.

Billeter, J. C., Rideout, E. J., Dornan, A. J., & Goodwin, S. F. (2006). Control of male sexual behavior in *Drosophila* by the sex determination pathway. *Current Biology, 16*, R766–R776.

Brand, A. H., & Perrimon, N. (1993). Targeted gene expression as a means of altering cell fates and generating dominant phenotypes. *Development, 118*, 401–415.

Bray, S., & Amrein, H. (2003). A putative *Drosophila* pheromone receptor expressed in male-specific taste neurons is required for efficient courtship. *Neuron, 39*, 1019–1029.

Broughton, S. J., Kitamoto, T., & Greenspan, R. J. (2004). Excitatory and inhibitory switches for courtship in the brain of *Drosophila melanogaster. Current Biology, 14*, 538–547.

Butterworth, F. M. (1969). Lipids of *Drosophila*: A newly detected lipid in the male. *Science, 163*, 1356–1357.

Cachero, S., Ostrovsky, A. D., Yu, J. Y., Dickson, B. J., & Jefferis, G. S. (2010). Sexual dimorphism in the fly brain. *Current Biology, 20*, 1589–1601.

Cande, J., Stern, D. L., Morita, T., Prud'homme, B., & Gompel, N. (2014). Looking under the lamp post: Neither fruitless nor *doublesex* has evolved to generate divergent male courtship in *Drosophila. Cell Reports, 8*, 363–370.

Chiang, A. S., Lin, C. Y., Chuang, C. C., Chang, H. M., Hsieh, C. H., Yeh, C. W., et al. (2011). Three-dimensional reconstruction of brain-wide wiring networks in *Drosophila* at single-cell resolution. *Current Biology, 21*, 1–11.

Clyne, J. D., & Miesenböck, G. (2008). Sex-specific control tuning of the pattern generator for courtship song in *Drosophila. Cell, 133*, 354–363.

Coen, P., Clemens, J., Weinstein, A. J., Pacheco, D. A., Deng, Y., & Murthy, M. (2014). Dynamic sensory cues shape song structure in *Drosophila. Nature, 507*, 233–237.

Dalton, J. E., Fear, J. M., Knott, S., Baker, B. S., McIntyre, L. M., & Arbeitman, M. N. (2013). Male-specific fruitless isoforms have different regulatory roles conferred by distinct zinc finger DNA binding domains. *BMC Genomics, 14*, 659.

Demir, E., & Dickson, B. J. (2005). *fruitless* splicing specifies male courtship behavior in *Drosophila. Cell, 121*, 785–794.

Dudai, Y., Jan, Y. N., Byers, D., Quinn, W. G., & Benzer, S. (1976). *dunce*, a mutant of *Drosophila* deficient in learning. *Proceedings of the National Academy of Sciences of the United States of America, 73*, 1684–1688.

Edwards, A. C., Ayroles, J. F., Stone, E. A., Carbone, M. A., Lyman, R. F., & Mackay, T. F. (2009). A transcriptional network associated with natural variation in *Drosophila* aggressive behavior. *Genome Biology, 10*, R76.

Ejima, A., & Griffith, L. C. (2008). Courtship initiation is stimulated by acoustic signals in *Drosophila melanogaster. PLoS ONE, 3*, e3246.

Fan, P., Manoli, D. S., Ahmed, O. M., Chen, Y., Agarwal, N., Kwong, S., et al. (2013). Genetic and neural mechanisms that inhibit *Drosophila* from mating with other species. *Cell, 154*, 89–102.

Ferveur, J. F. (1997). The pheromonal role of cuticular hydrocarbons in *Drosophila melanogaster. Bioessays, 19*, 353–358.

Ferveur, J. F., & Greenspan, R. J. (1998). Courtship behavior of brain mosaics in *Drosophila. Journal of Neurogenetics, 12*, 205–226.

Gailey, D. A., & Hall, J. C. (1989). Behavior and cytogenetics of *fruitless* in *Drosophila melanogaster*: Different courtship defects caused by separate, closely linked lesions. *Genetics, 121*, 773–785.

Gailey, D. A., Taylor, B. J., & Hall, J. C. (1991). Elements of the *fruitless* locus regulate development of the muscle of Lawrence, a male-specific structure in the abdomen of *Drosophila melanogaster* adults. *Development, 113*, 879–890.

Gardiner, A., Barker, D., Butlin, R. K., Jordan, W. C., & Ritchie, M. G. (2008). Evolution of a complex locus: Exon gain, loss and divergence at the Gr39a locus in *Drosophila. PLoS One, 3*, e1513.

Gill, K. S. (1963). A mutation causing abnormal courtship and mating behavior in males of *Drosophila melanogaster. American Zoologist, 3*, 507.

Gleason, J. M., & Ritchie, M. G. (2004). Do quantitative trait loci (QTL) for a courtship song difference between *Drosophila simulans* and *D. sechellia* coincide with candidate genes and intraspecific QTL? *Genetics, 166*, 1303–1311.

Goto, J., Mikawa, Y., Koganezawa, M., Ito, H., & Yamamoto, D. (2011). Sexually dimorphic shaping of interneuron dendrites involves the Hunchback transcription factor. *Journal of Neuroscience, 31*, 5454–5459.

Ha, T. S., & Smith, D. P. (2006). A pheromone receptor mediates 11-cis-vaccenyl acetate-induced responses in *Drosophila. Journal of Neuroscience, 26*, 8727–8733.

Hall, J. C. (1978). Courtship among males due to a male-sterile mutation in *Drosophila melanogaster. Behavior Genetics, 8*, 125–141.

Hall, J. C. (1979). Control of male reproductive behavior by the central nervous system of *Drosophila*: Dissection of a courtship pathway by genetic mosaics. *Genetics, 92*, 437–457.

Hamada, F. N., Rosenzweig, M., Kang, K., Pulver, S. R., Ghezzi, A., Jegla, T. J., et al. (2008). An internal thermal sensor controlling temperature preference in *Drosophila. Nature, 454*, 217–220.

Hotta, Y., & Benzer, S. (1970). Genetic dissection of the *Drosophila* nervous system by means of mosaics. *Proceedings of the National Academy of Sciences of the United States of America, 67*, 1156–1163.

Hotta, Y., & Benzer, S. (1972). Mapping of behaviour in *Drosophila mosaics. Nature, 240*, 527–535.

Ikeda, K., & Kaplan, W. D. (1970a). Patterned neural activity of a mutant *Drosophila melanogaster. Proceedings of the National Academy of Sciences of the United States of America, 66*, 765–772.

Ikeda, K., & Kaplan, W. D. (1970b). Unilaterally patterned neural activity of gynandromorphs, mosaic for a neurological mutant of *Drosophila melanogaster. Proceedings of the National Academy of Sciences of the United States of America, 67*, 1480–1487.

Ikeda, K., Ozawa, S., & Hagiwara, S. (1976). Synaptic transmission reversibly conditioned by single-gene mutation in *Drosophila melanogaster. Nature, 259*, 489–491.

Inagaki, H. K., Jung, Y., Hoopfer, E. D., Wong, A. M., Mishra, N., Lin, J. Y., et al. (2013). Optogenetic control of *Drosophila* using a red-shifted channelrhodopsin reveals experience-dependent influences on courtship. *Nature Methods, 11*, 325–332.

Isono, K., & Morita, H. (2010). Molecular and cellular designs of insect taste receptor system. *Frontiers in Cellular Neuroscience, 4*, 20.

Ito, H., Fujitani, K., Usui, K., Shimizu-Nishikawa, K., Tanaka, S., & Yamamoto, D. (1996). Sexual orientation in *Drosophila* is altered by the *satori* mutation in the sex-determination gene *fruitless* that encodes a zinc finger protein with a BTB domain. *Proceedings of the National Academy of Sciences of the United States of America, 93*, 9687–9692.

Ito, H., Sato, K., Koganezawa, M., Ote, M., Matsumoto, K., Hama, C., et al. (2012). Fruitless recruits two antagonistic chromatin factors to establish single-neuron sexual dimorphism. *Cell, 149*, 1327–1338.

Jallon, J.-M. (1984). A few chemical words exchanged by *Drosophila* during courtship and mating. *Behavior Genetics, 14*, 441–478.

Jallon, J.-M., Antony, C., & Benamar, O. (1981). Un anti-aphrodisiaque produit par les males *Drosophila melanogaster* et transfere aux femelles lors de la copulation. *Comptes Rendus. Académie des Sciences, 292*, 1147–1149.

Jefferis, G. S., Potter, C. J., Chan, A. M., Marin, E. C., Rohlfing, T., Maurer, C. R., Jr., et al. (2007). Comprehensive maps of *Drosophila* higher olfactory centers: Spatially segregated fruit and pheromone representation. *Cell, 128*, 1187–1203.

Jin, X., Ha, T. S., & Smith, D. P. (2008). SNMP is a signaling component required for pheromone sensitivity in *Drosophila. Proceedings of the National Academy of Sciences of the United States of America, 105*, 10996–11001.

Kamikouchi, A., Shimada, T., & Ito, K. (2006). Comprehensive classification of the auditory sensory projections in the brain of the fruit fly *Drosophila melanogaster. Journal of Comparative Neurology, 499*, 317–356.

Kandel, E. R. (1976). *Cellular basis of behavior*. New York: W. H. F. Freeman and Company.

Keleman, K., Vrontou, E., Kruttner, S., Yu, J. Y., Kurtovic-Kozaric, A., & Dickson, B. J. (2012). Dopamine neurons modulate pheromone responses in *Drosophila* courtship learning. *Nature, 489*, 145–149.

Kim, W. J., Jan, L. Y., & Jan, Y. N. (2013). A PDF/NPF neuropeptide signaling circuitry of male *Drosophila melanogaster* controls rival-induced prolonged mating. *Neuron, 80*, 1190–1205.

Kimura, K.-I., Hachiya, T., Koganezawa, M., Tazawa, T., & Yamamoto, D. (2008). Fruitless and Doublesex coordinate to generate male-specific neurons that can initiate courtship. *Neuron, 59*, 759–769.

Kimura, K.-I., Ote, M., Tazawa, T., & Yamamoto, D. (2005). *fruitless* specifies sexually dimorphic neural circuitry in the *Drosophila* brain. *Nature, 438*, 229–233.

Kitamoto, T. (2002). Conditional disruption of synaptic transmission induces male-male courtship behavior in *Drosophila. Proceedings of the National Academy of Sciences of the United States of America, 99*, 13232–13237.

Koganezawa, M., Haba, D., Matsuo, T., & Yamamoto, D. (2010). The shaping of male courtship posture by lateralized gustatory inputs to male-specific interneurons. *Current Biology, 20*, 1–8.

Koh, T. W., He, Z., Gorur-Shandilya, S., Menuz, K., Larter, N. K., Stewart, S., et al. (2014). The *Drosophila* IR20a clade of ionotropic receptors are candidate taste and pheromone receptors. *Neuron, 83*, 850–865.

Kohatsu, S., Koganezawa, M., & Yamamoto, D. (2011). Female contact activates male-specific interneurons that trigger stereotypic courtship behavior in *Drosophila. Neuron, 69*, 498–508.

Kohatsu, S., & Yamamoto, D. (2015). Visually induced initiation of *Drosophila* innate courtship-like following pursuit is mediated by central excitatory state. *Nature Communications, 6*, 6457.

Kohl, J., Ostrovsky, A. D., Frechter, S., & Jefferis, G. S. (2013). A bidirectional circuit switch reroutes pheromone signals in male and female brains. *Cell, 155*, 1610–1623.

Kondoh, Y., Kaneshiro, K. Y., Kimura, K.-I., & Yamamoto, D. (2003). Evolution of sexual dimorphism in the olfactory brain of Hawaiian *Drosophila. Proceedings of the Biological Sciences, 270*, 1005–1013.

Konopka, R., & Benzer, S. (1971). Clock mutants of *Drosophila melanogaster. Proceedings of the National Academy of Sciences of the United States of America, 68*, 2112–2116.

Krstic, D., Boll, W., & Noll, M. (2009). Sensory integration regulating male courtship behavior in *Drosophila. PLoS ONE, 4*, e4457.

Kurtovic, A., Widmer, A., & Dickson, B. J. (2007). A single class of olfactory neurons mediates behavioural responses to a *Drosophila* sex pheromone. *Nature, 446*, 542–546.

Kyriacou, C. P., & Hall, J. C. (1980). Circadian rhythm mutations in *Drosophila melanogaster* affect short-term fluctuations in the male's courtship song. *Proceedings of the National Academy of Sciences of the United States of America, 77*, 6729–6733.

Kyriacou, C. P., & Hall, J. C. (1984). Learning and memory mutations impair acoustic priming of mating behaviour in *Drosophila. Nature, 308*, 62–65.

Kyriacou, C. P., & Hall, J. C. (1986). Interspecific genetic control of courtship song production and reception in *Drosophila. Science, 232*, 494–497.

Lacaille, F., Hiroi, M., Twele, R., Inoshita, T., Umemoto, D., Manière, G., et al. (2007). An inhibitory sex pheromone tastes bitter for *Drosophila* males. *PLoS ONE, 2*, e661.
Laughlin, J. D., Ha, T. S., Jones, D. N., & Smith, D. P. (2008). Activation of pheromone-sensitive neurons is mediated by conformational activation of pheromone-binding protein. *Cell, 133*, 1255–1265.
Lawrence, P. A., & Johnston, P. (1984). The genetic specification of pattern in a *Drosophila* muscle. *Cell, 36*, 775–782.
Lawrence, P. A., & Johnston, P. (1986). The muscle pattern of a segment of *Drosophila* may be determined by neurons and not by contributing myoblasts. *Cell, 45*, 505–513.
Lebreton, S., Grabe, V., Omondi, A. B., Ignell, R., Becher, P. G., Hansson, B. S., et al. (2014). Love makes smell blind: Mating suppresses pheromone attraction in *Drosophila* females via Or65a olfactory neurons. *Scientific Reports, 4*, 7119.
Lee, G., Foss, M., Goodwin, S. F., Carlo, T., Taylor, B. J., & Hall, J. C. (2000). Spatial, temporal, and sexually dimorphic expression patterns of the *fruitless* gene in the *Drosophila* central nervous system. *Journal of Neurobiology, 43*, 404–426.
Lee, G., Hall, J. C., & Park, J. H. (2002). *doublesex* gene expression in the central nervous system of *Drosophila melanogaster. Journal of Neurogenetics, 16*, 229–248.
Lee, T., & Luo, L. (1999). Mosaic analysis with a repressible cell marker for studies of gene function in neuronal morphogenesis. *Neuron, 22*, 451–461.
Lin, H., Mann, K. J., Starostina, E., Kinser, R. D., & Pikielny, C. W. (2005). A *Drosophila* DEG/ENaC channel subunit is required for male response to female pheromones. *Proceedings of the National Academy of Sciences of the United States of America, 102*, 12831–12836.
Liu, W., Liang, X., Gong, J., Yang, Z., Zhang, Y. H., Zhang, J. X., et al. (2011). Social regulation of aggression by pheromonal activation of Or65a olfactory neurons in *Drosophila. Nature Neuroscience, 14*, 896–902.
Liu, T., Starostina, E., Vijayan, V., & Pikielny, C. W. (2012). Two Drosophila DEG/ENaC channel subunits have distinct functions in gustatory neurons that activate male courtship. *Journal of Neuroscience, 32*, 11879–11889.
Lu, B., LaMora, A., Sun, Y., Welsh, M. J., & Ben-Shahar, Y. (2012). *ppk23*-Dependent chemosensory functions contribute to courtship behavior in *Drosophila melanogaster. PLoS Genetics, 8*, e1002587.
Manoli, D. S., Foss, M., Villella, A., Taylor, B. J., Hall, J. C., & Baker, B. S. (2005). Male-specific fruitless specifies the neural substrates of Drosophila courtship behaviour. *Nature, 436*, 395–400.
Marín, I., & Baker, B. S. (1998). The evolutionary dynamics of sex determination. *Science, 281*, 1990–1994.
Masuyama, K., Zhang, Y., Rao, Y., & Wang, J. W. (2012). Mapping neural circuits with activity-dependent nuclear import of a transcription factor. *Journal of Neurogenetics, 26*, 89–102.
Mellert, D. J., Knapp, J. M., Manoli, D. S., Meissner, G. W., & Baker, B. S. (2010). Midline crossing by gustatory receptor neuron axons is regulated by *fruitless, doublesex* and the Roundabout receptors. *Development, 137*, 323–332.
Miyamoto, T., & Amrein, H. (2008). Suppression of male courtship by a *Drosophila* pheromone receptor. *Nature Neuroscience, 11*, 874–876.
Moon, S. J., Lee, Y., Jiao, Y., & Montell, C. (2009). A *Drosophila* gustatory receptor essential for aversive taste and inhibiting male-to-male courtship. *Current Biology, 19*, 1623–1627.
Morgan, T. H. (1910). Sex limited inheritance in *Drosophila. Science, 32*, 120–122.
Neville, M. C., Nojima, T., Ashley, E., Parker, D. J., Walker, J., Southall, T., et al. (2014). Male-specific fruitless isoforms target neurodevelopmental genes to specify a sexually dimorphic nervous system. *Current Biology, 24*, 229–241.
Nojima, T., Kimura, K., Koganezawa, M., & Yamamoto, D. (2010). Neuronal synaptic outputs determine the sexual fate of postsynaptic targets. *Current Biology, 20*, 836–840.
Pan, Y., & Baker, B. S. (2014). Genetic identification and separation of innate and experience-dependent courtship behaviors in *Drosophila. Cell, 156*, 236–248.
Pan, Y., Meissner, G. W., & Baker, B. S. (2012). Joint control of *Drosophila* male courtship behavior by motion cues and activation of male-specific P1 neurons. *Proceedings of the National Academy of Sciences of the United States of America, 109*, 10065–10070.

Pan, Y., Robinett, C. C., & Baker, B. S. (2011). Turning males on: Activation of male courtship behavior in *Drosophila melanogaster. PLoS ONE, 6*, e21144.

Park, S. K., Mann, K. J., Lin, H., Starostina, E., Kolski-Andreaco, A., & Pikielny, C. W. (2006). A *Drosophila* protein specific to pheromone-sensing gustatory hairs delays males' copulation attempts. *Current Biology, 16*, 1154–1159.

Possidente, D. R., & Murphey, R. K. (1989). Genetic control of sexually dimorphic axon morphology in *Drosophila* sensory neurons. *Developmental Biology, 132*, 448–457.

Riabinina, O., Dai, M., Duke, T., & Albert, J. T. (2011). Active process mediates species-specific tuning of *Drosophila* ears. *Current Biology, 21*, 658–664.

Rideout, E. J., Billeter, J. C., & Goodwin, S. F. (2007). The sex-determination genes *fruitless* and *doublesex* specify a neural substrate required for courtship song. *Current Biology, 17*, 1473–1478.

Rideout, E. J., Dornan, A. J., Neville, M. C., Eadie, S., & Goodwin, S. F. (2010). Control of sexual differentiation and behavior by the *doublesex* gene in *Drosophila melanogaster. Nature Neuroscience, 13*, 458–466.

Robinett, C. C., Vaughan, A. G., Knapp, J. M., & Baker, B. S. (2010). Sex and the single cell. II. There is a time and place for sex. *PLoS Biology, 8*, e1000365.

Rubinstein, C. D., Rivlin, P. K., & Hoy, R. R. (2010). Genetic feminization of the thoracic nervous system disrupts courtship song in male *Drosophila melanogaster. Journal of Neurogenetics, 24*, 234–245.

Ruta, V., Datta, S. R., Vasconcelos, M. L., Freeland, J., Looger, L. L., & Axel, R. (2010). A dimorphic pheromone circuit in *Drosophila* from sensory input to descending output. *Nature, 468*, 686–690.

Ryner, L. C., Goodwin, S. F., Castrillon, D. H., Anand, A., Villella, A., Baker, B. S., et al. (1996). Control of male sexual behavior and sexual orientation in *Drosophila* by the *fruitless* gene. *Cell, 87*, 1079–1089.

Sakurai, A., Koganezawa, M., Yasunaga, K., Emoto, K., & Yamamoto, D. (2013). Select interneuron clusters determine female sexual receptivity in *Drosophila. Nature Communications, 4*, 1825.

Sato, K., Pellegrino, M., Nakagawa, T., Vosshall, L. B., & Touhara, K. (2008). Insect olfactory receptors are heteromeric ligand-gated ion channels. *Nature, 452*, 1002–1006.

Schneiderman, A. M., Matsumoto, S. G., & Hidebrand, J. G. (1982). Trans-sexually grafted antennae influence development of sexually dimorphic neurons in moth brain. *Nature, 298*, 844–846.

Shirangi, T. R., Stern, D. L., & Truman, J. W. (2013). Motor control of *Drosophila* courtship song. *Cell Reports, 5*, 678–686.

Shirangi, T. R., Taylor, B. J., & McKeown, M. (2006). A double-switch system regulates male courtship behavior in male and female *Drosophila melanogaster. Nature Genetics, 38*, 1435–1439.

Siegel, R. W., & Hall, J. C. (1979). Conditioned responses in courtship behavior of normal and mutant *Drosophila. Proceedings of the National Academy of Sciences of the United States of America, 76*, 3430–3434.

Song, H. J., Billeter, J. C., Reynaud, E., Carlo, T., Spana, E. P., Perrimon, N., et al. (2002). The *fruitless* gene is required for the proper formation of axonal tracts in the embryonic central nervous system of *Drosophila. Genetics, 162*, 1703–1724.

Spieth, H. T. (1952). Mating behavior within the genus *Drosophila* (Diptera). *Bulletin of the American Museum of Natural History, 99*, 399–474.

Starostina, E., Liu, T., Vijayan, V., Zheng, Z., Siwicki, K. K., & Pikielny, C. W. (2012). A *Drosophila* deg/ENaC subunit functions specifically in gustatory neurons required for male courtship behavior. *Journal of Neuroscience, 32*, 4665–4674.

Stern, D. L. (2014). Reported *Drosophila* courtship song rhythms are artifacts of data analysis. *BMC Biology, 12*, 38.

Stockinger, P., Kvitsiani, D., Rotkopf, S., Tirián, L., & Dickson, B. J. (2005). Neural circuitry that governs *Drosophila* male courtship behavior. *Cell, 121*, 795–807.

Sturtevant, A. H. (1915). Experiments on sex recognition and the problem of sexual selection in *Drosophila. Journal of Animal Behavior, 5*, 351–366.

Taylor, B. J. (1992). Differentiation of a male-specific muscle in *Drosophila melanogaster* does not require the sex-determining genes *doublesex* or intersex. *Genetics, 132*, 179–191.

Taylor, B. J., Villella, A., Ryner, L. C., Baker, B. S., & Hall, J. C. (1994). Behavioral and neurobiological implications of sex-determining factors in *Drosophila. Developmental Genetics, 15*, 275–296.

Thistle, R., Cameron, P., Ghorayshi, A., Dennison, L., & Scott, K. (2012). Contact chemoreceptors mediate male-male repulsion and male-female attraction during *Drosophila* courtship. *Cell, 149*, 1140–1151.

Toda, H., Zhao, X., & Dickson, B. J. (2012). The Drosophila female aphrodisiac pheromone activates pp k23(+) sensory neurons to elicit male courtship behavior. *Cell Reports, 1*, 599–607.

Tompkins, L. (1984). Genetic analysis of sex appeal in *Drosophila. Behavior Genetics, 14*, 411–440.

Tompkins, L., & Hall, J. C. (1983). Identification of brain sites controlling female receptivity in mosaics of *Drosophila melanogaster. Genetics, 103*, 179–195.

Tompkins, L., Hall, J. C., & Hall, L. M. (1980). Courtship-stimulating volatile compounds from normal and mutant *Drosophila. Journal of Insect Physiology, 26*, 689–697.

Tootoonian, S., Coen, P., Kawai, R., & Murthy, M. (2012). Neural representations of courtship song in the *Drosophila* brain. *Journal of Neuroscience, 32*, 787–798.

Trott, A. R., Donelson, N. C., Griffith, L. C., & Ejima, A. (2012). Song choice is modulated by female movement in *Drosophila* males. *PLoS One, 7*, e46025.

Usui-Aoki, K., Ito, H., Ui-Tei, K., Takahashi, K., Lukacsovich, T., Awano, W., et al. (2000). Formation of the male-specific muscle in female *Drosophila* by ectopic *fruitless* expression. *Nature Cell Biology, 2*, 500–506.

van der Goes van Naters, W., & Carlson, J. R. (2007). Receptors and neurons for fly odors in *Drosophila. Current Biology, 17*, 606–612.

Vaughan, A. G., Zhou, C., Manoli, D. S., & Baker, B. S. (2014). Neural pathways for the detection and discrimination of conspecific song in *D. melanogaster. Current Biology, 24*, 1039–1049.

Vernes, S. C. (2014). Genome wide identification of *fruitless* targets suggests a role in upregulating genes important for neural circuit formation. *Scientific Reports, 4*, 4412.

Villela, A., Gailey, D. A., Berwald, B., Ohshima, S., Barnes, P. T., & Hall, J. C. (1997). Extended reproductive roles of the *fruitless* gene in *Drosophila melanogaster* revealed by behavioral analysis of new *fru* mutants. *Genetics, 147*, 1107–1130.

Villella, A., & Hall, J. C. (1996). Courtship anomalies caused by doublesex mutations in *Drosophila melanogaster. Genetics, 143*, 331–344.

von Philipsborn, A. C., Liu, T., Yu, J. Y., Masser, C., Bidaye, S. S., & Dickson, B. J. (2011). Neuronal control of *Drosophila* courtship song. *Neuron, 69*, 509–522.

von Schilcher, F. (1976). The role of auditory stimuli in the courtship of *Drosophila melanogaster. Animal Behaviour, 24*, 18–26.

von Schilcher, F., & Hall, J. C. (1979). Neural topography of courtship song in sex mosaics of *Drosophila melanogaster. Journal of Comparative Physiology, 129*, 85–95.

Vosshall, L. B. (2008). Scent of a fly. *Neuron, 59*, 685–689.

Wang, L., & Anderson, D. J. (2010). Identification of an aggression-promoting pheromone and its receptor neurons in *Drosophila. Nature, 463*, 227–231.

Wang, L., Han, X., Mehren, J., Hiroi, M., Billeter, J. C., Miyamoto, T., et al. (2011). Hierarchical chemosensory regulation of male-male social interactions in *Drosophila. Nature Neuroscience, 14*, 757–762.

Watanabe, H., Nishino, H., Nishikawa, M., Mizunami, M., & Yokohari, F. (2010). Complete mapping of glomeruli based on sensory nerve branching pattern in the primary olfactory center of the cockroach *Periplaneta americana. Journal of Comparative Neurology, 518*, 3907–3930.

Watanabe, K., Toba, G., Koganezawa, M., & Yamamoto, D. (2011). Gr39a, a highly diversified gustatory receptor in *Drosophila*, has a role in sexual behavior. *Behavior Genetics, 41*, 746–753.

Weiss, L. A., Dahanukar, A., Kwon, J. Y., Banerjee, D., & Carlson, J. R. (2011). The molecular and cellular basis of bitter taste in *Drosophila. Neuron, 69*, 258–272.

Wheeler, D. A., Kyriacou, C. P., Greenacre, M. L., Yu, Q., Rutila, J. E., Rosbash, M., et al. (1991). Molecular transfer of a species-specific behavior from *Drosophila simulans* to *Drosophila melanogaster. Science, 251*, 1082–1085.

Wu, C.-F., Ganetzky, B., Haugland, F. N., & Liu, A.-X. (1983). Potassium currents in *Drosophila*: Different components affected by mutations of two genes. *Science, 220*, 1076–1078.

Yamamoto, D., & Koganezawa, M. (2013). Genes and circuits of courtship behaviour in *Drosophila* males. *Nature Reviews Neuroscience, 14*, 681–692.

Yamamoto, D., Sano, Y., Ueda, R., Togashi, S., Tsurumura, S., & Sato, K. (1991). Newly isolated mutants of *Drosophila melanogaster* defective in mating behavior. *Journal of Neurogenetics, 7*, 152.

Yew, J. Y., Dreisewerd, K., Luftmann, H., Müthing, J., Pohlentz, G., & Kravitz, E. A. (2009). A new male sex pheromone and novel cuticular cues for chemical communication in *Drosophila*. *Current Biology, 19*, 1245–1254.

Yorozu, S., Wong, A., Fischer, B. J., Dankert, H., Kernan, M. J., Kamikouchi, A., et al. (2009). Distinct sensory representations of wind and near-field sound in the *Drosophila* brain. *Nature, 458*, 201–205.

Yu, Q., Colot, H. V., Kyriacou, C. P., Hall, J. C., & Rosbash, M. (1987). Behaviour modification by in vitro mutagenesis of a variable region within the *period* gene of *Drosophila. Nature, 326*, 765–769.

Yu, J. Y., Kanai, M. I., Demir, E., Jefferis, G. S., & Dickson, B. J. (2010). Cellular organization of the neural circuit that drives *Drosophila* courtship behavior. *Current Biology, 20*, 1602–1614.

Zawistowski, S. (1988). A replication demonstrating reduced courtship of *Drosophila-melanogaster* by associative learning. *Journal of Comparative Psychology, 102*, 174–176.

Chapter 2
A *Drosophila* Model for Aggression

Yong-Kyu Kim

There are many other structures and instincts which must have been developed through sexual selection—such as the weapons of offence and the means of defense—of the males for fighting with and driving away their rivals—their courage and pugnacity—their various ornaments—their contrivances for producing vocal or instrumental music—and their glands for emitting odors, most of these latter structures serving only to allure or excite the female. It is clear that these characters are the result of sexual and not of ordinary selection, since unarmed, unornamented, or unattractive males would succeed equally well in the battle for life and in leaving a numerous progeny, but for the presence of better endowed males.

Darwin (1871, p. 278)

Introduction

In animals, females make large, but few nutritious eggs and males produce small but many mobile sperm. There is tremendous competition between males over access to females, and females discriminate among their mating partners. Sexual selection arises from differences in reproductive success caused by competition for mates. Sexual selection is a mechanism by which conspicuous traits such as large body size, bright colors, songs, weapons, as well as behaviors are highly favored to attract

Y.-K. Kim (✉)
Janelia Research Campus, Howard Hughes Medical Institute,
19700 Helix Drive, Ashburn, VA 20147, USA
e-mail: kimy11@janelia.hhmi.org

J.C. Gewirtz, Y.-K. Kim (eds.), *Animal Models of Behavior Genetics*,
Advances in Behavior Genetics, DOI 10.1007/978-1-4939-3777-6_2

more mates and the traits enhance fitness of individuals (Fisher 1930; Lande 1981; Kirkpatrick 1987; Andersson 1994). There are two types of sexual selection (1) *intrasexual selection*, a competition within the same sex, usually males, for mates and (2) *intersexual selection*, mate selection by females. Intrasexual selection can occur in the form of competition for females without fighting with other males or in the form of contest between males. Morphological traits such as large body size, weaponry, and armor as well as aggressiveness are favored in the form of male–male competition. When males, however, are unable to monopolize either females or any resource vital to females, males advertise themselves for mates by displaying courtship, territory, songs, and ornaments. Intersexual selection occurs by choosing individuals with the best displays. Sexually dimorphic traits result from sexual selection by female choice. Males vigorously compete with each other for mates, and females show preference for males with the conspicuous traits. Female preferences have therefore evolved as a correlated response to selection on male traits, and genes both for attractive male traits and for female preferences for those attributes are inherited together from generation to generation. The two sets of genes are genetically correlated and evolution of the male traits drives a further change in the preference. This process leads to rapid coevolution of trait and preference until the exaggerated traits are opposed by natural selection due to selective pressure on high predation (Fisher 1930; Lande 1981) or until the traits are common in frequency in the population (O'Donald 1983).

Aggressive behavior is prevalent among animals, including fruit flies and is often observed at food resources (Jacob 1960, 1978; Spieth 1966, 1968; Dow and von Schilcher 1975; Ringo et al. 1983; Hoffmann 1987a). It serves for the acquisition or defense of food resources as well as for access to mates in nature. Despite its importance, little is known about the genetic and neural mechanisms that underlie aggressive behavior, other than that hormones play roles (e.g., Baier et al. 2002). Aggression is a complex behavior influenced by multiple genes and various environmental factors and has recently received attention from the perspectives of evolutionary biology, neuroscience, and psychology. *Drosophila* is a powerful model system for dissecting complex behaviors such as aggression because of (1) the availability of numerous genetic resources; (2) easy performance of genetic manipulations; (3) precise controls of environment and genetic background; and (4) large numbers of individuals of the same genotype.

In this chapter, studies of aggressive behavior in *Drosophila* will be reviewed from ecological, evolutionary, neurological, and genetic points of view. I discuss (1) how ecological or behavioral interactions between individuals that influence aggressive behavior in *Drosophila*; (2) relationships between social experience and the *Drosophila* central nervous system; and (3) the genetic architecture of aggressive behavior using both single gene analysis of induced mutations and quantitative genetic analysis. Recommended future research directions on aggressive behavior are addressed.

Current Research

Ecological and Evolutionary Perspectives

Insects, including fruit flies, search for food and mates for survival and reproduction daily. Food is unevenly distributed in nature. Males mate with multiple females and females discriminate among males. Such ecological and environment conditions lead to a development of aggressive behavior in *Drosophila*. Aggressive behavior functions to defend territories against intruders per se but is also exhibited during courtship. Male mating success is influenced by aggression as well as by courtship. Aggressive behavior may also contribute to reproductive isolation between species.

Territoriality

Most *Drosophila* matings occur at feeding and oviposition sites (reviewed in Wilkinson and Johns 2005), and there is a wide range of resources used by different *Drosophila* species (reviewed in Markow and O'Grady 2005). The distribution of food resources attracts females to feed and lay eggs. This, in turn, biases the males towards these same locations, which they visit to feed and obtain mates. Flies are usually observed near food patches or baits in nature as well as in the laboratory, and they interact with each other during territorial defenses of food resources. Territorial behavior has been reported for several *Drosophila* species (for *Drosophila melanogaster*, Jacob 1960, 1978; Dow and von Schilcher 1975; Hoffmann 1987a; for *D. pinicola*, Spieth and Heed 1975; for Hawaiian flies, Spieth 1966, 1974, 1981). For example, *D. melanogaster* males establish and defend patches of food as territories against intruding males (Hoffmann 1987a; Hoffmann & Cacoyianni 1990). A single male can defend a maximum diameter of 55–75 mm of food (Skrzipek et al. 1979; Hoffmann & Cacoyianni 1990). Old males are more successful at holding territories than younger males, although territorial behavior is initiated by as young as 2-day-old flies (Hoffmann 1990). Body size is positively correlated with territoriality (Hoffmann 1987b, 1991). In the *D. pinicola* species group, utilizing mushrooms for oviposition sites, *D. melanderi* males vigorously defend their territories and single males individually position on a mushroom cap (Spieth and Heed 1975). Males of the picture-winged Hawaiian *Drosophila* species occupy their mating territories within leks which consist of 5–10 males each and advertise their sexual readiness by displaying nonresource-based behavior, and females periodically visit these territories (Spieth 1982). Territorial males in *D. melanogaster* consequently have more encounters and are more successful in mating than nonterritorial males (Jacob 1960, 1978; Dow and von Schilcher 1975; Skrzipek et al. 1979; Hoffmann 1987a, 1991). However, Ringo et al. (1983) found no evidence of territoriality in three species they studied: *D. virilis, D. americana*, and *D. novamexicana*. Territoriality has been shown to be heritable in artificial selection experiments, and progeny of territorial parents have mating advantages (Hoffmann 1988, 1991; Hoffmann and Cacoyianni 1989). The incidence of territoriality is influenced by defended food types, male density in territories, female receptivity, as well as age. Higher territoriality is observed in natural breeding sites than in laboratory medium and is also detected when females are present in the territory regardless of their reproductive status (Hoffmann & Cacoyianni 1990). Territorial defense, however, ceases when benefits from mating advantages are overweighed by costs. Males are less likely to defend territories when food resources are abundant or territorial males are more common (Hoffmann & Cacoyianni 1990).

Resource defense is associated with fighting between individuals, usually males (see *Aggressive Behavior* for details). *Drosophila melanogaster* males vigorously defend their feeding or oviposition sites by displaying a series of aggressive behaviors: chasing, fencing, lunging, tussling, wing threat, boxing, holding, or head-to-head butting. Males of most Hawaiian *Drosophila* species—*heteroneura, silvestris*, and their close relatives—patrol their leks and vigorously defend mating territories exhibiting ritualized fighting such as curling and slashing which involve wings, legs, and body movements (Spieth 1966, 1968; see Ringo and Hodosh 1978 for *D. grimshawi* subgroup). *Drosophila melanderi* males approach and attempt to court any individuals invading their territories (Spieth and Heed 1975), and *D. subobscura* males are reported to display wing threat toward intruders (Milani 1956). It is hypothesized that forced copulation and patrolling at emergence sites are favored in *Drosophila* species possessing more male-biased operational sex ratio and intense competition for mates (Markow 1996). For example, Markow (2000) reported that *D. melanogaster* and *D. simulans* males patrolled pupation sites, waited for emerging females, and copulated with teneral females that were incapable of displaying rejection behavior; in addition, these mated teneral females successfully produced fertile offspring.

Mate Competition

Individuals of the same sex vigorously compete with each other to acquire mates. Males may pugnaciously fight with other males or have contests over females by exhibiting extravagant ornaments or conspicuous courtship signals toward females or elaborate courtship or mating behavior toward females. Traits that increase reproductive fitness will be favored by sexual selection (Fisher 1930; Lande 1981; Kirkpatrick 1987; Andersson 1994). Considerable variation in body size, head size, songs, pheromones, and courtship are observed among flies collected in nature and contribute to reproductive success. Adult body size is influenced by a variety of environmental factors such as nutrition, temperature, and larval density on food (David et al. 1983), and it is heritable in laboratory populations (Robertson and Reeve 1952). Female *Drosophila* are usually larger than males, and large females have fecundity advantages (Partridge and Farquhar 1983). Males consequently struggle over females who also compete for larger males. Larger males offer more courtship activities as well as agonistic behavior, and they win aggressive encounters (Spieth 1982; Partridge et al. 1987; Partridge et al. 1987; Markow 1988; Santos et al. 1988; but see Zamudio et al. 1995). Small males are less likely than are large males to hold territory (Hoffmann 1987b, 1991). Hawaiian *Drosophila* species also show considerable variation in morphology, sexual behavior, and pheromones (Spieth 1966; Carson et al. 1970; Ringo 1977; Ringo and Hodosh 1978; Kaneshiro and Boake 1987; Hoy et al. 1988; Carson 1997; Alves et al. 2010). These sexually dimorphic traits have evolved under sexual selection via female choice. For example, highly pugnacious *D. heteroneura* males have hammer-shaped heads with stalk eyes. Males with broader heads are more successful in male–male competition and there is a highly significant correlation between male mating success and head width (Spieth 1981; Boake et al. 1997). In *D. silvestris* males, however, body size is positively correlated with aggressive success but not with mating success (Boake 1989; Boake and Konigsberg 1998).

Drosophila males display elaborate courtship behavior to attract mates (e.g., see Spieth 1966, 1982 for the picture-winged Hawaiian flies) and females are subjected during courtship to a variety of stimuli—visual, acoustic, olfactory, and mechanical elements (Spiess 1987; Hall 1994; Greenspan and Ferveur 2000; Lasbleiz et al. 2006). The relative importance of these sensory modalities varies with species. Courtship is species specific and plays a major role in mate recognition (Paterson 1978, 1985), but males show individual variation in quantities of courtship, an indicator of their physical status. Females prefer to mate with males that vigorously display courtship. For example, Kim and Ehrman (1998) demonstrated that *D. paulistorum* females mate more frequently with males reared in isolation than those reared in groups, because the former are sexually more active in courting. During courtship, males also exhibit high levels of agonistic behavior to their sexual partners. Males intersperse between courtship behavior and aggressive behavior when they encounter females. When females are not receptive or display rejection behaviors, males often threaten females by raising both wings up in front of females or by lunging toward them (Jacob 1960), and females repel males by kicking and/or decamping. Unsuccessful males in mating are often observed to interrupt or terminate courtship of other males by attacking them.

Courtship songs influence male mating success and are sexually selected traits (Aspi and Hoikkala 1995; Hoikkala et al. 1998; Ritchie et al. 1998, 2001; Williams et al. 2001; Saarikettu et al. 2005; Snook et al. 2005). During courtship, *Drosophila* males produce two forms of courtship song via wing vibration: sine song and pulse song (see Tauber and Eberl 2003 for research approach). It functions as a means of communication from males to females and it serves to stimulate female receptivity and identify the species of the courting male. The wing vibration of many *Drosophila* species has been studied (see Gleason 2005; Hoikkala 2005 for review). The song of each species is unique in its acoustic characteristics such as interpulse interval (IPI) which is regarded as the critical parameter for discrimination between species. Such variable courtship songs may contribute to reproductive isolation between sympatric, closely related sibling species (Kyriacou and Hall 1982; Ritchie and Gleason 1995; Williams et al. 2001). There is considerable variation in courtship songs among *D. melanogaster* males (Wheeler et al. 1988). Females prefer to mate with males presenting high frequencies of IPI and courtship song traits are heritable (Ritchie et al. 1998; Ritchie and Kyriacou 1996). In competitive situations, however, males interrupt other males (Wallace 1974) by producing a rejection song linked to flicking behavior (Paillette et al. 1991) or by exhibiting aggressive behavior such as wing threat or lunging (Jacob 1960; Chen et al. 2002). These strategies can shorten the duration of courtship song as well as courtship duration or latency to copulation (Ewing and Ewing 1984; Crossley and Wallace 1987; Tauber and Eberl 2002).

During courtship, males and females exchange olfactory signals as well. Cuticular hydrocarbons present on insect cuticles primarily protect them from desiccation, but they also contribute to intraspecific mate recognition as well as to sexual isolation between species (Jallon 1984; Cobb and Jallon 1990; Coyne et al. 1994; Coyne and Charlesworth 1997; Blows and Allan 1998; Savarit et al. 1999; Marcillac et al. 2005; Billeter et al. 2009). Cuticular hydrocarbons, produced by oenocyte cells in the abdomen, are not too volatile and are perceived at a relatively short distance (1–2 mm) by olfactory organs or by contact with taste organs. Quantitative and qualitative variations in cuticular hydrocarbons exist according to sex, species, and geographic population (Ferveur 2005). Several studies have proven the roles of cuticular hydrocarbons in sexual selection. In *D. melanogaster*, females mate faster and more often with males carrying greater levels of (z)-7-tricosene, a principal male hydrocarbon (Grillet et al. 2006). *Drosophila serrata* females show significant sexual selection according to quantities of the male hydrocarbons (Blows 2002; Howard et al. 2003; Chenoweth and Blows 2005). Similarly, males that mated multiply in an array of *D. pseudoobscura* possessed higher relative abundances of (z)-5,9-pentacosadiene, a male-predominant hydrocarbon, than did males that were never accepted by females. They further mated more quickly with females (Kim et al. 2005).

Mate choice is influenced both intrinsically and extrinsically. For example, social constraints, such as intrasexual interference and intersexual coercion, significantly influence mate choice. Mate preference tests, in which social or ecological constraints are removed, have been designed to identify freely chosen preferred and nonpreferred partners, and the viability of offspring from matings to preferred and nonpreferred partners have been measured (Anderson et al. 2007; Gowaty et al. 2007). Matings with preferred partners, when either males or females were choosing, produced more offspring than matings with nonpreferred partners. During these trials, individual choosers were permitted to compare both visual and olfactory information concerning two partners of the opposite sex. Preferred partners were larger and produced greater quantities of male- or female-predominant hydrocarbons than nonpreferred partners (Kim et al. 2005). Quantities of hydrocarbons are also modulated with social experience. In the *D. paulistorum* species complex, males reared in isolation produced greater quantities of male hydrocarbons than did males reared in groups, and, as a result, they were more often accepted as mates (Kim et al. 2004; see Kent et al. 2008 for *D. melanogaster*).

Males often mate with several females per day in nature, and mating frequency among males varies. Mating frequency is positively correlated with mating speed, an important component of *Drosophila* fitness and one that is genetically controlled (Manning 1961; Spiess and Langer 1964; Kaul and Parsons 1965; Fulker 1966; Kessler 1969; Partridge et al. 1985). Carson (2002) demonstrated that approximately 30 % of Hawaiian *Drosophila* males did not mate in female choice tests. Kim et al. (2005) reported similar results with *D. melanogaster*, *D. pseudoobscura*, and *D. hydei*—males that mated multiply during observation periods showed significantly faster mating speeds than did males that were not accepted by females during the same intervals. The former also were more aggressive than the latter. Recently, Moehring and Mackay (2004) identified QTLs affecting *D. melanogaster* male mating speed and showed that seven candidate genes are associated with variations in mating behavior. Microarray analysis showed that 21 % of the genome is involved in regulating mating speed and that such genes also are likely to be genes involved in neurogenesis, metabolism, development, and cellular processes (Mackay et al. 2005).

Aggressive Behavior

Aggression is an adaptive behavior enabling access to more mates, among other goals. Individuals of the same sex, usually males, compete and fight for mates. They perform complex and stereotyped aggressive behaviors. Aggressive behavior has been reported in many *Drosophila* species (Jacob 1960, 1978; Spieth 1966, 1968; Dow and von Schilcher 1975; Ringo et al. 1983; Hoffmann 1987a). In *D. melanogaster*, males at feeding sites chase each other, push off with one of their middle legs, elevate both wings for a sustained period of time, rise on their hind legs and tussle with forelegs, hold other flies with forelegs in the back, raise the front part of body, and lunge toward the other (Jacob 1960; Hoffmann 1987b; Chen et al. 2002; see Table 2.1). Females also display aggressive though limited fighting behaviors (Chen et al. 2002; Ueda and Kidokoro 2002; Nilsen et al. 2004; Vrontou et al. 2006; Chan and Kravitz 2007). These sex differences in aggression are under the control of *fruitless* (*fru*), a regulator of sexual differentiation of the brain (Lee and Hall 2000; Vrontou et al. 2006). Hierarchical relationships are formed among *D. melanogaster* males, and winners lunge more while losers retreat more (Chen et al. 2002; Nilsen et al. 2004; Vrontou et al. 2006; Yurkovic et al. 2006; Penn et al. 2010; Trannoy et al. 2016). Thus, more aggressive males are shown to have a greater mating success than less aggressive males. In the majority of lekking Hawaiian species, males frequently intrude upon one another's leks, leading to aggressive behaviors (Spieth 1966, 1968, 1982; Boake 1989)—both males engage in head-on, wing waving, and bobbing. They elevate their heads and forebodies, then fully extend and slash their forelegs downward against the intruders. If the encounter continues, additional bobbing and slashing movements occur and terminate the encounter. *Drosophila heteroneura* in the *planitibia* subgroup of the Hawaiian species,

Table 2.1 Aggressive behavior in *Drosophila melanogaster*

Behavior	Description
Chasing	A male vigorously follows another male the short distance; often the male vibrates one wing during chasing (Jacob 1960)
Fencing	A male wards off another male by extending a middle of leg (Jacob 1960)
Wing threat	Both wings are extended to 30° and raised up to 30–40° in front of another male. This behavior is often observed in front of females (Dow and von Schilcher 1975)
Boxing	Two males approach each other, stand up and interlock their forelegs; it is infrequently displayed during fighting (Dow and von Schilcher 1975)
Holding	A male stays behind, and grabs and holds both wings of another male; the male often display this behavior in front of the opponent (Jacob 1960)
Lunging	A male raises the front part of his body and lunges down onto his opponent (Hoffmann 1987a)
Head-butting	Two males face each other in a straight-line position and push against each other using their heads (Lee and Hall 2000)
Tussling	Two males raise the front part of the body extending their hind legs and tussle with forelegs (Hoffmann 1987a)

however, display a unique fighting pattern (Spieth 1981). When they approach head-on, they keep their bodies depressed horizontally and face and push against each other, attempting to force the other backwards. Males with broader heads are more successful at mating success as well as in aggressive encounters with other males (Boake et al. 1997). Such fighting behavior likely provides a selection pressure that has led to the evolution of secondary sexual characters such as enhanced head size that attract more mates (Ringo 1977; Templeton 1977; Boake et al. 1997). However, aggression is not always positively correlated with mating success. Unlike *D. melanogaster* group species, Ringo et al. (1983) found that courtship and aggressive behavior were largely independent in the *D. virilis* group species; courtship was positively related to mating success, and aggressive males were not more successful at mating than unaggressive males. Boake (1989) also demonstrated that aggressive success in the picture-winged Hawaiian species, *D. silvestris*, was positively and highly correlated with body size but not correlated with mating success (see also Boake and Konigsberg 1998). Laboratory strains that have been reared under constant environments for many generations are less aggressive than field-caught flies (Dierick and Greenspan 2006). These authors also found that flies were less aggressive when food is not available in the observation chamber (see also Hoffmann and Cacoyianni 1989; Ueda and Kidokoro 2002).

Age affects aggressiveness in *Drosophila* too. The age at which males become sexually mature differs among species (see Markow 1996 for details). Old flies of both sexes are more territorial and aggressive than young ones (Hoffmann 1990; Papaj and Messing 1998). Shortly after eclosion young males are more often courted by mature males, but *Drosophila* aggressive behavior develops as fruit flies age along with the production of inhibitory pheromones, that is, 11-*cis*-vaccenyl acetate (cVA), especially in males (Jallon et al. 1981; Mane et al. 1983; Bartelt et al. 1985;

Wang and Anderson 2010; Wang, Han et al. 2011; Billeter and Levine 2015), and such male pheromones can inhibit male–male courtship but promote aggression among males. Anderson and his colleagues demonstrated that acute exposure to cVA promoted aggressive behavior by activating Or67d-expressing olfactory sensory neurons (Wang and Anderson 2010). In turn, a high density of cVA in a local population affects *Drosophila* males to regulate population density by displaying aggressive behavior. However, chronic exposure to cVA reduced aggression by activating Or65a OSN (Liu et al. 2011). This suggests that identification of signaling pathways in the central nervous system is essential to understanding mechanisms of aggressive behavior in *Drosophila*. Social experience also affects the expression of synaptic proteins and genes in the fly brain. Levels of several synaptic proteins increased during social interaction with conspecifics and decreased during sleep (Gilestro et al. 2009; see Donlea et al. 2009 for synaptic terminals).

Neurological and Genetic Perspectives

Aggressive behavior is affected by social experience, and therefore entails learning. The *Drosophila* mushroom bodies (MBs), which continue to grow in adults and whose growth is stimulated by social experience, are the sites of olfactory learning and memory. The products of several specific genes necessary for the acquisition and storage of olfactory memories have been identified. Therefore, we expect to find that the MBs are the site of learning during aggressive encounters, and that the genes involved in learning and memory are necessary for acquisition of some aspects of aggression, too.

Social Experience and Behavior

Social interactions or experiences with conspecifics or heterospecifics modulate *Drosophila* behaviors, including aggression (Hoffmann 1990; Kim et al. 1996a, b; Kim and Ehrman 1998; Papaj and Messing 1998; McRobert et al. 2003; Svetec and Ferveur 2005; Yurkovic et al. 2006; Wang, Dankert et al. 2008; Zhou et al. 2008; Wang and Anderson 2010). Individuals receive visual, olfactory, acoustic, and tactile signals during pre- and adult stages. These signals are used as templates and compared during subsequent encounters when adult (Lacy and Sherman 1983). But when flies are raised in isolation from conspecifics, they do not obtain any useful template information. In a series of investigations into the development of discriminatory abilities in *D. paulistorum*, Kim et al. (1992, 1996a, b) demonstrated that discriminatory abilities increased when flies had social contacts with conspecifics during development from egg to adult stages. However, when individuals were reared in total isolation during development, their discriminatory abilities were minimized and more heterogamic matings occurred. Socially isolated flies did not discriminate efficiently between sexes and homosexuality was frequently observed (Kim and Ehrman 1998).

Males reared in isolation have been shown to be more aggressive than those reared in groups (Hoffmann 1990; Kim and Ehrman 1998; Kamyshev et al. 2002; Svetec et al. 2005; Wang, Dankert et al. 2008). Wang and Anderson (2010) recently demonstrated that olfactory male pheromone, cVA, promotes aggression among males via olfactory receptor, Or67d (Kurtovic et al. 2007). When males were, however, group-housed for some periods, chronic exposure to the cVA suppressed male aggression via the olfactory receptor, Or65d (Liu et al. 2011; see Yuan et al. 2014 for female exposure). In addition, nonvolatile gustatory pheromones, that is, (z)-7-tricosene, promote aggression among males and act in a hierarchical manner with cVA through activation of Gr32a gustatory receptor where (z)-7-tricosene is required for the aggression-promoting effect of cVA (Wang, Han et al. 2011). The isolated males displayed lunging and tussling more frequently than did males held in groups. They also established territories more quickly (Hoffmann 1990). Consequently, the isolated males were more successful at mating than males in groups (Ellis and Kessler 1975; Kim and Ehrman 1998). Previous experience in fighting modified their social status when new hierarchical relationships were reestablished (Hoffmann 1990; Vrontou et al. 2006; Yurkovic et al. 2006; Penn et al. 2010; Trannoy et al. 2016)—winner males who were more territorial and aggressive tended to establish territories more readily and to attack more than did loser males. Loser males lost in subsequent fights when paired with familiar or unfamiliar winners. Such results indicate that *Drosophila* learn to recognize conspecifics through social experience, and learning about conspecifics is reinforced by the presence of heterospecifics (Kim et al. 1996a, b; Irwin and Price 1999). For example, when *D. pseudoobscura* and *D. persimilis* coexist in sympatry, the discriminatory abilities of *D. pseudoobscura* females increase (Noor 1995; Noor and Ortíz-Barrientos 2006) and more aggressive behaviors by males against heterospecifics are observed (Kim et al. 2008).

Mushroom Bodies as a Site for Associative Learning

Flies receive and process multisensory information, including olfactory stimuli in the brain. They detect odors through sensory receptors on antennae and maxillary palps, and olfactory inputs reach antennal lobes of the brain via the antennal nerve. Gustatory information, such as bitter tasting compounds, are detected by gustatory receptors in the mouth, proboscis, and legs and are delivered to subesophageal ganglions (SEG) that project to all regions in the brain. Larvae also have olfactory organs, the antenna-maxillary complex, at the anterior tip of the animal (Singh and Singh 1984). The mushroom bodies (MBs) of the brain receive olfactory inputs from antennal lobes and forward outputs to the central complex and lateral protocerebrum. The MBs are bilateral clusters of about 2500 Kenyon cells per hemisphere. They are crucial for olfactory associative learning and memory, but a few studies have demonstrated that MBs are also required for visual learning (Heisenberg et al. 1985; Davis 1993; de Belle and Heisenberg 1994; Strausfeld et al. 1998; Zars et al. 2000; Dubnau and Tully 2001; Pascual and Préat 2001; Roman and Davis 2001; Waddell and Quinn 2001; Heisenberg 2003; Busto et al. 2010). Flies whose MBs were disrupted chemically or genetically were impaired in olfactory associative learning (Heisenberg et al. 1985; de Belle and Heisenberg 1994; Connolly et al. 1996; Baier et al. 2002). As attempts to dissect the roles of MBs at different phases of memory processing—*acquisition, consolidation, and retrieval* (Dubnau et al. 2003)—McGuire et al. (2001) inactivated neurotransmission in this region and demonstrated that MB signaling is needed for olfactory memory retrieval. The MBs were also involved in sustaining courtship suppression after males were conditioned with mated females (McBride et al. 1999; Vrontou et al. 2006; Ejima et al. 2007; Keleman et al. 2012).

Genetic screening for learning-defective mutant flies identified *dunce, rutabaga, amnesiac*, and *radish* mutants (Dubai et al. 1976; Quinn et al. 1979; Folkers et al. 1993). The *dunce* mutant has a defect in the gene coding for cAMP-specific phosphodiesterase, and the *rutabaga* mutant has abnormal adenylyl cyclase, an enzyme activated by Ca^{2+}/calmodulin and G_s α proteins. These genes are expressed in the MBs. The *amnesiac* gene, encoding a prepronеuropeptide neurotransmitter, is expressed in dorsal paired medial (DPM) neurons. Cyclic AMP-dependent protein kinase (PKA) is also expressed in the MBs. Flies mutated in genes for the catalytic or regulatory subunits of PKA are deficient in learning. Transposon-induced mutant screening identified more new genes involved in olfactory learning: *latheo, nalyot, linotte, leonardo, volado*, and *fasciclin II* (see Waddell and Quinn 2001 for review).

The MBs have an embryonic origin and continue to grow during development to the adult stages (Technau and Heisenberg 1982; Armstrong et al. 1998). During metamorphosis, the MBs display a remarkable structural plasticity involving neural degeneration, birth, and regrowth. The volume of MBs is influenced by genetic and environmental factors during development (Technau 1984; Balling et al. 1987). Heisenberg et al. (1995) observed the effect of social experience on the volume of the *D. melanogaster* brain during adulthood and found that the volume of the MBs was significantly increased in the flies held in groups compared to those held in isolation (see also Barth and Heisenberg 1997; Barth et al. 1997 for visual experiences). In a series of investigations into the development of *D. paulistorum* discrimination abilities, Kim et al. (2009) also showed that flies reared in total isolation during development had a significantly smaller MB size than did flies reared in groups, and the isolates were more aggressive. Such results suggest that changes in MB structure result from social interactions during development and may play roles in drosophilid aggressiveness (Baier et al. 2002).

The neurotransmitters dopamine and octopamine are known to be involved in aggression, courtship conditioning, and olfactory learning and memory in *Drosophila* (Dubai et al. 1987; O'Dell 1994; Waddell 2010), and their receptors have been identified in the MBs (Han et al. 1996; Han et al. 1998). Baier et al. (2002) observed that the aggressiveness of *D. melanogaster* males was significantly reduced when dopamine and octopamine were genetically and pharmacologically depleted. In the same context, Alekseyenko et al. (2010) demonstrated that inactivation of neurons expressing Dopa decarboxylase (Ddc), an enzyme responsible for converting L-DOPA to dopamine (DA), reduced aggression. Subsequently, they identified two pairs of DA neurons in the T1 and PPM3 clusters that modulated aggression (Alekseyenko et al. 2013). Tyramine-β-hydroxylase (TβH), an enzyme responsible for converting tyramine (TA) to octopamine (OA), is involved in aggression. TβH-null mutant males display reduced aggression due to the decrease in OA. Further, tyrosine is converted to TA by tyrosine decarboxylase (Tdc). Tdc2 mutants also show decreased aggression due to a loss of OA, suggesting that OA is responsible for aggression. In contrast, activation of OA neurons enhanced aggression in adult males (Zhou et al. 2008; Hoyer et al. 2008). When synaptic outputs from MBs were blocked, males were not aggressive. Similarly, MB-ablated males were significantly less aggressive than controls with intact MBs (Kim et al. 2007). Such data indicate that the MBs may also be involved in *Drosophila* aggressive behavior. Meanwhile, serotonin (5-hydroxytryptophan, 5-HT) was initially reported to have no effects on aggression in *D. melanogaster* (Baier et al. 2002), although it has been known that 5-HT is associated with aggression and social dominance status in a variety of organisms (Kravitz 2000; Murakami and Itoh 2001, 2003; Kravitz and Huber 2003; Popova 2006). Dierick and Greenspan (2007) demonstrated that pharmacologically induced increases of 5-HT and overexpression of Tryptophan hydroxylase (Trh) enhanced aggression. Consistently, selective activation of 5-HT neurons with dTrpA1, a temperature-sensitive cation channel, using a Trh-GAL4 driver produced flies that escalated fighting faster and fought at higher intensity (Alekseyenko et al. 2010). However, pharmacological inhibition or silencing of 5-HT neurons did not affect aggression (Dierick and Greenspan 2007) while flies with acute inhibition of 5-HT neurons still fought but did not escalate fighting behavior (Alekseyenko et al. 2010). The complexity in the results of manipulation 5-HT activity is possibly due to differential regulation of aggressive behavior by different receptors. Johnson et al. (2009) demonstrated that activation of 5-HT2 receptors decreases overall aggression while activation of 5-HT1A-like receptors induces the opposite effect. Overall levels of 5-HT also influence male courtship behavior—a reduction in 5-HT levels in the brain induces male homosexual courtship (Zhang and Odenwald 1995; Hing and Carlson 1996; see Liu et al. 2008 for the effect of dopamine on male–male courtship).

Neuropeptides are one of the major regulators that control physiology and behavior in animals and are involved in a wide range of biological functions, including homeostasis, metabolism, reproduction, anxiety, stress, learning and memory, and alcohol addiction (reviewed in Nässel and Winther 2010). A large number of neuropeptides are structurally and functionally conserved between vertebrates and invertebrates. For instance, *Drosophila* neuropeptide F (NPF) is a homologue of vertebrate neuropeptide Y (NPY) and is detected in the brain and midgut (Brown et al. 1999). NPF is predominantly expressed in two pairs of neurons in the protocerebral lobes and one pair of neurons in the SEG of third instar larvae (Shen and Cai 2001). Additional interneurons are identified in adults, giving a total of 26 NPF neurons in the brain of males and 20 in females (Lee et al. 2006; Yoshill et al. 2008). NPF functions in foraging, feeding, alcohol sensitivity, as well as aggression (Shen and Cai 2001; Wu et al. 2003, 2005; Wen et al. 2005; Lee et al. 2006; Dierick and Greenspan 2007; Chen et al. 2008; Shohat-Ophir et al. 2012). NPF is involved in courtship and aggression. Lee et al. (2006) investigated regulatory mechanisms of *npf* expression in the adult fly brain. They determined that NPF expression was sex-specific in three neurons of the adult brain and that male-specific *npf* expression (ms-*npf*) was controlled by sex determination genes. Feminization of these neurons abolished NPF expression, and consequently males show less courtship behavior. ms-*nsf* expression was significantly reduced in the *fruitless* mutants as well. These results indicate that ms-*npf* expression is regulated by both sex-determinating and circadian genes. Genetically silencing or feminizing the NPF neurons of males with *tra*^{F}, Dierick and Greenspan (2007) found increased aggressive behavior. Therefore, the ms-*npf* neurons play roles in switching between courtship and aggression. In a search of peptidergic neurons that control aggression among males in *Drosophila*, Asahina et al. (2014) activated neurons labeled by a set of about 40 GAL4 drivers created from putative promotor regions of 20 different *Drosophila* neuropeptide genes in the combination of dTrpA1 at high temperature, and screened these lines for increases in aggressive behavior. They identified that Fru^{M+} neurons expressing the neuropeptide *tachykinin* controlled higher levels of aggression, and that activation and silencing of these neurons increased and decreased intermale aggression, without affecting male–female courtship. Consistently, a large-scale screening for aggression-promoting neurons in *D. melanogaster* has identified multiple P1 interneurons that enhance both aggression and courtship, and thermogenetic activation of Fru^{M} P1 neurons increases aggression in the absence of wing extension (Hoopfer et al. 2015).

Genetics of Aggressive Behavior

Jacob (1978) first reported that *ebony* mutants were more aggressive than wild types while *black* mutants were less aggressive and that aggressiveness was influenced by β-alanine. Boake et al. (1998) crossed *D. silvestris* (S) and *D. heteroneura* (H) and observed aggression (initial posture, wing extension, head position, and leg posture) in heterospecific pairs, a hybrid male and either an S or an H male. Parental males performed differently in regard to wing extensions and leg postures, and the hybrids resembled S males for early postures and extended leg postures, but were like H males for wing extensions. They hypothesized that aggressive behavior is controlled by a single gene with major effects.

Recent dissections of interactions between wild-type *D. melanogaster* males have identified multiple genes involved in *Drosophila* aggressiveness (see Robin et al. 2006 for review). Two genes involved in the sex-determination hierarchy, *fruitless (fru)* and *dissatisfaction (dsf)*, were reported to be associated with aggression (Lee and Hall 2000). The *fru* gene, which plays a prominent role in male courtship behavior, was associated with high levels of head-to-head interactions between males. Males homozygous for any of the *fru* mutations displayed vigorous head-buttings. The frequency of the head-buttings was significantly higher between *fru* mutant males than between *fru* male and a wild-type male or female. Aging *dsf* mutant males led to a high level of head-to-head interactions compared to *fru* mutant males. Generating alleles for *fru* that are constitutively spliced in the male (fru^M) or female (fru^F) mode (Demir and Dickson 2005), Vrontou et al. (2006) further analyzed the crucial role of the *fru* gene in aggression and found that fru^F males were more inclined to fight females than to court them, whereas fru^M females courted other females and fought males. In a pairing between fru^F males and fru^M females, the behavior of fru^M females was indistinguishable from that of control fru^F males; and fru^F males behaved like fru^C females. They also found that fru^F males, like normal females, did not establish strong dominance relationships; fights between fru^F males were significantly less frequent than those in fru^C males. See Chapter One in this volume for the roles of *fru* in courtship.

Manipulation of the sex determination factor *transformer* (*tra*) switched male and female patterns of aggression (Chan and Kravitz 2007). Using the GAL4/UAS binary expression system (Brand and Perrimon 1993), the female form of *transformer* (tra^F) was expressed in the male nervous system and a *transformer* dsRNA (tra^{IR}) in the female nervous system. Male patterns of fighting were switched to female patterns and vice versa. Similarly, masculinization of female pheromones or behaviors triggered male-to-female aggression (Fernandez et al. 2010). Using the GAL4/UAS system, the expression of tra^{IR} in pheromone-producing oenocytes of females switched males to display aggression to the females. Reciprocally, feminization of the oenocytes in males elicited courtship but no aggression from control males.

Utilizing classical selection experiment procedures widely employed in quantitative genetics (Falconer and Mackay 1996; Harshman and Hoffmann 2000), Dierick and Greenspan (2006) produced four lines of *D. melanogaster*, two selected for high aggression and two for "neutral" lines in which highly aggressive males were removed from the breeding pool of each generation. Aggressiveness in the selected lines consequently increased over generations. After 21 generations of selection, a fighting index increased more than 30-fold in the selected lines. This indicates that multiple genes contribute to aggressiveness. The neutral lines displayed decreased aggressiveness since selection was exercised against aggressive flies every generation. Selected lines dominated the neutral lines when paired with them. However, in mating competition assays, the selected lines mated significantly less than the neutral lines (but see Dow and von Schilcher 1975; Hoffmann 1987a, b). The selected lines were also lighter in weight than the neutral ones. Using microarray analysis, these authors evaluated gene expression in the brains of representative flies from their two lines and identified approximately 80 genes that were significantly differentially expressed. The selected lines expressed higher levels of these genes than did the neutral ones. Note that 42 genes showed differences in transcription abundances between these lines. Mutation analysis, inserting these candidate genes into wild types, showed that *Cyp6a20*, which encodes a cytochrome P450 (Feyereisen 2005), was directly involved in aggressive behavior during selection experiments. Social experience modulated expression levels of *Cyp6a20* and its expression levels were inversely related to the levels of aggression (Wang, Dankert et al. 2008). Subsequently, Dierick and Greenspan (2007) found that both pharmacological and genetic manipulations that increase 5-HT synthesis also increase aggression. This effect was the same in selected and neutral lines, indicating that selection had not affected 5-HT levels or receptors. They also found that silencing or feminizing brain cells that produce neuropeptide F (NPF) in a male-specific pattern increased aggression, which parallels what has been found with neuropeptide Y (NPY) in *Mus* (knocking out an NPY receptor gene increases territorial aggression in mice, which appears to be stimulated by 5-HT). Flies whose 5-HT synthesis in Ddc-producing cells was shut off nonetheless showed low levels of aggression, leading Dierick and Greenspan to postulate an "aggression circuit" in the brain, which does not require 5-HT but is modulated by both 5-HT and NPF. The modulatory roles of 5-HT (which stimulates aggression) and NPF (which inhibits aggression) appear to be additive, and this suggests that the two molecules act independently.

Edwards et al. (2006) used artificial selection as did Dierick and Greenspan (2006) selecting for high and low levels of male aggression. A realized heritability (h^2) of 0.10 over 28 generations was reported for aggressive behavior in *D. melanogaster*, indicating considerable environmental variance. Selection for aggressive behavior did not affect other *Drosophila* behaviors or physiology, such as mating, locomotion, stress resistance, temperature tolerance, and longevity. Using whole-genome microarrays, these authors assessed aggression levels of lines containing *P*-element insertional mutations in candidate genes and identified 15 novel genes that formed the genetic architecture of *Drosophila* aggressive behavior. More candidate genes affecting aggression were discovered increasing *P*-element mutant lines—57 genes with aberrant aggressive behavior were identified (Edwards et al. 2009). The authors also demonstrated that *P*-element insertional mutations in single genes exhibited pleiotropic phenotypes, such as aggression and brain anatomy (see also Rollmann et al. 2008). Subsequently, combining quantitative genetic analysis of variation in aggressive behavior and whole genome transcript profiling, Mackay and her colleagues quantified aggressive behavior of 40 inbred lines of wild-type *D. melanogaster.* Using a genome-wide association scan for quantitative trait transcripts (QTT) associated with aggressive behavior in natural populations of *D. melanogaster* (Edwards et al. 2009), and they reported that (1) many of candidate genes had pleiotropic effects on metabolism, development, and other behavioral traits; (2) the genetically correlated transcripts formed a transcriptional genetic network associated with natural variation in aggressive behavior; and (3) widespread epistatic interactions among these pleiotropic genes formed the genetic architecture of *D. melanogaster* aggressive behavior (Zwarts et al. 2011; Shorter et al. 2015).

Future Directions

The function of aggressive behavior in nature is to defend territories against intruders. Considerable variation in territorial behavior has been observed among *Drosophila* species and territoriality is positively correlated with male mating success. Aggressive behavior is reported in many picture-winged Hawaiian *Drosophila* species and *D. melanogaster* in the wild and in the laboratory, which demonstrates that more aggressive individuals are also more territorial and the beneficiaries of greater their reproductive fitness. However, these relationships are not always true and need to be studied more with different species. Aggressive behavior is also displayed during courtship but has received less attention from evolutionary biologists because its frequency is relatively low in the laboratory where flies are routinely placed with small amounts of food in *plastic* mating chambers. Role of aggressive behavior in sexual isolation between strains and species needs to be investigated.

Using single-gene studies of induced mutations, multiple genes have been reported to affect aggressive behavior in *D. melanogaster* (see Table 2.2). However, there are no genes controlling behavior directly. Rather, many genes have pleiotropic effects and influence other behaviors as well (e.g., Anholt and Mackay 2012). Aggressive behavior is polygenic and regulated by interactions between multiple genes with small effects as well as by brain neurotransmitters. Studies of quantitative genetic analysis for natural variation in aggression could be combined with single-gene studies of biochemical mutations providing additional candidate genes or gene regions (Greenspan 2004). Reproductive fitness components of candidate genes associated with Drosophila aggression need to be evaluated in an envolutionary perspective. Aggressive behavior is also significantly influenced by environment. In particular, various experiences that individuals garner during development will affect their subsequent behaviors including aggression. Studies of antisocial personality disorders in humans show that the magnitude of genetic and environmental influences on the traits substantially changes during development stages—these traits are more influenced by shared environments during adolescence but more genetically influenced during adulthood. Studies of social experience/learning in *Drosophila* aggressive behavior are limited to the observations of experience-dependent modification of behavior, memory, and structural plasticity in the brain. The availability of genetic tools for manipulating specific cell types in *D. melanogaster* enable explore neural circuitry of aggressive behavior and identify specific neurons that mediate and influence this behavior. In a combination of neurologic approach, studies of gene–environment interactions and correlations on aggressive behavior will provide insights into complex gene-neuron-behavior pathways in aggression.

Table 2.2 Factors involved in *Drosophila* aggressive behavior

Factor	References
1. *Genes*	
Ebony	Jacob (1960)
Black	
Fruitless (fru)	Lee and Hall (2000)
Dissatisfaction (dsf)	
fru^{M}	Vrontou et al. (2006)
fru^{F}	
Cyp6a20	Dierick and Greenspan (2006)
Neuropeptide F(npf)	Dierick and Greenspan (2007)
Muscleblind	Edwards et al. (2006)
tramtrack	
SP71	
Darkener of apricot	
Longitudinals lacking	
Scribbler	
Male-specific RNA 87F	
Kismet	
CG12292	
CG17154	
CG1623	
CG5966	
CG30015	
CG13512	
CG14478	
CG11448	Edwards et al. (2009a, b)
CG13760	
CG2556	
CG31038	
CG32425	
Gap1	
Late bloomer	
Schizo	
Skuld, etc.	
Tachykinin (Tk)	Asahina et al. (2014)
2. *Neurotransmitters*	
Dopamine	Baier et al. (2002), Alekseyenko et al. (2013)
Octopamine	Baier et al. (2002), Hoyer et al. (2008), Zhou et al. (2008), Certel et al. (2007, 2010), Andrews et al. (20102014)
Serotonin	Baier et al. (2002), Dierick and Greenspan (2007), Alekseyenko et al. (2010)

(continued)

Table 2.2 (continued)

Factor	References
3. *Environment*	
Food resource	Hoffmann (1990), Papaj and Messing (1998)
Social experience	Hoffmann (1990), Kim and Ehrman (1998), Kamyshev et al. (2002), Ueda and Kidokoro (2002), Svetec and Ferveur (2005), Yurkovic et al. (2006), Zhou et al. (2008), Wang, Dankert et al. (2008), Penn et al. (2010), Trannoy et al. (2016)

Acknowledgments I am very grateful to Drs. W. Anderson, L. Ehrman, J.-F., Ferveur, P. A. Gowaty, J.-M. Jallon, and J. Ringo for helpful comments on the earlier version of the manuscript.

References

Alekseyenko, O. V., Chan, Y.-B., Li, R., & Kravitz, E. A. (2013). Single dopaminergic neurons that modulate aggression in *Drosophila. Proceedings of the National Academy of Sciences of the United States of America, 110*, 6151–6156.

Alekseyenko, O. V., Lee, C., & Kravita, E. A. (2010). Targeted manipulation of serotonergic neurotransmission affects the escalation of aggression in adult male *Drosophila melanogaster. PLoS ONE, 5*, e10806.

Alves, H., Rouault, J.-D., Kondoh, Y., Nakano, Y., Yamamoto, D., Kim, Y.-K., et al. (2010). Evolution of cuticular hydrocarbons of Hawaiian Drosophilidae. *Behavior Genetics, 40*, 694–705.

Anderson, W. W., Kim, Y.-K., & Gowaty, P. A. (2007). Experimental constraints on mate preferences in *Drosophila pseudoobscura* decrease offspring viability and fitness of mated pairs. *Proceedings of the National Academy of Sciences of the United States of America, 104*, 4484–4488.

Andersson, M. (1994). *Sexual selection*. Princeton, NJ: Princeton University Press.

Andrews, J. C., Fernandez, M. P., Yu, Q., Leary, G. P., Leung, A. K., Kavanaugh, M. P., et al. (2014). Octopamine neuromodulation regulates Gr32a-linked aggression and courtship pathways in *Drosophila* males. *PLoS Genetics, 10*, e1004356.

Anholt, R. R. H., & Mackay, T. F. C. (2012). Genetics of aggression. *Annual Review of Genetics, 46*, 145–164.

Armstrong, J. D., de Belle, J. S., Wang, Z., & Kaiser, K. (1998). Meta-morphosis of the mushroom bodies: Large-scale rearrangements of the neural substrates for associative learning and memory in *Drosophila. Learning & Memory, 5*, 102–114.

Asahina, K., Watanabe, K., Duistermars, B. J., Hoopfer, E., Gonzalez, C. R., Eyjolfsdottir, E. A., et al. (2014). Tachykinin-expressing neurons control male-specific aggressive arousal in *Drosophila. Cell, 156*, 221–235.

Aspi, J., & Hoikkala, A. (1995). Male mating success and survival in the field with respect to size and courtship song character in *Drosophila littoralis* and *D. montana* (Diptera: Drosophilidae). *Journal of Insect Behavior, 8*, 67–87.

Baier, A., Wittek, B., & Brembs, B. (2002). *Drosophila* as a new model organisms for the neurobiology of aggression? *Journal of Experimental Biology, 205*, 1233–1240.

Balling, A., Technau, G. M., & Heisenberg, M. (1987). Are the structural changes in the adult *Drosophila* mushroom bodies memory traces? Studies on biochemical learning mutants. *Journal of Neurogenetics, 4*, 65–73.

Bartelt, R. J., Schaner, A. M., & Jackson, L. L. (1985). cis-vaccenyl acetate as an aggregation pheromone in *Drosophila melanogaster. Journal of Chemical Ecology, 11*, 1747–1756.

Barth, M., & Heisenberg, M. (1997). Vision affects mushroom bodies and central complex in *Drosophila melanogaster. Learning & Memory, 4*, 219–229.

Barth, M., Hirsch, H. V. B., Meinertzhagen, I. A., & Heisenberg, M. (1997). Experience-dependent developmental plasticity in the optic lobe of *Drosophila melanogaster. Journal of Neuroscience, 17*, 1493–1504.

Billeter, J.-C., & Levine, J. D. (2015). The role of cVA and the Odorant binding protein Lush in social and sexual behavior in *Drosophila melanogaster. Frontiers in Ecology and Evolution, 3*. doi:10.3389/fevo.2015.00075.

Billeter, J.-C., Atllah, J., Krupp, J. J., Millar, J. G., & Levine, J. D. (2009). Specialized cells tag sexual and species identity in *Drosophila melanogaster. Nature, 461*, 987–991.

Blows, M. W. (2002). Interaction between natural and sexual selection during the evolution of mate recognition. *Proceedings of the Royal Society of London, 269*, 1113–1118.

Blows, M. W., & Allan, R. A. (1998). Levels of mate recognition within and between two *Drosophila* species and their hybrids. *American Naturalist, 152*, 826–837.

Boake, C. R. B. (1989). Correlations between courtship success, aggressive success and body size in a picture-winged fly, *Drosophila sil- vestris. Ethology, 80*, 318–329.

Boake, C. R. B., DeAngelis, M. P., & Andreadis, D. K. (1997). Is sexual selection and species recognition a continuum? Mating behavior of the stalk-eyed *Drosophila heteroneura. Proceedings of the National Academy of Sciences of the United States of America, 94*, 12442–12445.

Boake, C. R. B., & Konigsberg, L. (1998). Inheritance of male courtship behavior, aggressive success, and body size in *Drosophila silvestris. Evolution, 52*, 1487–1492.

Boake, C. R. B., Price, D. K., & Andreadis, D. K. (1998). Inheritance of behavioural differences between two interfertile, sympatric species *Drosophila silvestris* and *D. heteroneura. Heredity, 80*, 642–650.

Brand, A. H., & Perrimon, N. (1993). Targeted gene expression as a means of altering cell fates and generating dominant phenotypes. *Development, 118*, 401–415.

Brown, M. R., Crim, J. W., Arata, R. C., Cai, H. N., et al. (1999). Identification of a *Drosophila* brain-gut peptide related to the neuropeptide Y family. *Peptides, 20*, 1035–1042.

Busto, G. U., Cervantes-Sandoval, I., & Davis, R. L. (2010). Olfactory learning in *Drosophila. Physiology, 25*, 338–346.

Carson, H. L. (1997). Sexual selection: A driver of genetic change in Hawaiian *Drosophila. Journal of Heredity, 88*, 343–352.

Carson, H. L. (2002). Female choice in *Drosophila*: Evidence from Hawaii and implications for evolutionary biology. *Genetica, 116*, 383–393.

Carson, H. L., Hardy, D. E., Spieth, H. T., & Stone, W. S. (1970). The evolutionary biology of the Hawaiian Drosophilidae. In M. K. Hecht & W. C. Steere (Eds.), *Essays in evolution and genetics in honor of Theodosius Dobzhansky* (pp. 437–543). New York: Appleton- Centry-Crofts.

Certel, S. J., Savella, M. G., Schlegel, D. C., & Kravitz, E. A. (2007). Modulation of *Drosophila* male behavioral choice. *Proceedings of the National Academy of Sciences of the United States of America, 104*, 4706–4711.

Certel, S. J., Leung, A., Lin, C. Y., Perez, P., Chiang, A.-S., & Kravitz, E. A. (2010). Octopamine neuromodulatory effects on a social behavior decision-making network in *Drosophila* males. *PLoS One, 5*, e13248.

Chan, Y.-K., & Kravitz, E. A. (2007). Specific subgroups of Fru^M neurons control sexually dimorphic patterns of aggression in *Drosophila melanogaster. Proceedings of the National Academy of Sciences of the United States of America, 104*, 19577–19582.

Chen, S., Lee, A. Y., Bowens, N., Huber, R., & Kravitz, E. A. (2002). Fighting fruit flies: A model system for the study of aggression. *Proceedings of the National Academy of Sciences of the United States of America, 99*, 5664–5668.

Chen, J., Zhang, Y., & Shen, P. (2008). A protein kinase C activity localized to neuropeptide Y-like neurons mediates ethanol intoxication in *Drosophila melanogaster. Neuroscience, 156*, 42–47.

Chenoweth, S. F., & Blows, M. W. (2005). Contrasting mutual sexual selection on homologous signal traits in *Drosophila serrata. American Naturalist, 165*, 281–289.
Cobb, M., & Jallon, J.-M. (1990). Pheromones, mate recognition and courtship stimulation in the *Drosophila melanogaster* species sub- group. *Animal Behaviour, 39*, 1058–1067.
Connolly, J. B., Roberts, I. J., Amstrong, J. D., Kaiser, K., Forte, M., Tully, T., & O'Kane, C. J. (1996). Associative learning disrupted by impaired G_s signaling in *Drosophila* mushroom bodies. *Science, 274*, 2104–2107.
Coyne, J. A., & Charlesworth, B. (1997). Genetics of a pheromonal difference affecting sexual isolation between *Drosophila mauritiana* and *D. sechellia. Genetics, 145*, 1015–1030.
Coyne, J. A., Crittenden, A. P., & Mah, K. (1994). Genetics of a pheromonal difference contributing to reproductive isolation in *Drosophila. Science, 265*, 1461–1464.
Crossley, S., & Wallace, B. (1987). The effects of crowding on courtship and mating success in *Drosophila melanogaster. Behavior Genetics, 17*, 513–522.
Darwin, C. (1871). *The descent of man, and selection in relation to sex.* London: J. Murray.
David, J. R., Allemand, R., Van Herrewege, J., & Cohert, Y. (1983). Ecophysiology: Abiotic factors. In M. Ashburner, H. L. Carson, & J. N. Thompson (Eds.), *The genetics and biology of Drosophila* (pp. 106–109). London: Academic.
Davis, R. L. (1993). Mushroom bodies and *Drosophila* learning. *Neuron, 11*, 1–14.
de Belle, J. S., & Heisenberg, M. (1994). Associative odor learning in *Drosophila* abolished by chemical ablation of mushroom bodies. *Science, 263*, 692–695.
Demir, E., & Dickson, B. J. (2005). *fruitless* splicing specifies male courtship behavior in *Drosophila. Cell, 121*, 785–794.
Dierick, H. A., & Greenspan, R. J. (2006). Molecular analysis of flies selected for aggressive behavior. *Nature Genetics, 38*, 1023–1031.
Dierick, H. A., & Greenspan, R. J. (2007). Serotonin and neuropeptide F have opposite modulatory effects on fly aggression. *Nature Genetics, 39*, 678–682.
Donlea, J. M., Ramanan, N., & Shaw, P. J. (2009). Use-dependent plasticity in clock neurons regulates sleep need in *Drosophila. Science, 324*, 105–108.
Dow, M. A., & von Schilcher, F. (1975). Aggression and mating success in *Drosophila melanogaster. Nature, 254*, 511–512.
Dubai, Y., Buxbaum, J., Corfas, G., & Ofarim, M. (1987). For- mamidines interact with *Drosophila* octopamine receptors alter the flies' behavior and reduce their learning ability. *Journal of Comparative Physiology, 161*, 739–746.
Dubai, Y., Jan, Y.-N., Byers, D., Quinn, W., & Benzer, S. (1976). *dunce*, a mutant of *Drosophila* deficient in learning. *Proceedings of the National Academy of Sciences of the United States of America, 73*, 1684–1688.
Dubnau, J., Chiang, A.-S., & Tully, T. (2003). Neural substrates of memory: From synapses to system. *Journal of Neurobiology, 54*, 238–253.
Dubnau, J., & Tully, T. (2001). Functional anatomy: From molecule to memory. *Current Biology, 11*, R24–R243.
Edwards, A. C., Ayroles, J. F., Stone, E. A., Carbone, M. A., Lyman, R. F., & Mackay, T. F. C. (2009a). A transcriptional network associated with natural variation in *Drosophila* aggressive behavior. *Gemone Biology, 10*, R76.
Edwards, A. C., Rollmann, S. M., Morgan, T. J., & Mackay, T. F. C. (2006). Quantitative genomics of aggressive behavior in *Drosophila melanogaster. PLoS Genetics, 2*, 1386–1395.
Edwards, A. C., Zwarts, L., Yamamoto, A., Callaerts, P., & Mackay, T. F. C. (2009b). Mutations in many genes affect aggressive behavior in *Drosophila melanogaster. BMC Biology, 7*, 29.
Ejima, A., Smith, B. P. C., Lucas, C., van der Goes van Naters, W., Miller, C. J., et al. (2007). Generalization of courtship learning in *Drosophila* is mediated by cis-vaccenyl acetate. *Current Biology, 17*, 599–605.
Ellis, L. B., & Kessler, S. (1975). Differential posteclosion housing experiences and reproduction in *Drosophila. Animal Behaviour, 23*, 949–952.

Ewing, L. S., & Ewing, A. W. (1984). Courtship in *Drosophila melanogaster*: Behaviour of mixed sex groups in large observation chambers. *Behaviour, 90*, 184–202.

Falconer, D. S., & Mackay, T. F. C. (1996). *Introduction to quantitative genetics* (4th ed.). London: Longmans Green.

Fernandez, M. P., Chan, Y.-B., Yew, J. Y., Billester, J.-C., Driesewerd, K., et al. (2010). Pheromonal and behavioral cues trigger male-to-female aggression in *Drosophila*. *PLoS Biology, 8*, e1000541.

Ferveur, J.-F. (2005). Cuticular hydrocarbons: The evolution and roles in *Drosophila* pheromonal communication. *Behavior Genetics, 35*, 279–295.

Feyereisen, R. (2005). Insect cytochrome P450. In L. I. Gilbert, K. Iatrou, & S. S. Gill (Eds.), *Comprehensive molecular insect science* (Vol. 4, pp. 1–77). Amsterdam: Elsevier.

Fisher, R. A. (1930). *The genetical theory of natural selection*. Oxford: Clarendon.

Folkers, E., Drain, P. F., & Quinn, W. G. (1993). *radish*, a *Drosophila* mutant deficient in consolidated memory. *Proceedings of the National Academy of Sciences of the United States of America, 90*, 8123–8127.

Fulker, D. W. (1966). Mating speed in male *Drosophila melanogaster*: A psychogenetic analysis. *Science, 153*, 203–205.

Gilestro, G. F., Tononi, G., & Cirelli, C. (2009). Widespread changes in synaptic markers as a function of sleep and wakefulness in *Drosophila*. *Science, 324*, 109–112.

Gleason, J. M. (2005). Mutations and natural genetic variation in the courtship song of *Drosophila*. *Behavior Genetics, 35*, 265–277.

Gowaty, P. A., Anderson, W. W., Bluhm, C. K., Drickamer, L. C., Kim, Y.-K., & Moore, A. (2007). The hypothesis of reproductive compensation for lowered offspring viability: Tests of assumptions and predictions. *Proceedings of the National Academy of Sciences of the United States of America, 104*, 15023–15027.

Greenspan, R. J. (2004). E pluribus unum, ex uno plura: Quantitative- and single-gene perspectives on the study of behavior. *Annual Reviews of Neuroscience, 27*, 79–105.

Greenspan, R. J., & Ferveur, J.-F. (2000). Courtship in *Drosophila*. *Annual Reviews of Genetics, 34*, 205–232.

Grillet, M., Dartevelle, L., & Ferveur, J.-F. (2006). A *Drosophila* male pheromone affects female sexual receptivity. *Proceedings of the Royal Society of London, 273*, 315–323.

Hall, J. C. (1994). The mating of a fly. *Science, 264*, 1702–1714.

Han, K.-A., Millar, N. S., Grotewiel, M. S., & Davis, R. L. (1996). DAMB, a novel dopamine receptor expressed specifically in *Drosophila* mushroom bodies. *Neuron, 16*, 1127–1135.

Han, K.-A., Millar, N. S., & Davis, R. L. (1998). A novel octopamine receptor with preferential expression in *Drosophila* mushroom bodies. *Journal of Neuroscience, 18*, 3650–3658.

Harshman, L. G., & Hoffmann, A. A. (2000). Laboratory selection experiments using *Drosophila*: What do they really tell us? *Trends in Ecology and Evolution, 15*, 32–36.

Heisenberg, M. (2003). Mushroom body memoir: From maps to models. *Nature Review/Neuroscience, 4*, 266–275.

Heisenberg, M., Borst, A., Wagner, S., & Byers, D. (1985). *Drosophila* mushroom body mutants are deficient in olfactory learning. *Journal of Neurogenetics, 2*, 1–30.

Heisenberg, M., Heusipp, M., & Wanke, C. (1995). Structural plasticity in the *Drosophila* brain. *Journal of Neuroscience, 15*, 1951–1960.

Hing, A. L., & Carlson, J. R. (1996). Male-male courtship behavior induced by ectopic expression of the *Drosophila* white gene: Role of sensory function and age. *Journal of Neurobiology, 30*, 454–464.

Hoffmann, A. A. (1987a). A laboratory study of male territoriality in the sibling species *Drosophila melanogaster and Drosophila simulans*. *Animal Behaviour, 35*, 807–818.

Hoffmann, A. A. (1987b). Territorial encounters between *Drosophila* males of different sizes. *Animal Behaviour, 35*, 1899–1901.

Hoffmann, A. A. (1988). Heritable variation for territorial success in two *Drosophila melanogaster* populations. *Animal Behaviour, 36*, 1180–1189.

Hoffmann, A. A. (1989). Georgraphic variation in the territorial success of *Drosophila melanogaster* males. *Behavior Genetics, 19*, 241–255.
Hoffmann, A. A. (1990). The influence of age and experience with con- specifics on territorial behavior in *Drosophila melanogaster. Journal of Insect Behavior, 3*, 1–12.
Hoffmann, A. A. (1991). Heritable variation for territorial success in field-collected *Drosophila melanogaster. American Naturalist, 138*, 668–679.
Hoffmann, A. A., & Cacoyianni, Z. (1989). Selection for territoriality in *Drosophila melanogaster*: Correlated responses in mating success and other fitness components. *Animal Behaviour, 38*, 23–34.
Hoffmann, A. A., & Cacoyianni, Z. (1990). Territoriality in *Drosophila melanogaster* as a conditional strategy. *Animal Behaviour, 40*, 526–537.
Hoikkala, A. (2005). Inheritance of male sound characteristics in *Drosophila* species. In S. Drosopoulos & M. F. Claridge (Eds.), *Insect sounds and communication: Physiology, behaviour, ecology and evolution* (pp. 167–177). Boca Raton, FL: CRC Taylor & Francis.
Hoikkala, A., Aspi, J., & Suvanto, L. (1998). Male courtship song frequency as an indicator of male genetic quality in an insect species, *Drosophila montana. Proceedings of the Royal Society of London, 265*, 503–508.
Hoopfer, E. D., Jung, Y., Inagaki, H. K., Rubin, G. M., & Anderson, D. J. (2015). P1 interneurons promote a persistent internal state that enhances inter-male aggression in *Drosophila. eLife, 4*, e11346.
Howard, R. W., Jackson, L. L., Banse, H., & Blows, M. W. (2003). Cuticular hydrocarbons of *Drosophila birchii* and *D. serrata*: Identification and role in mate choice in *D. serrata. Journal of Chemical Ecology, 29*, 961–976.
Hoy, R. R., Hoikkala, A., & Kaneshiro, K. Y. (1988). Hawaiian courtship songs: Evolutionary innovation in communication signals in *Drosophila. Science, 240*, 217–219.
Hoyer, S. C., Eckart, A., Herrel, A., Zars, T., Fischer, S. A., Hardie, S. L., et al. (2008). Octopamine in male aggression of *Drosophila. Current Biology, 18*, 159–167.
Irwin, D. E., & Price, T. (1999). Sexual imprinting, learning and speciation. *Heredity, 82*, 347–354.
Jacob, M. E. (1960). Influence of light on mating of *Drosophila melanogaster. Ecology, 41*, 182–188.
Jacob, M. E. (1978). Influence of β-alanine on mating and territorialism in *Drosophila melanogaster. Behavior Genetics, 8*, 487–502.
Jallon, J. M. (1984). A few chemical words exchanged by *Drosophila* during courtship and mating. *Behavio Genetics, 14*, 441–478.
Jallon, J.-M., Antony, C., & Benemar, O. (1981). Un antiaphrodisiac produit par les males de *Drosophila melanogaster* et transfere aux femelles lors de la copulation. *Comptes rendus de l'Acade'mie des sciences. Paris, 292*, 1147–1149.
Johnson, O., Becnel, J., & Nichols, C. D. (2009). Serotonin 5-HT(2) and 5-HT(1A)-like receptors differentially modulate aggressive behaviors in *Drosophila melanogaster. Neuroscience, 158*, 1292–1300.
Kamyshev, N. G., Smirnova, G. P., Kamysheva, E. A., Nikiforov, O. N., Parafenyuk, I. V., & Ponomarenko, V. V. (2002). Plasticity of social behavior in *Drosophila. Neuroscience and Behavioral Physiology, 32*, 401–408.
Kaneshiro, K. Y., & Boake, C. R. B. (1987). Sexual selection and speciation: Issue raised by *Hawaiian Drosophila. Trends in Ecology and Evolution, 2*, 207–212.
Kaul, D., & Parsons, P. A. (1965). The genotypic control of mating speed and duration of copulation in *Drosophila pseudoobscura. Heredity, 20*, 381–392.
Keleman, K., Vrontou, E., Krutter, S., Yu, J. Y., Kurtovic-Kozaric, A., & Dickson, B. J. (2012). Dopamine neurons modulate pheromone responses in *Drosophila* courtship learning. *Nature, 489*, 145–150.
Kent, C., Azanchi, R., Smith, B., Formosa, A., & Levine, J. D. (2008). Social context influences chemical communication in *D. melanogaster* males. *Current Biology, 18*, 1384–1389.

Kessler, S. (1969). The genetics of *Drosophila* mating behavior. II. The genetic architecture of mating speed in *Drosophila pseudoobscura. Genetics, 62*, 421–433.

Kim, Y.-K., Abramowicz, K., & Anderson, W. W. (2008). A role of aggressive behavior in sexual isolation between *Drosophila pseudoobscura* and *D. persimilis. Behavior Genetics, 38*, 632.

Kim, Y.-K., Alvarez, D., Barber, J., Brock, A., & Jeon, J. (2007). Genetic and environmental influence on the *Drosophila* aggressive behavior. *Behavior Genetics, 37*, 766.

Kim, Y.-K., Basset, C., Laverentz, J., & Anderson, W. W. (2005a). Female choice in sexual selection of *Drosophila pseudoobscura. Behavior Genetics, 35*, 808.

Kim, Y.-K., & Ehrman, L. (1998). Developmental isolation and sub- sequent adult behavior of *D. paulistorum*: IV. Courtship. *Behavior Genetics, 28*, 57–65.

Kim, Y.-K., Ehrman, L., & Koepfer, H. R. (1992). Developmental isolation and subsequent adult behavior of *Drosophila paulistorum*. I. Survey of the six semispecies. *Behavior Genetics, 22*, 545–556.

Kim, Y.-K., Ehrman, L., & Koepfer, H. R. (1996a). Developmental isolation and subsequent adult behavior of *Drosophila paulistorum*. II. Prior experience. *Behavior Genetics, 26*, 15–25.

Kim, Y.-K., Gowaty, P. A., & Anderson, W. W. (2005). Testing for mate preference and mate choice in *Drosophila pseudoobscura*. In L. Noldus, J. J. F. Grieco, L. W. S. Loijens, & P. H. Zimmerman (Eds.) *Proceedings of the 5th International Conference on Methods and Techniques in Behavioral Research* (pp. 533–535). Wageningen, The Netherlands.

Kim, Y.-K., Kim, A., & Tiemeyer, M. (2009). Social experience modulates *Drosophila* behavior and brain function. *Behavior Genetics, 39*, 661.

Kim, Y.-K., Koepfer, H. R., & Ehrman, L. (1996b). Developmental isolation and subsequent adult behavior of *Drosophila paulistorum*. III. Alternative rearing. *Behavior Genetics, 26*, 27–37.

Kim, Y.-K., Phillips, D., Chao, T., & Ehrman, L. (2004). Developmental isolation and subsequent adult behavior of *Drosophila paulistorum*. V. Quantitative variation of cuticular hydrocarbons. *Behavior Genetics, 34*, 385–394.

Kirkpatrick, M. (1987). Sexual selection by female choice in polygynous animals. *Annual Reviews of Ecology and Systematics, 18*, 43–70.

Kravitz, E. A. (2000). Serotonin and aggression: Insights gained from a lobster model system and speculations on the role of amine neurons in a complex behavior like aggressions. *Journal of Comparative Physiology, 186*, 221–238.

Kravitz, E. A., & Huber, R. (2003). Aggression in invertebrates. *Current Opinion in Neurobiology, 13*, 736–743.

Kurtovic, A., Widmer, A., & Dickson, B. J. (2007). A single class of olfactory neurons mediated behavioural responses to a *Drosophila* sex pheromone. *Nature, 446*, 542–546.

Kyriacou, C. P., & Hall, J. C. (1982). The function of courtship song rhythms in *Drosophila. Animal Behaviour, 30*, 794–801.

Lacy, R. C., & Sherman, P. W. (1983). Kin recognition by phenotype matching. *American Naturalist, 121*, 489–512.

Lande, R. (1981). Models of speciation by sexual selection on polygenic traits. *Proceedings of the National Academy of Sciences of the United States of America, 78*, 3721–3725.

Lasbleiz, C., Ferveur, J.-F., & Everaerts, C. (2006). Courtship behavior of *Drosophila melanogaster* revisited. *Animal Behaviour, 72*, 1001–1012.

Lee, G., Bahn, J. H., & Park, J. H. (2006). Sex- and clock-controlled expression of the *neuropeptide F* gene in *Drosophila. Proceedings of the National Academy of Sciences of the United States of America, 103*, 12580–12585.

Lee, G., & Hall, J. C. (2000). A newly uncovered phenotype associated with the fruitless gene of *Drosophila melanogaster*: Aggression- like head interactions between mutant males. *Behavior Genetics, 30*, 263–275.

Liu, T., Dartevelle, L., Yuan, C., Wei, H., Wang, Y., Ferveur, J.-F., et al. (2008). Increased dopamine level enhances male-male courtship in *Drosophila. Journal of Neuroscience, 28*, 5539–5546.

Liu, S., Liang, X., Gong, J., Yang, Z., et al. (2011). Social regulation of aggression by pheromonal activation of Or65a olfactory neurons in *Drosophila. Nature Neuroscience, 14*, 896–902.

Mackay, T. F., Heinsohn, S. L., Lyman, R. F., Moehring, A. J., Morgan, T. J., & Rollmann, S. M. (2005). Genetics and genomics of Drosophila mating behavior. *Proceedings of the National Academy of Sciences of the United States of America, 102*, 6622–6629.

Mane, S. D., Tompkins, L., & Richmond, R. C. (1983). Male esterase 6 catalyzes the synthesis of a sex pheromone in *Drosophila melanogaster* females. *Science, 222*, 419–421.

Manning, A. (1961). The effects of artificial selection for mating speed in *Drosophila melanogaster. Animal Behaviour, 9*, 82–92.

Marcillac, F., Bousquet, F., Alabouvette, J., Savarit, F., & Ferveur, J.-F. (2005). A mutation with major effects on *Drosophila melanogaster* sex pheromones. *Genetics, 171*, 1617–1628.

Markow, T. A. (1988). Reproductive behavior of *Drosophila melanogaster* and *D. nigrospiracula* in the field and in the laboratory. *Journal of Comparative Psychology, 102*, 169–173.

Markow, T. A. (1996). Evolution of *Drosophila* mating systems. *Evolutionary Biology, 29*, 73–106.

Markow, T. A. (2000). Forced matings in natural populations of *Drosophila. American Naturalist, 156*, 100–103.

Markow, T. A., & O'Grady, P. M. (2005). Evolutionary genetics of reproductive behavior in *Drosophila*: Connecting the dots. *Proceedings of the National Academy of Sciences of the United States of America, 39*, 263–291.

McBride, S. M. J., Giuliani, G., Choi, C., Krause, P., Correale, D., Watson, K., Baker, G., & Siwicki, K. K. (1999). Mushroom body ablation impairs short-term memory and long-term memory of courtship conditioning in *Drosophila melanogaster. Neuron, 24*, 967–977.

McGuire, S. E., Le, P. T., & Davis, R. L. (2001). The role of *Drosophila* mushroom body signaling in olfactory memory. *Science, 293*, 1330–1333.

McRobert, S. P., Tompkins, L., Barr, N. B., Bradner, J., Lucas, D., Rattigan, D. M., & Tannous, A. F. (2003). Mutations in *raised Drosophila melanogaster* affect experience-dependent aspects of sexual behavior in both sexes. *Behavior Genetics, 33*, 347–356.

Milani, R. (1956). Relations between courting and fighting behaviour in some *Drosophila* species (*obscura* group). *1st. Sup. di Sanita, 1*, 213–224.

Moehring, A. J., & Mackay, T. F. C. (2004). The quantitative genetic basis of male mating behavior in *Drosophila melanogaster. Genetics, 167*, 1249–1263.

Murakami, S., & Itoh, M. T. (2001). Effects of aggression and wing removal on brain serotonin levels in male crickets, *Gryllus bimacu- latus. Journal of Insect Physiology, 47*, 1309–1312.

Murakami, S., & Itoh, M. T. (2003). Removal of both antennae influences the courtship and aggressive behaviors in male crickets. *Journal of Neurobiology, 57*, 110–118.

Nässel, D. R., & Winther, A. M. E. (2010). *Drosophila* neuropeptides in regulation of physiology and behavior. *Progress in Neurobiology, 92*, 42–104.

Nilsen, S. P., Chan, Y.-B., Huber, R., & Kravitz, E. A. (2004). Gender-selective patterns of aggressive behavior in *Drosophila melanogaster. Proceedings of the National Academy of Sciences of the United States of America, 101*, 12342–12347.

Noor, M. A. (1995). Speciation driven by natural selection in*Drosophila. Nature, 375*, 674–675.

Noor, M. A. F., & Ortíz-Barrientos, D. (2006). Simulating natural conditions in the laboratory: A re-examination of sexual isolation between sympatric and allopatric populations of *Drosophila pseudoobscura* and *D. persimilis. Behavior Genetics, 36*, 322–327.

O'Dell, K. M. C. (1994). The inactive mutation leads to abnormal experience-dependent courtship modification in male *Drosophila melanogaster. Behavior Genetics, 24*, 381–388.

O'Donald, P. (1983). Sexual selection by female choice. In P. Bateson (Ed.), *Mate choice* (pp. 53–66). Cambridge: Cambridge University Press.

Paillette, M., Ikeda, H., & Jallon, J.-M. (1991). A new acoustic signal of the fruit-flies *Drosophila simulans* and *D. melanogaster. Bioacoustics, 3*, 247–254.

Papaj, D. R., & Messing, R. H. (1998). Asymmetries in physiological state as a possible cause of resident advantage in contests. *Behaviour, 135*, 1013–1030.

Partridge, L., Ewing, A., & Chandler, A. (1987a). Male size and mating success in *Drosophila melanogaster*: The roles of male and female behavior. *Animal Behaviour, 35*, 555–562.

Partridge, L., & Farquhar, M. (1983). Lifetime mating success of male fruitflies (*Drosophila melanogaster*) is related to their size. *Animal Behaviour, 31*, 871–877.

Partridge, L., Hoffmann, A., & Jones, S. (1987b). Male size and mating success in *Drosophila melanogaster* and *Drosophila pseudoobscura* under field conditions. *Animal Behaviour, 35*, 468–476.

Partridge, L., Mackay, T. F. C., & Aitken, S. (1985). Male mating success and fertility in *Drosophila melanogaster. Genetical Research Cambridge, 46*, 279–285.

Pascual, A., & Préat, T. (2001). Localization of long-term memory within the *Drosophila* mushroom body. *Science, 294*, 1115–1117.

Paterson, H. E. H. (1978). More evidence against speciation by reinforcement. *South African Journal of Science, 74*, 369–371.

Paterson, H. E. H. (1985). The recognition concept of species. In E. S. Vrba (Ed.), *Species and speciation. Transvaal Museum Monograph, 4* (pp. 21–29). Pretoria: Transvaal Museum.

Penn, J. K. M., Zito, M. F., & Kravitz, E. A. (2010). A single social defeat reduces aggression in a highly aggressive strain of *Drosophila. Proceedings of the National Academy of Sciences of the United States of America, 107*, 12682–12686.

Popova, N. K. (2006). From genes to aggressive behavior: The role of serotonin system. *BioEssays, 28*, 495–503.

Quinn, W. G., Sziber, P. P., & Booker, R. (1979). The *Drosophila* memory mutant amnesiac. *Nature, 277*, 212–214.

Ringo, J. M. (1977). Why 300 species of Hawaiian *Drosophila*? The sexual selection hypothesis. *Evolution, 31*, 694–696.

Ringo, J. M., & Hodosh, R. J. (1978). A multivariate analysis of behavioral divergence among closely related species of endemic Hawaiian *Drosophila. Evolution, 32*, 389–397.

Ringo, J., Kananen, M. K., & Wood, D. (1983). Aggression and mating success in three species of *Drosophila. Journal of Comparative Ethology, 61*, 341–350.

Ritchie, M. G., & Gleason, J. M. (1995). Rapid evolution of courtship song pattern in *Drosophila willistoni* sibling species. *Journal of Evolutionary Biology, 8*, 463–479.

Ritchie, M. G., & Kyriacou, C. P. (1996). Artificial selection for a courtship signal in *Drosophila melanogaster. Animal Behaviour, 52*, 603–611.

Ritchie, M. G., Saarikettu, M., Livingstone, S., & Hoikkala, A. (2001). Characterisation of female preference functions for a sexually selected acoustic signal in *D. montana*, and a test of the "temperature coupling" hypothesis. *Evolution, 55*, 721–727.

Ritchie, M. G., Townhill, R. M., & Hoikkala, A. (1998). Female preference for fly song: Playback experiments confirm the targets of sexual selection. *Animal Behaviour, 56*, 713–717.

Robertson, F. W., & Reeve, E. (1952). Studies in quantitative inheritance. I. The effects of selection on wing and thorax length in Drosophila melanogaster. *Journal of Genetics, 50*, 414–448.

Robin, C., Daborn, P. J., & Hoffmann, A. A. (2006). Fighting fly genes. *Trends in Genetics, 23*, 51–54.

Rollmann, S. M., Zwarts, L., Edwards, A. C., Yamamoto, A., Callaerts, P., et al. (2008). Pleiotropic effects of *Drosophila* neuralized on complex behaviors and brain structure. *Genetics, 179*, 1327–1336.

Roman, G., & Davis, R. L. (2001). Molecular biology and anatomy of *Drosophila* olfactory associative learning. *Bioessays, 23*, 571–581.

Saarikettu, M., Liimatainen, J., & Hoikkala, A. (2005). The role of male courtship song in species recognition in *Drosophila montana. Behavior Genetics, 35*, 257–263.

Santos, M., Ruiz, A., Barbadilla, A., Hasson, E., & Fontdevila, A. (1988). The evolutionary history of *Drosophila*. XIV. Larger flies mate more often in nature. *Heredity, 61*, 255–262.

Savarit, F., Sureau, G., Cobb, M., & Ferveur, J.-F. (1999). Genetic elimination of known pheromones reveals the fundamental chemical bases of mating and isolation in *Drosophila. Proceedings of the National Academy of Sciences of the United States of America, 96*, 9015–9020.

Shen, P., & Cai, H. N. (2001). *Drosophila* neuropeptide F mediates integration of chemosensory stimulation and conditioning of the nervous system by food. *Journal of Neuroscience, 47*, 16–25.

Shohat-Ophir, G., Kaun, K. R., Azanchi, R., Mohammed, H., & Heberlein, U. (2012). Sexual depriviation increases ethanol intake in *Drosophila*. *Science, 335*, 1351–1355.

Shorter, J., Couch, C., Huang, W., Carbone, M. A., Peiffer, J., Anholt, R. R. H., et al. (2015). Genetic architecture of natural variation in *Drosophila melanogaster* aggressive behavior. *Proceedings of the National Academy of Sciences of the United States of America, 112*, E3555. doi:10.1073/pnas.1510104112.

Singh, R. N., & Singh, K. (1984). Fine-structure of the sensory organs of *Drosophila melanogaster* meigen larva. *International Journal of Insect Morphology and Embryology, 13*, 255–273.

Skrzipek, K. H., Kroner, B., & Hager, H. (1979). Aggression bei *Drosophila melanogaster*—Laboruntersuchungen. *Journal of Comparative Psychology, 49*, 87–103.

Snook, R. R., Robertson, A., Crudgington, H. S., & Ritchie, M. G. (2005). Experimental manipulation of sexual selection and the evolution of courtship song in *Drosophila pseudoobscura*. *Behavior Genetics, 35*, 245–255.

Spiess, E. B. (1987). Discrimination among prospective mates in *Drosophila*. In D. J. C. Fletcher & C. D. Michener (Eds.), *Kin recognition in animals* (pp. 75–119). New York: John Wiley & Sons.

Spiess, E. B., & Langer, B. (1964). Mating speed control by gene arrangements in *Drosophila pseudoobscura* homokaryotypes. *Proceedings of the National Academy of Sciences of the United States of America, 51*, 1015–1019.

Spieth, H. T. (1966). Courtship behavior of endemic Hawaiian *Drosophila*. *Studies in Genetics III. University of Texas Publication 6615*, 245–313.

Spieth, H. T. (1968). Evolutionary implications of sexual behavior in *Drosophila*. *Evolutionary Biology, 2*, 157–193.

Spieth, H. T. (1974). Courtship behavior in *Drosophila*. *Annual Reviews of Entomology, 19*, 385–405.

Spieth, H. T. (1981). *Drosophila heteroneura* and *Drosophila silvestris*: Head shapes, behavior and evolution. *Evolution, 35*, 921–930.

Spieth, H. T. (1982). Behavioral biology and evolution of the Hawaiian picture-winged species group of *Drosophila*. *Evolutionary Biology, 14*, 351–437.

Spieth, H. T., & Heed, W. B. (1975). The *Drosophila pinicola* species group. *Pan-Pacific Entomologist, 51*, 287–295.

Strausfeld, N. J., Hansen, L., Li, Y., Gomez, R. S., & Ito, K. (1998). Evolution, discovery and interpretations of arthropod mushroom bodies. *Leaning & Memory, 5*, 11–37.

Svetec, N., Cobb, M., & Ferveur, J.-F. (2005). Chemical stimuli induce courtship dominance in *Drosophila*. *Current Biology, 15*, R790–R792.

Svetec, N., & Ferveur, J.-F. (2005). Social experience and pheromonal perception can exchange male-male interactions in *Drosophila melanogaster*. *Journal of Experimental Biology, 208*, 891–898.

Tauber, E., & Eberl, D. F. (2002). The effect of male competition on the courtship song of *Drosophila melanogaster*. *Journal of Insect Behavior, 15*, 109–120.

Tauber, E., & Eberl, D. F. (2003). Acoustic communication in *Drosophila*. *Behavioural Processes, 64*, 197–210.

Technau, G. M. (1984). Fiber number in the mushroom bodies of adult *Drosophila melanogaster* depends on age, sex and experience. *Journal of Neurogenetics, 1*, 113–126.

Technau, G. M., & Heisenberg, M. (1982). Neural reorganization during metamorphosis of the corpora pedunculata in *Drosophila melanogaster*. *Nautre, 295*, 405–407.

Templeton, A. R. (1977). Analysis of head shape differences between two interfertile species of Hawaiian *Drosophila*. *Evolution, 31*, 630–641.

Trannoy, S., Penn, J., Lucy, K., Popovic, D., & Kravitz, E. A. (2016). Short and long-lasting behavioral consequences of agonistic encounters between male *Drosophila melanogaster*. *Proceedings of the National Academy of Sciences of the United States of America, 113*, 4818–4823.

Ueda, A., & Kidokoro, Y. (2002). Aggressive behaviours of female *Drosophila melanogaster* are influenced by their social experience and food resources. *Phsiological Entomology, 27*, 21–28.

Vrontou, E., Nilsen, S. P., Kravitz, E. A., & Dickson, B. J. (2006). *fruitless* regulates aggression and dominance in *Drosophila. Nature Neuroscience, 9*, 1469–1471.

Waddell, S. (2010). Dopamine reveals neural circuit mechanisms of fly memory. *Trends in Neuroscience, 33,* 457–464.

Waddell, S., & Quinn, W. G. (2001). What can we teach *Drosophila*? What can they teach us? *Trends in Genetics, 17*, 719–726.

Wallace, B. (1974). Studies on intra- and inter-specific competition in *Drosophila. Ecology, 55*, 227–244.

Wang, L., & Anderson, D. J. (2010). Identification of an aggression-promoting pheromone and its receptor neurons in *Drosophila. Nature, 463*, 227–231.

Wang, L., Dankert, H., Perona, P., & Anderson, D. (2008). A common genetic target for environmental and heritable influences on aggressiveness in *Drosophila. Proceedings of the National Academy of Sciences of the United States of America, 105*, 5657–5663.

Wang, L., Han, X., Mehren, J., Hiroi, M., Billeter, J.-F., Miyamoto, T., et al. (2011). Hierarchical chemosensory regulation of male-male social interactions in *Drosophila. Nature Neuroscience, 14,* 757–762.

Wen, T., Parrish, C. A., Xu, D., Wu, Q., & Shen, P. (2005). *Drosophila* neuropeptide F and its receptor, NPFR1, define a signaling pathway that acutely modulates alcohol sensitivity. *Proceedings of the National Academy of Sciences of the United States of America, 102*, 2141–2146.

Wheeler, C. J., Fields, W. L., & Hall, J. C. (1988). Spectral analysis of *Drosophila* courtship songs: *D. melanogaster, D. simulans*, and their interspecific hybrids. *Behavior Genetics, 18*, 675–703.

Wilkinson, G. S., & Johns, P. M. (2005). Sexual selection and the evolution of mating systems in flies. In D. K. Yates & B. M. Wiegmann (Eds.), *The evolutionary biology of flies* (pp. 312–339). New York: Columbia University Press.

Williams, M. A., Blouin, A. G., & Noor, M. A. F. (2001). Courtship songs of *Drosophila pseudoobscura* and *D. persimilis*. II. Genetics of species differences. *Heredity, 86*, 68–77.

Wu, Q., Wen, T., Lee, G., Park, J. H., et al. (2003). Developmental control of foraging and social behavior by the *Drosophila* Neuropeptide Y-like system. *Neuron, 39*, 147–161.

Wu, Q., Zhao, Z., & Shen, P. (2005b). Regulation of aversion to noxious food by *Drosophila* neuropeptide Y- and insulin-like systems. *Nature Neuroscience, 8*, 1350–1355.

Yoshill, T., Todo, T., Wulbeck, C., Stanewsky, R., et al. (2008). Cryptochrome is present in the compound eyes and a subset of *Drosophila*'s clock neurons. *Journal of Comparative Neurology, 508*, 952–966.

Yuan, Q., Song, Y., Yang, C.-H., Jan, L. Y., & Jan, Y. N. (2014). Female contact modulates male aggression via a sexually dimorphic GABAergic circuit in *Drosophila. Nature Neuroscience, 17*, 81–88.

Yurkovic, A., Wang, O., Basu, A. C., & Kravitz, E. A. (2006). Learning and memory associated with aggression in *Drosophila melanogaster. Proceedings of the National Academy of Sciences of the United States of America, 103*, 17519–17524.

Zamudio, K. R., Huey, R. B., & Crill, W. D. (1995). Bigger isn't always better: Body size, developmental and parental temperature and male territorial success in *Drosophila melanogaster. Animal Behaviour, 49*, 671–677.

Zars, T., Fischer, M., Schulz, R., & Heisenberg, M. (2000). Localization of a short-term memory in *Drosophila. Science, 288*, 672–675.

Zhang, S.-D., & Odenwald, W. F. (1995). Misexpression of the white (w) gene triggers male-male courtship in *Drosophila. Proceedings of the National Academy of Sciences of the United States of America, 92*, 5525–5529.

Zhou, C., Rao, Y., & Rao, Y. (2008). A subset of octopaminergic neurons are important for *Drosophiila* aggression. *Nature Neuroscience, 11*, 1059–1067.

Zwarts, L., Magwire, M. M., Carbone, M. A., Versteven, M., Herteleer, L., et al. (2011). Complex genetic architecture of *Drosophila* aggressive behavior. *Proceedings of the National Academy of Sciences of the United States of America, 108*, 17070–17075.

Chapter 3
The Genetics of Impulsivity: A Synthesis of Findings in Humans and Rodent Models

Bianca Jupp and Jeffrey W. Dalley

Introduction

Impulsivity is a multidimensional behavioral trait encompassing an individual's tendency to act rapidly without appropriate foresight (Moeller et al. 2001; Evenden 1999). Within a "functional" range, where impulsiveness favors advantageous outcomes, an impulsive character is considered an essential aspect of human behavior where rapid, "gut" responses are often required (Dickman 1990). In the absence of "functional" impulsivity, individuals may fail to take acceptable risks or pursue unexpected opportunities. However, when excessively expressed, impulsive behavior is counterproductive and associated with a predisposition toward excessively risky and inappropriate actions, which can result in negative outcomes for the individual (Fernando and Robbins 2011). Indeed, in its extreme form, clinically relevant impulsivity is present in a wide range of neuropsychiatric morbidities including personality (e.g., Perry and Korner 2011) and mood disorder (e.g., Lombardo et al. 2012), drug addiction (e.g., Ersche et al. 2010; de Wit 2009; Kreek et al. 2005), schizophrenia (e.g., Kaladjian et al. 2011), problem gambling (e.g., Verdejo-Garcia et al. 2008), suicide (e.g., Dougherty et al. 2004), and attention deficit hyperactivity disorder (ADHD) (e.g., Geissler and Lesch 2011; Sonuga-Barke et al. 1992; Sergeant and Scholten 1985; Avila et al. 2004).

B. Jupp
Department of Psychology, University of Cambridge, Downing St, Cambridge CB2 3EB, UK

J.W. Dalley (✉)
Department of Psychology, University of Cambridge, Downing St, Cambridge CB2 3EB, UK

Department of Psychiatry, University of Cambridge, Cambridge CB2 2QQ, UK
e-mail: jwd20@cam.ac.uk

J.C. Gewirtz, Y.-K. Kim (eds.), *Animal Models of Behavior Genetics*,
Advances in Behavior Genetics, DOI 10.1007/978-1-4939-3777-6_3

Impulsivity spans a heterogeneous repertoire of behavioral processes mediated by distinct yet partially overlapping biological substrates (Winstanley et al. 2006; Dalley et al. 2011). Broadly described, these processes can be divided into two main components; impaired inhibition of motoric output (impulsive action) and a preference for immediate versus delayed rewards (impulsive choice). In reality however, impulsivity involves a highly complex myriad of behavioral processes including urgency, risk-taking, sensation-seeking, impaired planning, lack of premeditation, altered reward sensitivity, a disregard for future consequences, and insensitivity to punishment (Barratt 1985; Evenden 1999; Moeller et al. 2001; Monterosso and Ainslie 1999; Whiteside and Lynam 2003). This heterogeneity has hampered research on the precise contribution of impulsivity to various neuropsychiatric disorders.

Impulsivity can be assessed in humans using self-report questionnaires and experimental laboratory-based assays. Although self-report measures have obvious limitations (see Wilson and Dunn 2004), questionnaire based inventories provide a useful index of impulsivity and are widely used. These include the Eysenck personality questionnaire (EPQ; Eysenck and Eysenck 1975), Barratt Impulsivity Scale (BIS; Barratt 1985), Urgency, Premeditation, Perseverance Sensation-Seeking (UPSS) Impulsive Behaviour Scale (Whiteside and Lynam 2003) and Dickman's Impulsivity Inventory (DII; Dickman 1990). Experimental laboratory-based tests provide an objective assessment of impulsivity traits and states and can be broadly divided into tests of impulsive action (e.g., continuous performance tests, Voon et al. 2014; Dougherty et al. 2002; Sanchez-Roige et al. 2014), stop-signal reaction time (Aron et al. 2004) and go/no-go discrimination tasks (Fillmore 2003), and impulsive choice for immediate, small-magnitude or certain rewards (Mazur 1987). Importantly, many of these paradigms have been successfully translated to experimental animals, enabling the assessment of analogous behavioral processes in rodents (Winstanley 2011), non-human primates (Tobin and Logue 1994) and other species including zebrafish (Parker et al. 2013).

As mentioned above it is now widely accepted that impulsivity is a multifaceted behavioral construct with different measures of impulsivity assaying distinct neural and psychological processes (Dalley et al. 2011). Indeed, the degree of relatedness between laboratory-based tests and self-report measures is generally considered to be quite limited (Dalley et al. 2011), suggesting that psychometric and experimental approaches assay separate dimensions of impulsivity (i.e., behavioral vs. cognitive; trait vs. state). Similar distinctions have been noted for rodent-based impulsivity tasks, specifically between impulsive action as measured by either the stop-signal task or five-choice serial reaction time task, and impulsive choice as measured by performance on a delay discounting task (van den Bergh et al. 2006; Broos et al. 2012b), a finding which also extends to humans without impulsivity disorders (Broos et al. 2012b). However, some studies have found certain strains of rodents demonstrate both enhanced measures of impulsive action and choice as measured by these tasks (e.g., Moreno et al. 2010; Robinson et al. 2009). Given impulsivity is proposed to involve separable but partially overlapping neural mechanisms; it is therefore possible that an individual may express multiple types of impulsivity without these being directly correlated. To this effect, while enhanced impulsive action as measured by these tasks reflect deficits in inhibition of motoric output, tasks assessing impulsive choice also incorporate a value assessment and therefore performance may also be related to mechanisms involved in interpreting reward value, distinct from those involved in motor inhibition per se.

Heritability of Impulsivity

A number of lines of evidence suggest that impulsivity, in its many forms, is determined by genetic influences (e.g., Goldman and Fishbein 2000). For example, disorders that express high levels of impulsivity such as addiction (Bevilacqua and Goldman 2009) and ADHD (Franke et al. 2012; Sullivan and Rudnik-Levin 2001) have been shown to be highly heritable with average rates of heritability around 75 % for ADHD (Faraone et al. 2005), and 56 % across the various drug addictions (reviewed in Agrawal et al. 2012). A recent meta-analysis of twin, family and adoption studies suggests that approximately 50 % of the variance in the impulsivity trait is due to genetic influences, the degree of which appears not to be influenced by different assessment methods or impulsivity subtypes (Bezdjian et al. 2011). The identification of individual rodent strains with enhanced levels of impulsive behavior (e.g., Roman High Avoidance Rats,(Moreno et al. 2010); Spontaneous Hypertensive Rats, Adriani et al. 2003) as well as studies demonstrating the similarity within and variability between inbred strains of rodents have confirmed the heritability of impulsivity (e.g., Isles et al. 2004). Further, findings in rodents suggest a similar level of heritability to that observed in humans across measures of both impulsive action and impulsive choice (Richards et al. 2013; Wilhelm and Mitchell 2009; Loos et al. 2009). Interestingly, the heritability of this trait appears to vary across the lifespan with enhanced levels observed during adolescence (Niv et al. 2012; Bezdjian et al. 2011). Additionally, there is evidence for an influence of gender on this measure, with greater heritability observed in males (Bezdjian et al. 2011), which may relate to the sexual dimorphism observed in the expression of impulsivity and related neuropsychiatric disorders (reviewed in Trent and Davies 2012).

Genes Influencing Impulsivity

The heritability of impulsivity suggests that it may be a candidate endophenotype through which the genetic basis of more complex disease states may be investigated (Gottesman and Gould 2003; Berrettini 2005), a notion supported by recent research in addiction (Ersche et al. 2010). A number of clinical and preclinical studies have investigated the contribution of particular candidate genes to the expression of impulsivity; these have broadly implicated the monoaminergic systems, a consensus that is unsurprising given that the majority of candidate gene approaches have focused largely on genes within this category of neuromodulators (reviewed in Nielsen et al. 2012; Bevilacqua and Goldman 2009; Congdon and Canli 2008; Verdejo-Garcia et al. 2008; Brewer and Potenza 2008) (Table 3.1; Fig. 3.1). Increasingly, however, additional neurochemical systems are emerging as putative substrates underlying the expression of various impulsivity phenotypes, which we review below.

Dopamine Receptor Genes

Several lines of evidence implicate dysfunctional dopamine (DA) systems in impulsive behavior (reviewed in Dalley and Roiser 2012) drawing primarily from the efficacy of psychostimulant drugs such as methylphenidate (or Ritalin), which enhance levels of DA in the brain to reduce impulsivity symptoms in ADHD (e.g., Swanson and Volkow 2009). Pharmacological studies in rodents have implicated central D1 and D2-like receptors in mediating the effects of stimulant drugs on impulsivity (reviewed in Jupp and Dalley 2014). In general these studies accord with other findings demonstrating a correlation between reduced striatal D1 and D2-like receptor availability (e.g., Reeves et al. 2012; Clark et al. 2012; Lee et al. 2009; Dalley et al. 2007; Jupp et al. 2013), DA release (Buckholtz et al. 2010) and impulsivity levels in both humans and rodents.

Consistent with the hypothesized role of DA-ergic mechanisms in impulsivity, variants within genes encoding DA receptors (*DRD1-5*) and the DA transporter (*DAT1/SLC6A3*) have been shown to moderate risk for a number of impulsivity-related disorders including substance use disorder (reviewed in Le Foll et al. 2009), problem gambling (Perez de Castro et al. 1997; Comings et al. 1999), suicide (Suda et al. 2009) and ADHD (Wu et al. 2012). However, more relevant to understanding the genetic basis of impulsivity, and independent of confounding effects of disease susceptibility loci, sequence polymorphisms within DA-ergic genes have been reported to specifically predict enhanced levels of impulsivity, both clinically and in healthy individuals. For example, variation in the 40 base pair variable number tandem repeat (VNTR) polymorphism within the *DAT1* gene, which has been associated with altered binding potential and expression of the DA transporter (Faraone et al. 2014; Mill et al. 2002) has been shown to enhance impulsivity as measured by the BIS-11 (Paloyelis et al. 2010) and impulsive responding on continuous performance (Loo et al. 2003) and delay discounting tasks (Paloyelis et al. 2010). Similarly, the A2 allele of the rs1800497 variant in ankyrin repeat and kinase domain containing 1 gene (*ANKK1*), which has been demonstrated to be in linkage disequilibrium with the *DRD2* gene (Neville et al. 2004), and affects DA D2 receptor availability (e.g., Pohjalainen et al. 1998), is associated with increased measures of impulsivity on the BIS in alcohol-dependent individuals (Limosin et al. 2003). Polymorphisms in genes encoding DA receptors have also been found to account for variability in the expression of impulsivity within the general population with

Table 3.1 Examples of genes associated with impulsivity measured using psychometric and laboratory-based tasks in healthy subjects and clinical populations

Gene	Locus	Relevant physiological effects of variant	Population	Psychometric measure	Experimental measure	
					Impulsive action	Impulsive choice
Dopamine						
DRD2	rs6277	Altered striatal D2 binding Hirvonen et al. (2004)	HV	↑ DII Colzato et al. (2010)	↑ SSRT Colzato et al. (2010)	
ANKK1/DRD2	rs1800497	Altered striatal D2 binding Jonsson et al. (1999)	HV	– BIS-11 Eisenberg et al. (2007)	↑ GNG White et al. (2008)	↑ DD Eisenberg et al. (2007)
			AD	– EPQ Gullo et al. (2014)		
				↑ BIS-10 Limosin et al. (2003)		
DRD3	rs6280	Enhanced affinity of D3 receptors for DA Lundstrom and Turpin (1996)	Violent offenders	↑ EPQ Retz et al. (2003)		
			AD	↑ BIS-11 Limosin et al. (2005)		
			HV	– BIS-11 Limosin et al. (2005)		
DRD4	48 bp VNTR	Altered D4 binding Van Tol et al. (1992).	HV	– BIS-11 Eisenberg et al. (2007; Congdon et al. 2008)	SSRT Colzato et al. (2010)	– DD Eisenberg et al. (2007)
				↓ BIS-11 Varga et al. (2012)	↑ SSRT Congdon et al. (2008)	
				↑ NEO-PI-R Reiner and Spangler (2011)		
				↑ DII Colzato et al. (2010)		
	120 bp TR	Altered transcriptional regulation Paredes et al. (2013)	MDD, BP, AD	↑ TCI Rogers et al. (2004)		
DRD5	148 bp VNTR	–	ADHD		↑ CPT Manor et al. (2004)	
SLC6A3/DAT1	40 bp VNTR	Altered DAT binding Faraone et al. (2014) and expression Mill et al. (2002)	HV	– DII Colzato et al. (2010)	– SSRT Colzato et al. (2010)	
				– BIS-11 Congdon et al. (2008)		
			ADHD	↑ BIS-11 Paloyelis et al. (2010)	↑ CPT Loo et al. (2003)	↑ DD Paloyelis et al. (2010)
	rs37020 rs460000	Altered fronto-striatal activation during behavioral inhibition Cummins et al. (2012)	HV		↑ SSRT Cummins et al. (2012)	

(continued)

Table 3.1 (continued)

Gene	Locus	Relevant physiological effects of variant	Population	Psychometric measure	Experimental measure	
					Impulsive action	Impulsive choice
Serotonin						
HTR1A	rs6925	Altered 5HT1A expression Lemonde et al. (2003)	HV	– BIS-11 Varga et al. (2012)		
				↑ BIS-11 Benko et al. (2010)		
HTR1B	rs13212041	Altered microRNA binding site, likely affecting transcription Jensen et al. (2009)	HV	↓ BIS-11 Varga et al. (2012)		
GSK3β (modulates 5HT1B function)	rs1732170 rs334558	–	BP	↑ BIS-11 Jimenez et al. (2014)		
HTR2A	rs6313	Altered 5H2A binding Turecki et al. (1999)	AD	– BIS-11 Jakubczyk et al. (2012)	↑ SSRT Jakubczyk et al. (2012)	
				↑ BIS-11 Preuss et al. (2001)		
			Female HV	↑ BIS-11 Racine et al. (2009)		
			HV		↑ GNG Bjork et al. (2002)	
					↑ CPT Bjork et al. (2002)	
HTR2B	Q20* stop codon	Reduced 5HT2B expression Bevilacqua et al. (2010)	Violent offenders	↑ KSP Bevilacqua et al. (2010)		
SLC6A4	5HTTLPR	Altered 5HTT binding Little et al. (1998), transcription Lesch et al. (1994)	HV	– BIS-11 Varga et al. (2012)	– CPT Lage et al. (2011; Clark et al. 2005)	
				– EIQ (Retz et al. 2002)	↑ CPT Walderhaug et al. (2010)	
				↑ BIS-11 Sakado et al. (2003)		
			HV Tryptophan depletion		↑ CPT Walderhaug et al. (2010)	
					– CPT Clark et al. (2005)	
			Female HV	↑ BIS-11 Racine et al. (2009)		
			Suicide Attempters	– BIS-11 Baca-Garcia et al. (2004)		
			ED	– BIS-11 Steiger et al. (2005)	↑ GNG Steiger et al. (2005)	
			ADHD			↑ DD Sonuga-Barke et al. (2011)

Noradrenaline						
SLC6A2	rs5569	Altered NA metabolite levels, suggesting altered NA transporter function Jonsson et al. (1998)	ADHD		↑ CPT Song et al. (2011)	
Monoaminergic metabolism						
COMT	rs4680	Altered enzyme activity Chen et al. (2004)	HV	– BIS-11 Varga et al. (2012)	– SSRT Colzato et al. (2010)	↑ DD Smith and Boettiger (2012)
				↑ BIS-11 Soeiro-De-Souza et al. (2013)		
				– DII Colzato et al. (2010)		
			AD			↑ DD Boettiger et al. (2007)
			ADHD	– BIS-11 Paloyelis et al. (2010)	↑ GNG Bellgrove et al. (2005)	↑ DD Paloyelis et al. (2010)
MAOA	MAOA-uVNTR	Altered transcriptional activity Sabol et al. (1998)	HV	↑ BIS-11 Manuck et al. (2000)	↑ GNG Meyer-Lindenberg et al. (2006)	
DBH	rs1611115	Altered enzymatic activity Zabetian et al. (2001)	ADHD		↑ CPT Kieling et al. (2008)	
			PD	↑ NEO-PI-R Hess et al. (2009)		
TPH2	rs1473473	rs4570625 Altered transcriptional activity Scheuch et al. (2007)	ED	↑ DII Slof-Op't Landt et al. (2013)		
	rs1386483		HV		↑ SSRT Stoltenberg et al. (2006)	
	rs4570625 rs11178997		ADHD		↑ CPT Baehne et al. (2009)	
Other						

(continued)

Table 3.1 (continued)

Gene	Locus	Relevant physiological effects of variant	Population	Psychometric measure	Experimental measure	
					Impulsive action	Impulsive choice
GABRA2	rs1442060 rs10805145 rs426463	rs279858 altered GABRA2 mRNA Haughey et al. (2008)	AD	↑ NEO-PI-R Villafuerte et al. (2012)		
	rs279827					
	rs279826					
	rs279843					
	rs279847					
	rs279858 rs519270					
	rs693547					
	rs10805145					
	rs279826					
CNR1	rs806368 rs1535255 rs2023239 rs1049353 rs806368 (AAT)n triplet repeat	rs2023239 altered CB1 receptor binding Hirvonen et al. (2013)	SouthWest California Indians	↑ MPI Ehlers et al. (2007)		
	rs2023239		Regular cannabis users	↑ BIS-11 Bidwell et al. (2013)	– SSRT Bidwell et al. (2013)	– DD Bidwell et al. (2013)
NOS1	Ex1f-VNTR	Altered NOS expression Reif et al. (2009)	HV	↑ AMIS Reif et al. (2011)	↑ VCT Reif et al. (2011)	
			ADHD			↑ DD Hoogman et al. (2011)

FKBP5	rs1360780	rs1360780 Altered glucocorticoid receptor sensitivity Menke et al. (2013)	HV			↓ DD Kawamura et al. (2013)
	rs3800373 rs9296158 rs1360780 rs9470080		Incarcerated males	– BIS-11 Bevilacqua et al. (2012)		
AMBRA1	rs11819869	Altered cortical activity during behavioral inhibition Heinrich et al. (2013)	HV			↑ CGT Heinrich et al. (2013)
AR	(CAG)n triplet repeat	Altered receptor expression and sensitivity Ackerman et al. (2012)	Female AD	↑ BIS-11 Mettman et al. (2014)		

–/↑/↓ rare variant associated with no effect/enhanced/reduced measures of impulsivity

AD alcohol dependence, *ADHD* attention deficit/hyperactivity disorder, *AMIS* adaptive and maladaptive impulsivity scale, *BP* bipolar disorder, *BIS* barrett impulsivity scale, *CGT* Cambridge gambling task, *CPT* continuous performance task, *DII* dickman impulsivity inventory, *DAT* dopamine transporter, *DD* delay discounting, *ED* eating disorder, *EIQ* eysnick impulsivity questionaire, *GNG* Go/No-go task, *NA* noradrenaline, *NEO-PI-R* Revised NEO Personality Inventory, *HV* healthy volunteer, *KSP* Karolinska scales of personality, *MPI* Maudsley Personality Inventory, *PD* Parkinson's disease, *SSRT* stop-signal reaction time, *VCT* visual comparison test, *AMIS* adaptive and maladaptive impulsivity scale, *VNTR* variable number tandem repeat, *TR* tandem repeat

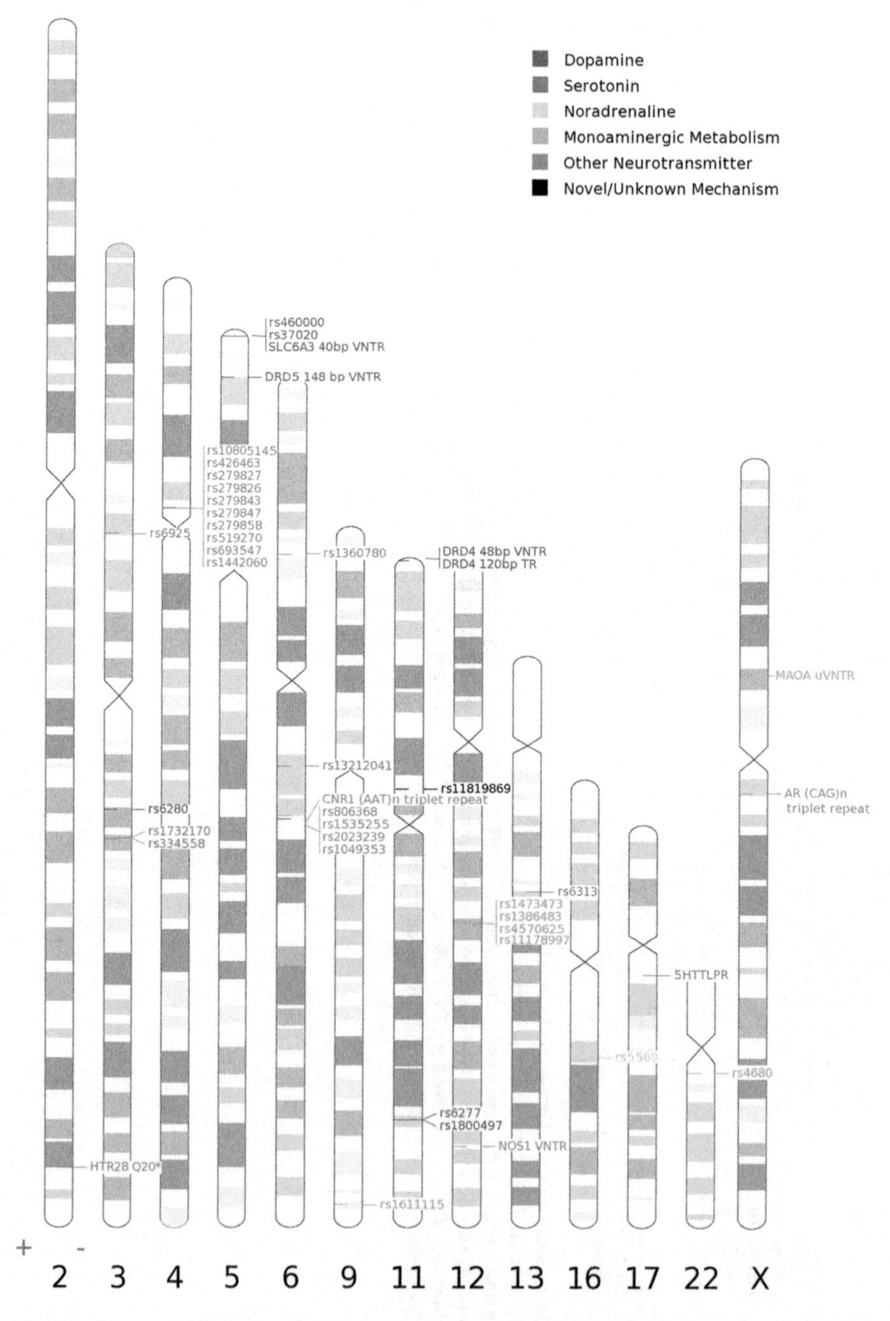

Fig. 3.1 Idiogram illustrating the chromosomic location of the genetic variants associated with impulsivity in humans (hg19, giemsa stain). Variants are color coded according to the neurotransmitter system/mechanism they affect. *Pink*, dopamine; *purple*, serotonin; *yellow*, noradrenaline; *gray*, other neurotransmitter; *black*, novel/unknown mechanism

the A1 allele of the rs1800497 *ANKK1/DAD2* variant (White et al. 2008; Eisenberg et al. 2007) and the 7-repeat allele of the *DRD4* 48 bp VNTR (Colzato et al. 2010; Reiner and Spangler 2011; Congdon et al. 2008) associated with increased psychometric and experimental measures of impulsivity. Interestingly, transgenic mice expressing a truncated D4 receptor are not overtly impulsive (Helms et al. 2008) suggesting important cross-species differences in the impact of D4 receptor dysfunction in the expression of impulsivity; these differences may of course simply reflect difficulties in interrelating different measures of impulsivity. Nevertheless, there is evidence to suggest that the effect of the VNTR on impulsivity is moderated by the co-expression of a polymorphism within the catechol-*O*-methyl transferase (*COMT*) gene, such that expression of this haplotype affected go/no-go performance, but no effect was observed when either variant was expressed independently (Heinzel et al. 2013). Thus it is possible the epistatic interaction between the *DRD4* VNTR and *COMT* genes may explain the null findings in Drd4 transgenic mice.

Serotonin Receptor Genes

Significant evidence implicates serotonin (5-HT) dysfunction in impulsivity (Dalley and Roiser 2012; Miyazaki et al. 2012; Soubrie et al. 1986; Fairbanks et al. 2001) and impulsivity-related disorders, including especially aggression, suicide and mood disorder (reviewed in Mann 2013; Olivier 2004). The involvement of 5-HT in impulsivity is thought to be mediated through its interaction with DA (Kapur and Remington 1996; Di Matteo et al. 2008; Di Giovanni et al. 2008), and other neurotransmitter systems (Fink and Gothert 2007). A reduced level of the principal 5-HT metabolite, 5-hydroxyindoleacetic acid (5-HIAA) in the cerebrospinal fluid has been associated with impulsivity in problem gamblers (Nordin and Eklundh 1999), suicide attempters (Chatzittofis et al. 2013) and alcoholics (Goldman et al. 1992). Further, experimentally-induced reduction in 5-HT is associated with enhanced measures of "waiting" impulsivity in both rodents and humans, as measured by premature responding on the 5-choice task (Harrison et al. 1997, 1999) and the go/no-go task (Harrison et al. 1999). However, findings in relation to "stopping" impulsivity, as measured by the SSRT (Eagle et al. 2009) and impulsive choice (Winstanley et al. 2004a; Mobini et al. 2000) have been more equivocal (Dalley and Roiser 2012). Nevertheless, pharmacological studies have consistently implicated the 5-HT transporter (5-HTT) and 5-HT1A, 2A, and 2C receptors in impulsivity (reviewed in Jupp and Dalley 2014). Further, there is evidence of altered binding and expression at these receptors in impulsivity and impulsivity-related disorders (e.g., Soloff et al. 2014; Witte et al. 2009; Di Narzo et al. 2014; Frankle et al. 2005).

Supporting the widely described relationship between 5-HT and impulsivity, polymorphisms within genes encoding 5HT1A, 2A and 2C receptors, and the 5-HT transporter have been associated with moderating risk for addiction (e.g., Cao et al. 2014; Gao et al. 2011), ADHD (e.g., Guimaraes et al. 2009; Zhao et al. 2005), suicide (e.g., Hung et al. 2011; Bah et al. 2008; Serretti et al. 2007), mood disorder (e.g., Jiang et al. 2013; Mansour et al. 2005) and problem gambling (e.g., Wilson et al. 2013). Similarly, variance within these genes has additionally been associated with individual levels of impulsivity.

Given the apparent role for altered 5-HT levels in impulsivity, a number of studies have focused on the 5HTTLPR polymorphism within the promoter region of the 5HTT gene (*SLC6A4*), which alters the transcriptional efficiency of the gene (Lesch et al. 1994; Heils et al. 1996) and reduces binding potential at the transporter (Little et al. 1998). The "S" allele of the 5HTTLPR polymorphism, resulting in reduced 5HTT mRNA expression (Heils et al. 1995), is associated with enhanced measures of impulsivity on the BIS-11 (Sakado et al. 2003; Racine et al. 2009; Paaver et al. 2007) and continuous performance task (Walderhaug et al. 2010) in healthy volunteers and impaired Go/No-Go performance in individuals with eating disorder (Steiger et al. 2005) and enhanced delay discounting in ADHD (Sonuga-Barke et al. 2011). Additionally, the expression of this allele has been found to moderate the effect of tryptophan depletion on continuous performance task performance (Walderhaug et al. 2010). There are conflicting reports, however, concerning the association between 5HTTLPR and impulsivity with studies demonstrating no effect of this variation on the expression of impulsivity (e.g., Varga et al. 2012; Lage et al. 2011; Clark et al. 2005). Further, rats with a genetic ablation of 5-HTT demonstrate reduced impulsive responding on the five-choice serial reaction time task (Homberg et al. 2007). There is evidence, however, for gender (Racine et al. 2009; Walderhaug et al. 2010; Guimaraes et al. 2007), environmental (Nishikawa et al. 2012; Stoltenberg et al. 2012; Lage et al. 2011) and epistatic influences (Stoltenberg et al. 2012) on the association between this polymorphism and impulsivity which may potentially explain the variability observed between these studies.

Expression of the "C" allele of the rs6313 variant within the promoter region of the *HTR2A* gene has been found to affect the binding potential of this receptor (Turecki et al. 1999), and in keeping with the suggested contribution of this receptor in impulsivity, the rs6313 polymorphism has been found to moderate levels of impulsivity in both alcohol-dependent (Preuss et al. 2001; Jakubczyk et al. 2012) and healthy individuals (Racine et al. 2009; Nomura and Nomura 2006; Bjork et al. 2002). Similarly, the rs6925 polymorphism within the *HTR1A* gene affects the expression of its encoded receptor (Lemonde et al. 2003) and is associated with impulsivity as measured by the BIS and EPQ scales (Benko et al. 2010). A stop codon within the *5HT2B* gene, which would reduce the availability of 5-HT2B receptors, has been associated with increased impulsivity scores on the Karolinska scales of personality questionnaire in a group of violent offenders (Bevilacqua et al. 2010). In support of these findings, 5-HT2B receptor knockout mice also demonstrate enhanced delay discounting impulsivity (Bevilacqua et al. 2010). In contrast, the rs13212041 polymorphism within the *HTR1B* gene has been suggested to confer protection against impulsivity (Varga et al. 2012). The expression of the less common "G" allele reduces microRNA binding and consequentially reduces silencing of receptor synthesis, likely increasing levels of this receptor (Jensen et al. 2009). This assumed mechanism of action accords with the findings that eltoprazine, a selective 5HT1B receptor agonist, reduces delay discounting impulsivity in rats (van den Bergh et al. 2006) and a finding suggesting behaviors representative of impulsivity are increased in mice with a genetic ablation of the 5HT1B receptor (Brunner and Hen 1997). Despite pharmacological evidence for the role of the 5HT2C receptor in impulsivity (e.g., Navarra et al. 2008a; Winstanley et al. 2004b), and some support for an association of polymorphisms within the gene encoding this receptor (*HTR2C*) with disorders of impulsivity (e.g., ADHD, Xu et al. 2009; suicide, Videtic et al. 2009; mood disorder, Massat et al. 2007) there is currently no evidence from human studies implicating genetic variation within the *HTR2C* gene and individual levels of impulsivity (e.g., Tochigi et al. 2006). This accords with findings from 5HT2C receptor null mice, which demonstrate no difference in basal levels of impulsive action when compared to wild-type animals (Fletcher et al. 2013). However the possibility of genetic compensation must be considered since these mice as well another strain (that also contains a deletion of a small nucleolar RNA resulting in increased 5HT2C receptor editing), showed differential effects on impulsivity relative to wild-type mice following the pharmacological manipulation of this receptor subtype (Fletcher et al. 2013; Doe et al. 2009).

Noradrenaline Receptor Genes

A role for noradrenaline (NA) in impulsivity is substantiated by the clinical efficacy of amphetamine and methylphenidate in ADHD, which act to enhance NA as well as DA transmission in the brain (reviewed in Del Campo et al. 2011) and more specifically by the effectiveness of the selective NA reuptake inhibitor (NARI) atomoxetine in this disorder (Simpson and Plosker 2004; Faraone et al. 2005) and animal models of impulsivity (Fernando et al. 2012; Blondeau and Dellu-Hagedorn 2007; Robinson et al. 2008; Navarra et al. 2008b; Tsutsui-Kimura et al. 2009). Despite evidence for polymorphisms within genes encoding NA receptors (*ADRA1A*, *ADRA1B*, *ADRA2B*) and the NA transporter (*SLC6A2*), which have been associated with ADHD (Hawi et al. 2013; Kim et al. 2008), only the *SLC6A2* gene has been investigated for any relationship with impulsivity. The "A" allele of the rs5569 variant was found to associate with increased errors of commission on the continuous performance task in healthy volunteers (Song et al. 2011). Other polymorphisms within the *SLC6A2* gene (e.g., rs2279805, rs3785143, rs28386840) have also been found to either potentiate or moderate the clinical efficacy of methylphenidate and atomoxetine in ADHD (Yang et al. 2013; Park et al. 2012).

Monoaminergic Metabolic Markers

The enzymes monoamine oxidase A and B (MAOA/B) and catechol-o-methyl transferase (COMT) are involved in the degradation of monoamines, while tryptophan hydroxylase (TPH) and DA-β-hydroxylase (DBH) contribute to the synthesis of 5-HT and the conversion of DA to NA, respectively. Several studies have found an association between the genes encoding these enzymes and various impulsivity-related disorders such as substance dependence (e.g., Gerra et al. 2004; Anney et al. 2004) and ADHD (e.g., Liu et al. 2011; Palmason et al. 2010; Roman et al. 2002). Polymorphisms within these genes have also been associated with enhanced measures of impulsivity in individuals with impulsivity disorders as well as the general population. While there is little evidence implicating the rs4680 polymorphism in the *COMT* gene with psychometric measures of impulsivity (Varga et al. 2012; Colzato et al. 2010; Paloyelis et al. 2010) this variant is associated with enhanced delay discounting behavior in individuals with ADHD (Paloyelis et al. 2010), alcohol dependence (Boettiger et al. 2007) and in healthy volunteers (Smith and Boettiger 2012). Interestingly, this effect appears to be moderated by age, given that in individuals with the valine allele, discounting behavior positively correlated with advancing age, while in individuals with the methionine allele, delay discounting negatively correlated with age (Smith and Boettiger 2012). The mechanism underlying this dissociation is unclear but may involve age-related reductions in prefrontal DA (e.g., Volkow et al. 2000). In keeping with human literature, genetic ablation of COMT in mice has been shown to increase impulsive responding on the five-choice serial reaction time task when compared to wild-type animals (Papaleo et al. 2012). Expression of the "low activity" allele of the VNTR within the *MAOA* gene, thought to result in enhanced monoamine levels (e.g., Ducci et al. 2006), is associated with increased psychometric and experimental measures of impulsivity within healthy populations (Manuck et al. 2000; Meyer-Lindenberg et al. 2006). In addition, the rs1473473, rs1386483, rs4570625, and rs11178997 polymorphisms within the *TPH2* gene and the rs1611115 polymorphism within the *DBH* gene, influencing levels of 5-HT and NA respectively within the brain, have been associated with both enhanced psychometric measures (Slof-Op't Landt et al. 2013; Hess et al. 2009) and experimental measures of impulsive action (Kieling et al. 2008; Baehne et al. 2009). Interestingly, transgenic mice carrying a polymorphism in the *Tph2* gene, which has the effect of reducing levels of 5-HT in the prefrontal cortex and ventral striatum, showed no alteration in delay discounting performance (Isles et al. 2005), although these mice did show enhanced impulsive–aggressive like social behaviours (Angoa-Perez et al. 2012).

Non-monoaminergic Substrates

Whilst research on impulsivity has tended to focus on the brain monoamine systems recent findings suggest that many other neurochemical systems also contribute to the etiology and modulation of impulsivity (Jupp and Dalley 2014). Given these emerging data few studies to date have investigated the contribution of gene polymorphisms in various non-monoaminergic systems implicated in impulsivity (Table 3.1; Fig. 3.1).

Gamma-aminobutyric acid (GABA) neurons make up the majority of inhibitory interneurons in the brain and form key projections both within and between relevant cortical and subcortical regions involved in the regulation and expression of impulsivity. Based on several lines of evidence, GABA has been strongly implicated in the expression and modulation of impulsivity (reviewed in Hayes et al. 2014). While a number of polymorphisms have been identified within genes encoding the GABA receptors and transporter and enzymes contributing to the metabolism of this neurotransmitter (e.g., GABA decarboxylase 65/67), and some of these have been linked to expression of impulsivity-related disorders (e.g., ADHD, Wang et al. 2012; addiction, Agrawal et al. 2012; Drgon et al. 2006; bipolar, Chen et al. 2009), very few studies to date to have investigated their relationship to the expression of impulsivity. Ten variants within the gene encoding the GABA receptor subunit 2A (*GABRA2*), predominantly located within intron 3, and one of which has been shown to alter expression levels of GABRA2 mRNA within the brain (rs279858, Haughey et al. 2008), are associated with higher impulsivity scores on NEO-PI-R questionnaire (Villafuerte et al. 2012, 2013) in alcohol-dependent individuals.

Endocannabinoids play a major role in regulating synaptic function and plasticity within the striatum (Mathur and Lovinger 2012). Increasingly, these lipid-derived substances have been implicated in the expression and modulation of impulsivity (reviewed in Moreira et al. 2015). A number of variants within the genes encoding the two receptor subtypes CB1 (*CNR1*) and CB2 (*CNR2*) have been associated with modifying risk for impulsivity-related disorders including drug dependence (Marcos et al. 2012; Clarke et al. 2013), ADHD (Lu et al. 2008) and mood disorder (Minocci et al. 2011; Monteleone et al. 2010). Again, few studies have specifically investigated the relationship of these variants to measures of impulsivity. However, there is evidence implicating at least six polymorphisms within or near the *CNR1* gene with enhanced impulsivity scores on the BIS-11 (Bidwell et al. 2013) and Maudsley Personality Inventory (MPI) (Ehlers et al. 2007).

Nitric oxide (NO) has also been reported to modulate activity within cortical striatal networks (reviewed in Pierucci et al. 2011). The main source of NO in the brain is the enzyme neuronal nitric oxide synthase, which is encoded by the neuronal nitric oxide (*NOS1*) gene. A variable tandem number repeat in the promoter region of the *NOS1* exon 1f (Ex1f-VNTR), which has been found to affect *NOS1* gene expression and the neuronal transcriptome (Reif et al. 2009; Rife et al. 2009), has been linked with a number of impulsivity-related disorders including mood disorder (Maziade et al. 2005), ADHD, suicide and violent behavior (Reif et al. 2009). This polymorphism has also been found to moderate psychometric (Reif et al. 2011) and experimental (Reif et al. 2011; Hoogman et al. 2011) measures of impulsivity, with enhanced measures associated with the short "S" allele, an effect which appears to be further moderated by the influence of traumatic life events and family environment (Reif et al. 2011).

FKBP5 (FK506 binding protein 5) is a protein that interacts with glucocorticoid receptors (GR) to modulate their sensitivity to cortisol (Storer et al. 2011) and thus is a key regulator of the hypothalamic pituitary adrenal (HPA) axis stress response (e.g., Ising et al. 2008; Binder et al. 2004). Cortisol levels and stress reactivity have been hypothesized to correlate with impulsivity (Magrys et al. 2013; Lovallo 2013;

Almeida et al. 2010) and in keeping with this relationship, variants within the 5KBP5 gene have been associated with impulsivity-related disorders of suicidality (Roy et al. 2010, 2012), aggression (Bevilacqua et al. 2012) and drug dependence (Levran et al. 2014; Bevilacqua et al. 2012). More recently the rs1360780 variant has been shown to predict enhanced measures of delay discounting impulsivity in healthy individuals (Kawamura et al. 2013). Interestingly, transgenic mice with impaired glucocorticoid function also show increased impulsivity on the five-choice serial reaction time task (Steckler et al. 2000).

Another novel gene product implicated in impulsive behavior is the pro-autophagic factor AMBRA1 (activating molecule in Beclin-1-regulated autophagy). This factor is strongly expressed in the cortex, hippocampus, and striatum and has been shown to play an important role in embryogenesis (Fimia et al. 2007). Further, autophagy has been implicated in the pathogenesis of neurodegenerative disease (Ghavami et al. 2014). A genome-wide association study (GWAS) originally identified a variant within the *AMBRA1* gene (rs11819869) with risk for schizophrenia and was additionally associated with reduced prefrontal activation during a Go/No-Go task (Rietschel et al. 2012). Further investigation of this polymorphism in a healthy population found it to predict enhanced delay aversion in a discounting task (Heinrich et al. 2013). This polymorphism is in strong linkage disequilibrium with genes for diacylglycerol kinase zeta (*DGKZ*), cholinergic receptor muscarinic 4 (*CHRM4*) and midkine (neurite growth-promoting factor 2, *MDK*) (Rietschel et al. 2012), thus potentially implicating an action at these genes in modulating impulsivity. However, mice with a heterozygous mutation of *Ambra1*, leading to a truncated, non-functional Ambra1 protein, express impaired social behaviors resembling autism (Dere et al. 2014).

Variation in the number of CAG repeats within the androgen receptor (*AR*) gene has also been shown to influence the expression and androgen sensitivity of this receptor (Ackerman et al. 2012). In keeping with the influence of the androgen hormone testosterone on impulsivity (e.g., Wood et al. 2013), variation within the CAG repeat has been associated with enhanced levels of impulsivity as measured by the BIS-11 in alcohol-dependent Caucasian women (Mettman et al. 2014) and a haplotype association has been observed in combination with another tandem repeat within this gene (GGN) and impulsivity on the Zuckerman-Khulman personality questionnaire in incarcerated males (Aluja et al. 2011).

Future Directions

To date, a number of genetic variants have been implicated in modulating individual variability in impulsivity. These include the widely recognized monoaminergic systems and increasingly a broader collection of neural substrates and mechanisms. Consistent with the acknowledged ethological and biological diversity of the impulsivity construct and tools for assessment, there are few genetic variants that influence impulsivity across multiple measures. Further there is evidence for disease and population-specific effects (e.g., Bellgrove et al. 2005 vs. Colzato et al. 2010). Given the approximate 50 % heritability of this behavioral trait, it is not surprising that the effect of many of these genetic variants on impulsivity is moderated by epistatic (e.g., Heinzel et al. 2013; Manuck et al. 2000; Stoltenberg et al. 2012) as well as contextual interactions including age (e.g., Smith and Boettiger 2012), gender (e.g., Racine et al. 2009) and environmental factors (e.g., Reif et al. 2011) and likely accounts for the variable influence of particular polymorphisms on the expression of impulsivity traits. For these reasons, while candidate gene approaches provide a valid basis for research, there are significant limitations in this approach in terms of identifying novel mechanisms and in characterizing epistatic, gene by environment interactions, age and gender-specific effects.

Studies in Animal Models of Impulsivity

Animal models of impulsivity have the advantage of overcoming many confounds relating to the genetic and environmental variability observed within human studies. Studies in rodents demonstrate both the stability of impulsive behavior within, and variability between, inbred strains of mice and rats and suggest a level of heritability similar to humans (Isles et al. 2004; Loos et al. 2009; Pena-Oliver et al. 2012; Gubner et al. 2010; Logue et al. 1998; Richards et al. 2013). Similarly, a study in our own laboratory has found significant evidence of heritability of impulsivity, as measured on the 5-choice task, in a multigenerational pedigree of an outbred strain of rats (Pitzoi et al., unpublished observations).

Transgenic approaches support the notion that genetic mechanisms contribute to impulsivity by confirming the contribution of genes previously linked to impulsive behavior in humans as discussed above (e.g., 5-HTT knockout rat, Homberg et al. 2007) as well as implied by selective pharmacological interventions (e.g., neurokinin 1 receptor, Yan et al. 2011; muscarinic actylcholine type 1 receptor, Bartko et al. 2011; mu or delta opioid receptors, Olmstead et al. 2009) (Table 3.2) and provide the advantage of comparing behaviours across controlled genetic and environmental backgrounds. In addition, this approach has revealed novel gene products not previously linked to alterations in individual levels of impulsivity; for example, subunit 1 of the AMPA receptor (Barkus et al. 2012), alpha 7 containing nicotinic actylcholine receptors (Keller et al. 2005), neuropeptide Y2 receptor (Greco and Carli 2006), steroid sulfatase (Davies et al. 2009; Trent et al. 2013), X-linked lymphocyte-regulated 3 complex (Davies et al. 2005), alpha-synuclein (Pena-Oliver et al. 2012), tau protein (Lambourne et al. 2007), fragile X mental retardation protein (Moon et al. 2006) and neuroregulin 3 (Loos et al. 2014). Mutagenesis screening approaches, affected, for example, using *N*-ethyl-*N*-nitrosourea have also provided insight into novel genes influencing behavior and impulsivity (reviewed in Oliver and Davies 2012).

Importantly, studies in transgenic mice have confirmed gene x environment interactions (Isles et al. 2004; Loos et al. 2009) and shed light on their underlying molecular mechanisms and how these contribute to impulsivity. Factors such as

Table 3.2 Examples of genes associated with impulsivity as assayed in transgenic rodents

Gene	Impulsive action	Impulsive choice
Monoaminergic neurotransmitter/function		
Drd4	– Helms et al. (2008)	– Helms et al. (2008)
Slc6a4 (5-HTT)	↓ Homberg et al. (2007)	
Htr2b		↑ Bevilacqua et al. (2010)
Htr2c	– Fletcher et al. (2013; Doe et al. 2009)	
Comt	↑ Papaleo et al. (2012)	
Tph2		– Isles et al. (2005)
Other neurotransmitters		
Gria1		↑ Barkus et al. (2012)
Nk1r	↑ Yan et al. (2011)	
Npy2r	↑ Greco and Carli (2006)	
Gal	– Wrenn et al. (2006)	
Chrna7	↑ Keller et al. (2005)	
Chrnb2	– Guillem et al. (2011; Serreau et al. 2011)	
Chrm1	↑ Bartko et al. (2011)	
Oprm1	↓ Olmstead et al. (2009)	
Oprd1	↑ Olmstead et al. (2009)	
Nr3c1	↑ Steckler et al. (2000)	
Other/novel mechanism		
Fmr1	↑ Moon et al. (2006)	
Sts	↑ Davies et al. (2009)	
Snca	↓ Pena-Oliver et al. (2012)	
Mapt	↑ Lambourne et al. (2007)	
Nrg3	↓ Loos et al. (2014)	

–/↑/↓ denotes no effect/increased/decreased measured impulsivity compared to "wild-type" respectively

early-life stress (e.g., maternal separation, Lovic et al. 2011; prenatal/adolescent exposure to alcohol, Banuelos et al. 2012; nicotine Schneider et al. 2011) and environmental conditions (e.g., enrichment, Perry et al. 2008) reportedly alter levels of impulsivity in adult rodents. While no studies have specifically investigated epistatic effects on impulsivity in animals, there are a number of studies demonstrating gene x gene interactions and downstream molecular effects of such interactions on genes associated with impulsivity (e.g., *COMT* Papaleo et al. 2014; *SLC6A4* Kerr et al. 2013; Murphy et al. 2003; *HTR2A* Auclair et al. 2004).

Genome-Wide Association Studies

While candidate gene approaches have provided significant insight into the molecular genetics of impulsivity, these studies require an a priori approach to investigate genes hypothesized to play a role in impulsivity. As such this approach makes the identification of novel genes unlikely. With the development of sophisticated genetic screening tools, providing the ability to scan the entire genome with genotypic arrays and massively parallel sequencing, it has become possible to conduct genome-wide analyses (GWA), independent of a priori assumptions on the biology of the trait of interest. As such a number of GWA studies have been conducted for some of the disorders associated with impulsivity; for example, addiction (reviewed in Hall et al. 2013; Nielsen et al. 2012); ADHD (reviewed in Li et al. 2014), suicidality (Willour et al. 2012; Perlis et al. 2010), and schizophrenia (Ripke et al. 2013; Shi et al. 2011). Given the apparent complexity of these disorders it is not surprising that a large number of functional genetic loci have been identified (see the National Human Genome Research Institute Catalogue of Published Genome-Wide Association Studies, http://www.genome.gov/gwastudies/). To date, however, no published GWA studies have specifically investigated individual levels of impulsivity, as an endophenotype, within a healthy or neuropsychiatric patient group.

Despite the apparent power of genome-wide approaches for identifying novel genetic contributions, they are still subject to issues associated with environmental and epistatic heterogeneity within the population, and as such most of the genetic variance observed in complex diseases remains largely unexplained (e.g., Maher 2008; Manolio et al. 2009; Schork et al. 2009). GWA studies within animal models may overcome some of these issues and to date, a few studies have utilized this approach with some success (e.g., Levey et al. 2014). Importantly, a number of inbred rodent strains have been identified demonstrating intrinsically elevated levels of impulsive behavior (e.g., Roman High Avoidance Rats, Moreno et al. 2010; spontaneously hypertensive rats, Adriani et al. 2003; and outbred populations with a natural distinction between "high" and "low" impulsivity phenotypes on the 5-choice task Dalley et al. 2007) and delay discounting task (Perry et al. 2005; Broos et al. 2012a); these represent ideal populations in which to conduct genome-wide studies to investigate the genetic substrates of impulsivity together with epistatic and context-dependent interactions that shape the expression of this trait.

Neuroimaging Genetics

Beyond issues associated with identifying novel mechanisms, including epistatic and environmental influences, the apparent heterogeneity of impulsivity further confounds the identification of genetic associations. As a consequence, the refinement of impulsivity constructs into "simpler," more discrete neural phenotypes are likely to provide more robust genetic links. Indeed it has been suggested that the identification of distinct neuroimaging endophenotypes, including specific patterns of brain activation during impulsivity tasks or neurochemical correlates of performance detected using positron emission tomography would inform this approach (for review see Bogdan et al. 2013). A recent study by Whelan and colleagues found that although performance on a SSRT task was similar between healthy individuals and those with ADHD or substance use disorder, patterns of activation during this task varied between these individual groups and further, these patterns varied with *SLC6A4* genotype (Whelan et al. 2012). This approach thus provides insight into the functional impact of genetic variants and therefore the underlying neurobiology of behavior.

Conclusions

Impulsivity is a heritable, polygenic, disease-associated behavioral trait, modifiable by epistatic and environmental influences. Despite the genetic complexity of impulsivity, candidate gene sequencing approaches have identified a number of variants involved in monoaminergic transmission (NA, DA, and 5-HT), which contribute to the variable expression of impulsive behaviors in both healthy and neuropsychiatric groups. Continued research is needed to elucidate the neurobiological bases of the various impulsivity subtypes, which also implicate GABA and other neurotransmitter systems. Advancing genetic tools such as GWA studies, transgenic animal models, mutagenesis screens and the further refinement of the impulsivity construct into distinct neural phenotypes, will undoubtedly facilitate our understanding of the biological origins of functional and pathological impulsivity.

Acknowledgements The author's research is supported by a Medical Research Council (MRC) grant to JD, Enrico Petretto, Tim Aitman, Barry Everitt, and Trevor Robbins (G0802729) and by a joint award from the MRC (G1000183) and Wellcome Trust (093875/Z/10/Z) in support of the Behavioural and Clinical Neuroscience Institute (BCNI) at Cambridge University. BJ is supported by a Fellowship from the AXA Research Fund.

References

Ackerman, C. M., Lowe, L. P., Lee, H., Hayes, M. G., Dyer, A. R., Metzger, B. E., et al. (2012). Ethnic variation in allele distribution of the androgen receptor (AR) (CAG)n repeat. *Journal of Andrology, 33*(2), 210–215. doi:10.2164/jandrol.111.013391.

Adriani, W., Caprioli, A., Granstrem, O., Carli, M., & Laviola, G. (2003). The spontaneously hypertensive-rat as an animal model of ADHD: Evidence for impulsive and non-impulsive subpopulations. *Neuroscience and Biobehavioral Reviews, 27*(7), 639–651.

Agrawal, A., Verweij, K. J., Gillespie, N. A., Heath, A. C., Lessov-Schlaggar, C. N., Martin, N. G., et al. (2012). The genetics of addiction-a translational perspective. *Transl Psychiatry, 2*, e140. doi:10.1038/tp.2012.54.

Almeida, M., Lee, R., & Coccaro, E. F. (2010). Cortisol responses to ipsapirone challenge correlate with aggression, while basal cortisol levels correlate with impulsivity, in personality disorder and healthy volunteer subjects. *Journal of Psychiatric Research, 44*(14), 874–880. doi:10.1016/j.jpsychires.2010.02.012.

Aluja, A., Garcia, L. F., Blanch, A., & Fibla, J. (2011). Association of androgen receptor gene, CAG and GGN repeat length polymorphism and impulsive-disinhibited personality traits in inmates: The role of short-long haplotype. *Psychiatric Genetics, 21*(5), 229–239. doi:10.1097/YPG.0b013e328345465e.

Angoa-Perez, M., Kane, M. J., Briggs, D. I., Sykes, C. E., Shah, M. M., Francescutti, D. M., et al. (2012). Genetic depletion of brain 5HT reveals a common molecular pathway mediating compulsivity and impulsivity. *Journal of Neurochemistry, 121*(6), 974–984. doi:10.1111/j.1471-4159.2012.07739.x.

Anney, R. J., Olsson, C. A., Lotfi-Miri, M., Patton, G. C., & Williamson, R. (2004). Nicotine dependence in a prospective population-based study of adolescents: The protective role of a functional tyrosine hydroxylase polymorphism. *Pharmacogenetics, 14*(2), 73–81.

Aron, A. R., Robbins, T. W., & Poldrack, R. A. (2004). Inhibition and the right inferior frontal cortex. *Trends in Cognitive Science, 8*(4), 170–177. doi:10.1016/j.tics.2004.02.010.

Auclair, A., Drouin, C., Cotecchia, S., Glowinski, J., & Tassin, J. P. (2004). 5-HT2A and alpha1b-adrenergic receptors entirely mediate dopamine release, locomotor response and behavioural sensitization to opiates and psychostimulants. *European Journal of Neuroscience, 20*(11), 3073–3084. doi:10.1111/j.1460-9568.2004.03805.x.

Avila, C., Cuenca, I., Felix, V., Parcet, M. A., & Miranda, A. (2004). Measuring impulsivity in school-aged boys and examining its relationship with ADHD and ODD ratings. *Journal of Abnormal Child Psychology, 32*(3), 295–304.

Baca-Garcia, E., Vaquero, C., Diaz-Sastre, C., Garcia-Resa, E., Saiz-Ruiz, J., Fernandez-Piqueras, J., et al. (2004). Lack of association between the serotonin transporter promoter gene polymorphism and impulsivity or aggressive behavior among suicide attempters and healthy volunteers. *Psychiatry Research, 126*(2), 99–106. doi:10.1016/j.psychres.2003.10.007.

Baehne, C. G., Ehlis, A. C., Plichta, M. M., Conzelmann, A., Pauli, P., Jacob, C., et al. (2009). Tph2 gene variants modulate response control processes in adult ADHD patients and healthy individuals. *Molecular Psychiatry, 14*(11), 1032–1039. doi:10.1038/mp.2008.39.

Bah, J., Lindstrom, M., Westberg, L., Manneras, L., Ryding, E., Henningsson, S., et al. (2008). Serotonin transporter gene polymorphisms: Effect on serotonin transporter availability in the brain of suicide attempters. *Psychiatry Research, 162*(3), 221–229. doi:10.1016/j.pscychresns.2007.07.004.

Banuelos, C., Gilbert, R. J., Montgomery, K. S., Fincher, A. S., Wang, H., Frye, G. D., et al. (2012). Altered spatial learning and delay discounting in a rat model of human third trimester binge ethanol exposure. *Behavioural Pharmacology, 23*(1), 54–65. doi:10.1097/FBP.0b013e32834eb07d.

Barkus, C., Feyder, M., Graybeal, C., Wright, T., Wiedholz, L., Izquierdo, A., et al. (2012). Do GluA1 knockout mice exhibit behavioral abnormalities relevant to the negative or cognitive symptoms of schizophrenia and schizoaffective disorder? *Neuropharmacology, 62*(3), 1263–1272. doi:10.1016/j.neuropharm.2011.06.005.

Barratt, E. S. (1985). Impulsiveness subtraits, arousal and information processing. In J. T. Spence & C. T. Itard (Eds.), *Motivation, emotion and personality* (pp. 137–146). North Holland: Elsevier Science.

Bartko, S. J., Romberg, C., White, B., Wess, J., Bussey, T. J., & Saksida, L. M. (2011). Intact attentional processing but abnormal responding in M1 muscarinic receptor-deficient mice using an automated touchscreen method. *Neuropharmacology, 61*(8), 1366–1378. doi:10.1016/j.neuropharm.2011.08.023.

Bellgrove, M. A., Domschke, K., Hawi, Z., Kirley, A., Mullins, C., Robertson, I. H., et al. (2005). The methionine allele of the COMT polymorphism impairs prefrontal cognition in children and adolescents with ADHD. *Experimental Brain Research, 163*(3), 352–360. doi:10.1007/s00221-004-2180-y.

Benko, A., Lazary, J., Molnar, E., Gonda, X., Tothfalusi, L., Pap, D., et al. (2010). Significant association between the C(-1019)G functional polymorphism of the HTR1A gene and impulsivity. *American Journal of Medical Genetics Part B: Neuropsychiatric Genetics, 153B*(2), 592–599. doi:10.1002/ajmg.b.31025.

Berrettini, W. H. (2005). Genetic bases for endophenotypes in psychiatric disorders. *Dialogues in Clinical Neuroscience, 7*(2), 95–101.

Bevilacqua, L., Carli, V., Sarchiapone, M., George, D. K., Goldman, D., Roy, A., et al. (2012). Interaction between FKBP5 and childhood trauma and risk of aggressive behavior. *Archives of General Psychiatry, 69*(1), 62–70. doi:10.1001/archgenpsychiatry.2011.152.

Bevilacqua, L., Doly, S., Kaprio, J., Yuan, Q., Tikkanen, R., Paunio, T., et al. (2010). A population-specific HTR2B stop codon predisposes to severe impulsivity. *Nature, 468*(7327), 1061–1066. doi:10.1038/nature09629.

Bevilacqua, L., & Goldman, D. (2009). Genes and addictions. *Clinical Pharmacology and Therapeutics, 85*(4), 359–361. doi:10.1038/clpt.2009.6.

Bezdjian, S., Baker, L. A., & Tuvblad, C. (2011). Genetic and environmental influences on impulsivity: A meta-analysis of twin, family and adoption studies. *Clinical Psychology Review, 31*(7), 1209–1223. doi:10.1016/j.cpr.2011.07.005.

Bidwell, L. C., Metrik, J., McGeary, J., Palmer, R. H., Francazio, S., & Knopik, V. S. (2013). Impulsivity, variation in the cannabinoid receptor (CNR1) and fatty acid amide hydrolase (FAAH) genes, and marijuana-related problems. *Journal of Studies on Alcohol and Drugs, 74*(6), 867–878.

Binder, E. B., Salyakina, D., Lichtner, P., Wochnik, G. M., Ising, M., Putz, B., et al. (2004). Polymorphisms in FKBP5 are associated with increased recurrence of depressive episodes and rapid response to antidepressant treatment. *Nature Genetics, 36*(12), 1319–1325. doi:10.1038/ng1479.

Bjork, J. M., Moeller, F. G., Dougherty, D. M., Swann, A. C., Machado, M. A., & Hanis, C. L. (2002). Serotonin 2a receptor T102C polymorphism and impaired impulse control. *American Journal of Medical Genetics, 114*(3), 336–339. doi:10.1002/ajmg.10206.

Blondeau, C., & Dellu-Hagedorn, F. (2007). Dimensional analysis of ADHD subtypes in rats. *Biological Psychiatry, 61*(12), 1340–1350. doi:10.1016/j.biopsych.2006.06.030.

Boettiger, C. A., Mitchell, J. M., Tavares, V. C., Robertson, M., Joslyn, G., D'Esposito, M., et al. (2007). Immediate reward bias in humans: fronto-parietal networks and a role for the catechol-O-methyltransferase 158(Val/Val) genotype. *Journal of Neuroscience, 27*(52), 14383–14391. doi:10.1523/JNEUROSCI.2551-07.2007.

Bogdan, R., Hyde, L. W., & Hariri, A. R. (2013). A neurogenetics approach to understanding individual differences in brain, behavior, and risk for psychopathology. *Molecular Psychiatry, 18*(3), 288–299. doi:10.1038/mp.2012.35.

Brewer, J. A., & Potenza, M. N. (2008). The neurobiology and genetics of impulse control disorders: Relationships to drug addictions. *Biochemical Pharmacology, 75*(1), 63–75. doi:10.1016/j.bcp.2007.06.043.

Broos, N., Diergaarde, L., Schoffelmeer, A. N., Pattij, T., & De Vries, T. J. (2012a). Trait impulsive choice predicts resistance to extinction and propensity to relapse to cocaine seeking: A bidirectional investigation. *Neuropsychopharmacology, 37*(6), 1377–1386. doi:10.1038/npp.2011.323.

Broos, N., Schmaal, L., Wiskerke, J., Kostelijk, L., Lam, T., Stoop, N., et al. (2012b). The relationship between impulsive choice and impulsive action: A cross-species translational study. *PloS One, 7*(5), e36781. doi:10.1371/journal.pone.0036781.

Brunner, D., & Hen, R. (1997). Insights into the neurobiology of impulsive behavior from serotonin receptor knockout mice. *Annals of the New York Academy of Sciences, 836*, 81–105.

Buckholtz, J. W., Treadway, M. T., Cowan, R. L., Woodward, N. D., Li, R., Ansari, M. S., et al. (2010). Dopaminergic network differences in human impulsivity. *Science, 329*(5991), 532. doi:10.1126/science.1185778.

Cao, J., Liu, X., Han, S., Zhang, C. K., Liu, Z., & Li, D. (2014). Association of the HTR2A gene with alcohol and heroin abuse. *Human Genetics, 133*(3), 357–365. doi:10.1007/s00439-013-1388-y.

Chatzittofis, A., Nordstrom, P., Hellstrom, C., Arver, S., Asberg, M., & Jokinen, J. (2013). CSF 5-HIAA, cortisol and DHEAS levels in suicide attempters. *European Neuropsychopharmacology, 23*(10), 1280–1287. doi:10.1016/j.euroneuro.2013.02.002.

Chen, J., Lipska, B. K., Halim, N., Ma, Q. D., Matsumoto, M., Melhem, S., et al. (2004). Functional analysis of genetic variation in catechol-O-methyltransferase (COMT): Effects on mRNA, protein, and enzyme activity in postmortem human brain. *American Journal of Human Genetics, 75*(5), 807–821. doi:10.1086/425589.

Chen, J., Tsang, S. Y., Zhao, C. Y., Pun, F. W., Yu, Z., Mei, L., et al. (2009). GABRB2 in schizophrenia and bipolar disorder: Disease association, gene expression and clinical correlations. *Biochemical Society Transactions, 37*(Pt 6), 1415–1418. doi:10.1042/BST0371415.

Clark, L., Roiser, J. P., Cools, R., Rubinsztein, D. C., Sahakian, B. J., & Robbins, T. W. (2005). Stop signal response inhibition is not modulated by tryptophan depletion or the serotonin transporter polymorphism in healthy volunteers: Implications for the 5-HT theory of impulsivity. *Psychopharmacology, 182*(4), 570–578. doi:10.1007/s00213-005-0104-6.

Clark, L., Stokes, P. R., Wu, K., Michalczuk, R., Benecke, A., Watson, B. J., et al. (2012). Striatal dopamine D(2)/D(3) receptor binding in pathological gambling is correlated with mood-related impulsivity. *NeuroImage, 63*(1), 40–46. doi:10.1016/j.neuroimage.2012.06.067.

Clarke, T. K., Bloch, P. J., Ambrose-Lanci, L. M., Ferraro, T. N., Berrettini, W. H., Kampman, K. M., et al. (2013). Further evidence for association of polymorphisms in the CNR1 gene with cocaine addiction: Confirmation in an independent sample and meta-analysis. *Addiction Biology, 18*(4), 702–708. doi:10.1111/j.1369-1600.2011.00346.x.

Colzato, L. S., van den Wildenberg, W. P., Van der Does, A. J., & Hommel, B. (2010). Genetic markers of striatal dopamine predict individual differences in dysfunctional, but not functional impulsivity. *Neuroscience, 170*(3), 782–788. doi:10.1016/j.neuroscience.2010.07.050.

Comings, D. E., Gonzalez, N., Wu, S., Gade, R., Muhleman, D., Saucier, G., et al. (1999). Studies of the 48 bp repeat polymorphism of the DRD4 gene in impulsive, compulsive, addictive behaviors: Tourette syndrome, ADHD, pathological gambling, and substance abuse. *American Journal of Medical Genetics, 88*(4), 358–368. doi:10.1002/(SICI)1096-8628(19990820)88:4<358::AID-AJMG13>3.0.CO;2-G.

Congdon, E., & Canli, T. (2008). A neurogenetic approach to impulsivity. *Journal of Personality, 76*(6), 1447–1484. doi:10.1111/j.1467-6494.2008.00528.x.

Congdon, E., Lesch, K. P., & Canli, T. (2008). Analysis of DRD4 and DAT polymorphisms and behavioral inhibition in healthy adults: Implications for impulsivity. *American Journal of Medical Genetics Part B: Neuropsychiatric Genetics, 147B*(1), 27–32. doi:10.1002/ajmg.b.30557.

Cummins, T. D., Hawi, Z., Hocking, J., Strudwick, M., Hester, R., Garavan, H., et al. (2012). Dopamine transporter genotype predicts behavioural and neural measures of response inhibition. *Molecular Psychiatry, 17*(11), 1086–1092. doi:10.1038/mp.2011.104.

Dalley, J. W., Everitt, B. J., & Robbins, T. W. (2011). Impulsivity, compulsivity, and top-down cognitive control. *Neuron, 69*(4), 680–694. doi:10.1016/j.neuron.2011.01.020.

Dalley, J. W., Fryer, T. D., Brichard, L., Robinson, E. S., Theobald, D. E., Laane, K., et al. (2007). Nucleus accumbens D2/3 receptors predict trait impulsivity and cocaine reinforcement. *Science, 315*(5816), 1267–1270. doi:10.1126/science.1137073.

Dalley, J. W., & Roiser, J. P. (2012). Dopamine, serotonin and impulsivity. *Neuroscience, 215*, 42–58. doi:10.1016/j.neuroscience.2012.03.065.

Davies, W., Humby, T., Kong, W., Otter, T., Burgoyne, P. S., & Wilkinson, L. S. (2009). Converging pharmacological and genetic evidence indicates a role for steroid sulfatase in attention. *Biological Psychiatry, 66*(4), 360–367. doi:10.1016/j.biopsych.2009.01.001.

Davies, W., Isles, A., Smith, R., Karunadasa, D., Burrmann, D., Humby, T., et al. (2005). Xlr3b is a new imprinted candidate for X-linked parent-of-origin effects on cognitive function in mice. *Nature Genetics, 37*(6), 625–629. doi:10.1038/ng1577.

de Wit, H. (2009). Impulsivity as a determinant and consequence of drug use: A review of underlying processes. *Addiction Biology, 14*(1), 22–31. doi:10.1111/j.1369-1600.2008.00129.x.

Del Campo, N., Chamberlain, S. R., Sahakian, B. J., & Robbins, T. W. (2011). The roles of dopamine and noradrenaline in the pathophysiology and treatment of attention-deficit/hyperactivity disorder. *Biological Psychiatry, 69*(12), e145–e157. doi:10.1016/j.biopsych.2011.02.036.

Dere, E., Dahm, L., Lu, D., Hammerschmidt, K., Ju, A., Tantra, M., et al. (2014). Heterozygous ambra1 deficiency in mice: A genetic trait with autism-like behavior restricted to the female gender. *Frontiers in Behavioral Neuroscience, 8*, 181. doi:10.3389/fnbeh.2014.00181.

Di Giovanni, G., Di Matteo, V., Pierucci, M., & Esposito, E. (2008). Serotonin-dopamine interaction: Electrophysiological evidence. *Progress in Brain Research, 172*, 45–71. doi:10.1016/S0079-6123(08)00903-5.

Di Matteo, V., Di Giovanni, G., Pierucci, M., & Esposito, E. (2008). Serotonin control of central dopaminergic function: Focus on in vivo microdialysis studies. *Progress in Brain Research, 172*, 7–44. doi:10.1016/S0079-6123(08)00902-3.

Di Narzo, A. F., Kozlenkov, A., Roussos, P., Hao, K., Hurd, Y., Lewis, D. A., et al. (2014). A unique gene expression signature associated with serotonin 2C receptor RNA editing in the prefrontal cortex and altered in suicide. *Human Molecular Genetics, 23*(18), 4801–4813. doi:10.1093/hmg/ddu195.

Dickman, S. J. (1990). Functional and dysfunctional impulsivity: Personality and cognitive correlates. *Journal of Personality and Social Psychology, 58*(1), 95–102.

Doe, C. M., Relkovic, D., Garfield, A. S., Dalley, J. W., Theobald, D. E., Humby, T., et al. (2009). Loss of the imprinted snoRNA mbii-52 leads to increased 5htr2c pre-RNA editing and altered 5HT2CR-mediated behaviour. *Human Molecular Genetics, 18*(12), 2140–2148. doi:10.1093/hmg/ddp137.

Dougherty, D. M., Marsh, D. M., & Mathias, C. W. (2002). Immediate and delayed memory tasks: A computerized behavioral measure of memory, attention, and impulsivity. *Behavior Research Methods, Instruments, & Computers, 34*(3), 391–398.

Dougherty, D. M., Mathias, C. W., Marsh, D. M., Papageorgiou, T. D., Swann, A. C., & Moeller, F. G. (2004). Laboratory measured behavioral impulsivity relates to suicide attempt history. *Suicide and Life-threatening Behavior, 34*(4), 374–385. doi:10.1521/suli.34.4.374.53738.

Drgon, T., D'Addario, C., & Uhl, G. R. (2006). Linkage disequilibrium, haplotype and association studies of a chromosome 4 GABA receptor gene cluster: Candidate gene variants for addictions. *American Journal of Medical Genetics Part B: Neuropsychiatric Genetics, 141B*(8), 854–860. doi:10.1002/ajmg.b.30349.

Ducci, F., Newman, T. K., Funt, S., Brown, G. L., Virkkunen, M., & Goldman, D. (2006). A functional polymorphism in the MAOA gene promoter (MAOA-LPR) predicts central dopamine function and body mass index. *Molecular Psychiatry, 11*(9), 858–866. doi:10.1038/sj.mp.4001856.

Eagle, D. M., Lehmann, O., Theobald, D. E., Pena, Y., Zakaria, R., Ghosh, R., et al. (2009). Serotonin depletion impairs waiting but not stop-signal reaction time in rats: Implications for theories of the role of 5-HT in behavioral inhibition. *Neuropsychopharmacology, 34*(5), 1311–1321. doi:10.1038/npp.2008.202.

Ehlers, C. L., Slutske, W. S., Lind, P. A., & Wilhelmsen, K. C. (2007). Association between single nucleotide polymorphisms in the cannabinoid receptor gene (CNR1) and impulsivity in southwest California Indians. *Twin Research and Human Genetics, 10*(6), 805–811. doi:10.1375/twin.10.6.805.

Eisenberg, D. T., Mackillop, J., Modi, M., Beauchemin, J., Dang, D., Lisman, S. A., et al. (2007). Examining impulsivity as an endophenotype using a behavioral approach: A DRD2 TaqI A and DRD4 48-bp VNTR association study. *Behavioral and Brain Functions, 3*, 2. doi:10.1186/1744-9081-3-2.

Ersche, K. D., Turton, A. J., Pradhan, S., Bullmore, E. T., & Robbins, T. W. (2010). Drug addiction endophenotypes: Impulsive versus sensation-seeking personality traits. *Biological Psychiatry, 68*(8), 770–773. doi:10.1016/j.biopsych.2010.06.015.

Evenden, J. L. (1999). Varieties of impulsivity. *Psychopharmacology, 146*(4), 348–361.

Eysenck, H. J., & Eysenck, S. B. G. (1975). *Manual of the Eysenck personality questionnaire (junior and adult)*. Kent, UK: Hodder & Stoughton.

Fairbanks, L. A., Melega, W. P., Jorgensen, M. J., Kaplan, J. R., & McGuire, M. T. (2001). Social impulsivity inversely associated with CSF 5-HIAA and fluoxetine exposure in vervet monkeys. *Neuropsychopharmacology, 24*(4), 370–378. doi:10.1016/S0893-133X(00)00211-6.

Faraone, S. V., Biederman, J., Spencer, T., Michelson, D., Adler, L., Reimherr, F., et al. (2005). Efficacy of atomoxetine in adult attention-deficit/hyperactivity disorder: A drug-placebo response curve analysis. *Behavioral and Brain Functions, 1*, 16. doi:10.1186/1744-9081-1-16.

Faraone, S. V., Spencer, T. J., Madras, B. K., Zhang-James, Y., & Biederman, J. (2014). Functional effects of dopamine transporter gene genotypes on in vivo dopamine transporter functioning: A meta-analysis. *Molecular Psychiatry, 19*(8), 880–889. doi:10.1038/mp.2013.126.

Fernando, A. B., Economidou, D., Theobald, D. E., Zou, M. F., Newman, A. H., Spoelder, M., et al. (2012). Modulation of high impulsivity and attentional performance in rats by selective direct and indirect dopaminergic and noradrenergic receptor agonists. *Psychopharmacology, 219*(2), 341–352. doi:10.1007/s00213-011-2408-z.

Fernando, A. B., & Robbins, T. W. (2011). Animal models of neuropsychiatric disorders. *Annual Review of Clinical Psychology, 7*, 39–61. doi:10.1146/annurev-clinpsy-032210-104454.

Fillmore, M. T. (2003). Drug abuse as a problem of impaired control: Current approaches and findings. *Behavioral and Cognitive Neuroscience Reviews, 2*(3), 179–197. doi:10.1177/1534582303257007.

Fimia, G. M., Stoykova, A., Romagnoli, A., Giunta, L., Di Bartolomeo, S., Nardacci, R., et al. (2007). Ambra1 regulates autophagy and development of the nervous system. *Nature, 447*(7148), 1121–1125. doi:10.1038/nature05925.

Fink, K. B., & Gothert, M. (2007). 5-HT receptor regulation of neurotransmitter release. *Pharmacological Reviews, 59*(4), 360–417. doi:10.1124/pr.107.07103.

Fletcher, P. J., Soko, A. D., & Higgins, G. A. (2013). Impulsive action in the 5-choice serial reaction time test in 5-HT(2)c receptor null mutant mice. *Psychopharmacology, 226*(3), 561–570. doi:10.1007/s00213-012-2929-0.

Franke, B., Faraone, S. V., Asherson, P., Buitelaar, J., Bau, C. H., Ramos-Quiroga, J. A., et al. (2012). The genetics of attention deficit/hyperactivity disorder in adults, a review. *Molecular Psychiatry, 17*(10), 960–987. doi:10.1038/mp.2011.138.

Frankle, W. G., Lombardo, I., New, A. S., Goodman, M., Talbot, P. S., Huang, Y., et al. (2005). Brain serotonin transporter distribution in subjects with impulsive aggressivity: A positron emission study with [11C]McN 5652. *The American Journal of Psychiatry, 162*(5), 915–923. doi:10.1176/appi.ajp.162.5.915.

Gao, F., Zhu, Y. S., Wei, S. G., Li, S. B., & Lai, J. H. (2011). Polymorphism G861C of 5-HT receptor subtype 1B is associated with heroin dependence in Han Chinese. *Biochemical and Biophysical Research Communications, 412*(3), 450–453. doi:10.1016/j.bbrc.2011.07.114.

Geissler, J., & Lesch, K. P. (2011). A lifetime of attention-deficit/hyperactivity disorder: Diagnostic challenges, treatment and neurobiological mechanisms. *Expert Review of Neurotherapeutics, 11*(10), 1467–1484. doi:10.1586/ern.11.136.

Gerra, G., Garofano, L., Bosari, S., Pellegrini, C., Zaimovic, A., Moi, G., et al. (2004). Analysis of monoamine oxidase A (MAO-A) promoter polymorphism in male heroin-dependent subjects: Behavioural and personality correlates. *Journal of Neural Transmission, 111*(5), 611–621. doi:10.1007/s00702-004-0129-8.

Ghavami, S., Shojaei, S., Yeganeh, B., Ande, S. R., Jangamreddy, J. R., Mehrpour, M., et al. (2014). Autophagy and apoptosis dysfunction in neurodegenerative disorders. *Progress in Neurobiology, 112*, 24–49. doi:10.1016/j.pneurobio.2013.10.004.

Goldman, D., Dean, M., Brown, G. L., Bolos, A. M., Tokola, R., Virkkunen, M., et al. (1992). D2 dopamine receptor genotype and cerebrospinal fluid homovanillic acid, 5-hydroxyindoleacetic acid and 3-methoxy-4-hydroxyphenylglycol in alcoholics in Finland and the United States. *Acta Psychiatrica Scandinavica, 86*(5), 351–357.

Goldman, D., & Fishbein, D. H. (2000). Genetic bases for impulsive and antisocial behaviors—Can their course be altered? In D. H. Fishbein (Ed.), *The science, treatment, and prevention of antisocial behaviors: Application to the criminal justice system* (pp. 9.1–9.18). Kingston, NJ: Civic Research Institute.

Gottesman, I. I., & Gould, T. D. (2003). The endophenotype concept in psychiatry: Etymology and strategic intentions. *The American Journal of Psychiatry, 160*(4), 636–645.

Greco, B., & Carli, M. (2006). Reduced attention and increased impulsivity in mice lacking NPY Y2 receptors: Relation to anxiolytic-like phenotype. *Behavioural Brain Research, 169*(2), 325–334. doi:10.1016/j.bbr.2006.02.002.

Gubner, N. R., Wilhelm, C. J., Phillips, T. J., & Mitchell, S. H. (2010). Strain differences in behavioral inhibition in a Go/No-go task demonstrated using 15 inbred mouse strains. *Alcoholism: Clinical and Experimental Research, 34*(8), 1353–1362. doi:10.1111/j.1530-0277.2010.01219.x.

Guillem, K., Bloem, B., Poorthuis, R. B., Loos, M., Smit, A. B., Maskos, U., et al. (2011). Nicotinic acetylcholine receptor beta2 subunits in the medial prefrontal cortex control attention. *Science, 333*(6044), 888–891. doi:10.1126/science.1207079.

Guimaraes, A. P., Schmitz, M., Polanczyk, G. V., Zeni, C., Genro, J., Roman, T., et al. (2009). Further evidence for the association between attention deficit/hyperactivity disorder and the serotonin receptor 1B gene. *Journal of Neural Transmission, 116*(12), 1675–1680. doi:10.1007/s00702-009-0305-y.

Guimaraes, A. P., Zeni, C., Polanczyk, G. V., Genro, J. P., Roman, T., Rohde, L. A., et al. (2007). Serotonin genes and attention deficit/hyperactivity disorder in a Brazilian sample: Preferential transmission of the HTR2A 452His allele to affected boys. *American Journal of Medical Genetics Part B: Neuropsychiatric Genetics, 144B*(1), 69–73. doi:10.1002/ajmg.b.30400.

Gullo, M. J., St John, N., Mc, D. Y. R., Saunders, J. B., Noble, E. P., & Connor, J. P. (2014). Impulsivity-related cognition in alcohol dependence: Is it moderated by DRD2/ANKK1 gene status and executive dysfunction? *Addictive Behaviors, 39*(11), 1663–1669. doi:10.1016/j.addbeh.2014.02.004.

Hall, F. S., Drgonova, J., Jain, S., & Uhl, G. R. (2013). Implications of genome wide association studies for addiction: Are our a priori assumptions all wrong? *Pharmacology and Therapeutics, 140*(3), 267–279. doi:10.1016/j.pharmthera.2013.07.006.

Harrison, A. A., Everitt, B. J., & Robbins, T. W. (1997). Central 5-HT depletion enhances impulsive responding without affecting the accuracy of attentional performance: Interactions with dopaminergic mechanisms. *Psychopharmacology, 133*(4), 329–342.

Harrison, A. A., Everitt, B. J., & Robbins, T. W. (1999). Central serotonin depletion impairs both the acquisition and performance of a symmetrically reinforced go/no-go conditional visual discrimination. *Behavioural Brain Research, 100*(1-2), 99–112.

Haughey, H. M., Ray, L. A., Finan, P., Villanueva, R., Niculescu, M., & Hutchison, K. E. (2008). Human gamma-aminobutyric acid A receptor alpha2 gene moderates the acute effects of alcohol and brain mRNA expression. *Genes, Brain, and Behavior, 7*(4), 447–454. doi:10.1111/j.1601-183X.2007.00369.x.

Hawi, Z., Matthews, N., Barry, E., Kirley, A., Wagner, J., Wallace, R. H., et al. (2013). A high density linkage disequilibrium mapping in 14 noradrenergic genes: Evidence of association between SLC6A2, ADRA1B and ADHD. *Psychopharmacology, 225*(4), 895–902. doi:10.1007/s00213-012-2875-x.

Hayes, D. J., Jupp, B., Sawiak, S. J., Merlo, E., Caprioli, D., & Dalley, J. W. (2014). Brain gamma-aminobutyric acid: A neglected role in impulsivity. *European Journal of Neuroscience, 39*(11), 1921–1932. doi:10.1111/ejn.12485.

Heils, A., Teufel, A., Petri, S., Seemann, M., Bengel, D., Balling, U., et al. (1995). Functional promoter and polyadenylation site mapping of the human serotonin (5-HT) transporter gene. *Journal of Neural Transmission: General Section, 102*(3), 247–254.

Heils, A., Teufel, A., Petri, S., Stober, G., Riederer, P., Bengel, D., et al. (1996). Allelic variation of human serotonin transporter gene expression. *Journal of Neurochemistry, 66*(6), 2621–2624.

Heinrich, A., Nees, F., Lourdusamy, A., Tzschoppe, J., Meier, S., Vollstadt-Klein, S., et al. (2013). From gene to brain to behavior: Schizophrenia-associated variation in AMBRA1 alters impulsivity-related traits. *European Journal of Neuroscience, 38*(6), 2941–2945. doi:10.1111/ejn.12201.

Heinzel, S., Dresler, T., Baehne, C. G., Heine, M., Boreatti-Hummer, A., Jacob, C. P., et al. (2013). COMT x DRD4 epistasis impacts prefrontal cortex function underlying response control. *Cerebral Cortex, 23*(6), 1453–1462. doi:10.1093/cercor/bhs132.

Helms, C. M., Gubner, N. R., Wilhelm, C. J., Mitchell, S. H., & Grandy, D. K. (2008). D4 receptor deficiency in mice has limited effects on impulsivity and novelty seeking. *Pharmacology, Biochemistry and Behavior, 90*(3), 387–393. doi:10.1016/j.pbb.2008.03.013.

Hess, C., Reif, A., Strobel, A., Boreatti-Hummer, A., Heine, M., Lesch, K. P., et al. (2009). A functional dopamine-beta-hydroxylase gene promoter polymorphism is associated with impulsive personality styles, but not with affective disorders. *Journal of Neural Transmission, 116*(2), 121–130. doi:10.1007/s00702-008-0138-0.

Hirvonen, M., Laakso, A., Nagren, K., Rinne, J. O., Pohjalainen, T., & Hietala, J. (2004). C957T polymorphism of the dopamine D2 receptor (DRD2) gene affects striatal DRD2 availability in vivo. *Molecular Psychiatry, 9*(12), 1060–1061. doi:10.1038/sj.mp.4001561.

Hirvonen, J., Zanotti-Fregonara, P., Umhau, J. C., George, D. T., Rallis-Frutos, D., Lyoo, C. H., et al. (2013). Reduced cannabinoid CB1 receptor binding in alcohol dependence measured with positron emission tomography. *Molecular Psychiatry, 18*(8), 916–921. doi:10.1038/mp.2012.100.

Homberg, J. R., Pattij, T., Janssen, M. C., Ronken, E., De Boer, S. F., Schoffelmeer, A. N., et al. (2007). Serotonin transporter deficiency in rats improves inhibitory control but not behavioural flexibility. *European Journal of Neuroscience, 26*(7), 2066–2073. doi:10.1111/j.1460-9568.2007.05839.x.

Hoogman, M., Aarts, E., Zwiers, M., Slaats-Willemse, D., Naber, M., Onnink, M., et al. (2011). Nitric oxide synthase genotype modulation of impulsivity and ventral striatal activity in adult ADHD patients and healthy comparison subjects. *The American Journal of Psychiatry, 168*(10), 1099–1106. doi:10.1176/appi.ajp.2011.10101446.

Hung, C. F., Lung, F. W., Chen, C. H., O'Nions, E., Hung, T. H., Chong, M. Y., et al. (2011). Association between suicide attempt and a tri-allelic functional polymorphism in serotonin transporter gene promoter in Chinese patients with schizophrenia. *Neuroscience Letters, 504*(3), 242–246. doi:10.1016/j.neulet.2011.09.036.

Ising, M., Depping, A. M., Siebertz, A., Lucae, S., Unschuld, P. G., Kloiber, S., et al. (2008). Polymorphisms in the FKBP5 gene region modulate recovery from psychosocial stress in healthy controls. *European Journal of Neuroscience, 28*(2), 389–398. doi:10.1111/j.1460-9568.2008.06332.x.

Isles, A. R., Hathway, G. J., Humby, T., de la Riva, C., Kendrick, K. M., & Wilkinson, L. S. (2005). An mTph2 SNP gives rise to alterations in extracellular 5-HT levels, but not in performance on a delayed-reinforcement task. *European Journal of Neuroscience, 22*(4), 997–1000. doi:10.1111/j.1460-9568.2005.04265.x.

Isles, A. R., Humby, T., Walters, E., & Wilkinson, L. S. (2004). Common genetic effects on variation in impulsivity and activity in mice. *Journal of Neuroscience, 24*(30), 6733–6740. doi:10.1523/JNEUROSCI.1650-04.2004.

Jakubczyk, A., Wrzosek, M., Lukaszkiewicz, J., Sadowska-Mazuryk, J., Matsumoto, H., Sliwerska, E., et al. (2012). The CC genotype in HTR2A T102C polymorphism is associated with behavioral impulsivity in alcohol-dependent patients. *Journal of Psychiatric Research, 46*(1), 44–49. doi:10.1016/j.jpsychires.2011.09.001.

Jensen, K. P., Covault, J., Conner, T. S., Tennen, H., Kranzler, H. R., & Furneaux, H. M. (2009). A common polymorphism in serotonin receptor 1B mRNA moderates regulation by miR-96 and associates with aggressive human behaviors. *Molecular Psychiatry, 14*(4), 381–389. doi:10.1038/mp.2008.15.

Jiang, H. Y., Qiao, F., Xu, X. F., Yang, Y., Bai, Y., & Jiang, L. L. (2013). Meta-analysis confirms a functional polymorphism (5-HTTLPR) in the serotonin transporter gene conferring risk of bipolar disorder in European populations. *Neuroscience Letters, 549*, 191–196. doi:10.1016/j.neulet.2013.05.065.

Jimenez, E., Arias, B., Mitjans, M., Goikolea, J. M., Roda, E., Ruiz, V., et al. (2014). Association between GSK3beta gene and increased impulsivity in bipolar disorder. *European Neuropsychopharmacology, 24*(4), 510–518. doi:10.1016/j.euroneuro.2014.01.005.

Jonsson, E. G., Nothen, M. M., Grunhage, F., Farde, L., Nakashima, Y., Propping, P., et al. (1999). Polymorphisms in the dopamine D2 receptor gene and their relationships to striatal dopamine receptor density of healthy volunteers. *Molecular Psychiatry, 4*(3), 290–296.

Jonsson, E. G., Nothen, M. M., Gustavsson, J. P., Neidt, H., Bunzel, R., Propping, P., et al. (1998). Polymorphisms in the dopamine, serotonin, and norepinephrine transporter genes and their relationships to monoamine metabolite concentrations in CSF of healthy volunteers. *Psychiatry Research, 79*(1), 1–9.

Jupp, B., Caprioli, D., Saigal, N., Reverte, I., Shrestha, S., Cumming, P., et al. (2013). Dopaminergic and GABA-ergic markers of impulsivity in rats: evidence for anatomical localisation in ventral striatum and prefrontal cortex. *European Journal of Neuroscience, 37*(9), 1519–1528. doi:10.1111/ejn.12146.

Jupp, B., & Dalley, J. W. (2014). Convergent pharmacological mechanisms in impulsivity and addiction: Insights from rodent models. *British Journal of Pharmacology*. doi:10.1111/bph.12787.

Kaladjian, A., Jeanningros, R., Azorin, J. M., Anton, J. L., & Mazzola-Pomietto, P. (2011). Impulsivity and neural correlates of response inhibition in schizophrenia. *Psychological Medicine, 41*(2), 291–299. doi:10.1017/S0033291710000796.

Kapur, S., & Remington, G. (1996). Serotonin-dopamine interaction and its relevance to schizophrenia. *The American Journal of Psychiatry, 153*(4), 466–476.

Kawamura, Y., Takahashi, T., Liu, X., Nishida, N., Tokunaga, K., Ukawa, K., et al. (2013). DNA polymorphism in the FKBP5 gene affects impulsivity in intertemporal choice. *Asia Pac Psychiatry, 5*(1), 31–38. doi:10.1111/appy.12009.

Keller, J. J., Keller, A. B., Bowers, B. J., & Wehner, J. M. (2005). Performance of alpha7 nicotinic receptor null mutants is impaired in appetitive learning measured in a signaled nose poke task. *Behavioural Brain Research, 162*(1), 143–152. doi:10.1016/j.bbr.2005.03.004.

Kerr, T. M., Muller, C. L., Miah, M., Jetter, C. S., Pfeiffer, R., Shah, C., et al. (2013). Genetic background modulates phenotypes of serotonin transporter Ala56 knock-in mice. *Molecular Autism, 4*(1), 35. doi:10.1186/2040-2392-4-35.

Kieling, C., Genro, J. P., Hutz, M. H., & Rohde, L. A. (2008). The -1021 C/T DBH polymorphism is associated with neuropsychological performance among children and adolescents with ADHD. *American Journal of Medical Genetics Part B: Neuropsychiatric Genetics, 147B*(4), 485–490. doi:10.1002/ajmg.b.30636.

Kim, J. W., Biederman, J., McGrath, C. L., Doyle, A. E., Mick, E., Fagerness, J., et al. (2008). Further evidence of association between two NET single-nucleotide polymorphisms with ADHD. *Molecular Psychiatry, 13*(6), 624–630. doi:10.1038/sj.mp.4002090.

Kreek, M. J., Nielsen, D. A., Butelman, E. R., & LaForge, K. S. (2005). Genetic influences on impulsivity, risk taking, stress responsivity and vulnerability to drug abuse and addiction. *Nature Neuroscience, 8*(11), 1450–1457. doi:10.1038/nn1583.

Lage, G. M., Malloy-Diniz, L. F., Matos, L. O., Bastos, M. A., Abrantes, S. S., & Correa, H. (2011). Impulsivity and the 5-HTTLPR polymorphism in a non-clinical sample. *PloS One, 6*(2), e16927. doi:10.1371/journal.pone.0016927.

Lambourne, S. L., Humby, T., Isles, A. R., Emson, P. C., Spillantini, M. G., & Wilkinson, L. S. (2007). Impairments in impulse control in mice transgenic for the human FTDP-17 tauV337M mutation are exacerbated by age. *Human Molecular Genetics, 16*(14), 1708–1719. doi:10.1093/hmg/ddm119.

Le Foll, B., Gallo, A., Le Strat, Y., Lu, L., & Gorwood, P. (2009). Genetics of dopamine receptors and drug addiction: A comprehensive review. *Behavioural Pharmacology, 20*(1), 1–17. doi:10.1097/FBP.0b013e3283242f05.

Lee, B., London, E. D., Poldrack, R. A., Farahi, J., Nacca, A., Monterosso, J. R., et al. (2009). Striatal dopamine d2/d3 receptor availability is reduced in methamphetamine dependence and is linked to impulsivity. *Journal of Neuroscience, 29*(47), 14734–14740. doi:10.1523/JNEUROSCI.3765-09.2009.

Lemonde, S., Turecki, G., Bakish, D., Du, L., Hrdina, P. D., Bown, C. D., et al. (2003). Impaired repression at a 5-hydroxytryptamine 1A receptor gene polymorphism associated with major depression and suicide. *Journal of Neuroscience, 23*(25), 8788–8799.

Lesch, K. P., Balling, U., Gross, J., Strauss, K., Wolozin, B. L., Murphy, D. L., et al. (1994). Organization of the human serotonin transporter gene. *Journal of Neural Transmission: General Section, 95*(2), 157–162.

Levey, D. F., Le-Niculescu, H., Frank, J., Ayalew, M., Jain, N., Kirlin, B., et al. (2014). Genetic risk prediction and neurobiological understanding of alcoholism. *Transcultural Psychiatry, 4*, e391. doi:10.1038/tp.2014.29.

Levran, O., Peles, E., Randesi, M., Li, Y., Rotrosen, J., Ott, J., et al. (2014). Stress-related genes and heroin addiction: A role for a functional FKBP5 haplotype. *Psychoneuroendocrinology, 45*, 67–76. doi:10.1016/j.psyneuen.2014.03.017.

Li, Z., Chang, S. H., Zhang, L. Y., Gao, L., & Wang, J. (2014). Molecular genetic studies of ADHD and its candidate genes: A review. *Psychiatry Research, 219*(1), 10–24. doi:10.1016/j.psychres.2014.05.005.

Limosin, F., Loze, J. Y., Dubertret, C., Gouya, L., Ades, J., Rouillon, F., et al. (2003). Impulsiveness as the intermediate link between the dopamine receptor D2 gene and alcohol dependence. *Psychiatric Genetics, 13*(2), 127–129. doi:10.1097/01.ypg.0000066963.66429.00.

Limosin, F., Romo, L., Batel, P., Ades, J., Boni, C., & Gorwood, P. (2005). Association between dopamine receptor D3 gene BalI polymorphism and cognitive impulsiveness in alcohol-dependent men. *European Psychiatry, 20*(3), 304–306. doi:10.1016/j.eurpsy.2005.02.004.

Little, K. Y., McLaughlin, D. P., Zhang, L., Livermore, C. S., Dalack, G. W., McFinton, P. R., et al. (1998). Cocaine, ethanol, and genotype effects on human midbrain serotonin transporter binding sites and mRNA levels. *The American Journal of Psychiatry, 155*(2), 207–213.

Liu, L., Guan, L. L., Chen, Y., Ji, N., Li, H. M., Li, Z. H., et al. (2011). Association analyses of MAOA in Chinese Han subjects with attention-deficit/hyperactivity disorder: Family-based association test, case-control study, and quantitative traits of impulsivity. *American Journal of Medical Genetics Part B: Neuropsychiatric Genetics, 156B*(6), 737–748. doi:10.1002/ajmg.b.31217.

Logue, S. F., Swartz, R. J., & Wehner, J. M. (1998). Genetic correlation between performance on an appetitive-signaled nosepoke task and voluntary ethanol consumption. *Alcoholism: Clinical and Experimental Research, 22*(9), 1912–1920.

Lombardo, L. E., Bearden, C. E., Barrett, J., Brumbaugh, M. S., Pittman, B., Frangou, S., et al. (2012). Trait impulsivity as an endophenotype for bipolar I disorder. *Bipolar Disorders, 14*(5), 565–570. doi:10.1111/j.1399-5618.2012.01035.x.

Loo, S. K., Specter, E., Smolen, A., Hopfer, C., Teale, P. D., & Reite, M. L. (2003). Functional effects of the DAT1 polymorphism on EEG measures in ADHD. *Journal of the American Academy of Child and Adolescent Psychiatry, 42*(8), 986–993. doi:10.1097/01.CHI.0000046890.27264.88.

Loos, M., Mueller, T., Gouwenberg, Y., Wijnands, R., van der Loo, R. J., Birchmeier, C., et al. (2014). Neuregulin-3 in the mouse medial prefrontal cortex regulates impulsive action. *Biological Psychiatry, 6*(8), 648–655. doi:10.1016/j.biopsych.2014.02.011.

Loos, M., van der Sluis, S., Bochdanovits, Z., van Zutphen, I. J., Pattij, T., Stiedl, O., et al. (2009). Activity and impulsive action are controlled by different genetic and environmental factors. *Genes, Brain, and Behavior, 8*(8), 817–828. doi:10.1111/j.1601-183X.2009.00528.x.

Lovallo, W. R. (2013). Early life adversity reduces stress reactivity and enhances impulsive behavior: Implications for health behaviors. *International Journal of Psychophysiology, 90*(1), 8–16. doi:10.1016/j.ijpsycho.2012.10.006.

Lovic, V., Keen, D., Fletcher, P. J., & Fleming, A. S. (2011). Early-life maternal separation and social isolation produce an increase in impulsive action but not impulsive choice. *Behavioral Neuroscience, 125*(4), 481–491. doi:10.1037/a0024367.

Lu, A. T., Ogdie, M. N., Jarvelin, M. R., Moilanen, I. K., Loo, S. K., McCracken, J. T., et al. (2008). Association of the cannabinoid receptor gene (CNR1) with ADHD and post-traumatic stress disorder. *American Journal of Medical Genetics Part B: Neuropsychiatric Genetics, 147B*(8), 1488–1494. doi:10.1002/ajmg.b.30693.

Lundstrom, K., & Turpin, M. P. (1996). Proposed schizophrenia-related gene polymorphism: Expression of the Ser9Gly mutant human dopamine D3 receptor with the Semliki Forest virus system. *Biochemical and Biophysical Research Communications, 225*(3), 1068–1072. doi:10.1006/bbrc.1996.1296.

Magrys, S. A., Olmstead, M. C., Wynne-Edwards, K. E., & Balodis, I. M. (2013). Neuroendocrinological responses to alcohol intoxication in healthy males: Relationship with impulsivity, drinking behavior, and subjective effects. *Psychophysiology, 50*(2), 204–209. doi:10.1111/psyp.12007.

Maher, B. (2008). Personal genomes: The case of the missing heritability. *Nature, 456*(7218), 18–21. doi:10.1038/456018a.

Mann, J. J. (2013). The serotonergic system in mood disorders and suicidal behaviour. *Philosophical Transactions of the Royal Society of London. Series B: Biological Sciences, 368*(1615), 20120537. doi:10.1098/rstb.2012.0537.

Manolio, T. A., Collins, F. S., Cox, N. J., Goldstein, D. B., Hindorff, L. A., Hunter, D. J., et al. (2009). Finding the missing heritability of complex diseases. *Nature, 461*(7265), 747–753. doi:10.1038/nature08494.

Manor, I., Corbex, M., Eisenberg, J., Gritsenkso, I., Bachner-Melman, R., Tyano, S., et al. (2004). Association of the dopamine D5 receptor with attention deficit hyperactivity disorder (ADHD) and scores on a continuous performance test (TOVA). *American Journal of Medical Genetics Part B: Neuropsychiatric Genetics, 127B*(1), 73–77. doi:10.1002/ajmg.b.30020.

Mansour, H. A., Talkowski, M. E., Wood, J., Pless, L., Bamne, M., Chowdari, K. V., et al. (2005). Serotonin gene polymorphisms and bipolar I disorder: Focus on the serotonin transporter. *Annals of Medicine, 37*(8), 590–602. doi:10.1080/07853890500357428.

Manuck, S. B., Flory, J. D., Ferrell, R. E., Mann, J. J., & Muldoon, M. F. (2000). A regulatory polymorphism of the monoamine oxidase-A gene may be associated with variability in aggression, impulsivity, and central nervous system serotonergic responsivity. *Psychiatry Research, 95*(1), 9–23.

Marcos, M., Pastor, I., de la Calle, C., Barrio-Real, L., Laso, F. J., & Gonzalez-Sarmiento, R. (2012). Cannabinoid receptor 1 gene is associated with alcohol dependence. *Alcoholism: Clinical and Experimental Research, 36*(2), 267–271. doi:10.1111/j.1530-0277.2011.01623.x.

Massat, I., Lerer, B., Souery, D., Blackwood, D., Muir, W., Kaneva, R., et al. (2007). HTR2C (cys-23ser) polymorphism influences early onset in bipolar patients in a large European multicenter association study. *Molecular Psychiatry, 12*(9), 797–798. doi:10.1038/sj.mp.4002018.

Mathur, B. N., & Lovinger, D. M. (2012). Endocannabinoid-dopamine interactions in striatal synaptic plasticity. *Frontiers in Pharmacology, 3*, 66. doi:10.3389/fphar.2012.00066.

Maziade, M., Roy, M. A., Chagnon, Y. C., Cliche, D., Fournier, J. P., Montgrain, N., et al. (2005). Shared and specific susceptibility loci for schizophrenia and bipolar disorder: A dense genome scan in Eastern Quebec families. *Molecular Psychiatry, 10*(5), 486–499. doi:10.1038/sj.mp.4001594.

Mazur, J. E. (1987). An adjusting procedure for studying delayed reinforcement. In M. L. Commons, J. E. Mazur, J. A. Nevin, & H. Rachlin (Eds.), *Quantitative analysis of behavior* (The effect of delay and of intervening events of reinforcement value, Vol. 5, pp. 55–73). Hillsdale, NJ: Erlbaum.

Menke, A., Klengel, T., Rubel, J., Bruckl, T., Pfister, H., Lucae, S., et al. (2013). Genetic variation in FKBP5 associated with the extent of stress hormone dysregulation in major depression. *Genes, Brain, and Behavior, 12*(3), 289–296. doi:10.1111/gbb.12026.

Mettman, D. J., Butler, M. G., Poje, A. B., Penick, E. C., & Manzardo, A. M. (2014). A preliminary case study of androgen receptor gene polymorphism association with impulsivity in women with alcoholism. *Advances in Genomics Genetics, 4*, 5–13. doi:10.2147/AGG.S57771.

Meyer-Lindenberg, A., Buckholtz, J. W., Kolachana, B., Hariri, R. A., Pezawas, L., Blasi, G., et al. (2006). Neural mechanisms of genetic risk for impulsivity and violence in humans. *Proceedings of the National Academy of Sciences of the United States of America, 103*(16), 6269–6274. doi:10.1073/pnas.0511311103.

Mill, J., Asherson, P., Browes, C., D'Souza, U., & Craig, I. (2002). Expression of the dopamine transporter gene is regulated by the 3′ UTR VNTR: Evidence from brain and lymphocytes using quantitative RT-PCR. *American Journal of Medical Genetics, 114*(8), 975–979. doi:10.1002/ajmg.b.10948.

Minocci, D., Massei, J., Martino, A., Milianti, M., Piz, L., Di Bello, D., et al. (2011). Genetic association between bipolar disorder and 524A>C (Leu133Ile) polymorphism of CNR2 gene, encoding for CB2 cannabinoid receptor. *Journal of Affective Disorders, 134*(1–3), 427–430. doi:10.1016/j.jad.2011.05.023.

Miyazaki, K., Miyazaki, K. W., & Doya, K. (2012). The role of serotonin in the regulation of patience and impulsivity. *Molecular Neurobiology, 45*(2), 213–224. doi:10.1007/s12035-012-8232-6.

Mobini, S., Chiang, T. J., Ho, M. Y., Bradshaw, C. M., & Szabadi, E. (2000). Effects of central 5-hydroxytryptamine depletion on sensitivity to delayed and probabilistic reinforcement. *Psychopharmacology, 152*(4), 390–397.

Moeller, F. G., Barratt, E. S., Dougherty, D. M., Schmitz, J. M., & Swann, A. C. (2001). Psychiatric aspects of impulsivity. *The American Journal of Psychiatry, 158*(11), 1783–1793.

Monteleone, P., Bifulco, M., Maina, G., Tortorella, A., Gazzerro, P., Proto, M. C., et al. (2010). Investigation of CNR1 and FAAH endocannabinoid gene polymorphisms in bipolar disorder and major depression. *Pharmacological Research, 61*(5), 400–404. doi:10.1016/j.phrs.2010.01.002.

Monterosso, J., & Ainslie, G. (1999). Beyond discounting: Possible experimental models of impulse control. *Psychopharmacology, 146*(4), 339–347.

Moon, J., Beaudin, A. E., Verosky, S., Driscoll, L. L., Weiskopf, M., Levitsky, D. A., et al. (2006). Attentional dysfunction, impulsivity, and resistance to change in a mouse model of fragile X syndrome. *Behavioral Neuroscience, 120*(6), 1367–1379. doi:10.1037/0735-7044.120.6.1367.

Moreira, F. A., Jupp, B., Belin, D., & Dalley, J. W. (2015). Endocannabinoids and basal ganglia striatal function: implications for addiction-related behaviours. *Behavioural Pharmacology, 26*, 59–72.

Moreno, M., Cardona, D., Gomez, M. J., Sanchez-Santed, F., Tobena, A., Fernandez-Teruel, A., et al. (2010). Impulsivity characterization in the Roman high- and low-avoidance rat strains: Behavioral and neurochemical differences. *Neuropsychopharmacology, 35*(5), 1198–1208. doi:10.1038/npp.2009.224.

Murphy, D. L., Uhl, G. R., Holmes, A., Ren-Patterson, R., Hall, F. S., Sora, I., et al. (2003). Experimental gene interaction studies with SERT mutant mice as models for human polygenic and epistatic traits and disorders. *Genes, Brain, and Behavior, 2*(6), 350–364.

Navarra, R., Comery, T. A., Graf, R., Rosenzweig-Lipson, S., & Day, M. (2008a). The 5-HT(2C) receptor agonist WAY-163909 decreases impulsivity in the 5-choice serial reaction time test. *Behavioural Brain Research, 188*(2), 412–415. doi:10.1016/j.bbr.2007.11.016.

Navarra, R., Graf, R., Huang, Y., Logue, S., Comery, T., Hughes, Z., et al. (2008b). Effects of atomoxetine and methylphenidate on attention and impulsivity in the 5-choice serial reaction time test. *Progress in Neuropsychopharmacology and Biological Psychiatry, 32*(1), 34–41. doi:10.1016/j.pnpbp.2007.06.017.

Neville, M. J., Johnstone, E. C., & Walton, R. T. (2004). Identification and characterization of ANKK1: A novel kinase gene closely linked to DRD2 on chromosome band 11q23.1. *Human Mutation, 23*(6), 540–545. doi:10.1002/humu.20039.

Nielsen, D. A., Proudnikov, D., & Kreek, M. J. (2012). The genetics of impulsivity. In J. E. Grant & M. N. Potenza (Eds.), *The oxford handbook of impulsive control disorders*. New York: Oxford University Press.

Nishikawa, S., Nishitani, S., Fujisawa, T. X., Noborimoto, I., Kitahara, T., Takamura, T., et al. (2012). Perceived parental rejection mediates the influence of serotonin transporter gene (5-HTTLPR) polymorphisms on impulsivity in Japanese adults. *PloS One, 7*(10), e47608. doi:10.1371/journal.pone.0047608.

Niv, S., Tuvblad, C., Raine, A., Wang, P., & Baker, L. A. (2012). Heritability and longitudinal stability of impulsivity in adolescence. *Behavior Genetics, 42*(3), 378–392. doi:10.1007/s10519-011-9518-6.

Nomura, M., & Nomura, Y. (2006). Psychological, neuroimaging, and biochemical studies on functional association between impulsive behavior and the 5-HT2A receptor gene polymorphism in humans. *Annals of the New York Academy of Sciences, 1086*, 134–143. doi:10.1196/annals.1377.004.

Nordin, C., & Eklundh, T. (1999). Altered CSF 5-HIAA disposition in pathologic male gamblers. *CNS Spectrums, 4*(12), 25–33.

Oliver, P. L., & Davies, K. E. (2012). New insights into behaviour using mouse ENU mutagenesis. *Human Molecular Genetics, 21*(R1), R72–R81. doi:10.1093/hmg/dds318.

Olivier, B. (2004). Serotonin and aggression. *Annals of the New York Academy of Sciences, 1036*, 382–392. doi:10.1196/annals.1330.022.

Olmstead, M. C., Ouagazzal, A. M., & Kieffer, B. L. (2009). Mu and delta opioid receptors oppositely regulate motor impulsivity in the signaled nose poke task. *PloS One, 4*(2), e4410. doi:10.1371/journal.pone.0004410.

Paaver, M., Nordquist, N., Parik, J., Harro, M., Oreland, L., & Harro, J. (2007). Platelet MAO activity and the 5-HTT gene promoter polymorphism are associated with impulsivity and cognitive style in visual information processing. *Psychopharmacology, 194*(4), 545–554. doi:10.1007/s00213-007-0867-z.

Palmason, H., Moser, D., Sigmund, J., Vogler, C., Hanig, S., Schneider, A., et al. (2010). Attention-deficit/hyperactivity disorder phenotype is influenced by a functional catechol-O-methyltransferase variant. *Journal of Neural Transmission, 117*(2), 259–267. doi:10.1007/s00702-009-0338-2.

Paloyelis, Y., Asherson, P., Mehta, M. A., Faraone, S. V., & Kuntsi, J. (2010). DAT1 and COMT effects on delay discounting and trait impulsivity in male adolescents with attention deficit/hyperactivity disorder and healthy controls. *Neuropsychopharmacology, 35*(12), 2414–2426. doi:10.1038/npp.2010.124.

Papaleo, F., Burdick, M. C., Callicott, J. H., & Weinberger, D. R. (2014). Epistatic interaction between COMT and DTNBP1 modulates prefrontal function in mice and in humans. *Molecular Psychiatry, 19*(3), 311–316. doi:10.1038/mp.2013.133.

Papaleo, F., Erickson, L., Liu, G., Chen, J., & Weinberger, D. R. (2012). Effects of sex and COMT genotype on environmentally modulated cognitive control in mice. *Proceedings of the National Academy of Sciences of the United States of America, 109*(49), 20160–20165. doi:10.1073/pnas.1214397109.

Paredes, U. M., Quinn, J. P., & D'Souza, U. M. (2013). Allele-specific transcriptional activity of the variable number of tandem repeats in 5′ region of the DRD4 gene is stimulus specific in human neuronal cells. *Genes, Brain, and Behavior, 12*(2), 282–287. doi:10.1111/j.1601-183X.2012.00857.x.

Park, S., Kim, J. W., Yang, Y. H., Hong, S. B., Park, M. H., Kim, B. N., et al. (2012). Possible effect of norepinephrine transporter polymorphisms on methylphenidate-induced changes in neuropsychological function in attention-deficit hyperactivity disorder. *Behavioral and Brain Functions, 8*, 22. doi:10.1186/1744-9081-8-22.

Parker, M. O., Ife, D., Ma, J., Pancholi, M., Smeraldi, F., Straw, C., et al. (2013). Development and automation of a test of impulse control in zebrafish. *Frontiers in Systems Neuroscience, 7*, 65. doi:10.3389/fnsys.2013.00065.

Pena-Oliver, Y., Buchman, V. L., Dalley, J. W., Robbins, T. W., Schumann, G., Ripley, T. L., et al. (2012). Deletion of alpha-synuclein decreases impulsivity in mice. *Genes, Brain, and Behavior, 11*(2), 137–146. doi:10.1111/j.1601-183X.2011.00758.x.

Perez de Castro, I., Ibanez, A., Torres, P., Saiz-Ruiz, J., & Fernandez-Piqueras, J. (1997). Genetic association study between pathological gambling and a functional DNA polymorphism at the D4 receptor gene. *Pharmacogenetics, 7*(5), 345–348.

Perlis, R. H., Huang, J., Purcell, S., Fava, M., Rush, A. J., Sullivan, P. F., et al. (2010). Genome-wide association study of suicide attempts in mood disorder patients. *The American Journal of Psychiatry, 167*(12), 1499–1507. doi:10.1176/appi.ajp.2010.10040541.

Perry, J. C., & Korner, A. C. (2011). Impulsive phenomena, the impulsive character (der Triebhafte Charakter) and DSM personality disorders. *Journal of Personality Disorders, 25*(5), 586–606. doi:10.1521/pedi.2011.25.5.586.

Perry, J. L., Larson, E. B., German, J. P., Madden, G. J., & Carroll, M. E. (2005). Impulsivity (delay discounting) as a predictor of acquisition of IV cocaine self-administration in female rats. *Psychopharmacology, 178*(2-3), 193–201. doi:10.1007/s00213-004-1994-4.

Perry, J. L., Stairs, D. J., & Bardo, M. T. (2008). Impulsive choice and environmental enrichment: Effects of d-amphetamine and methylphenidate. *Behavioural Brain Research, 193*(1), 48–54. doi:10.1016/j.bbr.2008.04.019.

Pierucci, M., Galati, S., Valentino, M., Di Matteo, V., Benigno, A., Pitruzzella, A., et al. (2011). Nitric oxide modulation of the basal ganglia circuitry: Therapeutic implication for Parkinson's disease and other motor disorders. *CNS & Neurological Disorders Drug Targets, 10*(7), 777–791.

Pohjalainen, T., Rinne, J. O., Nagren, K., Lehikoinen, P., Anttila, K., Syvalahti, E. K., et al. (1998). The A1 allele of the human D2 dopamine receptor gene predicts low D2 receptor availability in healthy volunteers. *Molecular Psychiatry, 3*(3), 256–260.

Preuss, U. W., Koller, G., Bondy, B., Bahlmann, M., & Soyka, M. (2001). Impulsive traits and 5-HT2A receptor promoter polymorphism in alcohol dependents: Possible association but no influence of personality disorders. *Neuropsychobiology, 43*(3), 186–191.

Racine, S. E., Culbert, K. M., Larson, C. L., & Klump, K. L. (2009). The possible influence of impulsivity and dietary restraint on associations between serotonin genes and binge eating. *Journal of Psychiatric Research, 43*(16), 1278–1286. doi:10.1016/j.jpsychires.2009.05.002.

Reeves, S. J., Polling, C., Stokes, P. R., Lappin, J. M., Shotbolt, P. P., Mehta, M. A., et al. (2012). Limbic striatal dopamine D2/3 receptor availability is associated with non-planning impulsivity in healthy adults after exclusion of potential dissimulators. *Psychiatry Research, 202*(1), 60–64. doi:10.1016/j.pscychresns.2011.09.011.

Reif, A., Jacob, C. P., Rujescu, D., Herterich, S., Lang, S., Gutknecht, L., et al. (2009). Influence of functional variant of neuronal nitric oxide synthase on impulsive behaviors in humans. *Archives of General Psychiatry, 66*(1), 41–50. doi:10.1001/archgenpsychiatry.2008.510.

Reif, A., Kiive, E., Kurrikoff, T., Paaver, M., Herterich, S., Konstabel, K., et al. (2011). A functional NOS1 promoter polymorphism interacts with adverse environment on functional and dysfunctional impulsivity. *Psychopharmacology, 214*(1), 239–248. doi:10.1007/s00213-010-1915-7.

Reiner, I., & Spangler, G. (2011). Dopamine D4 receptor exon III polymorphism, adverse life events and personality traits in a nonclinical German adult sample. *Neuropsychobiology, 63*(1), 52–58. doi:10.1159/000322291.

Retz, W., Rosler, M., Supprian, T., Retz-Junginger, P., & Thome, J. (2003). Dopamine D3 receptor gene polymorphism and violent behavior: Relation to impulsiveness and ADHD-related psychopathology. *Journal of Neural Transmission, 110*(5), 561–572. doi:10.1007/s00702-002-0805-5.

Retz, W., Thome, J., Blocher, D., Baader, M., & Rosler, M. (2002). Association of attention deficit hyperactivity disorder-related psychopathology and personality traits with the serotonin transporter promoter region polymorphism. *Neuroscience Letters, 319*(3), 133–136.

Richards, J. B., Lloyd, D. R., Kuehlewind, B., Militello, L., Paredez, M., Solberg Woods, L., et al. (2013). Strong genetic influences on measures of behavioral-regulation among inbred rat strains. *Genes, Brain, and Behavior, 12*(5), 490–502. doi:10.1111/gbb.12050.

Rietschel, M., Mattheisen, M., Degenhardt, F., Muhleisen, T. W., Kirsch, P., Esslinger, C., et al. (2012). Association between genetic variation in a region on chromosome 11 and schizophrenia in large samples from Europe. *Molecular Psychiatry, 17*(9), 906–917. doi:10.1038/mp.2011.80.

Rife, T., Rasoul, B., Pullen, N., Mitchell, D., Grathwol, K., & Kurth, J. (2009). The effect of a promoter polymorphism on the transcription of nitric oxide synthase 1 and its relevance to Parkinson's disease. *Journal of Neuroscience Research, 87*(10), 2319–2325. doi:10.1002/jnr.22045.

Ripke, S., O'Dushlaine, C., Chambert, K., Moran, J. L., Kahler, A. K., Akterin, S., et al. (2013). Genome-wide association analysis identifies 13 new risk loci for schizophrenia. *Nature Genetics, 45*(10), 1150–1159. doi:10.1038/ng.2742.

Robinson, E. S., Eagle, D. M., Economidou, D., Theobald, D. E., Mar, A. C., Murphy, E. R., et al. (2009). Behavioural characterisation of high impulsivity on the 5-choice serial reaction time task: Specific deficits in 'waiting' versus 'stopping'. *Behavioural Brain Research, 196*(2), 310–316. doi:10.1016/j.bbr.2008.09.021.

Robinson, E. S., Eagle, D. M., Mar, A. C., Bari, A., Banerjee, G., Jiang, X., et al. (2008). Similar effects of the selective noradrenaline reuptake inhibitor atomoxetine on three distinct forms of impulsivity in the rat. *Neuropsychopharmacology, 33*(5), 1028–1037. doi:10.1038/sj.npp.1301487.

Rogers, G., Joyce, P., Mulder, R., Sellman, D., Miller, A., Allington, M., et al. (2004). Association of a duplicated repeat polymorphism in the 5′-untranslated region of the DRD4 gene with novelty seeking. *American Journal of Medical Genetics Part B: Neuropsychiatric Genetics, 126B*(1), 95–98. doi:10.1002/ajmg.b.20133.

Roman, T., Schmitz, M., Polanczyk, G. V., Eizirik, M., Rohde, L. A., & Hutz, M. H. (2002). Further evidence for the association between attention-deficit/hyperactivity disorder and the dopamine-beta-hydroxylase gene. *American Journal of Medical Genetics, 114*(2), 154–158. doi:10.1002/ajmg.10194.

Roy, A., Gorodetsky, E., Yuan, Q., Goldman, D., & Enoch, M. A. (2010). Interaction of FKBP5, a stress-related gene, with childhood trauma increases the risk for attempting suicide. *Neuropsychopharmacology, 35*(8), 1674–1683. doi:10.1038/npp.2009.236.

Roy, A., Hodgkinson, C. A., Deluca, V., Goldman, D., & Enoch, M. A. (2012). Two HPA axis genes, CRHBP and FKBP5, interact with childhood trauma to increase the risk for suicidal behavior. *Journal of Psychiatric Research, 46*(1), 72–79. doi:10.1016/j.jpsychires.2011.09.009.

Sabol, S. Z., Hu, S., & Hamer, D. (1998). A functional polymorphism in the monoamine oxidase A gene promoter. *Human Genetics, 103*(3), 273–279.

Sakado, K., Sakado, M., Muratake, T., Mundt, C., & Someya, T. (2003). A psychometrically derived impulsive trait related to a polymorphism in the serotonin transporter gene-linked polymorphic region (5-HTTLPR) in a Japanese nonclinical population: Assessment by the Barratt impulsiveness scale (BIS). *American Journal of Medical Genetics Part B: Neuropsychiatric Genetics, 121B*(1), 71–75. doi:10.1002/ajmg.b.20063.

Sanchez-Roige, S., Baro, V., Trick, L., Pena-Oliver, Y., Stephens, D. N., & Duka, T. (2014). Exaggerated waiting impulsivity associated with human binge drinking, and high alcohol consumption in mice. *Neuropsychopharmacology, 39*(13), 2919–2927. doi:10.1038/npp.2014.151.

Scheuch, K., Lautenschlager, M., Grohmann, M., Stahlberg, S., Kirchheiner, J., Zill, P., et al. (2007). Characterization of a functional promoter polymorphism of the human tryptophan hydroxylase 2 gene in serotonergic raphe neurons. *Biological Psychiatry, 62*(11), 1288–1294. doi:10.1016/j.biopsych.2007.01.015.

Schneider, T., Ilott, N., Brolese, G., Bizarro, L., Asherson, P. J., & Stolerman, I. P. (2011). Prenatal exposure to nicotine impairs performance of the 5-choice serial reaction time task in adult rats. *Neuropsychopharmacology, 36*(5), 1114–1125.

Schork, N. J., Murray, S. S., Frazer, K. A., & Topol, E. J. (2009). Common vs. rare allele hypotheses for complex diseases. *Current Opinion in Genetics and Development, 19*(3), 212–219. doi:10.1016/j.gde.2009.04.010.

Sergeant, J. A., & Scholten, C. A. (1985). On resource strategy limitations in hyperactivity: Cognitive impulsivity reconsidered. *The Journal of Child Psychology and Psychiatry, 26*(1), 97–109.

Serreau, P., Chabout, J., Suarez, S. V., Naude, J., & Granon, S. (2011). Beta2-containing neuronal nicotinic receptors as major actors in the flexible choice between conflicting motivations. *Behavioural Brain Research, 225*(1), 151–159. doi:10.1016/j.bbr.2011.07.016.

Serretti, A., Mandelli, L., Giegling, I., Schneider, B., Hartmann, A. M., Schnabel, A., et al. (2007). HTR2C and HTR1A gene variants in German and Italian suicide attempters and completers. *American Journal of Medical Genetics Part B: Neuropsychiatric Genetics, 144B*(3), 291–299. doi:10.1002/ajmg.b.30432.

Shi, Y., Li, Z., Xu, Q., Wang, T., Li, T., Shen, J., et al. (2011). Common variants on 8p12 and 1q24.2 confer risk of schizophrenia. *Nature Genetics, 43*(12), 1224–1227. doi:10.1038/ng.980.

Simpson, D., & Plosker, G. L. (2004). Atomoxetine: A review of its use in adults with attention deficit hyperactivity disorder. *Drugs, 64*(2), 205–222.

Slof-Op't Landt, M. C., Bartels, M., Middeldorp, C. M., van Beijsterveldt, C. E., Slagboom, P. E., Boomsma, D. I., et al. (2013). Genetic variation at the TPH2 gene influences impulsivity in addition to eating disorders. *Behavior Genetics, 43*(1), 24–33. doi:10.1007/s10519-012-9569-3.

Smith, C. T., & Boettiger, C. A. (2012). Age modulates the effect of COMT genotype on delay discounting behavior. *Psychopharmacology, 222*(4), 609–617. doi:10.1007/s00213-012-2653-9.

Soeiro-De-Souza, M. G., Stanford, M. S., Bio, D. S., Machado-Vieira, R., & Moreno, R. A. (2013). Association of the COMT Met(1)(5)(8) allele with trait impulsivity in healthy young adults. *Molecular Medicine Reports, 7*(4), 1067–1072. doi:10.3892/mmr.2013.1336.

Soloff, P. H., Chiappetta, L., Mason, N. S., Becker, C., & Price, J. C. (2014). Effects of serotonin-2A receptor binding and gender on personality traits and suicidal behavior in borderline personality disorder. *Psychiatry Research, 222*(3), 140–148. doi:10.1016/j.pscychresns.2014.03.008.

Song, D. H., Jhung, K., Song, J., & Cheon, K. A. (2011). The 1287 G/A polymorphism of the norepinephrine transporter gene (NET) is involved in commission errors in Korean children with attention deficit hyperactivity disorder. *Behavioral and Brain Functions, 7*, 12. doi:10.1186/1744-9081-7-12.

Sonuga-Barke, E. J., Kumsta, R., Schlotz, W., Lasky-Su, J., Marco, R., Miranda, A., et al. (2011). A functional variant of the serotonin transporter gene (SLC6A4) moderates impulsive choice in attention-deficit/hyperactivity disorder boys and siblings. *Biological Psychiatry, 70*(3), 230–236. doi:10.1016/j.biopsych.2011.01.040.

Sonuga-Barke, E. J., Taylor, E., Sembi, S., & Smith, J. (1992). Hyperactivity and delay aversion—I. The effect of delay on choice. *The Journal of Child Psychology and Psychiatry, 33*(2), 387–398.

Soubrie, P., Martin, P., el Mestikawy, S., Thiebot, M. H., Simon, P., & Hamon, M. (1986). The lesion of serotonergic neurons does not prevent antidepressant-induced reversal of escape failures produced by inescapable shocks in rats. *Pharmacology, Biochemistry and Behavior, 25*(1), 1–6.

Steckler, T., Sauvage, M., & Holsboer, F. (2000). Glucocorticoid receptor impairment enhances impulsive responding in transgenic mice performing on a simultaneous visual discrimination task. *European Journal of Neuroscience, 12*(7), 2559–2569.

Steiger, H., Joober, R., Israel, M., Young, S. N., Ng Ying Kin, N. M., Gauvin, L., et al. (2005). The 5HTTLPR polymorphism, psychopathologic symptoms, and platelet [3H-] paroxetine binding in bulimic syndromes. *International Journal of Eating Disorders, 37*(1), 57–60. doi:10.1002/eat.20073.

Stoltenberg, S. F., Christ, C. C., & Highland, K. B. (2012). Serotonin system gene polymorphisms are associated with impulsivity in a context dependent manner. *Progress in Neuropsychopharmacology and Biological Psychiatry, 39*(1), 182–191. doi:10.1016/j.pnpbp.2012.06.012.

Stoltenberg, S. F., Glass, J. M., Chermack, S. T., Flynn, H. A., Li, S., Weston, M. E., et al. (2006). Possible association between response inhibition and a variant in the brain-expressed tryptophan hydroxylase-2 gene. *Psychiatric Genetics, 16*(1), 35–38.

Storer, C. L., Dickey, C. A., Galigniana, M. D., Rein, T., & Cox, M. B. (2011). FKBP51 and FKBP52 in signaling and disease. *Trends in Endocrinology and Metabolism, 22*(12), 481–490. doi:10.1016/j.tem.2011.08.001.

Suda, A., Kawanishi, C., Kishida, I., Sato, R., Yamada, T., Nakagawa, M., et al. (2009). Dopamine D2 receptor gene polymorphisms are associated with suicide attempt in the Japanese population. *Neuropsychobiology, 59*(2), 130–134. doi:10.1159/000213566.

Sullivan, M. A., & Rudnik-Levin, F. (2001). Attention deficit/hyperactivity disorder and substance abuse. Diagnostic and therapeutic considerations. *Annals of the New York Academy of Sciences, 931*, 251–270.

Swanson, J. M., & Volkow, N. D. (2009). Psychopharmacology: Concepts and opinions about the use of stimulant medications. *The Journal of Child Psychology and Psychiatry, 50*(1–2), 180–193. doi:10.1111/j.1469-7610.2008.02062.x.

Tobin, H., & Logue, A. W. (1994). Self-control across species (Columba livia, Homo sapiens, and Rattus norvegicus). *Journal of Comparative Psychology, 108*(2), 126–133.

Tochigi, M., Hibino, H., Otowa, T., Ohtani, T., Ebisawa, T., Kato, N., et al. (2006). No association of 5-HT2C, 5-HT6, and tryptophan hydroxylase-1 gene polymorphisms with personality traits in the Japanese population. *Neuroscience Letters, 403*(1–2), 100–102. doi:10.1016/j.neulet.2006.04.020.

Trent, S., & Davies, W. (2012). The influence of sex-linked genetic mechanisms on attention and impulsivity. *Biological Psychology, 89*(1), 1–13. doi:10.1016/j.biopsycho.2011.09.011.

Trent, S., Dean, R., Veit, B., Cassano, T., Bedse, G., Ojarikre, O. A., et al. (2013). Biological mechanisms associated with increased perseveration and hyperactivity in a genetic mouse model of neurodevelopmental disorder. *Psychoneuroendocrinology, 38*(8), 1370–1380. doi:10.1016/j.psyneuen.2012.12.002.

Tsutsui-Kimura, I., Ohmura, Y., Izumi, T., Yamaguchi, T., Yoshida, T., & Yoshioka, M. (2009). The effects of serotonin and/or noradrenaline reuptake inhibitors on impulsive-like action assessed by the three-choice serial reaction time task: A simple and valid model of impulsive action using rats. *Behavioural Pharmacology, 20*(5–6), 474–483. doi:10.1097/FBP.0b013e3283305e65.

Turecki, G., Briere, R., Dewar, K., Antonetti, T., Lesage, A. D., Seguin, M., et al. (1999). Prediction of level of serotonin 2A receptor binding by serotonin receptor 2A genetic variation in post-mortem brain samples from subjects who did or did not commit suicide. *The American Journal of Psychiatry, 156*(9), 1456–1458.

van den Bergh, F. S., Bloemarts, E., Groenink, L., Olivier, B., & Oosting, R. S. (2006). Delay aversion: Effects of 7-OH-DPAT, 5-HT1A/1B-receptor stimulation and D-cycloserine. *Pharmacology, Biochemistry and Behavior, 85*(4), 736–743. doi:10.1016/j.pbb.2006.11.007.

Van Tol, H. H., Wu, C. M., Guan, H. C., Ohara, K., Bunzow, J. R., Civelli, O., et al. (1992). Multiple dopamine D4 receptor variants in the human population. *Nature, 358*(6382), 149–152. doi:10.1038/358149a0.

Varga, G., Szekely, A., Antal, P., Sarkozy, P., Nemoda, Z., Demetrovics, Z., et al. (2012). Additive effects of serotonergic and dopaminergic polymorphisms on trait impulsivity. *American Journal of Medical Genetics Part B: Neuropsychiatric Genetics, 159B*(3), 281–288. doi:10.1002/ajmg.b.32025.

Verdejo-Garcia, A., Lawrence, A. J., & Clark, L. (2008). Impulsivity as a vulnerability marker for substance-use disorders: Review of findings from high-risk research, problem gamblers and genetic association studies. *Neuroscience and Biobehavioral Reviews, 32*(4), 777–810. doi:10.1016/j.neubiorev.2007.11.003.

Videtic, A., Peternelj, T. T., Zupanc, T., Balazic, J., & Komel, R. (2009). Promoter and functional polymorphisms of HTR2C and suicide victims. *Genes, Brain, and Behavior, 8*(5), 541–545. doi:10.1111/j.1601-183X.2009.00505.x.

Villafuerte, S., Heitzeg, M. M., Foley, S., Yau, W. Y., Majczenko, K., Zubieta, J. K., et al. (2012). Impulsiveness and insula activation during reward anticipation are associated with genetic variants in GABRA2 in a family sample enriched for alcoholism. *Molecular Psychiatry, 17*(5), 511–519. doi:10.1038/mp.2011.33.

Villafuerte, S., Strumba, V., Stoltenberg, S. F., Zucker, R. A., & Burmeister, M. (2013). Impulsiveness mediates the association between GABRA2 SNPs and lifetime alcohol problems. *Genes, Brain, and Behavior, 12*(5), 525–531. doi:10.1111/gbb.12039.

Volkow, N. D., Logan, J., Fowler, J. S., Wang, G. J., Gur, R. C., Wong, C., et al. (2000). Association between age-related decline in brain dopamine activity and impairment in frontal and cingulate metabolism. *The American Journal of Psychiatry, 157*(1), 75–80.

Voon, V., Irvine, M. A., Derbyshire, K., Worbe, Y., Lange, I., Abbott, S., et al. (2014). Measuring "waiting" impulsivity in substance addictions and binge eating disorder in a novel analogue of rodent serial reaction time task. *Biological Psychiatry, 75*(2), 148–155. doi:10.1016/j.biopsych.2013.05.013.

Walderhaug, E., Herman, A. I., Magnusson, A., Morgan, M. J., & Landro, N. I. (2010). The short (S) allele of the serotonin transporter polymorphism and acute tryptophan depletion both increase impulsivity in men. *Neuroscience Letters, 473*(3), 208–211. doi:10.1016/j.neulet.2010.02.048.

Wang, G. X., Ma, Y. H., Wang, S. F., Ren, G. F., & Guo, H. (2012). Association of dopaminergic/GABAergic genes with attention deficit hyperactivity disorder in children. *Molecular Medicine Reports, 6*(5), 1093–1098. doi:10.3892/mmr.2012.1028.

Whelan, R., Conrod, P. J., Poline, J. B., Lourdusamy, A., Banaschewski, T., Barker, G. J., et al. (2012). Adolescent impulsivity phenotypes characterized by distinct brain networks. *Nature Neuroscience, 15*(6), 920–925. doi:10.1038/nn.3092.

White, M. J., Morris, C. P., Lawford, B. R., & Young, R. M. (2008). Behavioral phenotypes of impulsivity related to the ANKK1 gene are independent of an acute stressor. *Behavioral and Brain Functions, 4*, 54. doi:10.1186/1744-9081-4-54.

Whiteside, S. P., & Lynam, D. R. (2003). Understanding the role of impulsivity and externalizing psychopathology in alcohol abuse: Application of the UPPS impulsive behavior scale. *Experimental and Clinical Psychopharmacology, 11*(3), 210–217.

Wilhelm, C. J., & Mitchell, S. H. (2009). Strain differences in delay discounting using inbred rats. *Genes, Brain, and Behavior, 8*(4), 426–434. doi:10.1111/j.1601-183X.2009.00484.x.

Willour, V. L., Seifuddin, F., Mahon, P. B., Jancic, D., Pirooznia, M., Steele, J., et al. (2012). A genome-wide association study of attempted suicide. *Molecular Psychiatry, 17*(4), 433–444. doi:10.1038/mp.2011.4.

Wilson, D., da Silva Lobo, D. S., Tavares, H., Gentil, V., & Vallada, H. (2013). Family-based association analysis of serotonin genes in pathological gambling disorder: Evidence of vulnerability risk in the 5HT-2A receptor gene. *Journal of Molecular Neuroscience, 49*(3), 550–553. doi:10.1007/s12031-012-9846-x.

Wilson, T. D., & Dunn, E. W. (2004). Self-knowledge: Its limits, value, and potential for improvement. *Annual Review of Psychology, 55*, 493–518. doi:10.1146/annurev.psych.55.090902.141954.

Winstanley, C. A. (2011). The utility of rat models of impulsivity in developing pharmacotherapies for impulse control disorders. *British Journal of Pharmacology, 164*(4), 1301–1321. doi:10.1111/j.1476-5381.2011.01323.x.

Winstanley, C. A., Dalley, J. W., Theobald, D. E., & Robbins, T. W. (2004a). Fractionating impulsivity: Contrasting effects of central 5-HT depletion on different measures of impulsive behavior. *Neuropsychopharmacology, 29*(7), 1331–1343. doi:10.1038/sj.npp.1300434.

Winstanley, C. A., Theobald, D. E., Dalley, J. W., Glennon, J. C., & Robbins, T. W. (2004b). 5-HT2A and 5-HT2C receptor antagonists have opposing effects on a measure of impulsivity: Interactions with global 5-HT depletion. *Psychopharmacology (Berl), 176*(3-4), 376–385. doi:10.1007/s00213-004-1884-9.

Winstanley, C. A., Eagle, D. M., & Robbins, T. W. (2006). Behavioral models of impulsivity in relation to ADHD: Translation between clinical and preclinical studies. *Clinical Psychology Review, 26*(4), 379–395. doi:10.1016/j.cpr.2006.01.001.

Witte, A. V., Floel, A., Stein, P., Savli, M., Mien, L. K., Wadsak, W., et al. (2009). Aggression is related to frontal serotonin-1A receptor distribution as revealed by PET in healthy subjects. *Human Brain Mapping, 30*(8), 2558–2570. doi:10.1002/hbm.20687.

Wood, R. I., Armstrong, A., Fridkin, V., Shah, V., Najafi, A., & Jakowec, M. (2013). 'Roid rage in rats? Testosterone effects on aggressive motivation, impulsivity and tyrosine hydroxylase. *Physiology and Behavior, 110–111*, 6–12. doi:10.1016/j.physbeh.2012.12.005.

Wrenn, C. C., Turchi, J. N., Schlosser, S., Dreiling, J. L., Stephenson, D. A., & Crawley, J. N. (2006). Performance of galanin transgenic mice in the 5-choice serial reaction time attentional task. *Pharmacology, Biochemistry and Behavior, 83*(3), 428–440. doi:10.1016/j.pbb.2006.03.003.

Wu, J., Xiao, H., Sun, H., Zou, L., & Zhu, L. Q. (2012). Role of dopamine receptors in ADHD: A systematic meta-analysis. *Molecular Neurobiology, 45*(3), 605–620. doi:10.1007/s12035-012-8278-5.

Xu, X., Brookes, K., Sun, B., Ilott, N., & Asherson, P. (2009). Investigation of the serotonin 2C receptor gene in attention deficit hyperactivity disorder in UK samples. *BMC Research Notes, 2*, 71.

Yan, T. C., Dudley, J. A., Weir, R. K., Grabowska, E. M., Pena-Oliver, Y., Ripley, T. L., et al. (2011). Performance deficits of NK1 receptor knockout mice in the 5-choice serial reaction-time task: Effects of d-amphetamine, stress and time of day. *PloS One, 6*(3), e17586. doi:10.1371/journal.pone.0017586.

Yang, L., Qian, Q., Liu, L., Li, H., Faraone, S. V., & Wang, Y. (2013). Adrenergic neurotransmitter system transporter and receptor genes associated with atomoxetine response in attention-deficit hyperactivity disorder children. *Journal of Neural Transmission, 120*(7), 1127–1133. doi:10.1007/s00702-012-0955-z.

Zabetian, C. P., Anderson, G. M., Buxbaum, S. G., Elston, R. C., Ichinose, H., Nagatsu, T., et al. (2001). A quantitative-trait analysis of human plasma-dopamine beta-hydroxylase activity: Evidence for a major functional polymorphism at the DBH locus. *American Journal of Human Genetics, 68*(2), 515–522. doi:10.1086/318198.

Zhao, A. L., Su, L. Y., Zhang, Y. H., Tang, B. S., Luo, X. R., Huang, C. X., et al. (2005). Association analysis of serotonin transporter promoter gene polymorphism with ADHD and related symptomatology. *International Journal of Neuroscience, 115*(8), 1183–1191. doi:10.1080/00207450590914545.

Chapter 4
The Neurobiology and Genetics of Affiliation and Social Bonding in Animal Models

Zoe R. Donaldson and Larry J. Young

Introduction

Two basic processes are required for social affiliation. First and foremost, social relationships require the ability to process social stimuli and to distinguish familiar and unfamiliar individuals. The neuropeptides oxytocin (OXT) and arginine vasopressin (AVP) have been widely implicated in the regulation of social recognition in rodents (Bielsky and Young 2004). In parallel, motivational systems drive the "desire" to engage in social relationships. It has been suggested that social motivation is regulated by two processes, one generating the reinforcing effects of social interactions and the other generating negative emotional states after social isolation or social bond disruption (Panksepp et al. 1997; Nelson and Panksepp 1998; Bosch et al. 2009; Burkett and Young 2012; Resendez and Aragona 2013). Neurochemically, both of these processes appear to be mediated in part by endogenous opioids and the mesocorticolimbic dopamine system (Panksepp et al. 1980; Kelley and Berridge 2002; Dolen et al. 2013). The interaction between these systems modulates both reward and distress (Resendez and Aragona 2013). Ultimately, a complete understanding of social attachments will require an understanding of the interaction between those

Z.R. Donaldson Ph.D. (✉)
Department of Molecular, Cellular and Developmental Biology and Department of Psychology and Neuroscience, University of Colorado, Boulder, CO 80309, USA
e-mail: zoe.donaldson@gmail.com

L.J. Young Ph.D.
Department of Psychiatry and Behavioral Sciences, Center for Translational Social Neuroscience, Yerkes National Primate Research Center, Emory University, 954 Gatewood Rd., Atlanta, GA 30322, USA
e-mail: lyoun03@emory.edu

J.C. Gewirtz, Y.-K. Kim (eds.), *Animal Models of Behavior Genetics*, Advances in Behavior Genetics, DOI 10.1007/978-1-4939-3777-6_4

systems regulating social recognition and those mediating social distress (Johnson and Young 2015).

While affiliation and social bonding has received much attention among studies in human and nonhuman primates, this chapter focuses on more developed mammalian models that have significantly informed our knowledge of the neural substrates of these processes. First we will discuss the neurobiology and genetics of social recognition (the ability to discriminate individuals based on social cues) in rodents. We will then explore three types of social relationships: (1) infant–mother attachment, (2) maternal nurturing, and (3) male–female pair bonding. While a number of molecules, including sex steroids, prolactin, serotonin, and norepinephrine, have been implicated in various forms of affiliative behavior, this review focuses on molecules involved in modulating social memories and mediating the rewarding aspects of bonding. Specifically, we will focus on the role of OXT, AVP, opioids, dopamine, and their respective receptors in animal models of affiliation. We will further highlight genetic mechanisms involved in these processes as well as those generating diversity in affiliative behaviors. The human implications and potential translational applications of this research are discussed briefly at the end of the chapter.

Current Research

The Molecules of Affiliation and Social Bonding

Social attachment requires the motivation to interact with another individual, neural processing of social cues, and the ability to recognize specific individuals. *OXT, AVP, endogenous opioids, and dopamine* mediate different aspects of the neurobiological processes involved in social bonding. AVP and OXT impact social recognition, pup–infant attachment, parental care, and pair bonding in monogamous species. Peripherally, OXT is well known for its role in inducing uterine contractions during parturition and milk ejection during lactation (Burbach et al. 2006). AVP, also known as antidiuretic hormone, stimulates water reabsorption in the kidney and regulates vascular tone. There is a single known OXT receptor (OXTR), which is expressed both in the periphery and the brain. The effects of AVP are mediated by three receptor subtypes: AVPR1A, AVPR1B, and AVPR2. The AVPR1A and AVPR1B subtypes are localized within the brain and are implicated in regulating social behaviors. These receptors are members of the G-protein coupled receptor family. Importantly, OXT and AVP can activate each other's receptors, which has important considerations when interpreting the results of pharmacological and other studies (Chini et al. 2008).

Endogenous opioids have been most thoroughly studied within the context of drug addiction and pain modulation, which closely parallel natural reward and reinforcement underlying the positive affect associated with a specific individual during bonding (Panksepp 1981). Three classes of endogenous opioid peptides—endorphins, enkephalins, and dynorphins—are expressed extensively throughout the brain and act with differing affinities on three main receptor subtypes, designated κ, μ, and δ (Milligan 2005). Recent research suggests that these different receptors play distinct roles in social bonding (Resendez and Aragona 2013).

The mesocorticolimbic dopamine (DA) system has been implicated in drug dependence, and is associated with motivation, reward, reward saliency, and learning (Koob and Nestler 1997; Di Chiara 2002; Kelley and Berridge 2002; Wise 2002; Pierce and Kumaresan 2006). Dopamine is a catecholamine neurotransmitter synthesized within the ventral tegmental area (VTA), substantia nigra, and hypothalamus. Dopaminergic projections from the VTA innervate numerous brain regions, including the nucleus accumbens (NAcc), a brain region that has been heavily studied for its role in reward and reinforcement. VTA-derived dopamine exerts its behavioral effects primarily through the G-protein coupled dopamine D1 and D2 receptors (Wise 2002). The involvement of the opioid and dopamine systems in both addiction and social attachment suggests that there are many common underlying neural pathways between these processes, leading some to suggest that social bonding is a form of addiction (Young et al. 2011; Burkett and Young 2012).

Social Recognition

Social recognition is the ability to distinguish individuals of the same species and is therefore essential for the establishment and maintenance of any social relationship. Social recognition in rodents is mediated by olfactory signals (Beynon and Hurst 2003; Luo et al. 2003). Social recognition behavior tests are based on the phenomenon that rats and mice will spend less time investigating a previously encountered animal than a novel one. Rodents with impaired social recognition behavior do not show differences in attention towards previously encountered versus novel individuals (Thor and Holloway 1982; Engelmann et al. 1995). When interpreting the results of this habituation/dishabituation task, it is important to consider that the task does not distinguish between deficits in sensory discrimination and social recognition memory. While original work pioneered by de Wied and colleagues focused on the role of AVP in nonsocial learning and memory, further refinement suggested a specific role for the peptide in social learning and memory (van Wimersma Greidanus et al. 1983; Engelmann et al. 1996). Enhancing central AVP levels results in a potentiation of social memory in male rats (Dantzer et al. 1987). Conversely, interruptions in AVP signaling in rats abolish social recognition abilities (van Wimersma Greidanus 1982; Le Moal et al. 1987; Landgraf et al. 1995; van Wimersma Greidanus and Maigret 1996). Additionally, pharmacological and genetic manipulation studies suggest that AVPR1A serves as a primary site of action for AVP's role in social recognition. Central infusions of selective AVPR1A antagonists inhibit social recognition, and *Avpr1a* knockout mice display social amnesia but normal memory for nonsocial odorants (Fig. 4.1a) (Engelmann and Landgraf 1994; Everts and Koolhaas 1999; Bielsky et al. 2004, 2005a, b).

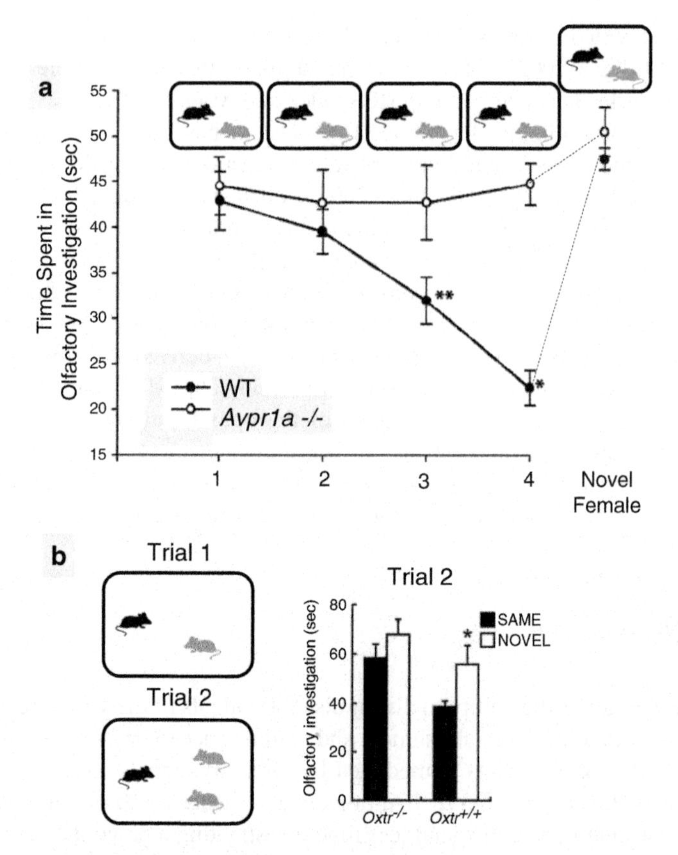

Fig. 4.1 Social recognition is disrupted in *Avpr1a* and *Oxtr* knockout mice. Rodents display a preference to spend time investigating a previously unencountered individual, which is used as a metric for social recognition. (**a**) Habituation/dishabituation task: Shown is the duration of olfactory investigation of male experimental mice toward an ovariectomized female stimulus mouse over four 1-min exposures separated by 10 min intervals. In the final exposure a novel stimulus female is presented. Note that wild-type mice exhibit decreased interest in the stimulus animals over the four trials, indicating that it is recognized as familiar, while *Avpr1a* –/– mice do not show a decrease, indicating a lack of recognition. Data redrawn from (Bielsky et al. 2004). (**b**) Social discrimination test. Male experimental mice are exposed to a single ovariectomized female mouse. A 30 min intertrial interval, the male is re-exposed to the original female (SAME) and a new ovariectomized female (NOVEL). Direct olfactory investigation of male experimental mice toward each female is measured for 5 min. WT male mice spent significantly more time investigating the novel female while *Oxtr* –/– failed to discriminate between novel and familiar females. Data redrawn from (Takayanagi et al. 2005)

Site-specific manipulations have identified the lateral septum as a key region for the action of AVP's effects on social recognition. AVP infused into the lateral septum prolongs social memory whereas selective AVPR1A antagonist and anti-sera infusions in this area block social recognition in rats (Engelmann and Landgraf 1994; Everts and Koolhaas 1999). Likewise, AVPR1A antisense oligonucleotide infused into the lateral septum abolishes social recognition, while viral vector-mediated upregulation of AVPR1A in the lateral septum results in prolonged social memory (Landgraf et al. 1995, 2003), or in the case of *Avpr1a* knockout mice, rescues social recognition in these animals (Bielsky et al. 2005a).

OXT is also involved in social recognition in both male and female rats and mice (Popik and Vetulani 1991; Popik et al. 1992). An OXTR antagonist blocks social recognition abilities (Benelli et al. 1995), and *Oxt* null mutant mice lack social recognition abilities (Ferguson et al. 2000). *Oxtr* null mutant mice display similar impairments in a social discrimination task (Fig. 4.1b) (Takayanagi et al. 2005). *Oxt* null mutant female mice are also unable to differentiate between parasite ridden and healthy males (Choleris et al. 2003; Kavaliers et al. 2003, 2004a, b, 2005a, b, 2006; Temple et al. 2003). The role of the OXT system in social recognition appears to be conserved from rodent to humans (Young 2015). A common polymorphism in the human *OXTR* predicts face recognition abilities in family members with an autistic child (Skuse et al. 2014).

OXTR affects social memory via its actions in multiple brain regions. Infusion of *Oxt* within the medial amygdala of OXT mutant mice rescues normal social recognition, while OXTR antagonist into the medial amygdala of wild-type mice abolishes social recognition (Ferguson et al. 2001). Likewise, deletion of *Oxtr* selectively within the lateral septum abolishes social recognition (Mesic et al. 2015). Septal OXTR also appear to enhance the memory of social interactions regardless of their valence. For instance, OXTR in the lateral septum are required for enhanced fear conditioning in socially defeated mice and social buffering of fear responses (Guzman et al. 2013, 2014). In addition, OXTR on projections from raphe to NAcc are required for the formation of a preference in a social conditioned place preference task through modulation of serotonergic signaling, although it is unclear whether this is due to effects on social memory or on reward (Dolen et al. 2013).

Infant–Mother Attachment

Infant attachment elicits care from a primary caregiver, which increases the offspring's likelihood of survival (Bowlby 1980). When pre-weanling rodents are removed from their mother and littermates, they undergo a number of physiological, distress-related changes including increased heart and respiratory rate, corticosterone release, and the production of ultrasonic vocalizations (Noirot 1972; Hofer 1984; Stuckey et al. 1989). While all three aspects of separation distress have been characterized in a number of species, the most commonly quantified measures of distress are changes in ultrasonic vocalizations (USV) in rodents or distress vocalizations in other species including chicks and puppies (Panksepp et al. 1980; Shair 2014).

There are few studies examining the potential role of AVP in infant–mother attachment. Infant Brattleboro rats, which harbor a mutated *Avp* gene, have an attenuated preference for maternal odors and fail to develop a preference for a maternally paired, nonsocial odor (Nelson and Panksepp 1998). AVP administration results in a decrease in ultrasonic vocalization frequency in isolated rat pups, which is thought to be mediated by AVPR1A (Winslow and Insel 1993). AVPR1B are required for

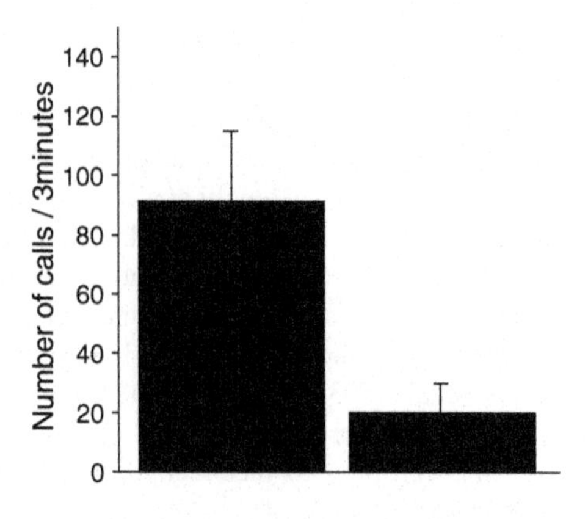

Fig. 4.2 Mouse pups lacking *Oxt* show reduced separation-induced ultrasonic vocalizations. Mouse pups will typically emit ultrasonic vocalizations when separated from the dam. This behavior is significantly reduced in *Oxt* −/− mice. Adapted from (Winslow et al. 2000)

maternal potentiation of USVs but not for isolation-induced calls (Scattoni et al. 2008). More recent work suggests that central administration of AVP in infant mice, unlike rats, blocks the orienting bias to maternally paired, nonsocial odors. However, AVP administered to *Avpr1a* KO mice did not block orienting bias. These differences in behavioral outcome may be related to known species differences in receptor distributions during development, underscoring the importance of AVP in modulating *species-typical* social behaviors (Hammock et al. 2013).

In contrast, there are several studies linking OXT with social attachment in rodent pups. OXT administration attenuates reactions to social separation and decreases milk intake by rat pups (Insel and Winslow 1991; Nelson et al. 1998). As would be expected, mice lacking either *Oxt* or *Oxtr* show less vocal distress during maternal separation (Fig. 4.2) (Winslow et al. 2000; Winslow and Insel 2002; Takayanagi et al. 2005). Finally, an OXTR antagonist blocks the acquisition of maternally associated odor preferences in young rats (Nelson and Panksepp 1996). Together, the results of these experiments suggest that OXT and perhaps AVP are important for modulating the affective aspects of social reward and distress in infants and may also be involved in forming associations between maternal and nonsocial stimuli (Hammock et al. 2013).

Opioids were initially hypothesized to play a role in infant–mother attachment because of the strong similarities between opioid addiction and social dependence. Both display an initial attachment phase, a development of tolerance, and similar withdrawal patterns (Panksepp et al. 1978, 1980). The administration of opioids powerfully attenuates the reaction to social separation, decreasing the rate of isolation-induced distress calls (Kehoe and Blass 1986; Winslow and Insel 1991). Similarly, mutant mice lacking the μ-opioid receptor fail to exhibit signs of distress due to maternal deprivation but show normal levels of distress in response to other stressors (Moles et al. 2004). Milk transfer and somatosensory contact are coupled with opioid release and endogenous opioids facilitate preferences for nonsocial odors paired with tactile stimulation (Blass and Fitzgerald 1988; Smotherman and Robinson 1992; Nelson and Panksepp 1998; Roth and Sullivan 2006). Opioid-mediated social dependence has been further validated by similar findings in a number of different species at different developmental time points including rats, mice, chicks, guinea pigs, and primates (Herman and Panksepp 1978; Panksepp et al. 1978, 1980; Kalin et al. 1988).

Our current knowledge of the role of opioids, OXT, and AVP in infant attachment support the idea of a previously hypothesized socially directed motivational system. The components of this system seem to be important for both mediating the affective state associated with maternal presence or absence, as well as learning maternally- and nest-associated cues. As suggested previously, the use of similar molecular and neural components in other affiliative behaviors discussed below suggests a common brain circuitry that may have its ontogenetic roots in infant attachment (Panksepp et al. 1997; Rilling and Young 2014). Accordingly, it has been hypothesized that disruption of infant attachment may lead to attachment disorders throughout development and later life (Nelson and Panksepp 1998). A variety of early life manipulations in monogamous prairie voles support this hypothesis. Specifically, differences in parental care and pharmacological manipulations of the abovementioned systems impact subsequent social behavior in adulthood, including the propensity to form pair bonds (Bales et al. 2007, 2011; Carter et al. 2009; Ahern et al. 2011; Hostetler et al. 2011; Greenberg et al. 2012; Barrett et al. 2014).

Tactile stimulation mimicking maternal licking and grooming activates OXT neurons in prairie vole infants (Barrett et al. 2015). Daily neonatal social isolations (3 h per day for the first 2 weeks of life) disrupt the ability to form partner preferences later in life in socially monogamous prairie voles. However there is remarkable individual variation in how these short duration social isolations affect later life bonding. Prairie vole pups that express high densities of OXTR in the NAcc are resilient to this paradigm of neglect and form social attachments as adults normally, while those with naturally low densities of receptors who experience social isolations are incapable of forming bonds (Barrett et al. 2015). A melanocortin agonist, which activates OXT neurons and stimulates endogenous OXT release, enhances the ability to form social attachments later in life (Barrett et al. 2014) and rescues the effects of neonatal isolations in pups on later life partner preference formation (Barrett et al. 2015).

Maternal Nurturing

Models of rodent maternal behavior include dams and, in some species, alloparenting behavior where nulliparous animals will display maternal-like behaviors towards pups. Virgin rats typically ignore or avoid pups, but around the time of parturition they begin to show high levels of maternal behavior (Rosenblatt 1969; Bridges 1975; Ross and Young 2009). Maternal behavior in rats and mice is not a form of social bonding, per se, since dams do not discriminate their own pups from novel pups and typically will care for any pup presented to them. Nevertheless, maternal nurturing in rodents is a highly motivated affiliative behavior and is likely to share common neural pathways with maternal behavior in species where mothers do form selective bonds with their offspring. Elucidating the neurobiology underlying the regulation of maternal behavior has been approached from a variety of perspectives, with a focus on various hormones and neurotransmitters released during pregnancy, parturition, and/or lactation (Rosenblatt et al. 1988; Bridges 1990; Keverne and Kendrick 1994; Rosenblatt 1994; McCarthy 1995; Mann and Bridges 2001; Rilling and Young 2014). While steroids and prolactin play a significant role in regulating the onset of maternal behavior, here we focus on the role of OXT/AVP and the classical reward system in accordance with the general aims of this chapter.

The first studies linking OXT with the onset of maternal behavior used rats and were initiated because of the known role of OXT in the stimulation of parturition and lactation. Central OXT administration induces maternal behavior in estrogen-primed virgin female rats (Pedersen and Prange 1979; Pedersen et al. 1992). Conversely, OXT antagonist administration blocks the transition to maternal behavior in parturient rats (Numan 1994 and references therein, Numan 2014). The facilitation of maternal behaviors by OXT involves multiple brain regions. OXT acts at the level of the medial preoptic area (MPOA) to facilitate maternal care. Within the paraventricular nucleus (PVN) and central amygdala (CeA), it facilitates maternal aggression, whereas OXT in the bed nucleus of the stria terminalis (BNST) inhibits maternal aggression (Fig. 4.5) (Bosch and Neumann 2012).

Genetic manipulation of *Oxt* and its receptor in mice have also been used to investigate the role of OXT in maternal behavior. Mutation of *Oxtr* results in a severe disruption in maternal behavior in mice (Takayanagi et al. 2005). In contrast, mutation of *Oxt* leads to only modest deficits in maternal behavior in nulliparous females in one report and no change in maternal behavior in other reports (Winslow and Insel 2002; Ragnauth et al. 2005; Takayanagi et al. 2005; Pedersen et al. 2006). This suggests that in the absence of OXT, AVP may be able to stimulate the OXTR or otherwise independently compensate for the lack of OXT to produce nearly normal levels of maternal behavior in this mouse strain. In addition to these studies on the OXT system genes, CD38 knockout mice, which display a disruption in OXT release from axon terminals, also have impaired maternal behavior, which can be rescued by exogenous OXT (Jin et al. 2007).

Recent studies in voles also suggest a role for OXTR signaling in regulating maternal and alloparental behavior. Juvenile rats, mice, prairie, and meadow voles vary in their propensity to display spontaneous maternal behavior towards pups, or alloparental behavior. Olazabal and Young (2006a, b) found that species with low densities of OXTR in the NAcc display low levels of spontaneous alloparental care, while species with high densities of receptor in the striatum display high levels of alloparental care. Among prairie voles (Fig. 4.3), individual differences in alloparental responsiveness are also positively correlated with OXTR density within the striatum (Fig. 4.4a, b). In addition, infusion of an OXTR antagonist into the NAcc of female prairie voles temporarily blocks spontaneous maternal behavior (Olazabal and Young 2006a, b). These studies suggest that variation in OXTR density in the NAcc may underlie both species and individual differences in pup responsiveness in rodents (Olazabal and Young 2006a, b). This appears to be a developmentally mediated phenotype since overexpression of OXTR in the NAcc of juvenile female prairie voles increases alloparental behaviors towards novel pups (Fig. 4.4c, d), but the same manipulation in adulthood does not (Ross et al. 2009; Keebaugh and Young

Fig. 4.3 Image of prairie voles. Photo credit: Todd Ahern

2011). Furthermore, silencing OXTR in the NAcc beginning at weaning using a viral vector-mediated RNAi approach inhibits alloparental behavior in prairie voles (Keebaugh et al. 2015). Interestingly, the natural variation in OXTR levels in the striatum appears to be robustly regulated by a cis-regulatory element in the *Oxtr* gene since single nucleotide polymorphisms in the prairie vole *Oxtr* account for the majority of variation in OXTR expression (King et al. 2014).

Unlike most rodents, maternal behavior in sheep involves selective social bonding. Precocial lambs are born at the same time of year within a flock. Therefore, the mother must bond with her lamb in the few hours after birth. Ewes will nurse only their own lamb and will reject foreign lambs. OXT infusion can induce acceptance of an unfamiliar lamb even in a nonpregnant ewe (Kendrick et al. 1987). Additionally, OXT release into the olfactory bulb increases GABA, that in conjunction with changes in norepinephrine and acetylcholine signaling appears to tune the olfactory mitral cells to the odor of a specific lamb, which may be responsible for the selective bonding between the mother and her lamb (reviewed by Kendrick et al. 1997). Similar tuning has been observed via oxytocin in the auditory cortex of mouse dams, suggesting that OXT may play a broad role in modulating sensory information processing in the context of maternal behavior (reference PMID 25874674).

Although less well studied, vasopressin is also an important factor in maternal behavior and maternal aggression (Bosch and Neumann 2008, 2012; Bosch 2011, 2013). In rodents, AVP infusion increases maternal care, and elevated levels of AVP are detected in the medial preoptic area and the bed nucleus of the stria terminalis in dams engaged in maternal care (Bosch et al. 2010; Bosch and Neumann 2012). AVPR1A in the MPOA is upregulated during lactation, and blockade or reduction of these receptors impairs maternal care (Pedersen et al. 1994; Bosch and Neumann 2008). In addition, AVPR1A in the central amygdala modulate maternal aggression (Fig. 4.5) (Bosch and Neumann 2010). Finally, antagonism of AVPR1B in rats also impairs maternal care but does not impact maternal aggression (Bayerl et al. 2014).

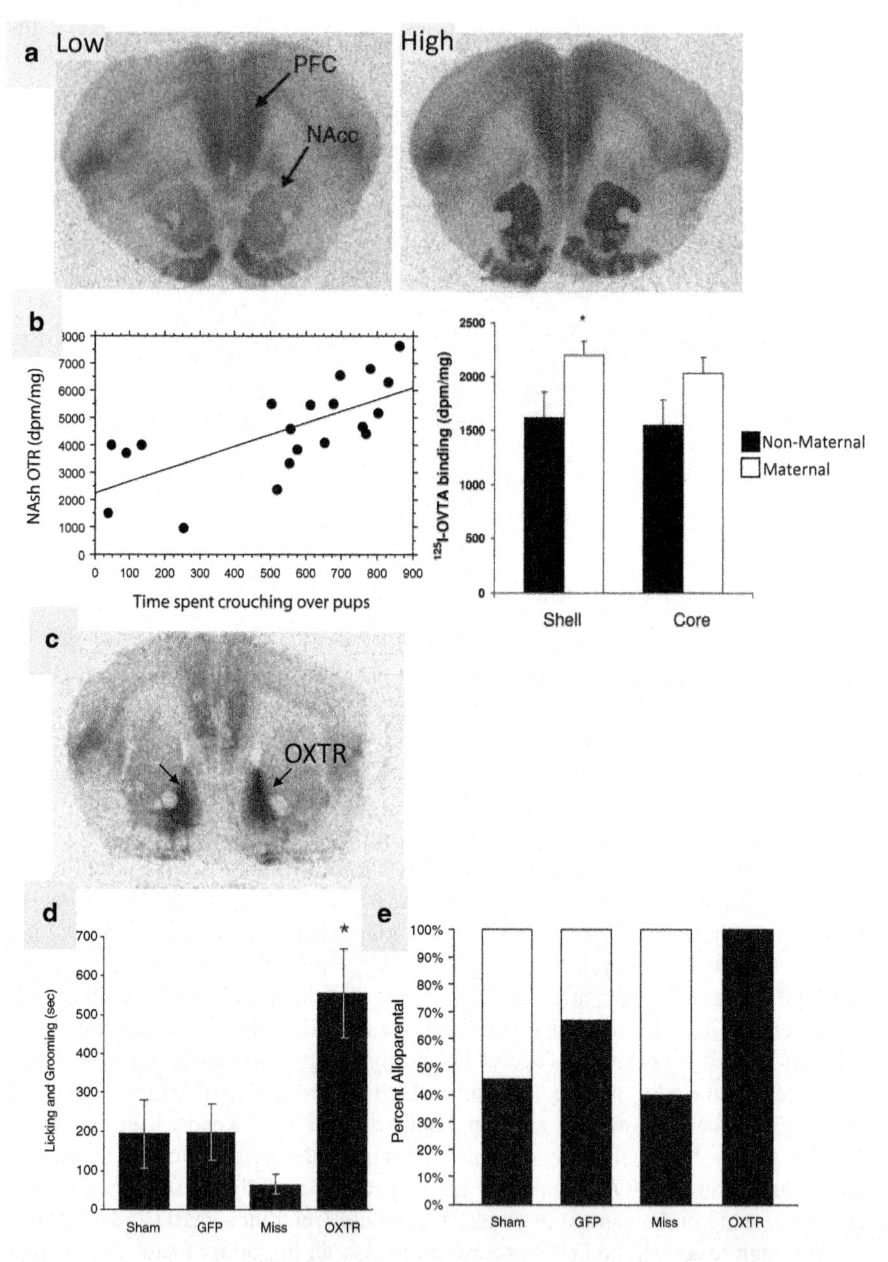

Fig. 4.4 Individual differences in NAcc OXTR modulate alloparental behavior. (**a**) Prairie voles show individual differences in OXTR levels in the nucleus accumbens (NAcc) but not the prefrontal cortex (PFC). (**b**) Differences in spontaneous alloparental behavior in female prairie voles are correlated with NAcc OXTR levels. Virgin females with higher levels of OXTR in the NAcc shell spend more time crouching over pups, a behavior typically associated with parenting. (**c**–**e**) Virally mediated overexpression of OXTR in the NAcc facilitates alloparental behavior. (**d**) Animals with virally increased NAcc OXTR exhibit significantly higher levels of pup-directed licking and grooming. (**e**) Approximately 50 % of virgin female prairie voles will exhibit alloparental behavior. However, when OXTR levels are virally increased in the NAcc shell, 100 % of virgin females engage in alloparental behavior. Adapted from (Olazabal and Young 2006a, b; Keebaugh and Young 2011)

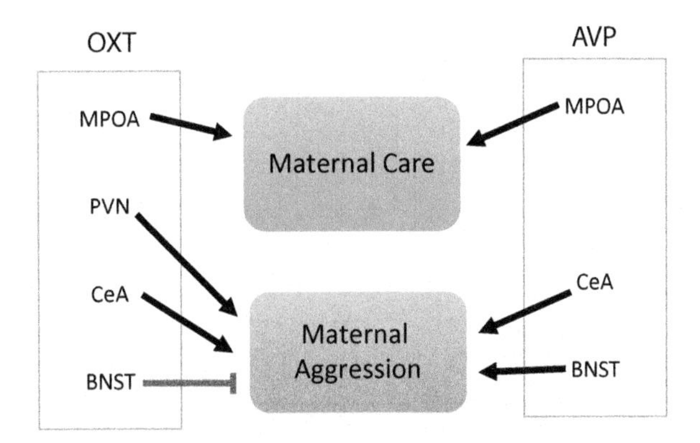

Fig. 4.5 Schematic representation of OXT and AVP actions on maternal behavior via hypothalamic and limbic brain regions. OXT and AVP systems are upregulated while rodent dams are engaged in maternal behaviors and maternal aggression. Both OXT and AVP facilitate maternal nurturing via the MPOA. OXT facilitates maternal aggression via the paraventricular nucleus (PVN) and central amygdala (CeA) whereas OXT signaling in the bed nucleus of the stria terminalis (BNST) inhibits maternal aggression. AVP facilitates maternal aggression via its activity within the CeA and BNST. *Black arrow*: increasing the behavior; *gray line*: reducing the behavior. Adapted from (Bosch 2011, 2013)

The interactions between OXTR, AVPR1A, and AVPR1B as they relate to maternal behavior and pup defense are an ongoing area of research (Fig. 4.5).

Another major area of research has focused on the role of the endogenous reward system in mother–pup bonding (Insel 2003). Post-parturient rats will choose exposure to pups over cocaine, suggesting that they find pups more rewarding than cocaine (Mattson et al. 2001). Similar to the brain's response to drugs of abuse, dopamine is released into the NAcc of a mother rat while interacting with a pup (Hansen et al. 1993). Conversely, dopaminergic-selective lesions of either the dopamine projecting VTA or of its downstream contact, the NAcc, is sufficient to disrupt maternal behavior, specifically impairing motivation to retrieve and care for pups (Keer and Stern 1999; Vernotica et al. 1999). Peripheral administration of a nonselective DA antagonist decreases the performance of active maternal behaviors, but facilitates a switch to nursing (Giordano et al. 1990; Stern and Taylor 1991; Fleming et al. 1994; Stern and Keer 1999; Silva et al. 2001; Byrnes et al. 2002; Li et al. 2004). Neuroimaging studies have examined the neural basis of the intensity of pup-associated reward. They showed that while pup exposure activates the same reward-associated brain regions that cocaine activates in virgin rats, cocaine administered to lactating rats actually suppressed activity within these regions (Ferris et al. 2005).

Unlike the primary role for endogenous opioids in infant–mother attachment, the role of opioids in maternal behavior remains controversial. Female mice mutant for the μ-opioid receptor display normal maternal behaviors (Moles et al. 2004). In rats a μ-opioid receptor agonist decreases levels of maternal behavior (Stafisso-Sandoz et al. 1998). However, the antagonist naltrexone has been found to inhibit the onset of maternal behavior in sheep but restore maternal behavior in rats (Caba et al. 1995; Stafisso-Sandoz et al. 1998).

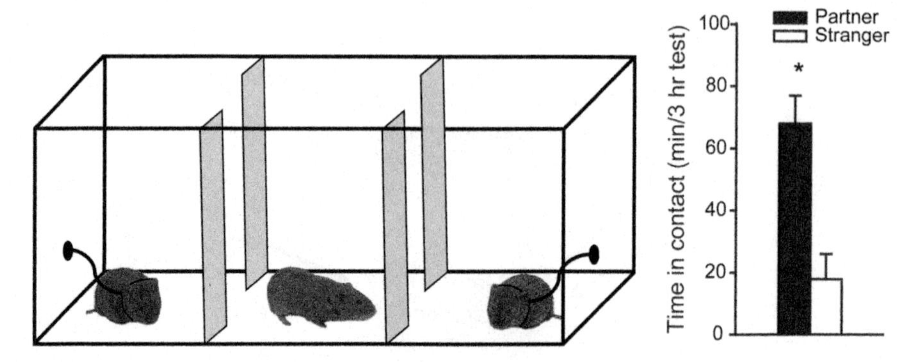

Fig. 4.6 Partner preference test. In the partner preference test, the test animal is allowed to freely roam among 3 chambers with two stimulus animals, typically the partner and a stranger, tethered at opposite ends of the apparatus. An animal that has formed a pair bond will spend more time in direct contact with its partner than the stranger over a 3 h test period. An animal that has failed to form a bond will not show a partner preference

Pair Bonding

A monogamous social structure is characterized by selective affiliation exhibited towards a particular partner, aggression towards opposite gender conspecifics, shared territories and resources, and often biparental care of offspring (Kleiman 1977). The core component of monogamy is the formation of a selective pair bond between a male–female pair, which can be assessed in rodents using a partner preference test (Fig. 4.6). While only 3–5 % of mammalian species are monogamous, the intense social bonds formed between mates of these species have particular relevance to human male–female bonding.

Comparative studies of socially distinct microtine rodents, or voles, have been used as a model system for understanding the neural substrates underlying pair bonding (Carter et al. 1995; Insel and Young 2001; Young and Wang 2004; McGraw et al. 2010; Johnson and Young 2015). In monogamous prairie voles, mating facilitates pair bond formation, although mating is not necessary. Nearly all neuromodulatory systems that have been implicated in pair bonding appear to have a role in pair bond formation and maintenance. However, vasopressinergic and oxytocinergic systems primarily contribute to the processing of social cues necessary for individual recognition in males and females, respectively, while mesolimbic dopamine and striatal opioid signaling are involved in the reward, reinforcement, and motivational aspects of bonding (Young and Wang 2004; Johnson and Young 2015).

OXT and AVP were initially hypothesized to be involved in pair bonding because of their role in maternal behavior, social recognition, and other social behaviors. While both peptides are capable of influencing pair bonding in both males and females when administered exogenously, OXT is often considered to be more important for females and AVP more so for males (Cho et al. 1999). This suggests that evolutionarily, female bonding utilized systems already in place for maternal care while male bonding expanded the role of a peptide that is involved in a variety of male social behaviors including territoriality and aggression (Goodson and Bass 2001). However, recent evidence suggests that endogenous OXT signaling does play a critical role in male pair bonding as well (Johnson et al. 2016). Central infusion of OXT in female prairie voles accelerates pair bonding even in the absence of mating, while an OXTR antagonist prevents mating-induced pair bond formation (Williams et al. 1994; Insel and Hulihan 1995). Comparison of OXT receptor distribution within the brains of prairie voles and the non-monogamous montane voles revealed higher densities of OXTR within certain brain regions including the prefrontal cortex and NAcc (Insel and Shapiro 1992). Mating-induced pair bond formation is blocked by site-specific injection of OXTR antagonist into the prefrontal cortex and NAcc, suggesting that OXTR signaling within the prefrontal cortex and/or NAcc is important for pair bonding (Fig. 4.7) (Young et al. 2001). Virally increased OXTR in the NAcc also enhances pair bonding in female prairie voles (Ross et al. 2009), while silencing OXTR in the NAcc using a virally mediated RNAi inhibits pair bonding (Keebaugh et al. 2015).

In male prairie voles, infusion of AVP facilitates pair bonding, while a selective AVPR1A antagonist blocks pair bond formation after mating (Winslow et al. 1993). Prairie voles have higher levels of AVPR1A binding in the ventral pallidum than do

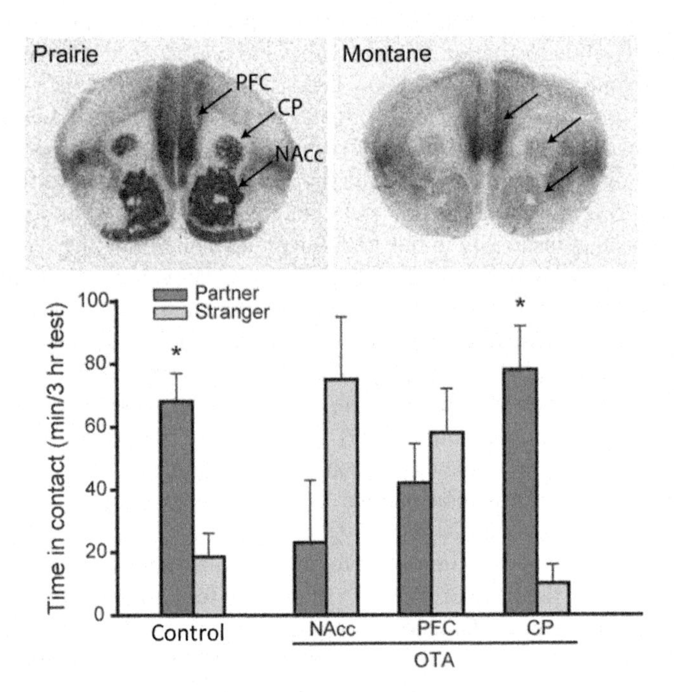

Fig. 4.7 Role of OXTR in partner preference formation. Prairie voles have higher levels of OXTR in the NAcc as compared with meadow voles. Blockade of OXTR in the NAcc or prefrontal cortex (PFC) but not the caudate/putamen (CP) during the initial mating period inhibits pair bond formation. Adapted from (Young et al. 2001)

non-monogamous montane or meadow voles, and blockade of these receptors, as well as those in the lateral septum, blocks pair bond formation (Fig. 4.8a, b) (Insel et al. 1994; Liu et al. 2001; Lim and Young 2004). Increasing *Avpr1a* gene expression in prairie vole males using viral vector gene transfer enhances partner preference formation (Pitkow et al. 2001), while reducing *Avpr1a* mRNA in the ventral pallidum using RNAi inhibits pair bonding (Fig. 4.8c, d) (Barrett et al. 2013). Furthermore, increasing AVPR1A protein levels in the ventral pallidum of the non-monogamous meadow vole results in the manifestation of partner preferences (Lim et al. 2004). Interestingly, the ventral pallidum is a major output of the NAcc, suggesting that different nodes of the same neural circuit regulate pair bond formation in males and females. The role of septal AVP is also consistent with its known role in social recognition.

In addition to the large species differences in V1aR distribution in the brain, there is significant individual variation in V1aR distribution among prairie voles (Phelps and Young 2003), which correlates with natural differences in a number of social behaviors including individual differences in propensities to form a pair bond (Hammock et al. 2005; Hammock and Young 2005). A polymorphic microsatellite consisting of a series of highly repetitive DNA sequences upstream of the vole *Avpr1a* varies strikingly between prairie and montane voles and more subtly amongst individual prairie voles (Fig. 4.9a) (Hammock and Young 2002). These microsatellite-containing polymorphisms in the promoter drive different levels of reporter gene expression in cell culture and are also correlated with individual patterns of AVPR1A and variation in social behavior both between and within species (Fig. 4.9b) (Hammock and Young 2004, 2005). Further, mice in which the endogenous *Avpr1a* promoter was replaced with the corresponding vole sequence and different microsatellite polymorphisms confirmed that these elements directly contribute to differences in receptor patterns in vivo, but in a fairly restricted set of brain regions (Fig. 4.9c) (Donaldson and Young 2013). While these studies support the hypothesis that instability in this highly repetitive microsatellite sequence contributes to some variability in receptor expression in the brain, and consequently to behavior, a survey of other vole species does not support a strict relationship between the length of the microsatellite and social monogamy (Fink et al. 2006). The previously mentioned experiment in mice also suggests that other genetic regulatory elements beyond the proximal promoter are also important modulators of AVPR1A expression (Donaldson and Young 2013). Interestingly, transgenic mice carrying a bacterial artificial chromosome (BAC) containing the human *AVPR1A* gene display a receptor binding pattern more similar to primate expression patterns than the mouse pattern (Charles et al. 2014).

AVPR1A activation is also important for maintenance of the bond via at least two behavioral effects. Blockade of AVPR1A after bond formation inhibits the expression of partner preference (Donaldson et al. 2010), perhaps by acting on memory systems. Second, AVPR1A binding is upregulated in the anterior hypothalamus following pair bond formation, and selective receptor antagonists and virally enhanced expression demonstrate that this is both necessary and sufficient for selective aggression (Gobrogge et al. 2009).

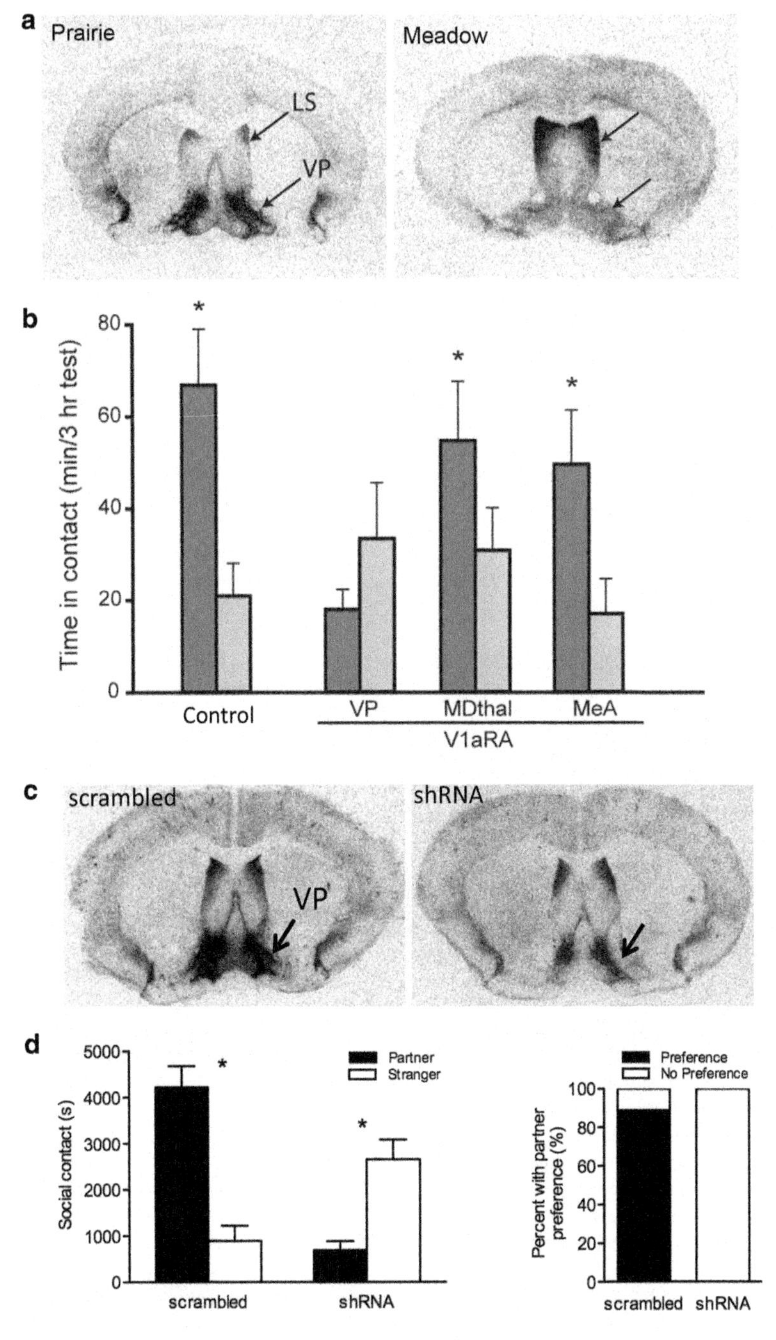

Fig. 4.8 Role of AVPR1A in partner preference formation. (**a**) Prairie voles have higher levels of AVPR1A in the ventral pallidum (VP) as compared with meadow voles. The converse is observed in the lateral septum (LS). (**b**) Blockade of AVPR1A in the VP but not the caudate/putamen (CP) during the initial mating period inhibits pair bond formation. (**c, d**) shRNA-mediated suppression of AVPR1A in the VP inhibits partner preference formation. Adapted from (Lim and Young 2004; Barrett et al. 2013)

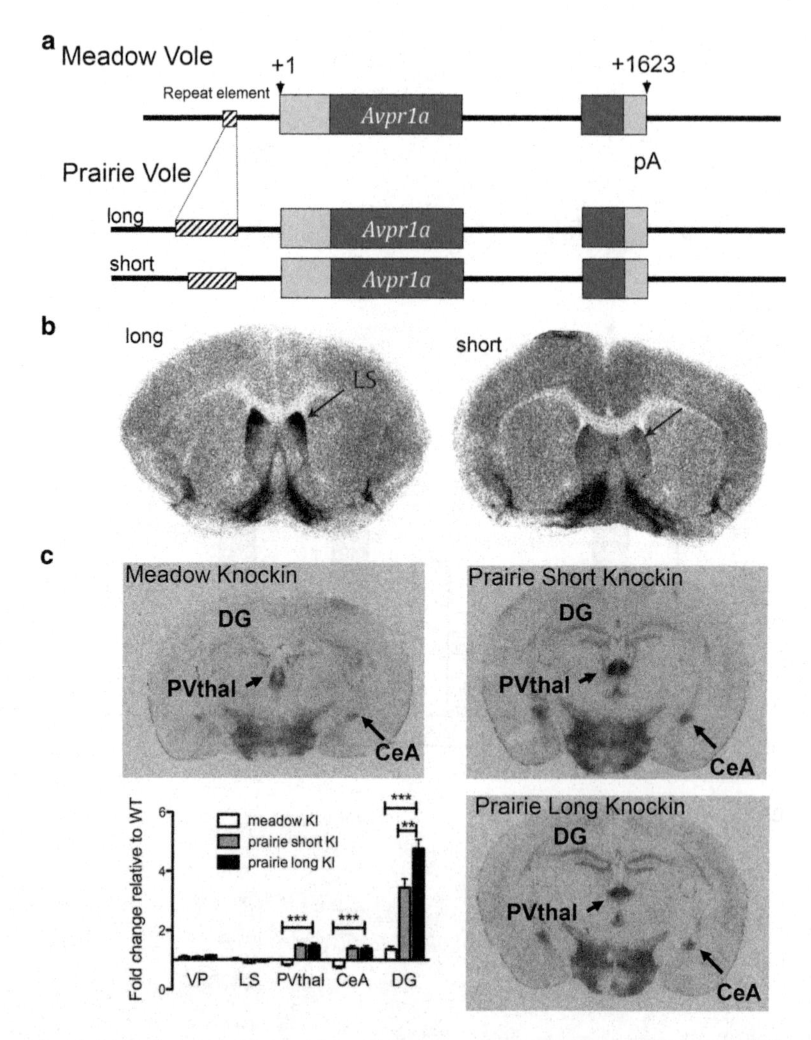

Fig. 4.9 Variation in noncoding repetitive elements modulate species and individual differences in AVPR1A levels. (**a**) Prairie and meadow voles differ in the content of a repeat containing element located upstream of the vole *Avpr1a* transcription start site. This element also exhibits allelic variation across prairie voles. (**b**) Prairie voles bred to homozygosity for longer than average and shorter than average versions of this element demonstrate region-specific differences in AVPR1A levels. (**c**) Mice in which the endogenous *Avpr1a* promoter has been replaced with corresponding vole sequence and different versions of the repeat element demonstrate that diversity in this sequence directly contributes to both species and individual differences in AVPR1A expression. In particular, mice carrying the prairie vole version of the repeat element exhibit increased levels of AVPR1A in the dentate gyrus (DG), paraventricular nucleus of the thalamus (PVthal), and central amygdala (CeA) compared with mice carrying the meadow vole version of the repeat element. In addition, intraspecies differences in the length of the prairie vole repeat element confer variation in AVPR1A levels in the dentate gyrus. This indicates that the repeat element directly contributes to both inter- and intra-species variation in neural AVPR1A levels. Adapted from (Hammock and Young 2005; Donaldson and Young 2013)

Dopamine also acts in the NAcc to impact pair bond formation and maintenance. Mating induces dopaminergic activity in the NAcc and a nonspecific dopamine receptor antagonist blocks pair bond formation (Gingrich et al. 2000; Aragona et al. 2003). D1 and D2 dopamine receptors within the NAcc contribute differently to the formation and maintenance of pair bonds. Activation of D2 receptors facilitates pair bonding while D1 receptor activation blocks it (Fig. 4.10a, b). Furthermore, pair bonding results in an increase in D1 receptor density within the NAcc, which may help to maintain pair bonding by inhibiting pair bond formation with another individual (Fig. 4.10c, d). Behaviorally, the increase in D1 receptors is associated with increased aggressive behavior towards novel individuals of the opposite sex, which can be blocked by a D1 receptor antagonist (Fig. 4.10e, f) (Aragona et al. 2006). Thus, increased D1 receptor in the NAcc and increased AVPR1A in the anterior hypothalamus represent two examples of neuroplastic changes that support pair bonding.

In addition, endogenous opioid signaling is an important modulator of pair bond formation and maintenance via parallel processing through different receptors in different parts of the striatum. Preliminary reports showed a decrease in affiliative behaviors of prairie voles after morphine administration (Shapiro et al. 1989), and subsequent studies also showed that systemic application of the nonselective opioid agonist, naltrexone, during an initial mating period showed a dose-dependent effect on partner preference formation, with the higher doses leading to a preference for the stranger (Burkett et al. 2011). Subsequent work suggests that blockade of μ-opioid receptors (MOR) in the dorsal striatum inhibits mating and subsequent partner preference formation while MORs in the dorsomedial NAcc shell inhibit partner preference formation without affecting mating behavior. In addition, blockade of κ-opioid receptors (KOR) in the NAcc shell blocks selective aggression. This leads to a model in which MORs in the dorsal striatum may be involved in the motivational aspects of pair bonding via their effects on mating, while MORs in the dorsomedial NAcc shell mediate pair bonding through the positive hedonics associated with mating, and KORs subsequently act to maintain bond formation via their effects on selective aggression (Burkett et al. 2011; Resendez et al. 2012, 2013; Johnson and Young 2015). Thus parallel processing by multiple endogenous opioid systems coordinate different aspects of pair bonding.

Finally, pair bonding is maintained not only via the hedonic aspects of partner interaction and an increase in selective aggression, but also by the emotionally aversive aspects of being separated from the partner. Prairie voles who have been separated from their mate exhibit increased levels of passive stress-coping in the forced swim test, a classical test of depression-like behavior in rodents (Fig. 4.11a), as well as autonomic imbalance and altered cardiac function (Bosch et al. 2009; McNeal et al. 2014). These behavioral effects are mediated at least in part by changes in the hypothalamic-pituitary axis (Fig. 4.11b). Ventricular infusion of a nonselective corticotropin release factor receptor (CRFR) antagonist blocks the depression-related behavioral changes without affecting the expression of partner preference (Fig. 4.11c) (Bosch et al. 2009). More recent studies suggest that the CRFR2 receptor-mediated social loss-induced depression occurs in concert with changes in the OXT system. CRFR2 is abundantly expressed on OXT neurons in the hypothalamus (Dabrowska et al. 2011).

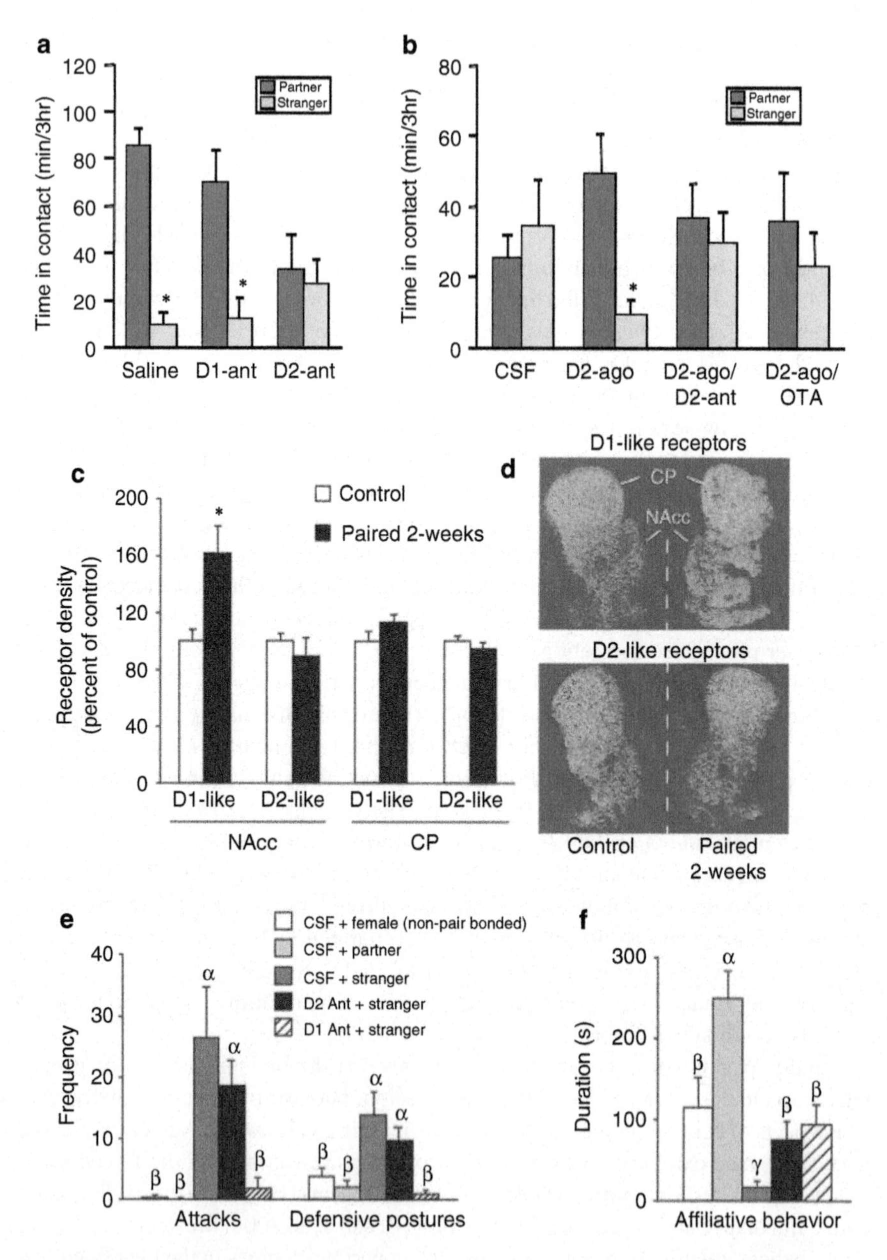

Fig. 4.10 Role of dopaminergic receptors in pair bond formation and maintenance. (**a, b**) D2 receptors facilitate while D1 receptors inhibit pair bond formation. (**c, d**) Animals that have been paired for 2 weeks have increased D1 receptor expression in the NAcc. (**e, f**) Blockade of D1 receptors blocks selective aggression and increases affiliative behaviors towards a novel opposite conspecific. Adapted from (Aragona et al. 2003, 2006)

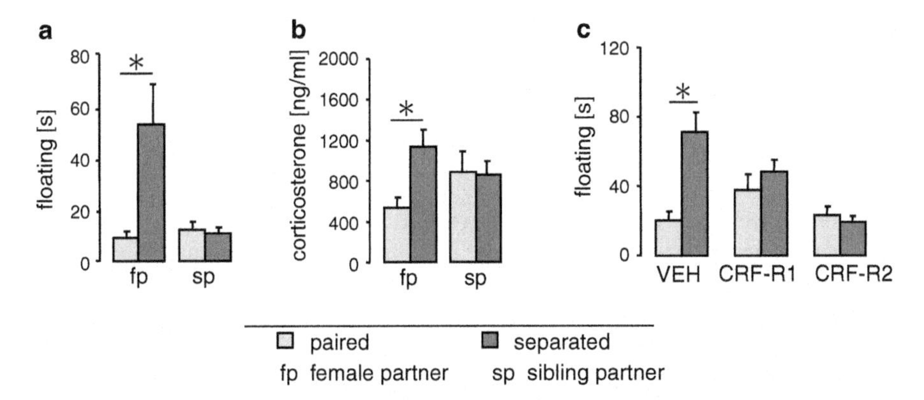

Fig. 4.11 Effects of partner separation in prairie voles. (**a**, **b**) Male prairie voles separated from their female partner but not those separated from a same-sex sibling exhibit increased floating in the forced swim test and elevated corticosterone levels. (**c**) Application of a CRF-R1 or CRF-R2 antagonist blocks separation induced increases in passive coping (Bosch et al. 2009)

CRFR2 agonists, mimicking endogenous release during separation, inhibit OXT release in the shell of the NAcc, while CRFR2 antagonists elevate OXT in the NAcc shell. Finally, direct infusion of OXT into the NAcc abolishes increased passive coping following social loss (Bosch et al. 2016).

Human Implications and Translational Opportunities

Human society is characterized by an incredible complexity and range of social relationships. The bonds we form amongst family members, spouses, caregivers, friends, and coworkers shape our day-to-day interactions and play a key role in our development and overall health. Developmentally, the ability and motivation to seek out social encounters and form bonds is critical for the formation of social skills; early severe social deprivation is associated with attachment disorders (O'Connor et al. 1999; Rutter et al. 1999). As adults, studies have suggested that social support affects both the rate of disease progression and general cardiovascular health (Leserman et al. 1999, 2000; Frasure-Smith et al. 2000). The power of social bonds has been demonstrated both anecdotally by stories of the intense power of love that fill our book and movie shelves, as well as by the emotional significance of social loss. Bereavement or grief due to social loss exhibits many of the same behavioral and physiological changes that characterize depression (Reite and Boccia 1994). In addition, grief is accompanied by intense yearning for the loss of a loved one, suggesting a potential dysregulation of reward processing (Zisook and Shear 2009).

Functional neuroimaging studies have elucidated some of the neural substrates underlying the powerful emotions associated with human social bonding and its loss. Functional imaging of the brains of people viewing pictures of someone they claimed to be desperately in love with revealed activation of many of the same regions that are activated by cocaine administration (Bartels and Zeki 2000; Fisher et al. 2005). Conversely, social exclusion or loss of a loved one activates brain regions typically associated with physical pain, similar to the animal experiments previously discussed in this chapter (Bartels and Zeki 2000; Eisenberger et al. 2003; Najib et al. 2004). In conjunction, they support the recurring hypothesis that bonds are reinforced through activation of brain regions that likely originally evolved as an endogenous reward system and that the dissolution of bonds results in negative affect due to activation of pain-mediating brain regions.

Alternatively, manipulations involving both intranasal OXT and AVP have suggested a general role for these peptides in species-appropriate social behavior. Intranasal OXT impacts social interactions as well as perceptions of social situations (Zink and Meyer-Lindenberg 2012). For instance, OXT increases trust (Kosfeld et al. 2005), facilitates eye gaze (Guastella et al. 2008), and enhances ability to infer another's emotional state (Domes et al. 2007). OXT also increases striatal activation in men during cooperation (Rilling et al. 2012), but not in women (Feng et al. 2015). In relation to pair bonding, polymorphisms in the OXTR gene predict aspects of pair bonding behavior in women (Walum et al. 2012). Intranasal OXT administration in men in monogamous relationships increases the comfortable social distances of an attractive female, perhaps maintaining pair bonds by avoiding close contact with potential mates (Scheele et al. 2012). Furthermore, for men in a relationship, intranasal OXT selectively increases the reported attractiveness of their partner but not other women, and enhances the activation of the NAcc when viewing photographs of their partner (Scheele et al. 2013). Thus there are intriguing parallels between OXT's role in vole pair bonding and human romantic relationships. However, OXT does not necessarily have prosocial effects across all contexts as it increases aggressive actions directed towards an out group and facilitates the sensation of social stress (De Dreu et al. 2010; Eckstein et al. 2014). This reflects that fact that human social behavior is more complex than the relatively simple social behaviors typically assayed in rodents in the laboratory. As such, the underlying neuromolecular systems that influence these behaviors may also exhibit significantly more nuance in humans.

Likewise, there are interesting parallels between animal and human parenting behaviors and the systems described above (Rilling and Young 2014). Regions of the brain that encompass the mesocorticolimbic DA system are activated in both fathers and mothers in response to videos or pictures of their infant. These responses are correlated with positive parenting behaviors (Glocker et al. 2009; Mascaro et al. 2013; Rilling 2013; Michalska et al. 2014). Infant cries activate the insula and prefrontal cortex, which may represent mechanisms underlying aversive reinforcement of parental care (Landi et al. 2011; Rilling 2013; Mascaro et al. 2014). Multiple lines of evidence suggest that OXT modulates aspects of human parenting. Peripheral measures of OXT in mothers and fathers largely correlate with positive parent–infant interactions, although the relationship between peripheral and central OXT release remains unclear (Ross and Young 2009; Feldman et al. 2010; Feldman 2012; Apter-Levi et al. 2014). Intranasal OXT also increases paternal attention, maternal defensiveness, and improves the perceived relationship between depressed mothers and their infant (Naber et al. 2010; Feldman 2012; Mah et al. 2013, 2015; Weisman et al. 2013, 2014). Finally, variation in *OXTR* and *CD38* have been linked with differences in maternal sensitivity (Riem et al. 2011) and levels of parental touch (Feldman 2012) although the samples sizes in these studies are small and require replication.

Less work has been done using intranasal AVP, but the limited studies show that it alters both facial motor patterns and interpretation of facial expressions in a sexually dimorphic fashion. The authors of this work postulate that AVP may mediate sexually dimorphic social strategies within social contexts (Thompson et al. 2006). Polymorphisms in a microsatellite in the human *AVPR1A*, analogous to those found in voles, also predict pair bonding behaviors in men, but not women (Walum et al. 2008). These findings support a broader and potentially nuanced role for these peptides in human social interaction and complex social constructs, such as culture. Interestingly, chimpanzees are polymorphic for the presences of the microsatellite in the *AVPR1A* that has been linked to pair bonding behavior in humans (Donaldson et al. 2008). Genetic studies in chimpanzees reveal that this RS3 microsatellite contributes to variation in dominance personality traits and joint attention in a sex-dependent manner in chimpanzees (Hopkins et al. 2012, 2014).

Further insight into the role of both sociomodulatory peptides and reward systems in human bonding have been garnered from studies of psychiatric diseases that manifest symptoms of altered or absent social interactions. Patients with autism spectrum disorders (ASD) are particularly noted for their deficits in social interaction (DiCicco-Bloom et al. 2006). The apparent lack of interest in social interactions observed amongst autistics has led to a number of hypotheses ranging from excess brain opioids to OXT and AVP deficits. Many of these hypotheses have been experimentally investigated with mixed results. While numerous studies have manipulated endogenous opioid levels, the results of these studies are conflicting and the most substantial improvements were observed in activity levels and attention rather than social interest (reviewed by Gillberg 1995; King and Bostic 2006).

Much recent work has focused on variation in the human *OXTR* gene as a risk factor for autism and the potential use of OXT to treat the social aspects of a number of disorders, most notably high functioning autism (Young and Barrett 2015). Variation in *OXTR* had previously been associated with a number of sociobehavioral traits in humans (Ebstein et al. 2012). A more recent meta-analysis of 3941 individuals with ASD also indicates that at least four SNPs in the *OXTR* locus are overrepresented in ASD (LoParo and Waldman 2015). These genetic polymorphisms may also interact with epigenetic modification of the locus; increased methylation of the *OXTR* locus, along with decreased *OXTR* expression, was observed in both peripheral and prefrontal tissue of ASD patients (Gregory et al. 2009). A remaining question is whether childhood abuse, which is associated with decreased OXT levels in the CSF (Heim et al. 2009), may impact the epigenetic status of the *OXT* or *OXTR* gene (Kumsta et al. 2013). Initial studies also showed that autistic children have lowered peripheral levels of OXT (Green et al. 2001), and OXT infusion reduces repetitive behaviors in adults with autism (Hollander et al. 2003). Across a number of small clinical trials, there is some evidence to suggest that OXT may be valuable for increasing eye gaze and emotion recognition. However, the effect sizes are quite small, and additional work is needed to determine the optimal potential use of OXT as a therapeutic (Preti et al. 2014; Guastella et al. 2015; Young and Barrett 2015). Underpowered studies represent a limitation of many intranasal OXT studies to date, not just those focused on autism (Walum et al. 2016), which should be factored in when interpreting this growing and complex literature.

Conclusions and Future Directions

The research outlined in this chapter suggests that two distinct sets of modulatory factors play a role in social affiliation and bonding across a wide array of social contexts and species. In particular, converging lines of evidence suggest that OXT and AVP modulate aspects of social memory while dopamine and endogenous opioids impact motivation and reward circuits to facilitate social attachment (Fig. 4.12). Given that social interactions play a significant role in species survival for all animals, it is unsurprising that these systems are also highly conserved at a broad level (O'Connell and Hofmann 2012). However, it is important to note that these systems modulate species-typical behaviors, and what is normative social behavior for one species may be very different than what is typical for another. As such, it is important to keep in mind individual, species-level, developmental, and ethological differences when considering the emerging literature on social behavior and its relevance to humans.

There are a number of important questions of future research. For instance, to what extent can we identify general principles versus species-specific adaptations of these systems when considering different social systems (Anacker and Beery 2013)? To what extent and how do these systems exhibit plasticity across behavioral context? This question has particular relevance for understanding the ways in which the human brain may be impacted by bond formation and bond loss.

Finally, it remains to be seen how these systems can be harnessed to provide potential therapeutics. ASD and other disorders that manifest social symptoms are complex diseases with multiple intermediate phenotypes and variable penetrance in different individuals. It is not hard to imagine that while these disorders may result from alterations in the neural components discussed in this paper, different alterations may ultimately result in a similar disease presentation. Therefore, it is important that we understand the various endophenotypes of these diseases and also seek ways to provide individualized treatments, perhaps partially through the use of pharmacogenetic strategies.

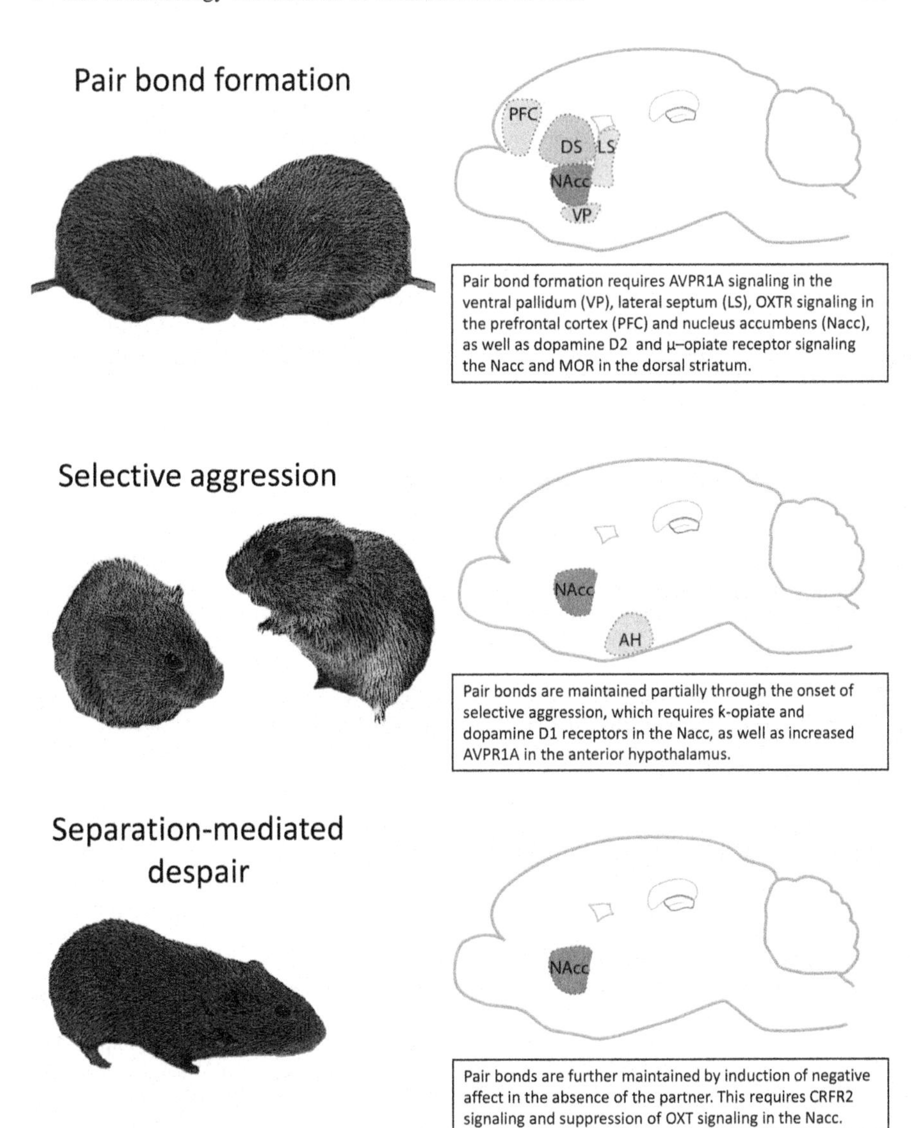

Fig. 4.12 Components of pair bonding in prairie voles

References

Ahern, T. H., Hammock, E. A., & Young, L. J. (2011). Parental division of labor, coordination, and the effects of family structure on parenting in monogamous prairie voles (Microtus ochrogaster). *Developmental Psychobiology, 53*(2), 118–131.

Anacker, A. M., & Beery, A. K. (2013). Life in groups: The roles of oxytocin in mammalian sociality. *Frontiers in Behavioral Neuroscience, 7*, 185.

Apter-Levi, Y., Zagoory-Sharon, O., & Feldman, R. (2014). Oxytocin and vasopressin support distinct configurations of social synchrony. *Brain Research, 1580*, 124–132.

Aragona, B. J., Liu, Y., Curtis, J. T., Stephan, F. K., & Wang, Z. (2003). A critical role for nucleus accumbens dopamine in partner-preference formation in male prairie voles. *The Journal of Neuroscience, 23*(8), 3483–3490.

Aragona, B. J., Liu, Y., Yu, Y. J., Curtis, J. T., Detwiler, J. M., Insel, T. R., et al. (2006). Nucleus accumbens dopamine differentially mediates the formation and maintenance of monogamous pair bonds. *Nature Neuroscience, 9*(1), 133–139.

Bales, K. L., Boone, E., Epperson, P., Hoffman, G., & Carter, C. S. (2011). Are behavioral effects of early experience mediated by oxytocin? *Frontiers in Psychiatry, 2*, 24.

Bales, K. L., van Westerhuyzen, J. A., Lewis-Reese, A. D., Grotte, N. D., Lanter, J. A., & Carter, C. S. (2007). Oxytocin has dose-dependent developmental effects on pair-bonding and alloparental care in female prairie voles. *Hormones and Behavior, 52*(2), 274–279.

Barrett, C. E., Arambula, S. E., & Young, L. J. (2015). The oxytocin system promotes resilience to the effects of neonatal isolation on adult social attachment in female prairie voles. *Translational Psychiatry, 5*, e606.

Barrett, C. E., Keebaugh, A. C., Ahern, T. H., Bass, C. E., Terwilliger, E. F., & Young, L. J. (2013). Variation in vasopressin receptor (Avpr1a) expression creates diversity in behaviors related to monogamy in prairie voles. *Hormones and Behavior, 63*(3), 518–526.

Barrett, C. E., Modi, M. E., Zhang, B. C., Walum, H., Inoue, K., & Young, L. J. (2014). Neonatal melanocortin receptor agonist treatment reduces play fighting and promotes adult attachment in prairie voles in a sex-dependent manner. *Neuropharmacology, 85*, 357–366.

Bartels, A., & Zeki, S. (2000). The neural basis of romantic love. *Neuroreport, 11*(17), 3829–3834.

Bayerl, D. S., Klampfl, S. M., & Bosch, O. J. (2014). Central V1b receptor antagonism in lactating rats: Impairment of maternal care but not of maternal aggression. *Journal of Neuroendocrinology, 26*(12), 918–926.

Benelli, A., Bertolini, A., Poggioli, R., Menozzi, B., Basaglia, R., & Arletti, R. (1995). Polymodal dose-response curve for oxytocin in the social recognition test. *Neuropeptides, 28*(4), 251.

Beynon, R. J., & Hurst, J. L. (2003). Multiple roles of major urinary proteins in the house mouse, Mus domesticus. *Biochemical Society Transactions, 31*(Part 1), 142.

Bielsky, I. F., Hu, S. B., Ren, X., Terwilliger, E. F., & Young, L. J. (2005a). The V1a vasopressin receptor is necessary and sufficient for normal social recognition: A gene replacement study. *Neuron, 47*(4), 503–513.

Bielsky, I. F., Hu, S. B., & Young, L. J. (2005b). Sexual dimorphism in the vasopressin system: Lack of an altered behavioral phenotype in female V1a receptor knockout mice. *Behavioural Brain Research, 164*, 132–136.

Bielsky, I. F., Hu, S.-B., Szegda, K. L., Westphal, H., & Young, L. J. (2004). Profound impairment in social recognition and reduction in anxiety-like behavior in vasopressin V1a receptor knockout mice. *Neuropsychopharmacology, 29*(3), 483.

Bielsky, I. F., & Young, L. J. (2004). Oxytocin, vasopressin, and social recognition in mammals. *Peptides, 25*(9), 1565.

Blass, E. M., & Fitzgerald, E. (1988). Milk-induced analgesia and comforting in 10-day-old rats: Opioid mediation. *Pharmacology, Biochemistry, and Behavior, 29*(1), 9–13.

Bosch, O. J. (2011). Maternal nurturing is dependent on her innate anxiety: The behavioral roles of brain oxytocin and vasopressin. *Hormones and Behavior, 59*(2), 202–212.

Bosch, O. J. (2013). Maternal aggression in rodents: Brain oxytocin and vasopressin mediate pup defence. *Philosophical Transactions of the Royal Society of London: Series B, Biological Sciences, 368*(1631), 20130085.

Bosch, O. J., Nair, H. P., Ahern, T. H., Neumann, I. D., & Young, L. J. (2009). The CRF system mediates increased passive stress-coping behavior following the loss of a bonded partner in a monogamous rodent. *Neuropsychopharmacology, 34*(6), 1406–1415.

Bosch, O. J., & Neumann, I. D. (2008). Brain vasopressin is an important regulator of maternal behavior independent of dams' trait anxiety. *Proceedings of the National Academy of Sciences, 105*(44), 17139–17144.

Bosch, O. J., & Neumann, I. D. (2010). Vasopressin released within the central amygdala promotes maternal aggression. *The European Journal of Neuroscience, 31*(5), 883–891.
Bosch, O. J., & Neumann, I. D. (2012). Both oxytocin and vasopressin are mediators of maternal care and aggression in rodents: From central release to sites of action. *Hormones and Behavior, 61*(3), 293–303.
Bosch, O. J., Pfortsch, J., Beiderbeck, D. I., Landgraf, R., & Neumann, I. D. (2010). Maternal behaviour is associated with vasopressin release in the medial preoptic area and bed nucleus of the stria terminalis in the rat. *Journal of Neuroendocrinology, 22*(5), 420–429.
Bosch, O. J., Dabrowska, J., Modi, M. E., Johnson, Z. V., Keebaugh, A. C., Barrett, C. E., et al. (2016). Oxytocin in the nucleus accumbens shell reverses CRFR2-evoked passive stress-coping after partner loss in monogamous male prairie voles. *Psychoneuroendocrinology, 64*, 66–78.
Bowlby, J. (1980). *Attachment*. New York: Basic.
Bridges, R. S. (1975). Long-term effects of pregnancy and parturition upon maternal responsiveness in the rat. *Physiology & Behavior, 14*(3), 245–249.
Bridges, R. S. (1990). *Endocrine regulation of parental behavior in rodents*. Oxford: Oxford University Press.
Burbach, P., Young, L. J., & Russell, J. (2006). Oxytocin: Synthesis, secretion and reproductive functions. In J. D. Neill (Ed.), *Knobil and Neill's physiology of reproduction* (pp. 3055–3127). St Louis, MO: J. D. Neill, Elsevier.
Burkett, J. P., Spiegel, L. L., Inoue, K., Murphy, A. Z., & Young, L. J. (2011). Activation of mu-opioid receptors in the dorsal striatum is necessary for adult social attachment in monogamous prairie voles. *Neuropsychopharmacology, 36*(11), 2200–2210.
Burkett, J. P., & Young, L. J. (2012). The behavioral, anatomical and pharmacological parallels between social attachment, love and addiction. *Psychopharmacology, 224*(1), 1–26.
Byrnes, E. M., Rigero, B. A., & Bridges, R. S. (2002). Dopamine antagonists during parturition disrupt maternal care and the retention of maternal behavior in rats. *Pharmacology, Biochemistry, and Behavior, 73*(4), 869–875.
Caba, M., Poindron, P., Krehbiel, D., Levy, F., Romeyer, A., & Venier, G. (1995). Naltrexone delays the onset of maternal behavior in primiparous parturient ewes. *Pharmacology, Biochemistry, and Behavior, 52*(4), 743.
Carter, C. S., Boone, E. M., Pournajafi-Nazarloo, H., & Bales, K. L. (2009). Consequences of early experiences and exposure to oxytocin and vasopressin are sexually dimorphic. *Developmental Neuroscience, 31*(4), 332–341.
Carter, C. S., DeVries, A. C., & Getz, L. L. (1995). Physiological substrates of mammalian monogamy: The prairie vole model. *Neuroscience and Biobehavioral Reviews, 19*(2), 303–314.
Charles, R., Sakurai, T., Takahashi, N., Elder, G. A., Gama Sosa, M. A., Young, L. J., et al. (2014). Introduction of the human AVPR1A gene substantially alters brain receptor expression patterns and enhances aspects of social behavior in transgenic mice. *Disease Models & Mechanisms, 7*(8), 1013–1022.
Chini, B., Manning, M., & Guillon, G. (2008). Affinity and efficacy of selective agonists and antagonists for vasopressin and oxytocin receptors: An "easy guide" to receptor pharmacology. *Progress in Brain Research, 170*, 513–517.
Cho, M. M., DeVries, A. C., Williams, J. R., & Carter, C. S. (1999). The effects of oxytocin and vasopressin on partner preferences in male and female prairie voles (Microtus ochrogaster). *Behavioral Neuroscience, 113*, 1071–1079.
Choleris, E., Gustafsson, J.-A., Korach, K. S., Muglia, L. J., Pfaff, D. W., & Ogawa, S. (2003). An estrogen-dependent four-gene micronet regulating social recognition: A study with oxytocin and estrogen receptor-alpha and -beta knockout mice. *Proceedings of the National Academy of Sciences, 100*(10), 6192.
Dabrowska, J., Hazra, R., Ahern, T. H., Guo, J. D., McDonald, A. J., Mascagni, F., et al. (2011). Neuroanatomical evidence for reciprocal regulation of the corticotrophin-releasing factor and oxytocin systems in the hypothalamus and the bed nucleus of the stria terminalis of the rat: Implications for balancing stress and affect. *Psychoneuroendocrinology, 36*(9), 1312–1326.

Dantzer, R., Bluthe, R. M., Koob, G. F., & Le Moal, M. (1987). Modulation of social memory in male rats by neurohypophyseal peptides. *Psychopharmacologia, 91*(3), 363.
De Dreu, C. K., Greer, L. L., Handgraaf, M. J., Shalvi, S., Van Kleef, G. A., Baas, M., et al. (2010). The neuropeptide oxytocin regulates parochial altruism in intergroup conflict among humans. *Science, 328*(5984), 1408–1411.
Di Chiara, G. (2002). Nucleus accumbens shell and core dopamine: Differential role in behavior and addiction. *Behavioural Brain Research, 137*(1–2), 75.
DiCicco-Bloom, E., Lord, C., Courchesne, E., Dager, S., Schmitz, C., Schultz, R., et al. (2006). Developmental neurobiology of autism spectrum disorder: Clinical phenotypes, neurobiologic abnormalities and animal models. *Journal of Neuroscience, 26*, 6897–6906.
Dolen, G., Darvishzadeh, A., Huang, K. W., & Malenka, R. C. (2013). Social reward requires coordinated activity of nucleus accumbens oxytocin and serotonin. *Nature, 501*(7466), 179–184.
Domes, G., Heinrichs, M., Michel, A., Berger, C., & Herpertz, S. C. (2007). Oxytocin improves "mind-reading" in humans. *Biological Psychiatry, 61*(6), 731–733.
Donaldson, Z. R., Kondrashov, F. A., Putnam, A., Bai, Y., Stoinski, T. L., Hammock, E. A., et al. (2008). Evolution of a behavior-linked microsatellite-containing element in the 5′ flanking region of the primate AVPR1A gene. *BMC Evolutionary Biology, 8*, 180.
Donaldson, Z. R., Spiegel, L., & Young, L. J. (2010). Central vasopressin V1a receptor activation is independently necessary for both partner preference formation and expression in socially monogamous male prairie voles. *Behavioral Neuroscience, 124*(1), 159–163.
Donaldson, Z. R., & Young, L. J. (2013). The relative contribution of proximal 5′ flanking sequence and microsatellite variation on brain vasopressin 1a receptor (Avpr1a) gene expression and behavior. *PLoS Genetics, 9*(8), e1003729.
Ebstein, R. P., Knafo, A., Mankuta, D., Chew, S. H., & Lai, P. S. (2012). The contributions of oxytocin and vasopressin pathway genes to human behavior. *Hormones and Behavior, 61*(3), 359–379.
Eckstein, M., Scheele, D., Weber, K., Stoffel-Wagner, B., Maier, W., & Hurlemann, R. (2014). Oxytocin facilitates the sensation of social stress. *Human Brain Mapping, 35*(9), 4741–4750.
Eisenberger, N. I., Lieberman, M. D., & Williams, K. D. (2003). Does rejection hurt? An FMRI study of social exclusion. *Science, 302*(5643), 290–292.
Engelmann, M., & Landgraf, R. (1994). Microdialysis administration of vasopressin into the septum improves social recognition in Brattleboro rats. *Physiology & Behavior, 55*(1), 145.
Engelmann, M., Wotjak, C. T., Neumann, I., Ludwig, M., & Landgraf, R. (1996). Behavioral consequences of intracerebral vasopressin and oxytocin: Focus on learning and memory. *Neuroscience & Biobehavioral Reviews, 20*(3), 341.
Engelmann, M., Wotjak, C. T., & Landgraf, R. (1995). Social discrimination procedure: An alternative method to investigate juvenile recognition abilities in rats. *Physiology & Behavior, 58*(2), 315.
Everts, H. G. J., & Koolhaas, J. M. (1999). Differential modulation of lateral septal vasopressin receptor blockade in spatial learning, social recognition, and anxiety-related behaviors in rats. *Behavioural Brain Research, 99*(1), 7.
Feldman, R. (2012). Oxytocin and social affiliation in humans. *Hormones and Behavior, 61*(3), 380–391.
Feldman, R., Gordon, I., Schneiderman, I., Weisman, O., & Zagoory-Sharon, O. (2010). Natural variations in maternal and paternal care are associated with systematic changes in oxytocin following parent-infant contact. *Psychoneuroendocrinology, 35*(8), 1133–1141.
Feng, C., Hackett, P. D., DeMarco, A. C., Chen, X., Stair, S., Haroon, E., et al. (2015). Oxytocin and vasopressin effects on the neural response to social cooperation are modulated by sex in humans. *Brain Imaging and Behavior, 9*(4), 754–764.
Ferguson, J. N., Aldag, J. M., Insel, T. R., & Young, L. J. (2001). Oxytocin in the medial amygdala is essential for social recognition in the mouse. *The Journal of Neuroscience, 21*(20), 8278.
Ferguson, J. N., Young, L. J., Hearn, E. F., Matzuk, M. M., Insel, T. R., & Winslow, J. T. (2000). Social amnesia in mice lacking the oxytocin gene. *Nature Genetics, 25*(3), 284.

Ferris, C. F., Kulkarni, P., Sullivan, J. M., Harder, J. A., Messenger, T. L., & Febo, M. (2005). Pup suckling is more rewarding than cocaine: Evidence from functional magnetic resonance Imaging and three-dimensional computational analysis. *Journal of Neuroscience, 25*(1), 149–156.
Fink, S., Excoffier, L., & Heckel, G. (2006). Mammalian monogamy is not controlled by a single gene. *Proceedings of the National Academy of Sciences, 103*(29), 10956–10960.
Fisher, H., Aron, A., & Brown, L. L. (2005). Romantic love: An fMRI study of a neural mechanism for mate choice. *The Journal of Comparative Neurology, 493*(1), 58–62.
Fleming, A. S., Korsmit, M., & Deller, M. (1994). Rat pups are potent reinforcers to the maternal animal: Effects of experience, parity, hormones, and dopamine function. *Psychobiology, 22*, 44–53.
Frasure-Smith, N., Lesperance, F., Gravel, G., Masson, A., Juneau, M., Talajic, M., et al. (2000). Social support, depression, and mortality during the first year after myocardial infarction. *Circulation, 101*(16), 1919–1924.
Gillberg, C. (1995). Endogenous opioids and opiate antagonists in autism: Brief review of empirical findings and implications for clinicians. *Developmental Medicine and Child Neurology, 37*(3), 239–245.
Gingrich, B., Liu, Y., Cascio, C., Wang, Z., & Insel, T. R. (2000). Dopamine D2 receptors in the nucleus accumbens are important for social attachment in female prairie voles (Microtus ochrogaster). *Behavioral Neuroscience, 114*(1), 173–183.
Giordano, A. L., Johnson, A. E., & Rosenblatt, J. S. (1990). Haloperidol-induced disruption of retrieval behavior and reversal with apomorphine in lactating rats. *Physiology & Behavior, 48*(1), 211–214.
Glocker, M. L., Langleben, D. D., Ruparel, K., Loughead, J. W., Valdez, J. N., Griffin, M. D., et al. (2009). Baby schema modulates the brain reward system in nulliparous women. *Proceedings of the National Academy of Sciences, 106*(22), 9115–9119.
Gobrogge, K. L., Liu, Y., Young, L. J., & Wang, Z. (2009). Anterior hypothalamic vasopressin regulates pair-bonding and drug-induced aggression in a monogamous rodent. *Proceedings of the National Academy of Sciences, 106*(45), 19144–19149.
Goodson, J. L., & Bass, A. H. (2001). Social behavior functions and related anatomical characteristics of vasotocin/vasopressin systems in vertebrates. *Brain Research Reviews, 35*(3), 246.
Green, L., Fein, D., Modahl, C., Feinstein, C., Waterhouse, L., & Morris, M. (2001). Oxytocin and autistic disorder: Alterations in peptide forms. *Biological Psychiatry, 50*(8), 609–613.
Greenberg, G. D., van Westerhuyzen, J. A., Bales, K. L., & Trainor, B. C. (2012). Is it all in the family? The effects of early social structure on neural-behavioral systems of prairie voles (Microtus ochrogaster). *Neuroscience, 216*, 46–56.
Gregory, S. G., Connelly, J. J., Towers, A. J., Johnson, J., Biscocho, D., Markunas, C. A., et al. (2009). Genomic and epigenetic evidence for oxytocin receptor deficiency in autism. *BMC Medicine, 7*, 62.
Guastella, A. J., Gray, K. M., Rinehart, N. J., Alvares, G. A., Tonge, B. J., Hickie, I. B., et al. (2015). The effects of a course of intranasal oxytocin on social behaviors in youth diagnosed with autism spectrum disorders: A randomized controlled trial. *Journal of Child Psychology and Psychiatry, 56*(4), 444–452.
Guastella, A. J., Mitchell, P. B., & Dadds, M. R. (2008). Oxytocin increases gaze to the eye region of human faces. *Biological Psychiatry, 63*(1), 3–5.
Guzman, Y. F., Tronson, N. C., Jovasevic, V., Sato, K., Guedea, A. L., Mizukami, H., et al. (2013). Fear-enhancing effects of septal oxytocin receptors. *Nature Neuroscience, 16*(9), 1185–1187.
Guzman, Y. F., Tronson, N. C., Sato, K., Mesic, I., Guedea, A. L., Nishimori, K., et al. (2014). Role of oxytocin receptors in modulation of fear by social memory. *Psychopharmacology, 231*(10), 2097–2105.
Hammock, E. A., Law, C. S., & Levitt, P. (2013). Vasopressin eliminates the expression of familiar odor bias in neonatal female mice through V1aR. *Hormones and Behavior, 63*(2), 352–360.
Hammock, E. A., Lim, M. M., Nair, H. P., & Young, L. J. (2005). Association of vasopressin 1a receptor levels with a regulatory microsatellite and behavior. *Genes, Brain, and Behavior, 4*(5), 289–301.

Hammock, E. A., & Young, L. J. (2002). Variation in the vasopressin V1a receptor promoter and expression: Implications for inter- and intraspecific variation in social behaviour. *The European Journal of Neuroscience, 16*(3), 399–402.

Hammock, E. A., & Young, L. J. (2004). Functional microsatellite polymorphism associated with divergent social structure in vole species. *Molecular Biology and Evolution, 21*(6), 1057–1063.

Hammock, E. A., & Young, L. J. (2005). Microsatellite instability generates diversity in brain and sociobehavioral traits. *Science, 308*(5728), 1630–1634.

Hansen, S., Bergvall, A. H., & Nyiredi, S. (1993). Interaction with pups enhances dopamine release in the ventral striatum of maternal rats: A microdialysis study. *Pharmacology, Biochemistry, and Behavior, 45*(3), 673.

Heim, C., Young, L. J., Newport, D. J., Mletzko, T., Miller, A. H., & Nemeroff, C. B. (2009). Lower CSF oxytocin concentrations in women with a history of childhood abuse. *Molecular Psychiatry, 14*(10), 954–958.

Herman, B. H., & Panksepp, J. (1978). Effects of morphine and naloxone on separation distress and approach attachment: Evidence for opiate mediation of social affect. *Pharmacology, Biochemistry, and Behavior, 9*(2), 213.

Hofer, M. A. (1984). Relationships as regulators: A psychobiologic perspective on bereavement. *Psychosomatic Medicine, 46*(3), 183–197.

Hollander, E., Novotny, S., Hanratty, M., Yaffe, R., DeCaria, C. M., Aronowitz, B. R., et al. (2003). Oxytocin infusion reduces repetitive behaviors in adults with autistic and Asperger's disorders. *Neuropsychopharmacology, 28*(1), 193–198.

Hopkins, W. D., Donaldson, Z. R., & Young, L. J. (2012). A polymorphic indel containing the RS3 microsatellite in the 5′ flanking region of the vasopressin V1a receptor gene is associated with chimpanzee (Pan troglodytes) personality. *Genes, Brain, and Behavior, 11*(5), 552–558.

Hopkins, W. D., Keebaugh, A. C., Reamer, L. A., Schaeffer, J., Schapiro, S. J., & Young, L. J. (2014). Genetic influences on receptive joint attention in chimpanzees (Pan troglodytes). *Scientific Reports, 4*, 3774.

Hostetler, C. M., Harkey, S. L., Krzywosinski, T. B., Aragona, B. J., & Bales, K. L. (2011). Neonatal exposure to the D1 agonist SKF38393 inhibits pair bonding in the adult prairie vole. *Behavioural Pharmacology, 22*(7), 703–710.

Insel, T. R. (2003). Is social attachment an addictive disorder? *Physiology & Behavior, 79*(3), 351–357.

Insel, T. R., & Hulihan, T. J. (1995). A gender-specific mechanism for pair bonding: Oxytocin and partner preference formation in monogamous voles. *Behavioral Neuroscience, 109*(4), 782.

Insel, T. R., & Shapiro, L. E. (1992). Oxytocin receptor distribution reflects social organization in monogamous and polygamous voles. *Proceedings of the National Academy of Sciences, 89*(13), 5981.

Insel, T. R., Wang, Z. X., & Ferris, C. F. (1994). Patterns of brain vasopressin receptor distribution associated with social organization in microtine rodents. *The Journal of Neuroscience, 14*(9), 5381–5392.

Insel, T. R., & Winslow, J. T. (1991). Central administration of oxytocin modulates the infant rats response to social isolation. *European Journal of Pharmacology, 203*(1), 149.

Insel, T. R., & Young, L. J. (2001). The neurobiology of attachment. *Nature Reviews Neuroscience, 2*(2), 129–136.

Jin, D., Liu, H. X., Hirai, H., Torashima, T., Nagai, T., Lopatina, O., et al. (2007). CD38 is critical for social behaviour by regulating oxytocin secretion. *Nature, 446*(7131), 41–45.

Johnson, Z. V., & Young, L. J. (2015). Neurobiological mechanisms of social attachment and pair bonding. *Current Opinion in Behavioral Sciences, 3*, 38–44.

Johnson, Z. V., Walum, H., Jamal, Y. A., Xiao, Y., Keebaugh, A. C., Inoue, K., et al. (2016). Central oxytocin receptors mediate mating-induced partner preferences and enhance correlated activation across forebrain nuclei in male prairie voles. *Hormones and Behavior*, 79, 8–17.

Kalin, N. H., Shelton, S. E., & Barksdale, C. M. (1988). Opiate modulation of separation-induced distress in non-human primates. *Brain Research, 440*(2), 285.

Kavaliers, M., Agmo, A., Choleris, E., Gustafsson, J. A., Korach, K. S., Muglia, L. J., et al. (2004a). Oxytocin and estrogen receptor alpha and beta knockout mice provide discriminably different odor cues in behavioral assays. *Genes, Brain, and Behavior, 3*(4), 189–195.
Kavaliers, M., Choleris, E., Agmo, A., Braun, W. J., Colwell, D. D., Muglia, L. J., et al. (2006). Inadvertent social information and the avoidance of parasitized male mice: A role for oxytocin. *Proceedings of the National Academy of Sciences, 103*(11), 4293–4298.
Kavaliers, M., Choleris, E., Agmo, A., & Pfaff, D. W. (2004b). Olfactory-mediated parasite recognition and avoidance: Linking genes to behavior. *Hormones and Behavior, 46*(3), 272–283.
Kavaliers, M., Choleris, E., & Pfaff, D. W. (2005a). Genes, odours and the recognition of parasitized individuals by rodents. *Trends in Parasitology, 21*(9), 423–429.
Kavaliers, M., Choleris, E., & Pfaff, D. W. (2005b). Recognition and avoidance of the odors of parasitized conspecifics and predators: Differential genomic correlates. *Neuroscience and Biobehavioral Reviews, 29*(8), 1347–1359.
Kavaliers, M., Colwell, D. D., Choleris, E., Agmo, A., Muglia, L. J., Ogawa, S., et al. (2003). Impaired discrimination of and aversion to parasitized male odors by female oxytocin knockout mice. *Genes, Brain, and Behavior, 2*(4), 220.
Keebaugh, A. C., Barrett, C. E., LaPrairie, J. L., Jenkins, J. J., & Young, L. J. (2015). RNAi knockdown of oxytocin receptor in the nucleus accumbens inhibits social attachment and parental care in monogamous female prairie voles. *Social Neuroscience, 10*(5), 561–570.
Keebaugh, A. C., & Young, L. J. (2011). Increasing oxytocin receptor expression in the nucleus accumbens of pre-pubertal female prairie voles enhances alloparental responsiveness and partner preference formation as adults. *Hormones and Behavior, 60*(5), 498–504.
Keer, S. E., & Stern, J. M. (1999). Dopamine receptor blockade in the nucleus accumbens inhibits maternal retrieval and licking, but enhances nursing behavior in lactating rats. *Physiology & Behavior, 67*(5), 659.
Kehoe, P., & Blass, E. M. (1986). Opioid-mediation of separation distress in 10-day-old rats: Reversal of stress with maternal stimuli. *Developmental Psychobiology, 19*(4), 385.
Kelley, A. E., & Berridge, K. C. (2002). The neuroscience of natural rewards: Relevance to addictive drugs. *Journal of Neuroscience, 22*(9), 3306–3311.
Kendrick, K. M., Da Costa, A. P., Broad, K. D., Ohkura, S., Guevara, R., Levy, F., et al. (1997). Neural control of maternal behaviour and olfactory recognition of offspring. *Brain Research Bulletin, 44*(4), 383–395.
Kendrick, K. M., Keverne, E. B., & Baldwin, B. A. (1987). Intracerebroventricular oxytocin stimulates maternal behaviour in the sheep. *Neuroendocrinology, 46*(1), 56–61.
Keverne, E. B., & Kendrick, K. M. (1994). Maternal behaviour in sheep and its neuroendocrine regulation. *Acta Paediatrica Supplement, 397*, 47–56.
King, B. H., & Bostic, J. Q. (2006). An update on pharmacologic treatments for autism spectrum disorders. *Child and Adolescent Psychiatric Clinics of North America, 15*(1), 161–175.
King, L. B., Inoue, K., & Young, L. J. (2014). *Genetic variation in the oxytocin receptor exerts robust regionally-selective control of expression to facilitate prairie vole pair bond formation*. Washington, DC: Society for Neuroscience Annual Meeting.
Kleiman, D. (1977). Monogamy in mammals. *Quarterly Review of Biology, 52*, 39–69.
Koob, G. F., & Nestler, E. J. (1997). The neurobiology of drug addiction. *Journal of Neuropsychiatry and Clinical Neurosciences, 9*(3), 482–497.
Kosfeld, M., Heinrichs, M., Zak, P. J., Fischbacher, U., & Fehr, E. (2005). Oxytocin increases trust in humans. *Nature, 435*(7042), 673–676.
Kumsta, R., Hummel, E., Chen, F. S., & Heinrichs, M. (2013). Epigenetic regulation of the oxytocin receptor gene: Implications for behavioral neuroscience. *Frontiers in Neuroscience, 7*, 83.
Landgraf, R., Frank, E., Aldag, J. M., Neumann, I. D., Sharer, C. A., Ren, X., et al. (2003). Viral vector-mediated gene transfer of the vole V1a vasopressin receptor in the rat septum: Improved social discrimination and active social behaviour. *The European Journal of Neuroscience, 18*(2), 403.
Landgraf, R., Gerstberger, R., Montkowski, A., Probst, J. C., Wotjak, C. T., Holsboer, F., et al. (1995). V1 vasopressin receptor antisense oligodeoxynucleotide into septum reduces vasopressin

binding, social discrimination abilities, and anxiety-related behavior in rats. *The Journal of Neuroscience, 15*(6), 4250.

Landi, N., Montoya, J., Kober, H., Rutherford, H. J., Mencl, W. E., Worhunsky, P. D., et al. (2011). Maternal neural responses to infant cries and faces: Relationships with substance use. *Frontiers in Psychiatry, 2*, 32.

Le Moal, M., Dantzer, R., Michaud, B., & Koob, G. F. (1987). Centrally injected arginine vasopressin (AVP) facilitates social memory in rats. *Neuroscience Letters, 77*(3), 353.

Leserman, J., Jackson, E. D., Petitto, J. M., Golden, R. N., Silva, S. G., Perkins, D. O., et al. (1999). Progression to AIDS: The effects of stress, depressive symptoms, and social support. *Psychosomatic Medicine, 61*(3), 397–406.

Leserman, J., Petitto, J. M., Golden, R. N., Gaynes, B. N., Gu, H., Perkins, D. O., et al. (2000). Impact of stressful life events, depression, social support, coping, and cortisol on progression to AIDS. *The American Journal of Psychiatry, 157*(8), 1221–1228.

Li, M., Davidson, P., Budin, R., Kapur, S., & Fleming, A. S. (2004). Effects of typical and atypical antipsychotic drugs on maternal behavior in postpartum female rats. *Schizophrenia Research, 70*(1), 69–80.

Lim, M. M., Wang, Z., Olazabal, D. E., Ren, X., Terwilliger, E. F., & Young, L. J. (2004). Enhanced partner preference in a promiscuous species by manipulating the expression of a single gene. *Nature, 429*(6993), 754–757.

Lim, M. M., & Young, L. J. (2004). Vasopressin-dependent neural circuits underlying pair bond formation in the monogamous prairie vole. *Neuroscience, 125*(1), 35–45.

Liu, Y., Curtis, J. T., & Wang, Z. (2001). Vasopressin in the lateral septum regulates pair bond formation in male prairie voles (Microtus ochrogaster). *Behavioral Neuroscience, 115*(4), 910–919.

LoParo, D., & Waldman, I. D. (2015). The oxytocin receptor gene (OXTR) is associated with autism spectrum disorder: A meta-analysis. *Molecular Psychiatry, 20*(5), 640–646.

Luo, M., Fee, M. S., & Katz, L. C. (2003). Encoding pheromonal signals in the accessory olfactory bulb of behaving mice. *Science, 299*(5610), 1196.

Mah, B. L., Bakermans-Kranenburg, M. J., Van, I. M. H., & Smith, R. (2015). Oxytocin promotes protective behavior in depressed mothers: A pilot study with the enthusiastic stranger paradigm. *Depression and Anxiety, 32*(2), 76–81.

Mah, B. L., Van Ijzendoorn, M. H., Smith, R., & Bakermans-Kranenburg, M. J. (2013). Oxytocin in postnatally depressed mothers: Its influence on mood and expressed emotion. *Progress in Neuropsychopharmacology and Biological Psychiatry, 40*, 267–272.

Mann, P. E., & Bridges, R. S. (2001). Lactogenic hormone regulation of maternal behavior. *Progress in Brain Research, 133*, 251–262.

Mascaro, J. S., Hackett, P. D., Gouzoules, H., Lori, A., & Rilling, J. K. (2014). Behavioral and genetic correlates of the neural response to infant crying among human fathers. *Social Cognitive and Affective Neuroscience, 9*(11), 1704–1712.

Mascaro, J. S., Hackett, P. D., & Rilling, J. K. (2013). Testicular volume is inversely correlated with nurturing-related brain activity in human fathers. *Proceedings of the National Academy of Sciences, 110*(39), 15746–15751.

Mattson, B. J., Williams, S., Rosenblatt, J. S., & Morrell, J. I. (2001). Comparison of two positive reinforcing stimuli: Pups and cocaine throughout the postpartum period. *Behavioral Neuroscience, 115*(3), 683.

McCarthy, M. M. (1995). Estrogen modulation of oxytocin and its relation to behavior. *Advances in Experimental Medicine and Biology, 395*, 235–245.

McGraw, L. A., Davis, J. K., Lowman, J. J., ten Hallers, B. F., Koriabine, M., Young, L. J., et al. (2010). Development of genomic resources for the prairie vole (Microtus ochrogaster): Construction of a BAC library and vole-mouse comparative cytogenetic map. *BMC Genomics, 11*, 70.

McNeal, N., Scotti, M. A., Wardwell, J., Chandler, D. L., Bates, S. L., Larocca, M., et al. (2014). Disruption of social bonds induces behavioral and physiological dysregulation in male and female prairie voles. *Autonomic Neuroscience, 180*, 9–16.

Mesic, I., Guzman, Y. F., Guedea, A. L., Jovasevic, V., Corcoran, K. A., Leaderbrand, K., et al. (2015). Double dissociation of the roles of metabotropic glutamate receptor 5 and oxytocin receptor in discrete social behaviors. *Neuropsychopharmacology, 40*(10), 2337–2346.

Michalska, K. J., Decety, J., Liu, C., Chen, Q., Martz, M. E., Jacob, S., et al. (2014). Genetic imaging of the association of oxytocin receptor gene (OXTR) polymorphisms with positive maternal parenting. *Frontiers in Behavioral Neuroscience, 8*, 21.

Milligan, G. (2005). Opioid receptors and their interacting proteins. *Neuromolecular Medicine, 7*(1–2), 51–59.

Moles, A., Kieffer, B. L., & D'Amato, F. R. (2004). Deficit in attachment behavior in mice lacking the mu-opioid receptor gene. *Science, 304*(5679), 1983–1986.

Naber, F., van Ijzendoorn, M. H., Deschamps, P., van Engeland, H., & Bakermans-Kranenburg, M. J. (2010). Intranasal oxytocin increases fathers' observed responsiveness during play with their children: A double-blind within-subject experiment. *Psychoneuroendocrinology, 35*(10), 1583–1586.

Najib, A., Lorberbaum, J. P., Kose, S., Bohning, D. E., & George, M. S. (2004). Regional brain activity in women grieving a romantic relationship breakup. *The American Journal of Psychiatry, 161*(12), 2245–2256.

Nelson, E. E., Alberts, J. R., Tian, Y., & Verbalis, J. G. (1998). Oxytocin is elevated in plasma of 10-day-old rats following gastric distension. *Developmental Brain Research, 111*(2), 301.

Nelson, E., & Panksepp, J. (1996). Oxytocin mediates acquisition of maternally associated odor preferences in preweanling rat pups. *Behavioral Neuroscience, 110*(3), 583–592.

Nelson, E. E., & Panksepp, J. (1998). Brain substrates of infant-mother attachment: Contributions of opioids, oxytocin, and norepinephrine. *Neuroscience and Biobehavioral Reviews, 22*(3), 437–452.

Noirot, E. (1972). Ultrasounds and maternal behavior in small rodents. *Developmental Psychobiology, 5*(4), 371.

Numan, M. (1994). *Maternal behavior*. New York: Raven.

Numan, M. (2014). *Neurobiology of social behavior*. Oxford: Academic.

O'Connell, L. A., & Hofmann, H. A. (2012). Evolution of a vertebrate social decision-making network. *Science, 336*(6085), 1154–1157.

O'Connor, T. G., Bredenkamp, D., & Rutter, M. (1999). Attachment disturbances and disorders in children exposed to early severe deprivation. *Infant Mental Health Journal, 20*(1), 10–29.

Olazabal, D. E., & Young, L. J. (2006a). Oxytocin receptors in the nucleus accumbens facilitate "spontaneous" maternal behavior in adult female prairie voles. *Neuroscience, 141*, 559–568.

Olazabal, D. E., & Young, L. J. (2006b). Species and individual differences in juvenile female alloparental care are associated with oxytocin receptor density in the striatum and the lateral septum. *Hormones and Behavior, 49*(5), 681–687.

Panksepp, J. (1981). *Brain opioids: A neurochemical substrate for narcotic and social dependence*. London: Academic.

Panksepp, J., Herman, B., Conner, R., Bishop, P., & Scott, J. P. (1978). The biology of social attachments: Opiates alleviate separation distress. *Biological Psychiatry, 13*(5), 607–618.

Panksepp, J., Herman, B. H., Vilberg, T., Bishop, P., & DeEskinazi, F. G. (1980). Endogenous opioids and social behavior. *Neuroscience & Biobehavioral Reviews, 4*(4), 473.

Panksepp, J., Nelson, E., & Bekkedal, M. (1997). Brain systems for the mediation of social separation-distress and social-reward. Evolutionary antecedents and neuropeptide intermediaries. *Annals of the New York Academy of Sciences, 807*, 78–100.

Pedersen, C. A., Caldwell, J. D., Peterson, G., Walker, C. H., & Mason, G. A. (1992). Oxytocin activation of maternal behavior in the rat. *Annals of the New York Academy of Sciences, 652*, 58–69.

Pedersen, C. A., Caldwell, J. D., Walker, C., Ayers, G., & Mason, G. A. (1994). Oxytocin activates the postpartum onset of rat maternal behavior in the ventral tegmental and medial preoptic areas. *Behavioral Neuroscience, 108*(6), 1163–1171.

Pedersen, C. A., & Prange, A. J., Jr. (1979). Induction of maternal behavior in virgin rats after intracerebroventricular administration of oxytocin. *Proceedings of the National Academy of Science, 76*(12), 6661–6665.

Pedersen, C. A., Vadlamudi, S. V., Boccia, M. L., & Amico, J. A. (2006). Maternal behavior deficits in nulliparous oxytocin knockout mice. *Genes, Brain, and Behavior, 5*(3), 274–281.

Phelps, S. M., & Young, L. J. (2003). Extraordinary diversity in vasopressin (V1a) receptor distributions among wild prairie voles (Microtus ochrogaster): Patterns of variation and covariation. *The Journal of Comparative Neurology, 466*(4), 564–576.

Pierce, R. C., & Kumaresan, V. (2006). The mesolimbic dopamine system: The final common pathway for the reinforcing effect of drugs of abuse? *Neuroscience & Biobehavioral Reviews, 30*(2), 215.

Pitkow, L. J., Sharer, C. A., Ren, X., Insel, T. R., Terwilliger, E. F., & Young, L. J. (2001). Facilitation of affiliation and pair-bond formation by vasopressin receptor gene transfer into the ventral forebrain of a monogamous vole. *The Journal of Neuroscience, 21*(18), 7392–7396.

Popik, P., & Vetulani, J. (1991). Opposite action of oxytocin and its peptide antagonists on social memory in rats. *Neuropeptides, 18*(1), 23.

Popik, P., Vetulani, J., & van Ree, J. M. (1992). Low doses of oxytocin facilitate social recognition in rats. *Psychopharmacology, 106*(1), 71–74.

Preti, A., Melis, M., Siddi, S., Vellante, M., Doneddu, G., & Fadda, R. (2014). Oxytocin and autism: A systematic review of randomized controlled trials. *Journal of Child and Adolescent Psychopharmacology, 24*(2), 54–68.

Ragnauth, A. K., Devidze, N., Moy, V., Finley, K., Goodwillie, A., Kow, L. M., et al. (2005). Female oxytocin gene-knockout mice, in a semi-natural environment, display exaggerated aggressive behavior. *Genes, Brain, and Behavior, 4*(4), 229–239.

Reite, M., & Boccia, M. L. (1994). *Attachment in adults*. New York: Guilford.

Resendez, S. L., & Aragona, B. J. (2013). Aversive motivation and the maintenance of monogamous pair bonding. *Reviews in the Neurosciences, 24*(1), 51–60.

Resendez, S. L., Dome, M., Gormley, G., Franco, D., Nevarez, N., Hamid, A. A., et al. (2013). mu-Opioid receptors within subregions of the striatum mediate pair bond formation through parallel yet distinct reward mechanisms. *The Journal of Neuroscience, 33*(21), 9140–9149.

Resendez, S. L., Kuhnmuench, M., Krzywosinski, T., & Aragona, B. J. (2012). kappa-Opioid receptors within the nucleus accumbens shell mediate pair bond maintenance. *The Journal of Neuroscience, 32*(20), 6771–6784.

Riem, M. M., Pieper, S., Out, D., Bakermans-Kranenburg, M. J., & van Ijzendoorn, M. H. (2011). Oxytocin receptor gene and depressive symptoms associated with physiological reactivity to infant crying. *Social Cognitive and Affective Neuroscience, 6*(3), 294–300.

Rilling, J. K. (2013). The neural and hormonal bases of human parental care. *Neuropsychologia, 51*(4), 731–747.

Rilling, J. K., DeMarco, A. C., Hackett, P. D., Thompson, R., Ditzen, B., Patel, R., et al. (2012). Effects of intranasal oxytocin and vasopressin on cooperative behavior and associated brain activity in men. *Psychoneuroendocrinology, 37*(4), 447–461.

Rilling, J. K., & Young, L. J. (2014). The biology of mammalian parenting and its effect on offspring social development. *Science, 345*(6198), 771–776.

Rosenblatt, J. S. (1969). The development of maternal responsiveness in the rat. *The American Journal of Orthopsychiatry, 39*(1), 36–56.

Rosenblatt, J. S. (1994). Psychobiology of maternal behavior: Contribution to the clinical understanding of maternal behavior among humans. *Acta Paediatrica Supplement, 397*, 3–8.

Rosenblatt, J. S., Mayer, A. D., & Giordano, A. L. (1988). Hormonal basis during pregnancy for the onset of maternal behavior in the rat. *Psychoneuroendocrinology, 13*(1–2), 29–46.

Ross, H. E., Freeman, S. M., Spiegel, L. L., Ren, X., Terwilliger, E. F., & Young, L. J. (2009). Variation in oxytocin receptor density in the nucleus accumbens has differential effects on affiliative behaviors in monogamous and polygamous voles. *The Journal of Neuroscience, 29*(5), 1312–1318.

Ross, H. E., & Young, L. J. (2009). Oxytocin and the neural mechanisms regulating social cognition and affiliative behavior. *Frontiers in Neuroendocrinology, 30*(4), 534–547.

Roth, T. L., & Sullivan, R. M. (2006). Examining the role of endogenous opioids in learned odor-stroke associations in infant rats. *Developmental Psychobiology, 48*(1), 71–78.

Rutter, M., Andersen-Wood, L., Beckett, C., Bredenkamp, D., Castle, J., Groothues, C., et al. (1999). Quasi-autistic patterns following severe early global privation. English and Romanian Adoptees (ERA) Study Team. *Journal of Child Psychology and Psychiatry, 40*(4), 537–549.

Scattoni, M. L., McFarlane, H. G., Zhodzishsky, V., Caldwell, H. K., Young, W. S., Ricceri, L., et al. (2008). Reduced ultrasonic vocalizations in vasopressin 1b knockout mice. *Behavioural Brain Research, 187*(2), 371–378.

Scheele, D., Striepens, N., Gunturkun, O., Deutschlander, S., Maier, W., Kendrick, K. M., et al. (2012). Oxytocin modulates social distance between males and females. *The Journal of Neuroscience, 32*(46), 16074–16079.

Scheele, D., Wille, A., Kendrick, K. M., Stoffel-Wagner, B., Becker, B., Gunturkun, O., et al. (2013). Oxytocin enhances brain reward system responses in men viewing the face of their female partner. *Proceedings of the National Academy of Sciences, 110*(50), 20308–20313.

Shair, H. N. (2014). Parental potentiation of vocalization as a marker for filial bonds in infant animals. *Developmental Psychobiology, 56*(8), 1689–1697.

Shapiro, L. E., Meyer, M. E., & Dewsbury, D. A. (1989). Affiliative behavior in voles: effects of morphine, naloxone, and cross-fostering. *Physiology & Behavior, 46*(4), 719–723.

Silva, M. R., Bernardi, M. M., & Felicio, L. F. (2001). Effects of dopamine receptor antagonists on ongoing maternal behavior in rats. *Pharmacology, Biochemistry, and Behavior, 68*(3), 461–468.

Skuse, D. H., Lori, A., Cubells, J. F., Lee, I., Conneely, K. N., Puura, K., et al. (2014). Common polymorphism in the oxytocin receptor gene (OXTR) is associated with human social recognition skills. *Proceedings of the National Academy of Sciences, 111*(5), 1987–1992.

Smotherman, W. P., & Robinson, S. R. (1992). Kappa opioid mediation of fetal responses to milk. *Behavioral Neuroscience, 106*(2), 396–407.

Stafisso-Sandoz, G., Polley, D., Holt, E., Lambert, K. G., & Kinsley, C. H. (1998). Opiate disruption of maternal behavior: Morphine reduces, and naloxone restores, c-fos activity in the medial preoptic area of lactating rats. *Brain Research Bulletin, 45*(3), 307.

Stern, J. M., & Keer, S. E. (1999). Maternal motivation of lactating rats is disrupted by low dosages of haloperidol. *Behavioural Brain Research, 99*(2), 231–239.

Stern, J. M., & Taylor, L. A. (1991). Haloperidol inhibits maternal retrieval and licking, but facilitates nursing behavior and milk ejection in lactating rats. *Journal of Neuroendocrinology, 3*, 591–596.

Stuckey, J., Marra, S., Minor, T., & Insel, T. R. (1989). Changes in mu opiate receptors following inescapable shock. *Brain Research, 476*(1), 167.

Takayanagi, Y., Yoshida, M., Bielsky, I. F., Ross, H. E., Kawamata, M., Onaka, T., et al. (2005). Pervasive social deficits, but normal parturition, in oxytocin receptor-deficient mice. *Proceedings of the National Academy of Sciences, 102*(44), 16096–16101.

Temple, J. L., Young, W. S. I., & Wersinger, S. R. (2003). *Disruption of the genes for either oxytocin or vasopressin 1B receptor alters male-induced pregnancy block (the Bruce effect)*. New Orleans: Society for Neuroscience Meeting.

Thompson, R. R., George, K., Walton, J. C., Orr, S. P., & Benson, J. (2006). Sex-specific influences of vasopressin on human social communication. *Proceedings of the National Academy of Sciences, 103*(20), 7889–7894.

Thor, D. H., & Holloway, W. R. (1982). Social memory of the male laboratory rat. *Journal of Comparative and Physiological Psychology, 96*, 1000–1006.

van Wimersma Greidanus, T. B. (1982). Disturbed behavior and memory of the Brattleboro rat. *Annals of the New York Academy of Sciences, 394*, 655.

van Wimersma Greidanus, T. B., & Maigret, C. (1996). The role of limbic vasopressin and oxytocin in social recognition. *Brain Research, 713*(1–2), 153.

van Wimersma Greidanus, T. B., van Ree, J. M., & de Wied, D. (1983). Vasopressin and memory. *Pharmacology & Therapeutics, 20*(3), 437.

Vernotica, E. M., Rosenblatt, J. S., & Morrell, J. I. (1999). Microinfusion of cocaine into the medial preoptic area or nucleus accumbens transiently impairs maternal behavior in the rat. *Behavioral Neuroscience, 113*(2), 377.

Walum, H., Lichtenstein, P., Neiderhiser, J. M., Reiss, D., Ganiban, J. M., Spotts, E. L., et al. (2012). Variation in the oxytocin receptor gene is associated with pair-bonding and social behavior. *Biological Psychiatry, 71*(5), 419–426.

Walum, H., Waldman, I. D., & Young, L. J. (2016). Statistical and methodological considerations for the interpretation of intranasal oxytocin studies. *Biological Psychiatry, 79*(3), 251–257.

Walum, H., Westberg, L., Henningsson, S., Neiderhiser, J. M., Reiss, D., Igl, W., et al. (2008). Genetic variation in the vasopressin receptor 1a gene (AVPR1A) associates with pair-bonding behavior in humans. *Proceedings of the National Academy of Sciences, 105*(37), 14153–14156.

Weisman, O., Delaherche, E., Rondeau, M., Chetouani, M., Cohen, D., & Feldman, R. (2013). Oxytocin shapes parental motion during father-infant interaction. *Biology Letters, 9*(6), 20130828.

Weisman, O., Zagoory-Sharon, O., & Feldman, R. (2014). Oxytocin administration, salivary testosterone, and father-infant social behavior. *Progress in Neuropsychopharmacology and Biological Psychiatry, 49*, 47–52.

Williams, J. R., Insel, T. R., Harbaugh, C. R., & Carter, C. S. (1994). Oxytocin administered centrally facilitates formation of a partner preference in female prairie voles (Microtus ochrogaster). *Journal of Neuroendocrinology, 6*(3), 247–250.

Winslow, J. T., Hastings, N., Carter, C. S., Harbaugh, C. R., & Insel, T. R. (1993). A role for central vasopressin in pair bonding in monogamous prairie voles. *Nature, 365*(6446), 545–548.

Winslow, J. T., Hearn, E. F., Ferguson, J., Young, L. J., Matzuk, M. M., & Insel, T. R. (2000). Infant vocalization, adult aggression, and fear behavior of an oxytocin null mutant mouse. *Hormones and Behavior, 37*(2), 145–155.

Winslow, J. T., & Insel, T. R. (1991). Endogenous opioids: Do they modulate the rat Pup's response to social isolation? *Behavioral Neuroscience, 105*(2), 253.

Winslow, J. T., & Insel, T. R. (1993). Effects of central vasopressin administration to infant rats. *European Journal of Pharmacology, 233*(1), 101.

Winslow, J. T., & Insel, T. R. (2002). The social deficits of the oxytocin knockout mouse. *Neuropeptides, 36*(2–3), 221–229.

Wise, R. A. (2002). Brain reward circuitry: Insights from unsensed incentives. *Neuron, 36*(2), 229–240.

Young, L. J. (2015). Oxytocin, social cognition and psychiatry. *Neuropsychopharmacology, 40*(1), 243–244.

Young, L. J., & Barrett, C. E. (2015). Neuroscience. Can oxytocin treat autism? *Science, 347*(6224), 825–826.

Young, K. A., Gobrogge, K. L., & Wang, Z. (2011). The role of mesocorticolimbic dopamine in regulating interactions between drugs of abuse and social behavior. *Neuroscience and Biobehavioral Reviews, 35*(3), 498–515.

Young, L. J., Lim, M. M., Gingrich, B., & Insel, T. R. (2001). Cellular mechanisms of social attachment. *Hormones and Behavior, 40*(2), 133.

Young, L. J., & Wang, Z. (2004). The neurobiology of pair bonding. *Nature Neuroscience, 7*(10), 1048–1054.

Zink, C. F., & Meyer-Lindenberg, A. (2012). Human neuroimaging of oxytocin and vasopressin in social cognition. *Hormones and Behavior, 61*(3), 400–409.

Zisook, S., & Shear, K. (2009). Grief and bereavement: What psychiatrists need to know. *World Psychiatry, 8*(2), 67–74.

Part II
Animal Models of Cognitive Processes and Cognitive Decline

Chapter 5
Intellectual Disability

Pierre L. Roubertoux and Michèle Carlier

Introduction

Up to January 2015, more than 84,811 PubMed entries cited intellectual disability (ID). The high prevalence (total number of cases in the population at a given time) of ID and the cost of managing persons with disability explain this high level of interest and also the upsurge in the development of organism models. ID prevalence should be in the vicinity of 21.4 per 1000 assuming that IQ score distribution follows a normal curve, with an average of 100, a standard deviation of 15, and a cutoff at 70 (i.e., 2 standard deviations below the mean). A prevalence of ID at 10–20 per 1000 was reported in a qualitative review of studies most of which were published before 2000 (McDermott et al. 2007; Oeseburg et al. 2011). Lower and higher estimates could be found depending on the populations surveyed and methods used (nationality and age of the population, national registry or not, cross-sectional data on children in mainstream public schools, data from special education schools, etc.). Variability in data collected may also be attributable to revisions of the classification systems. In fact, many epidemiological studies did not take adaptive behavior into account and the sole criterion used to estimate the prevalence of ID was IQ. Some studies that included adaptive behavior as one of the criteria suggested that the prevalence of ID would go down from 2 to 1 % if adaptive behavior were included (Leonard and Wen 2002). A recent meta-analysis (Maulik et al. 2011)

P.L. Roubertoux (✉)
INSERM UMR S910, Génétique Médicale et Génomique Fonctionnelle,
Aix-Marseille Université, Marseille, France
e-mail: Pierre.ROUBERTOUX@univ-amu.fr

M. Carlier
CNRS UMR 7290 Psychologie Cognitive, Fédération de Recherche,
Aix-Marseille Université, Marseille, France
e-mail: michele.carlier@univ-amu.fr

J.C. Gewirtz, Y.-K. Kim (eds.), *Animal Models of Behavior Genetics*,
Advances in Behavior Genetics, DOI 10.1007/978-1-4939-3777-6_5

found a prevalence of 10.37/1000 (95 %CI 9.55–11.18 per 1000) across 52 studies published between 1981 and 2008. The highest estimates were in low- and middle-income countries, and in child and adolescent populations, confirming earlier observational data. The estimates have stabilized around 11 per 1000 in studies published over the past decade (i.e., slightly higher than in earlier studies). A higher overall estimate (14.3/1000) was found in studies based on psychometric scales.

While there are differences in nomenclature and debate about the merits of some terms rather than others, there is a general consensus on the definition of lower levels of intelligence (Fisch 2011). The term "intellectual disability" is used throughout this chapter according to its most recent definition: it is a developmental disability characterized by significant limitations in both intellectual functioning and adaptive behavior (AAIDD 2010; Tasse et al. 2013). The cutoff score for ID on the IQ scale is around 70, although sometimes as high as 75. Developmental disability means that the onset of the impairment must be during the developmental period, i.e., before the age of 18 in western societies. Adaptive behavior includes conceptual skills (e.g., language and number concepts), social skills, and practical skills.

The etiology of ID does not come down to simply to genetics (Carlier and Roubertoux 2014). Many cases of ID of differing degrees of severity are due to infection, to physical, social, or emotional trauma, or to undernutrition. A number of authors have focused on the effects of undernutrition, and this has been confirmed by more recent reviews (Benton 2010; Benton et al. 2010). Here we are considering ID in a genetic context, within the scope of this book. The chapter will first examine the role of organism models in the translational approach to ID. ID nosography covers a wide and heterogeneous range of cognitive profiles with equally heterogeneous brain correlates. The last section will propose a comprehensive strategy for exploring the behavioral neurogenetics of ID.

Contribution of Organism Models to a Translational Strategy for ID

Figure 5.1 summarizes the translational strategy. The expression *translational strategy* means the combination of both the bench and the medical approach, of animal modeling and clinical investigation, with continuous and reciprocal adjustments. The strategy starts by observing the patient, then goes on to identify the genetic correlates of the disease, to find abnormal cell trafficking mechanisms and the pathophysiological processes of the disease, and ends with the discovery of molecules able to cure or alleviate the condition. A translational strategy is a two-way movement between clinical investigations and the experimental or bench approach. The *animal model* strategy, now referred to as the *organism model* strategy, plays a role in translational research which is of crucial importance in the genetic analysis of diseases associated with ID.

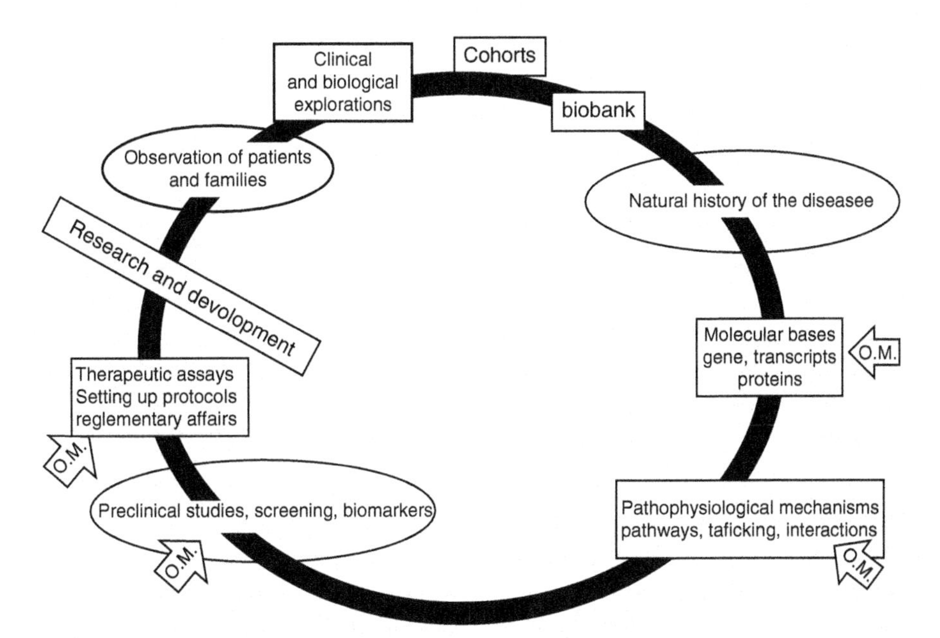

Fig. 5.1 Translational strategy, adapted from Lévy and Roubertoux (2015). O.M. indicates the integration of organism models in the strategy

The study of a disease begins with a careful observation of the patient, and even of his/her relatives, and this is the first stage in the translational approach. The patient can be assigned to a nosography group if he/she has a known disorder or may not yet have a clinical diagnosis if the disease has only recently been described. A cohort must be set up to test the consistency/variability of the observations and is essential for any endeavors to find therapeutic options. The difficulty is to find an adequate number of patients for these purposes. ID is often associated with rare diseases; a disease is said to be rare when prevalence is less than 1:2000 births. Angelman disease, a syndrome with ID, is very rare with a birth incidence of 1 per 12,000 live births (Mertz et al. 2013). Hirschsprung's disease has an ID component of 0.63 per 10,000 (Best et al. 2012), and for Smith–Lemli–Opitz syndrome it is 0.18 per 10,000 live births (Witsch-Baumgartner et al. 2008), hence the difficulty in finding the numbers for research cohorts. The development of international consortiums may help overcome this, but with more refined medical diagnoses through the advances of brain imaging and the widespread use of high-throughput sequencing technologies, several sub-diseases may be defined instead of the single disease initially diagnosed. One example is Smith–Lemli–Opitz syndrome which is associated with six different mutations in the same gene (Nowaczyk and Irons 2012) and that would require sub-cohorts. The situation is even more difficult with newly discovered diseases requiring a long time to establish a cohort. Organism models can be considered as a substitute for cohorts, but under strict conditions that will be specified in the section "Criteria for Developing

Organism Models of ID." It might be argued that it is too time-consuming to develop a mouse model of ID. New experimental models (zebrafish, drosophila, *Caenorhabditis elegans*, yeast, and Induced Pluripotent Stem Cells—IPSCs) are now available and are cheaper and less time-consuming to develop. The second stage in the translational approach is the documentary recording of the natural history of the disease, including the detection of infectious, traumatic, or other underlying events. Molecular events occurring between the DNA template and the functional protein (modification of gene transcription and gene expression) and any deregulation of other genes triggered by genetic modifications are part of this natural history. It is an essential prerequisite to have a description of the pathophysiological mechanisms for a proper understanding of the disease, and this requires investigations at different levels: molecular (pathway, cell trafficking mechanisms and their interactions), biochemical, and physiological. Brain imaging technologies are still not of sufficiently high resolution to show subtle biochemical dysregulation in the human brain. Electrophysiological data (average evoked potentials) is rarely specific to a disease. Further analysis is required and must be sought in nonhuman models. The description of the natural history of a condition includes the developmental aspects of the disease, and this is of particular relevance for ID as many syndromes with ID follow unusual courses. With Rett syndrome, for instance, development seems typical over the first three or so years, but suddenly deteriorates in the third or fourth year (Villard 2007). People with trisomy 21 (Down syndrome) are capable of acquiring knowledge, but around the age of 40, involution occurs for most of them (Teipel and Hampel 2006). Many studies report heterogeneous changes in cognitive performances of patients with William–Beuren syndrome (Karmiloff-Smith et al. 2012). The biological causes of atypical development need to be understood as they may provide a therapeutic target. When a cohort is formed, a biobank (tissue bank) needs to be set up. This is not so important for diseases affecting organs on which biopsies can be performed (e.g., muscle, liver or kidney), but is essential for studying ID-associated brain diseases where it is obviously inconceivable to conduct a biopsy. While banks of cryopreserved brains do exist (Lévy and Roubertoux 2015), logically the availability of brain tissue is proportional to the prevalence of the disease, so brain tissue for most diseases of the brain involving ID is not available. The brains of organism models may be sufficiently similar to patient brains and could thus compensate for the scarce supply of cryopreserved brain tissue, but for the translational approach, postmortem examinations of patient brains are essential. For example, postmortem examination of the brains of patients with Angelman syndrome has observed a reduction in the size of the cortex, but such brains are rarely available (Jay et al. 1991). A mouse model generated with a *UBE3A* maternal-deficient allele, that causes the syndrome in about 10 % of cases, confirmed the observation of reduced spine density of basal dendrites in pyramidal neurons of the visual cortex and pyramidal neurons (hippocampus and cortex) (Dindot et al. 2008; Sato and Stryker 2010). The same neuronal abnormalities (reduced dendrite branching of sensory neurons) appear in *Drosophila* carrying an *UBE3A*-null mutant (Lu et al. 2009). Many examples show that it is possible to overcome the problem of not having a brain bank by developing organism models, under strict conditions that will be discussed below.

The introduction of organism models in the second stage can be important in helping confirm the etiology of the disease. Human genetics is limited to gene–phenotype correlations. Correlation becomes causation when the transfer of the putative gene to another species produces a human-like phenotype. The story of the *SRY* gene (sex determining region of the Y chromosome) illustrates the decisive role played by the animal model in confirming the involvement of a gene in a given physiological function and consequently its contribution to the etiology of the disease in question. Molecular analysis (Quintana-Murci and Fellous 2001; Sinclair et al. 1990) suggested that the *SRY* gene could be involved in abnormal development, structure and function of the testes. Lovell-Badge group (Koopman et al. 1990) developed an organism model that provided convincing evidence of the contribution of *SRY* to the development of the testes. Transgenic mice overexpressing SRY appeared to have testes even if they were identified as females with an XX karyotype. The testes were smaller than in wild-type males and the carriers were sterile. The mouse model thus demonstrated that *Sry* was needed, but was not sufficient for the development of functional testes. Organism models can also shed light on the function of genes or groups of genes thought to be involved in complex syndromes. The subtractive approach performed with segmental models of trisomy 21 (Down syndrome) gives a better understanding of the functions of the HSA21 genes (Roubertoux and Carlier 2010). A comparison of a well-selected panel of segmental trisomy strains of mice shows the region around the *Dyrk1a* gene to be involved in brain and cognitive disorders associated with trisomy 21.

The breeding and observation of organism models confirm their usefulness for the third stage, which includes the identification of biomarkers that could be used to screen new molecules. In most countries, applications for marketing authorization for a new molecule must include reports of the findings of experiments conducted on different species. Evidence must be provided proving the therapeutic effect of the molecule and its nontoxicity. The level for teratogenic effects is closely examined, but the appearance of such effects can sometimes be species-dependent. For example, the mouse proved to be a poor model for detecting the teratogenic effects of stilbestrol which, when administered to pregnant females, did not induce any organ malformations, whereas in humans it caused abnormalities of the reproductive tract.

An organism model makes four key contributions to the translational strategy: (1) it answers the need for a cohort, (2) confirms etiology and identifies pathophysiological processes, (3) tests the safety and efficacy of drugs, and in so doing, (4) provides a means of addressing the complex issues in real time.

What are the characteristics of the organism models of ID?

Criteria for Developing Organism Models of ID

We make a distinction between two meanings of the word "model" (Lévy and Roubertoux 2015), the first being the human model of characteristic features seen in patients, and the second being the organism model achieved through modeling. For clarity, we have chosen to use the term "paragon" to denote the pattern of features combining to form a human model, and to reserve the word "model" for the nonhuman organism with features produced as a result of modeling. As stated in a previous publication, "the term paragon does not imply any superiority, but simply the abstract case of the patient" (Roubertoux 2015).

Previous papers attempted to provide criteria for accepting an organism construct as a "model" (McKinney 2001). With improved diagnostic tools and the opportunity for molecular intervention in an organism model, these criteria need to be revised. Tordjman et al. (2007) and Lévy and Roubertoux (2015) proposed four criteria that need to be postulated as evidence that the organism model tallies with the paragon (Tordjman et al. 2007): (1) identical etiology, (2) identical metabolic signature, (3) similar pathophysiological mechanisms, and (4) a similar clinical description.

The First Criterion: Identical Etiology

Through comparative genomics it is possible to establish that genes in different species are similar and to compute base-pair homologies and/or amino acid homologies. In a nonhuman organism the overexpression or deletion of a gene that is homologous in humans has been considered sufficient evidence for studying the overexpression or the null expression of the gene in the model, and this has been the case for decades. Evidence provided by the mouse model confirming the role played by *Sry* in masculinization shows the power of the technique, but is not sufficient for establishing the genetic etiology of most genetic diseases.

A single gene can be targeted by several mutations and each mutation has specific molecular, biochemical, and cellular effects, as is illustrated by the two mutations of neuroligin 4 gene. The deletion of two base pairs in codon nucleotide 418 causes a shift resulting in a premature stop codon 429 causing a truncated protein, a mutation associated with ID. Another mutation in the same gene (Coding nucleotide 396:396 G>T) induces a shift and transforms GAC TTC into TGA CTT TGA being a stop codon, the resulting protein is truncated. The two proteins do not have the same truncation and different protein domains are affected. The first mutation is associated with ID (Laumonnier et al. 2004), but the second mutation is associated with Autistic Syndrome Disorders (ASD) (Jamain et al. 2003), and independently of cognitive abilities as both high- and low-functioning patients share the mutation. Another illustration shows that targeting a whole gene is not always the right strategy for understanding the etiology of a disease. Two genes (*TSC1* and *TSC2*) contribute to tuberous sclerosis associated with ID. Howe group (Serfontein et al. 2011) noted that the majority of the mutations identified in *TSC1* are nonsense or frameshift mutations that will lead to premature termination of the protein. It was reported that a few splice site mutations and missense mutations are associated with *TSC1* (Hoogeveen-Westerveld et al. 2011). The conditions are quite different for *TSC2* in which 25 % are missense mutations or short in-frame insertions or deletions. To generate a *TSC1* model with a similar etiology to the paragon would require several different genes to be targeted, but a *TSC2* knockout would not be appropriate for deciphering the effects of genetic events with 25 % occurrence.

The deletion of a whole gene gives information on the function of the gene except when the deletion also deletes miRNAs (posttranscriptional regulators) that are carried by intron regions. More than 1000 miRNAs contribute to brain functioning and modulate learning in mice (Dias et al. 2014; Fiorenza et al. 2016; van Spronsen et al. 2013), and other species such as bees (Cristino et al. 2014), *Drosophila* (Li et al. 2013), and zebrafish (Kelley et al. 2012). The impact of miRNA on human ID cannot be overestimated. Sixty percent of the human genes are modulated by miRNAs and several miRNAs are known to be associated with ID (Green et al. 2013; Provost 2010a, b; Willemsen et al. 2011). The risk of deleting the entire sequence is that it may delete the exonic sequences plus the miRNA and attribute an effect to the gene when it is partially due to the neighbor miRNA. The deletion of a whole gene rarely generates a suitable model of genetic disease even if there is no miRNA involvement. (1) An incomplete protein is not always the genetic event underlying the disease. (2) The truncated proteins are not the same on the same gene, when protein truncation is the cause of the disease. (3) Truncated proteins therefore have specific and different effects. (4) Incomplete proteins interact differentially with the genetic background or proteins and contribute differently to signaling pathways.

Replacement techniques are available to generate a model in which the causes of the disease are as close as possible to the causes observed in the paragon. These techniques have been described extensively in relation to brain disorders (Bouabe and Okkenhaug 2013, see also Lévy and Roubertoux 2015). Most techniques used for generating complex etiologies are based on the *cre-loxp* system which provides the possibility of truncating a protein, skipping one exon or generating a specific genetic event. A clear illustration of this can be found with the complex etiology of Gilford-Progeria Syndrome. Nicolas Lévy and his team discovered that the syndrome is the result of a point mutation in *LMNA* gene (coding nucleotide 1824:c.1824 C>T) causing altered splicing in the gene's pre-mRNAs, and leads to the elimination of 150 bp encoding 50 amino acids in exon 11, resulting in a truncated and toxic form of prelamin A (De Sandre-Giovannoli et al. 2003). Using the *cre-loxp* system, it is possible to mimic the mutation and eliminate the same 50 amino acids in exon 11 of murine Lamin A (Osorio et al. 2011).

Diseases with polygenic etiology, e.g., those caused by chromosomal abnormalities (DiGeorge, trisomy 21 and William–Beuren syndromes), can be also modeled. Complex technologies such as MICER (Mutagenic Insertion and Chromosome Engineering Resource) or TAMERE (Targeted Meiotic Recombination) (Herault et al. 1998) can be used to develop mouse models of partial trisomies or partial monosomies. MICER has been widely used to generate segmental trisomies in trisomy 21 studies and could be of interest for generating mouse models of 2q37 deletion (Felder et al. 2009; Ingason et al. 2011a; Leroy et al. 2013; Mazzone et al. 2012a, b), 15q11-q13 duplication (Depienne et al. 2009; Ingason et al. 2011b; Madrigal et al. 2012; Urraca et al. 2013), 16p11.2 duplication (Barber et al. 2013; Bedoyan et al. 2010; Dittwald et al. 2013; Fernandez et al. 2010; Sanders et al.

2011), 16p11 deletion (Bassuk et al. 2013; Golzio et al. 2012; Grayton et al. 2012; Horev et al. 2011; Zufferey et al. 2012), and 22q13 deletion (Ahn et al. 2014; Aldinger et al. 2013; Chen et al. 2010a; Cohen et al. 1991; Denayer et al. 2012; Lo-Castro et al. 2009; McMichael et al. 2014; Mukaddes and Herguner 2007; Phelan 2008; Philippe et al. 2008; Yang et al. 2012), all of which have been reported to be associated with ID. Another solution consists in inserting fragments of the human genome encompassed in artificial chromosomes into the mouse genome. Segmental transgenic mice have been generated from HSA21 fragments carried by Yeast Artificial Chromosomes and inserted into the mouse genome for modeling.

With recent technical developments it is possible to fulfill the first criterion, but before using the refined tools of gene engineering, preliminary investigations must check and confirm the homology of the paragon and the model. References can be found through the National Center for Biotechnology Information (NCBI), Mouse Genome Informatics (MGI) or software programs such as FASTA (http://www.ebi.ac.uk/Tools/sss/fasta/) where there are several sections devoted to this.

The Second Criterion: Similar Molecular Signature

Identical etiology is not sufficient as the same trigger may produce different effects in two species. Several events separate the DNA template from the protein that determines cell functioning (Roubertoux and Carlier 2007). The story of a mouse model of Lesh–Nyhan disease which is associated with self-harm and cognitive impairment (Tordjman et al. 2007) provides a warning for anyone who might be tempted by modeling based solely on identical etiology. Lesh–Nyhan syndrome is caused by mutations, some causing truncated proteins, in the hypoxanthine phospho-ribosyltransferase (*HPRT*) gene. As the *Hprt* gene is homologous in humans and mice, an *Hprt*-knockout mouse was considered to be a suitable model (Hooper et al. 1987; Kuehn et al. 1987; Wu and Melton 1993), but knocking out *Hprt* proved to have no effect on the mouse phenotypes. However, knocking out the adenine phosphoribosyl-transferase (*Aprt*) gene that is also involved in purine metabolism in HPRT knockout mice did produce the expected Lesh–Nyhan phenotype. The fact that the paragon and the model differ in spite of *HPRT* and *APRT* homology in Humans and Mice can be explained by different compensation mechanisms in the two species. Strict paragon-model gene homology is required, but alone it is not sufficient.

The paragon-model homology of proteins, and more generally of the molecular signatures, is a prerequisite for modeling human ID. An overall estimate sets a ratio of 0.115 (non-synonymous mutations/synonymous mutations) for mouse–human homology (Mouse Genome Sequencing et al. 2002). A sizeable percentage of genes (21 %) are expressed differently in humans and mice (Zeng et al. 2012). Co-expression of genes is also important as interactions, cascades, and additive effects can contribute to the disease. Software programs have been developed and are available for detecting protein interactions in a wide range of species. *String* (Known and Predicted Protein–Protein Interactions) and *David* (bioinformatics resources) are two such programs that can help detect interactions. A gene may be homologous in two species, but there can be species-dependent variations in protein interactions. We can cite the example of the fragile X mental retardation 1 (*FMR1*) gene and its mutation which is often a cause of ID (1/3600). Using *String* software, we compared FMR1 protein interactions in three species frequently used for modeling diseases. Interactions in *Homo sapiens* are given first and contrasted with *Mus musculus* and *Danio rerio*. Modeling human brain disorders, we also selected *Drosophila melanogaster* as invertebrates have been successful models for this, as reported in a recent publication (Ruis-Rubio et al. 2015). The results are presented in Fig. 5.2. We only studied proteins with binding activity scoring above 0.70. Twenty proteins have been found to be associated with FMR1 in humans, eight in mice, eight in zebrafish, and seven in the fruitfly.

The result of gene targeting depends not only on the gene directly involved in the etiology but also on its interaction with other genes. Table 5.1 summarizes the findings and shows the patterns of associated proteins differing according to the species. The mouse, zebrafish, and fruitfly have, respectively, 7, 5, and 0 associated FMR1 proteins shared with humans. The cell component where the protein is found and the biological process suggest that the associated genes, *CYF1P1, FRX1, FRX2, or NUFIP1*, are involved in neuron function in the mouse. This does not preclude a

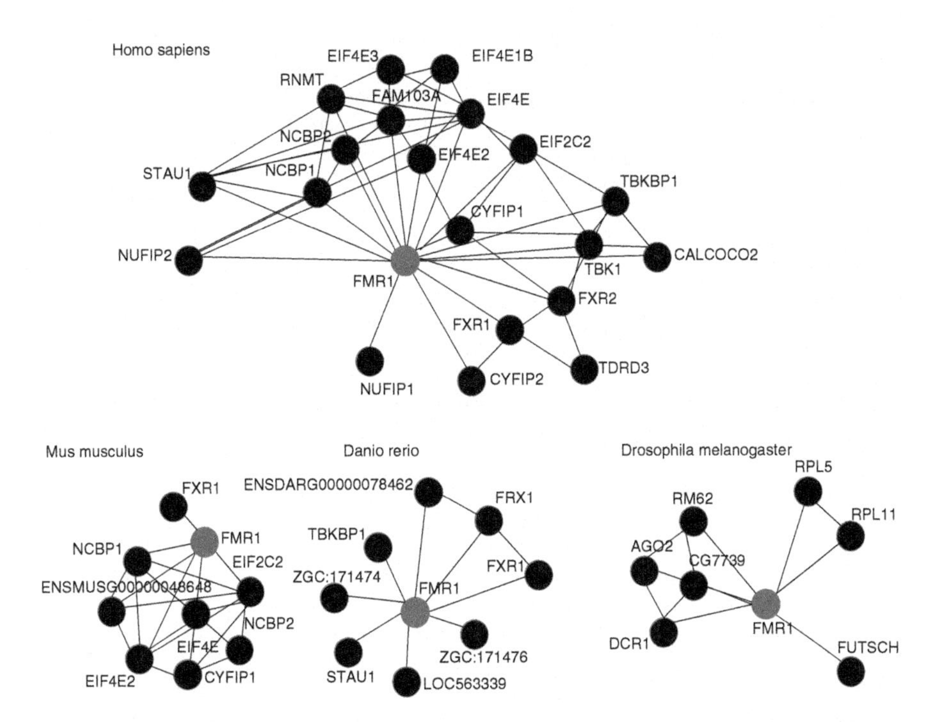

Fig. 5.2 FMR1 protein interactions in humans and organism models. Results using *String* (Franceschini et al. 2013), considered experimental interactions and a score of 0.70 at least. The names and functions of the genes corresponding to the proteins are indicated in Table 5.1

Table 5.1 Proteins interacting with FMR1 in four species: *Homo sapiens* (Hs), *Mus musculus* (Mm), *Danio rerio* (Dr), and *Drosophila melanogaster* (Dm)

Human protein name	Cell component	Biological process	Hs	Mm	Dr	Dm
FMR1	Dendritic spine	Regulates microtubule-dependent synaptic growth and function	X	X	X	X
CALCOCO2 (calcium binding and coiled-coil domain 2)	Cytoskeleton	Autophagy nervous vesicles under genetic conditions	X			
CYFIP1 (cytoplasmic FMR1 interacting protein 1)	Synapse	Axon outgrowth, dendrite formation	X	X		
EIF4E1B (Eukaryotic Translation Initiation Factor 4E Type 1B[2])	Cytoplasm	Contribution to neuron embryonic development suggested	X			
EIF4E1c (eukaryotic translation initiation factor 4E family member 1c)			X			
EIF4E1e (eukaryotic translation initiation factor 4E family member 1e)	RISC complex	Translational initiation	X	X		
EIF4E2 (eukaryotic translation initiation factor 4E family member 2)	mRNA cap binding complex	Interacts with the scaffolding molecule postsynaptic density 95 of excitatory synapses	X	X		
FAM103A1 (family with sequence similarity 103, member A1)	mRNA cap binding complex	Methylation	X			
FXR1 (fragile X mental retardation, autosomal homolog 1)	Dendrites	Negative regulation of translation	X	X	X	
FXR2 (fragile X mental retardation, autosomal homolog 2)	Dendrites	Negative regulation of translation	X		X	
NCBP1 (nuclear cap binding protein subunit 1)	Nuclear cap binding complex	Transport	X	X		
NCBP2 (nuclear cap binding protein subunit 2)	Nuclear cap binding complex	Regulation of translation	X	X		
NUFIP1 (nuclear fragile X mental retardation protein interacting protein 1)	Presynaptic active zone	Positive regulation of transcription from RNA polymerase II promoter	X			
NUFIP2 (nuclear fragile X mental retardation protein interacting protein 2)	Polysomal ribosome		X			

(continued)

Table 5.1 (continued)

Human protein name	Cell component	Biological process	Hs	Mm	Dr	Dm
STAU1 (staufen double-stranded RNA binding protein 1)	Neuronal cell body	Neuron projection, tubulin binding	X		X	
TBK1 (TANK-binding kinase 1)	Cytoplasm	Dendritic cell proliferation	X			
TBKBP1 (TBK1 binding protein 1)		Innate immune response	X		X	
TDRD3 (tudor domain containing 3)	Cytoplasm	Chromatin modification	X			

X indicate a significant interaction

possible more general effect of the protein. But there is a solution here too. The most straightforward option is to test the interaction between the main gene involved in the etiology and the other genes. In Rett mouse models, correction of the FXYD1 associated with Rett syndrome partially rescues the cognitive impairment (Matagne et al. 2013).

The molecular signature of a particular mutation may be explored using yeast models (Mason and Giorgini 2011). Yeast was successfully used to decipher the relationships between abnormal expansion of the polyglutamine tract and huntingtin protein toxicity. The complex gene expression profiling in mutant yeast for huntingtin and the chart produced are of fundamental importance for analyzing the transcription-related effects of huntingtin toxicity (Tauber et al. 2011).

Pathophysiological Pathways

Pathophysiology, molecular indicators, and clinical observations are part of the diagnosis itself and conversely, similar physiological pathways contribute to the phenotypes of the model and the paragon. The repertory of available species often limits any consistent comparison of model and paragon but it would be wrong, indeed an expression of prejudice, to believe that only species that are closely related to humans could provide models for brain function. Quite obviously the mouse brain is not a miniature human brain and while it has similarities, the organization of the brain is different. The efficacy of the model depends on the tissue. Several physiological processes are highly conserved across phylogenetically divergent organisms and match perfectly with human brain function or dysfunction. Differences between rodents and primates can be seen in the anatomy of the nervous system, in the bone marrow organization of motor neurons (Courtine et al. 2007), in prefrontal differentiation of the cortex, and in the organization of the cortical layers, but the connections in the rodent prefrontal cortex that are more similar to the connections in the primate median cortex (Muly et al. 2008; Reep 1984; Reep et al. 1984; Stevens 2010; Stevens et al. 2010). Mutations may induce modifications that are similar in the human paragon and in a phylogenetically remote model. Mutation of the *Minibrain* reduces the size of mushroom bodies, as does a mutation of *DYRK-1A*, located on the D21S17-ETS2 chromosomal region of chromosome 21, and contributes to the trisomy 21 phenotype (Guimera et al. 1996, 1999). Maternally deficient *UBE3A* decreases dendritic arborization and the number of dendritic spines in pyramidal neurons (visual cortex) (Jay et al. 1991), while *UBE3A*-null mutant drosophila have reduced dendritic branching of sensory neurons (Lu et al. 2009). The combination of these findings with comparative neuroanatomy may reveal unexpected pathophysiological models, but may also prove to be a total blunder. Genetic engineering can often help to overcome the difficulty by monitoring neuronal integration.

The brain is not a single organ but a federation of structures, and the most difficult task is to monitor the expression of the gene in the right structure, or even in the right neuron. Neuronal integration shows that a gene is not expressed similarly in different brain structures. As stated in a previous publication: "The brain differs from other organs, such as the liver, which have quite homogeneous cells with similar structure and function; the brain is a collection of structures that differ in their anatomy, morphology, neurochemistry, and in the functioning of the cells that compose the structures" (Roubertoux and Carlier 2007). The human brain has some 50 regions, each with a different cytoarchitecture (Brodmann 1909). The amygdala and hippocampus, for example, do not have many cytological, neurochemical, or functional similarities, but operate jointly for learning and exploration tasks. It is not unusual to have poor transcription or expression in the amygdala and strong transcription or expression of the same gene in the hippocampus. As a consequence, the same gene can have different functions in different brain structures. Clear evidence of this has been found in experiments rescuing dopamine in mice knocked out for the tyrosine hydroxylase gene. Dopamine rescue does not have the same effect on behavior in the nucleus accumbens, the lateral caudate putamen, or the central caudate putamen. Dopamine administered to the nucleus accumbens increases the preference for sucrose, but this is not the case for the putamen nuclei; and dopamine administered to the lateral or central caudate putamen increases nest building activity (Szczypka et al. 2001).

The challenge is to obtain Cre-recombinase driven by a reporter expressed in selected regions only. Lists of cre reporters expressed in the cortex, hippocampus, and striatum have been published (Madisen et al. 2010, 2012), and in finer detail for a subregion of the thalamus (Song et al. 2012). Seventeen cre reporters of differing degrees of specificity to GABAergic neurons were identified in the mouse (Taniguchi et al. 2011). Gene targeting in serotoninergic and in dopaminergic neurons (Zhuang et al. 2005) can be used to model Angelman disease (Farook et al. 2012; Kato et al. 2013; Staal et al. 2012). All the different *Cre-loxp* techniques are expensive and time-consuming, but more recent techniques are pushing back these limits.

The CRISPRs (clustered regularly interspaced palindromic repeat) abridge most of the stages required for performing targeted genes modifications (Swiech et al. 2015). The bacterial RNA-guided CRISPR-Cas9 system may be more versatile and more promising (Hsu et al. 2014). The CRISPR-Cas9 system is still being used to study gene function in the brain (Swiech et al. 2015), synaptic proteins (Incontro et al. 2014), and post-mitotic neurons (Straub et al. 2014). More insight on the CRISPR-Cas9 system is required.

Morpholino technology is being used in alternate organism models (mouse, chicken, zebrafish, Xenopus). Morpholino refers to a synthetic molecule that prevents gene expression by steroid-blocking and binding to a sequence of RNA that prevents the further stages developing towards the protein (Summerton et al. 1997; Summerton and Weller 1997); the gene is always present but the protein is absent. With the small size of morpholinos (around 25 bases), small RNA regions can be targeted. Morpholinos are used with cell culture and in particular with IPSCs, but they can be inserted into eggs or few-cell-stage embryos by electroporation or injection. A comparison across several species of the effects of morpholinos on homologous genes can provide further evidence of a mutation-phenotype correlation by demonstrating phylogenetic consistency. The implication of a mutation in the *DYX1C1* gene, which is associated with neuronal migration in the neo-cortex during embryonic development and in learning difficulties in humans, was confirmed in an interspecific comparison of the mouse and zebrafish. The contribution of a mutation of the acid retinoic (*RAI1*) gene to Smith–Magenis syndrome was confirmed in *Xenopus laevis* and *Xenopus tropicalis. Ra/1* morphants affect axon patterns, forebrain ventricle size, and forebrain apoptosis (Tarkar et al. 2013).

Another difficulty is that genes are not always expressed in one cell or one type of cell. It is known that a number of genes act during a limited time window and may sometimes trigger cascades (Ryan et al. 2012). A key issue for a rescue strategy is to generate organisms with one gene over- or under-expressed, but only at a specific age.

Two systems are commonly used, tet and cre-lox. A good description of tetracycline-dependent regulatory systems (abbreviated tet) was given in a paper published by Jaisser (Jaisser 2000). "The tet systems rely on two components, i.e., a tetracycline-controlled transactivator (tTA or rtTA) and a tTA/rtTA-dependent promoter that controls expression of a downstream cDNA, in a tetracycline-dependent manner (…) tTA is a fusion protein containing the repressor of the Tn10 tetracycline-resistance operon of Escherichia coli and a carboxyl-terminal portion of protein 16 of herpes simplex virus (VP16). The tTA-dependent promoter consists of a minimal RNA polymerase II promoter fused to tet operator (tetO) sequences (an array of seven cognate operator sequences). This fusion converts the tet repressor into a strong transcriptional activator in eukaryotic cells." The tTA binds to the tetO when tetracycline or similar molecules are missing in the organism carrying the system. tTa–tetO binding triggers the transcriptional activity of the tTA-dependent promoter. When tetracycline or doxycycline is added to drinking water or administered by gavage, transcription is stopped and the gene functions as a wild-type gene. Within-individual comparisons (with and without tetracycline in the drinking water) can study the same mouse under control and genetically modified conditions. Other systems based on the same rationale have been used (Jaisser 2000).

The combination of the cre-lox (tissue-dependent) system and the inducible gene expression system can generate the expression of a gene in an organ (brain structure or neuron family) but only within a certain time window. The use of a mutated form of the ligand-binding domain of the estrogen receptor makes Cre activity tamoxifen-dependent (Danielian et al. 1998). An injection of tamoxifen opens a time window of approximately 6 h (Hayashi and McMahon 2002). The technique has been used to observe the timing of gene expression in the neurons of the cortex (Weber et al. 2001; Erdmann et al. 2007; Sakamoto et al. 2011) and the hippocampus (Li et al. 2011) and in the synapses. The combination of cre-lox and of the cre-tamoxifen-dependent inducible system has also been used to show the role of 7miRNA in learning and in neuronal correlates (Konopka et al. 2010).

Stimulation and lesions have been used for decades to explore the function of the different brain structures. The advent of gene targeting, both gene deletion and gene overexpression, combined with tissue-specific expression, now stands as more than just an alternative to neurosurgery. It is possible to correlate the deletion/overexpression in a given structure with the appearance of the phenotype (ID). Co-occurrence of the deletion in the neural tissue and the expected phenotype supports the hypothesis that the paragon and model have similar pathophysiological pathways.

Clinical Similarity

The benefit that can be expected of the translational strategy is the possibility of having a chain of causality including etiology, physiological pathways, and clinical signs, with consistent observations and relevant prospects for curing or alleviating the disease. Researchers must always remember that our task is to "devise fine organism models that match the paragon, but [that] all of this would be to no avail without criteria to detect reactions to treatments" (Roubertoux 2015). Conditions required for the clinical features observed in the paragon to find a perfect match in the model are not always met.

1. Differences often exist in the behavioral repertoire of the species, making it unrealistic to attempt to model-specific anomalies in human behavior, e.g., amorality (Fisch 2015).
2. Very often clinical information on patients is not available, sometimes because of the small number of cases or because the disease has only recently been discovered.
3. A poor match between the paragon and the model may be the result of inadequate screening of the model. Cognitive tasks are used for modeling ID in organisms, and learning tests are available for different species, e.g., rodents, fish, and insects, but the features of a given disease are not all uniformly impaired, and there is no similarity of cognitive impairment across diseases. An IQ score is an overall estimate of intelligence covering a wide range of heterogeneous cognitive profiles. Researches have reported substantial differences in the skill levels of persons with trisomy 21 and with Williams–Beuren (Mervis and John 2008, 2010; Roubertoux and Kerdelhue 2006; Vicari 2006; Vicari and Carlesimo 2006). Simply citing

"intellectual disability" as a feature associated with a metabolic or a neurological disease is not sufficient for developing an organism model. For modeling, it is of prime importance to have detailed descriptions of behavioral and cognitive profiles as different profiles are observed across genetic syndromes. The profiles of PKU, fragile X, DiGeorge syndrome, and trisomy 21 all differ (Carlier and Roubertoux 2014), as do the fragile X, Wolf–Hirschhorn, and William–Beuren profiles (Carrasco et al. 2005; Fisch et al. 2012; Mervis and John 2010). Differences in developmental trajectories and developmental timing should also be taken into account, with their differential effects on cognitive and behavioral profiles (Fisch et al. 2012; Karmiloff-Smith et al. 2012; Paterson et al. 1999).

There are two challenges here. First a battery of tests covering the broadest possible field of cognitive skills in the model species must be developed. Evidence must be found to show that the model species matches the paragon species when considering its overall cognitive profile. And here we suggest a top-down transverse strategy for generating valid organism models of ID.

Top-Down Transverse Strategy for Generating Organism Models of ID

The broader approach to modeling suggested is what we call the "top-down transverse strategy" (T-DTS). T-DTS meets the criteria required of a model as stated above and can solve the question of validity which is so crucial in behavioral tests used to model ID.

When developing an ability test in human, there must be evidence proving that it is valid. Without going beyond the scope of the present chapter, we can briefly review the concept of test validity. In the early twentieth century, the validity of an intelligence test was mainly determined by its correlates with other measurements, designed as criterion measures, that measured, in part, the same psychological phenomenon. For example, intelligence was seen as essential to school achievement so, for an intelligence test to be considered as valid, it should be able to predict school achievement. But this basic approach was then thought to be too limited, one of the fundamental problems being the inherent tautology as the validity of the external criterion measures had not been determined. By the early twenty-first century, major advances had been made producing a number of types of validity (Suen and French 2003). More recently it has been argued that validity is a process for systematically developing arguments: the "argument-based approach to validation" (Kane 1992, 2013). "… test-score interpretations and uses that are clearly stated and are supported by appropriate evidence are considered to be valid. Conversely, interpretations or uses that are not well defined or that involve doubtful inferences or assumptions are not considered valid." (Kane 2013, p. 448) There are two prerequisites for validity: first a clear statement of the claims inherent in the interpretation of the test score, and second an evaluation of these claims.

The top-down strategy requires definition of external criteria. The principle of T-DTS is to include measured of cognition in a causal chain encompassing brain physiological and biochemical correlates as well as genetic and genomic data in a comparative (paragon-model) perspective. The transverse dimension consists of an additional comparative strategy across several organism models. By finding a consistently similar causal chain in a variety of organism models, conclusions drawn from data stand as valid evidence and may provide insights for therapeutic strategies. The T-DTS can also be used to annotate a gene and this is helpful for modeling ID when dealing with so many heterogeneous cognitive profiles. The cognitive dimensions impacted by gene impairment therefore need to be described.

The Top-Down Approach to Disease Associated with ID

The approach must cover all clinical data, including biochemical analyses, behavioral observations, brain imaging, academic achievement, developmental data, and, if available, anatomical and neurochemical data from postmortem examinations of the brain. Two illustrations help to clarify the T-D T approach.

In mammals, nonmotile (primary) cilia are thin cell membrane extensions found on most cells including brain neurons (Goetz and Anderson 2010); an impairment of cilium morphology and/or function causes ciliopathies. The basic function of a cilium is intraflagellar anterograde transport, mediated by kinesin, and retrograde transport mediated by dyneins. The process is involved in the regulation of signal transduction. Any impairment of the process modifies the Sonic hedgehog (Shh) signaling pathways subsequently causing abnormal development of neural tissues in the embryo (Goetz and Anderson 2010). The anterograde-retrograde process plays a critical role in the development of the nervous system, in CNS regeneration and stem cell regulation via the Shh and Wingless (Wnt) pathways. Primary cilia that contribute to neuron migration are involved in cerebellar granule neuron precursor proliferation (Breunig et al. 2008) and in cortical development (Guemez-Gamboa et al. 2014). Ciliopathies form a group of heterogeneous diseases including several syndromes, some of which are associated with ID (Joubert, Bardet–Biedl, nephronophthisis, and orofaciodigital syndromes) and sometimes with brain malformations as shown by MRI (Barnett et al. 2002; Bassuk et al. 2013; Bedoyan et al. 2010). The question is: "Is there a specific link with one particular gene, the structure where that gene is expressed, and the type of learning?" When the Intraflagellar transport 88 (*Ift88*) gene is deleted there are no cilia. *Ift88* knockout mice with deletion only in the cortex and the hippocampus were tested for learning skills, and the results of the cognitive indicators displayed great heterogeneity. Synaptic plasticity, measured by long-term potentiation after Schaffer collateral stimulation, was enhanced, while learning performance tested in the Morris water-maze and for novel object recognition had typical scores. Memory was tested in a fear conditioning task and was impaired (Berbari et al. 2014). When events, from the gene mutation through to cognitive performance, are found to be consistent, this is taken as compelling evidence validating the model. The validity in this case is not deduced from a correlation with an external criterion but is evidenced by the description of a mechanism.

With minor modifications, a similar strategy has been used to model trisomy 21 (Down syndrome) in mice. In mouse models of trisomy 21 HAS 21 is syntenic to MMU 16, MMU 10, and MMU 17. A number of studies have found that genes carried by MMU 10 and MMU 17 do not contribute to the neurological and cognitive disorders associated with Down syndrome, but that mice carrying a triple copy of MMU16 do present the deficits (Roubertoux and Carlier 2010). We explored the D21S17-ETS2 region located on MMU16, the region known to encompass 19 genes which, when triplicated, are involved in trisomy 21 (Chabert et al. 2004; Smith et al. 1995, 1997). Partial triplication of the D21S17-ETS2 region was performed using Smith's model using four different segmental trisomic strains of mice with a fragment covering the region. One strain with four triplicated genes was severely impaired, while the other strains triplicating a variable number of genes were less severely affected (Chabert et al. 2004; Roubertoux et al. 2005, 2006; Seregaza et al. 2006). The question is: "Does each trisomic strain contribute a significant percentage of the impairment produced by the triplication of MMU16?" To investigate this question, we compared the four segmental trisomies in the Ts65Dn mouse that triplicates the *Mrpl-39-Mrx1* region of MMU16. We selected features in patients with trisomy 21 that were recorded as stable features and taken from a review of published articles listing brain and cognitive characteristics (Roubertoux and Kerdelhue 2006) for modeling neurological and cognitive traits.

Brain characteristics observed by MRI or in postmortem analyses are: small brain, cerebellar hypoplasia, granular layer less developed than the cell layer, less developed hippocampus and cortex, and enlarged brain ventricles (Roubertoux and Carlier 2010). Motor difficulties were also reported (Jover et al. 2014) with poor motor coordination and hypotonia. The cognitive profile model took into account both the strong points and weak points of the syndrome; for example, associative learning is not impacted by trisomy 21 (Roubertoux and Kerdelhue 2006), but spatial learning, flexibility, and long-term memory are defective. By considering the homology of the genes carried by the D21S17-ETS2 region together with the cognitive and brain features, a consistent causal chain can be found.

Several top-down approaches exist in the field of ID, most of them using the mouse as the organism model. The attractiveness of other species, such as the zebrafish, could change the organism model landscape in the near future.

Complementarity of the Top-Down Approach and the Transverse Approach

The main argument for the transverse approach is that it provides confirmatory evidence of a causal link in a number of species. For modeling this is logical from a phylogenetic standpoint. A transverse complement may be found in a single link of the causal chain or may involve the entire chain.

The transverse approach has provided confirmation for a number of diseases. With the expansion of CGG (trinucleotide repeat) in the *FMR1* gene there is no production of the fragile X mental retardation protein (FMRP) which is needed for neuron development. The absence of FMRP causes methylation that silences the *FMR1* gene. Postmortem studies of fragile X patients describe immature morphology of dendritic spines in the cortex (Irwin et al. 2001). Different mouse models (deletion of *FMRP*, expanded CGG trinucleotide repeat) display morphological anomalies in dendrites, with a reduction of dendritic branches proximal to the soma, reduced total dendritic length, and a reduction in dendritic arborization (in the medial prefrontal cortex, basal lateral amygdala, and the hippocampus) (Berman et al. 2012; Jacobs et al. 2010; Qin et al. 2011; Wang et al. 2014). Given the number of modifications affecting dendrites, it may be hypothesized that synaptic function is modified too. Synaptic protein changes were observed in a mouse model of fragile X (Wang et al. 2014) and a large number of mouse studies have confirmed the link between the protein anomaly, dendritic and/or synaptic dysfunction and learning. The FMR1 protein has a number of isoforms and, studying the fly, correlations have been established between certain isoforms and certain learning tasks (Banerjee et al. 2010). Rescuing the protein modifies both the synaptic functions and the learning performance (Bilousova et al. 2009). Similar results were found in the zebrafish. Targeting the FMR1 homolog produced abnormal dendrite branching and abnormal LTD, indicating synaptic dysfunction plus spatial learning impairment (Ng et al. 2013). A recent study reported parallel rescue of both synaptic function and learning in mouse and drosophila models of fragile X syndrome given the same treatment (Choi et al. 2015).

Angelman patients with ID also have reduced dendritic arborization and a reduction in the number of dendritic spines in pyramidal neurons of the visual cortex as measured postmortem (Jay et al. 1991). This has been confirmed in models with a *UBE3A* maternal-deficient allele. The basal dendrites of the pyramidal neurons, of the visual cortex, of the cerebellar Purkinje cells, and of the pyramidal neurons in the hippocampus and cortex had reduced reduced or abnormal spine density in the *UBE3A* maternal-deficient allele mice (Dindot et al. 2008). The transverse approach used on *UBE3A*-null mutant drosophilas found reduced dendrite branching of sensory neurons (Lu et al. 2009).

Yeast model can be used to analyze protein interactions and this should be particularly relevant for examining the proteome of a range of different isoforms (Mason and Giorgini 2011). Expression profiling in yeast can shed light on the abnormal expansion of a polyglutamine tract and huntingtin protein toxicity (Tauber et al. 2011). Induced Pluripotent Stem Cell (iPSC) technology can be a further source of information for performing transverse analysis. After reprogramming cells to pluripotency, iPSC differentiation can be tracked and cell development can be guided towards a certain type of cell that has been selected, and iPSCs can be differentiated into neurons that reproduce the genetic defects of the cell that was reprogrammed. The technique was used for reprogramming Rett syndrome (Chen et al. 2010b; Cheng et al. 2012) and Fragile X (Eiges et al. 2007; Marchetto et al. 2010a, b) cells.

Future Directions: Recommendations and Reservations

Three key ideas are noted to conclude the chapter: (1) extension of the organism list, (2) better selection of cognitive tasks, and (3) better congruence between data and the use of statistics.

Finding New Organism Models

Rodents have been the most popular organism models for so long, even though other species have been suggested. A good illustration is provided by modeling diseases in the domestic dog. Inbreeding is currently practiced in dogs. It favors the appearance of spontaneous autosomal diseases that could be models of rare diseases occurring in consanguineous human populations—see Ellinwood et al. (2011) for a model of Hurler and Sanfilippo diseases in dog. Occasionally a different species may be used for research on a specific disease, e.g., the ferret for modeling lisencephaly.

Because of the strong response mammals have to their environment, there has been incentive to investigate non-mammalian organisms such as fish, worms, and insects. As most are oviparous, gene modification is easier. The zebrafish is currently being used as a model for brain disorders, including autism and ID (Glasgow et al. 1997; Kalueff et al. 2013). If we do not insist on having a holistic model of a disease and are willing to accept tissue-specific models, the list of organism models will obviously expand. One example is research on the *minibrain* gene and its impact on the development of mushroom bodies in the fly. The mushroom bodies of the Drosophila have the same functions as the mammalian hippocampus (Heisenberg et al. 1985). *Minibrain* mutations that reduce the size of mushroom bodies are the same as the mutations of the ortholog of Dyrk1a that induce a small hippocampus in the mouse. This modeling only applies to hippocampal tissues and is useful for understanding the contribution of *minibrain* to Down syndrome. Obviously, given the anatomy of the fly, this cannot be extended to any other nerve tissue. The worm Caenohabditis elegans is appropriate for modeling a wide range of neuron and/or synapse impairments found in many diseases, e.g., ciliopathies (Huang et al. 2011).

Better Selection of the Cognitive Tasks

There is a great deal of imagination being exercised in the development of new learning tests for different species, and yet researchers often select tests that are already established and widely used when working on models of disease. A survey of publications between 2006 and 2014 studying mouse models of ID found that less than 4 % used more than one test to annotate a model. All learning tests are not equivalent and conclusions drawn from studies using a single test may be flawed. A learning task not only has a "face-value" validity, but is also an indicator of the functioning of underlying physiological, cellular, and molecular processes. Learning tasks do not have the same neuronal and/or neurochemical correlates, even when the learning tasks appear to be identical. A number of tasks can be used to assess associative learning, also referred to as Pavlovian learning, and two tasks have been studied extensively: fear conditioning and eyeblink conditioning. For a long time, the two protocols were believed to bring associative processes and similar neurophysiological mechanisms into play. The amygdala is the connection center for information processing during fear conditioning, while the cerebellum is the processing center for eyeblink conditioning. Synaptic plasticity controls the associative processes in both of forms of conditioning. While the core of the two conditioning processes is synaptic plasticity, the cellular and molecular processes involved in plasticity are not the same for eyeblink and fear conditioning (Fanselow and Poulos 2005). Cognitive modeling to provide information should include a systematic investigation of the physiological processes behind the learning tests; this means having thorough knowledge of the neurophysiology of learning and conducting an exhaustive examination of both the paragon and model.

The Importance of Experimental Design and Statistical Analyses

It should be remembered that the first step before undertaking any experimental study is to ensure that the design is appropriate, i.e., that the tests have sound internal and external validity. Poor internal validity compromises the degree of certainty of findings stating that the dependent variable (e.g., the test score) was caused by the independent variable (e.g., the genotype) and not by another variable. Poor external validity limits the possibility of a wider application of the findings. These basic concepts lay the foundations of experimental research, but, unfortunately, recent critical reviews have found unreliable findings in studies published in the fields of neuroscience (Carp 2012; Button et al. 2013), biomedical research (Ioannidis 2005), and psychological research (Simmons et al. 2011). Even when the experimental design is sound, many studies published present findings that are unreliable because of the low power for statistical analysis (Button et al. 2013). "The power of a statistical test of a null hypothesis is the probability that it will lead to the rejection of the null hypothesis, i.e., the probability that it will result in the conclusion that the phenomenon exists" (Cohen 1988, p. 4). Statistical power depends on three parameters: the significance criterion (p level), the reliability of the data, and the size of the statistical effect. When the study is underpowered, statistical effects are difficult to detect and, unfortunately, the likelihood that a statistically significant result reflects a true effect is reduced (Button et al. 2013). Before undertaking an experiment, the power of the statistical analysis and the expected effect sizes should be taken into account, and power and effects sizes should be detailed in the description reporting the results. In many cases, particularly in studies of ID, the rejection of the null hypothesis is not informative. The most crucial element of information is to know whether the difference is noticeable.

Acknowledgements *Thanks are due to U 910, Génétique médicale, génomique fonctionnelle, to INSERM,* to *Aix-Marseille Université, and to the Fédération pour la Recherche sur le Cerveau.*

References

AAIDD. (2010). *Intellectual disability: definition, classification and system of supports/The AAIDD Ad Hoc Committee On terminology And Classification (11th ed.).* Washington, DC: American Association on Intellectual and Developmental Disabilities.

Ahn, K., Gotay, N., Andersen, T. M., Anvari, A. A., Gochman, P., Lee, Y., et al. (2014). High rate of disease-related copy number variations in childhood onset schizophrenia. *Molecular Psychiatry, 19*, 568–572.

Aldinger, K. A., Kogan, J., Kimonis, V., Fernandez, B., Horn, D., Klopocki, E., et al. (2013). Cerebellar and posterior fossa malformations in patients with autism-associated chromosome 22q13 terminal deletion. *American Journal of Medical Genetics Part A, 161A*, 131–136.

Banerjee, P., Schoenfeld, B. P., Bell, A. J., Choi, C. H., Bradley, M. P., Hinchey, P., et al. (2010). Short- and long-term memory are modulated by multiple isoforms of the fragile X mental retardation protein. *The Journal of Neuroscience: The Official Journal of the Society for Neuroscience, 30*, 6782–6792.

Barber, J. C., Hall, V., Maloney, V. K., Huang, S., Roberts, A. M., Brady, A. F., et al. (2013). 16p11.2-p12.2 duplication syndrome; a genomic condition differentiated from euchromatic variation of 16p11.2. *European Journal of Human Genetics: EJHG, 21*, 182–189.

Barnett, A., Mercuri, E., Rutherford, M., Haataja, L., Frisone, M. F., Henderson, S., et al. (2002). Neurological and perceptual-motor outcome at 5–6 years of age in children with neonatal encephalopathy: Relationship with neonatal brain MRI. *Neuropediatrics, 33*, 242–248.

Bassuk, A. G., Geraghty, E., Wu, S., Mullen, S. A., Berkovic, S. F., Scheffer, I. E., et al. (2013). Deletions of 16p11.2 and 19p13.2 in a family with intellectual disability and generalized epilepsy. *American Journal of Medical Genetics Part A, 161A*, 1722–1725.

Bedoyan, J. K., Kumar, R. A., Sudi, J., Silverstein, F., Ackley, T., Iyer, R. K., et al. (2010). Duplication 16p11.2 in a child with infantile seizure disorder. *American Journal of Medical Genetics Part A, 152A*, 1567–1574.

Benton, D. (2010). The influence of dietary status on the cognitive performance of children. *Molecular Nutrition & Food Research, 54*, 457–470.

Benton, M. J., Wagner, C. L., & Alexander, J. L. (2010). Relationship between body mass index, nutrition, strength, and function in elderly individuals with chronic obstructive pulmonary disease. *Journal of Cardiopulmonary Rehabilitation and Prevention, 30*, 260–263.

Berbari, N. F., Malarkey, E. B., Yazdi, S. M., McNair, A. D., Kippe, J. M., Croyle, M. J., et al. (2014). Hippocampal and cortical primary cilia are required for aversive memory in mice. *PLoS One, 9*, e106576.

Berman, R. F., Murray, K. D., Arque, G., Hunsaker, M. R., & Wenzel, H. J. (2012). Abnormal dendrite and spine morphology in primary visual cortex in the CGG knock-in mouse model of the fragile X premutation. *Epilepsia, 53*(Suppl 1), 150–160.

Best, K. E., Tennant, P. W., Addor, M. C., Bianchi, F., Boyd, P., Calzolari, E., et al. (2012). Epidemiology of small intestinal atresia in Europe: A register-based study. *Archives of Disease in Childhood. Fetal and Neonatal Edition, 97*, F353–F358.

Bilousova, T. V., Dansie, L., Ngo, M., Aye, J., Charles, J. R., Ethell, D. W., et al. (2009). Minocycline promotes dendritic spine maturation and improves behavioural performance in the fragile X mouse model. *Journal of Medical Genetics, 46*, 94–102.

Bouabe, H., & Okkenhaug, K. (2013). Gene targeting in mice: A review. *Methods in Molecular Biology, 1064*, 315–336.

Breunig, J. J., Sarkisian, M. R., Arellano, J. I., Morozov, Y. M., Ayoub, A. E., Sojitra, S., et al. (2008). Primary cilia regulate hippocampal neurogenesis by mediating sonic hedgehog signaling. *Proceedings of the National Academy of Sciences of the United States of America, 105*, 13127–13132.

Brodmann, C. (1909). *Lokalisationslehre der Grosshirnrinde*. Leipzig: JA Barth.

Button, K. S., Ioannidis, J. P., Mokrysz, C., Nosek, B. A., Flint, J., Robinson, E. S., et al. (2013). Empirical evidence for low reproducibility indicates low pre-study odds. *Nature Reviews Neuroscience, 14*, 877.

Carlier, M., & Ayoun, C. (2007). *Déficiences intellectuelles et intégration sociale*. Mardaga: Wavre (Belgique).

Carlier, M., & Roubertoux, P. L. (2014). Genetic and environmental influences on intellectual disability in childhood. In D. R. Finkel & C. A. Reynolds (Eds.), *Behavior genetics of cognition across the lifespan* (pp. 69–101). Springer: New York.

Carp, J. (2012). The secret lives of experiments: Methods reporting in the fMRI literature. *Euroimage, 3*, 89–300.

Carrasco, X., Castillo, S., Aravena, T., Rothhammer, P., & Aboitiz, F. (2005). Williams syndrome: Pediatric, neurologic, and cognitive development. *Pediatric Neurology, 32*, 166–172.

Chabert, C., Jamon, M., Cherfouh, A., Duquenne, V., Smith, D. J., Rubin, E., et al. (2004). Functional analysis of genes implicated in Down syndrome: 1. Cognitive abilities in mice transpolygenic for Down Syndrome Chromosomal Region-1 (DCR-1). *Behavior Genetics, 34*, 559–569.

Chen, C. P., Lin, S. P., Chern, S. R., Tsai, F. J., Wu, P. C., Lee, C. C., et al. (2010a). A de novo 7.9 Mb deletion in 22q13.2→qter in a boy with autistic features, epilepsy, developmental delay, atopic dermatitis and abnormal immunological findings. *European Journal of Medical Genetics, 53*, 329–332.

Chen, Y. J., Zhang, M., Yin, D. M., Wen, L., Ting, A., Wang, P., et al. (2010b). ErbB4 in parvalbumin-positive interneurons is critical for neuregulin 1 regulation of long-term potentiation. *Proceedings of the National Academy of Sciences of the United States of America, 107*, 21818–21823.

Cheng, C., Sourial, M., & Doering, L. C. (2012). Astrocytes and developmental plasticity in fragile X. *Neural Plasticity, 2012*, 197491.

Choi, C. H., Schoenfeld, B. P., Weisz, E. D., Bell, A. J., Chambers, D. B., Hinchey, J., et al. (2015). PDE-4 inhibition rescues aberrant synaptic plasticity in drosophila and mouse models of fragile X syndrome. *The Journal of Neuroscience: The Official Journal of the Society for Neuroscience, 35*, 396–408.

Cohen, J. (1988). *Statistical power analysis for the behavioral sciences* (2nd ed.). Hillsdale, NJ: L. Erlbaum Associates.

Cohen, I. L., Vietze, P. M., Sudhalter, V., Jenkins, E. C., & Brown, W. T. (1991). Effects of age and communication level on eye contact in fragile X males and non-fragile X autistic males. *American Journal of Medical Genetics, 38*, 498–502.

Courtine, G., Bunge, M. B., Fawcett, J. W., Grossman, R. G., Kaas, J. H., Lemon, R., et al. (2007). Can experiments in nonhuman primates expedite the translation of treatments for spinal cord injury in humans? *Nature Medicine, 13*, 561–566.

Cristino, A. S., Barchuk, A. R., Freitas, F. C., Narayanan, R. K., Biergans, S. D., Zhao, Z., et al. (2014). Neuroligin-associated microRNA-932 targets actin and regulates memory in the honeybee. *Nature Communications, 5*, 5529.

Danielian, P. S., Muccino, D., Rowitch, D. H., Michael, S. K., & McMahon, A. P. (1998). Modification of gene activity in mouse embryos in utero by a tamoxifen-inducible form of Cre recombinase. *Current Biology: CB, 8*, 1323–1326.

De Sandre-Giovannoli, A., Bernard, R., Cau, P., Navarro, C., Amiel, J., Boccaccio, I., et al. (2003). Lamin a truncation in Hutchinson-Gilford progeria. *Science, 300*, 2055.

Denayer, A., Van Esch, H., de Ravel, T., Frijns, J. P., Van Buggenhout, G., Vogels, A., et al. (2012). Neuropsychopathology in 7 Patients with the 22q13 Deletion Syndrome: Presence of bipolar disorder and progressive loss of skills. *Molecular Syndromology, 3*, 14–20.

Depienne, C., Moreno-De-Luca, D., Heron, D., Bouteiller, D., Gennetier, A., Delorme, R., et al. (2009). Screening for genomic rearrangements and methylation abnormalities of the 15q11-q13 region in autism spectrum disorders. *Biological Psychiatry, 66*, 349–359.

Dias, B. G., Goodman, J. V., Ahluwalia, R., Easton, A. E., Andero, R., & Ressler, K. J. (2014). Amygdala-dependent fear memory consolidation via miR-34a and Notch signaling. *Neuron, 83*, 906–918.

Dindot, S. V., Antalffy, B. A., Bhattacharjee, M. B., & Beaudet, A. L. (2008). The Angelman syndrome ubiquitin ligase localizes to the synapse and nucleus, and maternal deficiency results in abnormal dendritic spine morphology. *Human Molecular Genetics, 17*, 111–118.

Dittwald, P., Gambin, T., Szafranski, P., Li, J., Amato, S., Divon, M. Y., et al. (2013). NAHR-mediated copy-number variants in a clinical population: Mechanistic insights into both genomic disorders and Mendelizing traits. *Genome Research, 23*, 1395–1409.

Eiges, R., Urbach, A., Malcov, M., Frumkin, T., Schwartz, T., Amit, A., et al. (2007). Developmental study of fragile X syndrome using human embryonic stem cells derived from preimplantation genetically diagnosed embryos. *Cell Stem Cell, 1*, 568–577.

Ellinwood, N. M., Ausseil, J., Desmaris, N., Bigou, S., Liu, S., Jens, J. K., et al. (2011). Safe, efficient, and reproducible gene therapy of the brain in the dog models of Sanfilippo and Hurler syndromes. *Molecular Therapy: The Journal of the American Society of Gene Therapy, 19*, 251–259.

Erdmann, G., Schutz, G., & Berger, S. (2007). Inducible gene inactivation in neurons of the adult mouse forebrain. *BMC Neuroscience, 8*, 63.

Fanselow, M. S., & Poulos, A. M. (2005). The neuroscience of mammalian associative learning. *Annual Review of Psychology, 56*, 207–234.

Farook, M. F., DeCuypere, M., Hyland, K., Takumi, T., LeDoux, M. S., & Reiter, L. T. (2012). Altered serotonin, dopamine and norepinepherine levels in 15q duplication and Angelman syndrome mouse models. *PLoS One, 7*, e43030.

Felder, B., Radlwimmer, B., Benner, A., Mincheva, A., Todt, G., Beyer, K. S., et al. (2009). FARP2, HDLBP and PASK are downregulated in a patient with autism and 2q37.3 deletion syndrome. *American Journal of Medical Genetics Part A, 149A*, 952–959.

Fernandez, B. A., Roberts, W., Chung, B., Weksberg, R., Meyn, S., Szatmari, P., et al. (2010). Phenotypic spectrum associated with de novo and inherited deletions and duplications at 16p11.2 in individuals ascertained for diagnosis of autism spectrum disorder. *Journal of Medical Genetics, 47*, 195–203.

Fiorenza, A., Lopez-Atalaya, J. P., Rovira, V., Scandaglia, M., Geijo-Barrientos, E., & Barco, A. (2016). Blocking miRNA biogenesis in adult forebrain neurons enhances seizure susceptibility, fear memory, and food intake by increasing neuronal responsiveness. *Cerebral Cortex, 26*(4), 1619–1633.

Fisch, G. S. (2011). Mental retardation or intellectual disability? Time for a change. *American Journal of Medical Genetics Part A, 155A*, 2907–2908.

Fisch, G. S. (2015). Communication and language in animals. In P. L. Roubertoux (Ed.), *Organism models of autism spectrum disorders* (pp. 265–282). Springer: New York.

Fisch, G. S., Carpenter, N., Howard-Peebles, P. N., Holden, J. J., Tarleton, J., Simensen, R., et al. (2012). Developmental trajectories in syndromes with intellectual disability, with a focus on Wolf-Hirschhorn and its cognitive-behavioral profile. *American Journal on Intellectual and Developmental Disabilities, 117*, 167–179.

Franceschini, A., Szklarczyk, D., Frankild, S., Kuhn, M., Simonovic, M., Roth, A., et al. (2013). STRING v9.1: Protein-protein interaction networks, with increased coverage and integration. *Nucleic Acids Research, 41*, D808–D815.

Glasgow, E., Karavanov, A. A., & Dawid, I. B. (1997). Neuronal and neuroendocrine expression of lim3, a LIM class homeobox gene, is altered in mutant zebrafish with axial signaling defects. *Developmental Biology, 192*, 405–419.
Goetz, S. C., & Anderson, K. V. (2010). The primary cilium: A signalling centre during vertebrate development. *Nature Reviews Genetics, 11*, 331–344.
Golzio, C., Willer, J., Talkowski, M. E., Oh, E. C., Taniguchi, Y., Jacquemont, S., et al. (2012). KCTD13 is a major driver of mirrored neuroanatomical phenotypes of the 16p11.2 copy number variant. *Nature, 485*, 363–367.
Grayton, H. M., Fernandes, C., Rujescu, D., & Collier, D. A. (2012). Copy number variations in neurodevelopmental disorders. *Progress in Neurobiology, 99*, 81–91.
Green, M. J., Cairns, M. J., Wu, J., Dragovic, M., Jablensky, A., Tooney, P. A., et al. (2013). Genome-wide supported variant MIR137 and severe negative symptoms predict membership of an impaired cognitive subtype of schizophrenia. *Molecular Psychiatry, 18*, 774–780.
Guemez-Gamboa, A., Coufal, N. G., & Gleeson, J. G. (2014). Primary cilia in the developing and mature brain. *Neuron, 82*, 511–521.
Guimera, J., Casas, C., Estivill, X., & Pritchard, M. (1999). Human minibrain homologue (MNBH/DYRK1): Characterization, alternative splicing, differential tissue expression, and overexpression in Down syndrome. *Genomics, 57*, 407–418.
Guimera, J., Casas, C., Pucharcos, C., Solans, A., Domenech, A., Planas, A. M., et al. (1996). A human homologue of Drosophila minibrain (MNB) is expressed in the neuronal regions affected in Down syndrome and maps to the critical region. *Human Molecular Genetics, 5*, 1305–1310.
Hayashi, S., & McMahon, A. P. (2002). Efficient recombination in diverse tissues by a tamoxifen-inducible form of Cre: A tool for temporally regulated gene activation/inactivation in the mouse. *Developmental Biology, 244*, 305–318.
Heisenberg, M., Borst, A., Wagner, S., & Byers, D. (1985). Drosophila mushroom body mutants are deficient in olfactory learning. *Journal of Neurogenetics, 2*, 1–30.
Herault, Y., Rassoulzadegan, M., Cuzin, F., & Duboule, D. (1998). Engineering chromosomes in mice through targeted meiotic recombination (TAMERE). *Nature Genetics, 20*, 381–384.
Hoogeveen-Westerveld, M., Wentink, M., van den Heuvel, D., Mozaffari, M., Ekong, R., Povey, S., et al. (2011). Functional assessment of variants in the TSC1 and TSC2 genes identified in individuals with Tuberous Sclerosis Complex. *Human Mutation, 32*, 424–435.
Hooper, M., Hardy, K., Handyside, A., Hunter, S., & Monk, M. (1987). HPRT-deficient (Lesch-Nyhan) mouse embryos derived from germline colonization by cultured cells. *Nature, 326*, 292–295.
Horev, G., Ellegood, J., Lerch, J. P., Son, Y. E., Muthuswamy, L., Vogel, H., et al. (2011). Dosage-dependent phenotypes in models of 16p11.2 lesions found in autism. *Proceedings of the National Academy of Sciences of the United States of America, 108*, 17076–17081.
Hsu, P. D., Lander, E. S., & Zhang, F. (2014). Development and applications of CRISPR-Cas9 for genome engineering. *Cell, 157*, 1262–1278.
Huang, L., Szymanska, K., Jensen, V. L., Janecke, A. R., Innes, A. M., Davis, E. E., et al. (2011). TMEM237 is mutated in individuals with a Joubert syndrome related disorder and expands the role of the TMEM family at the ciliary transition zone. *American Journal of Human Genetics, 89*, 713–730.
Incontro, S., Asensio, C. S., Edwards, R. H., & Nicoll, R. A. (2014). Efficient, complete deletion of synaptic proteins using CRISPR. *Neuron, 83*, 1051–1057.
Ingason, A., Kirov, G., Giegling, I., Hansen, T., Isles, A. R., Jakobsen, K. D., et al. (2011a). Maternally derived microduplications at 15q11-q13: Implication of imprinted genes in psychotic illness. *The American Journal of Psychiatry, 168*, 408–417.
Ingason, A., Rujescu, D., Cichon, S., Sigurdsson, E., Sigmundsson, T., Pietilainen, O. P., et al. (2011b). Copy number variations of chromosome 16p13.1 region associated with schizophrenia. *Molecular Psychiatry, 16*, 17–25.
Ioannidis, J. P. (2005). Why most published research findings are false. *PLoS Medicine, 2*, e124.

Irwin, S. A., Patel, B., Idupulapati, M., Harris, J. B., Crisostomo, R. A., Larsen, B. P., et al. (2001). Abnormal dendritic spine characteristics in the temporal and visual cortices of patients with fragile-X syndrome: A quantitative examination. *American Journal of Medical Genetics, 98*, 161–167.

Jacobs, S., Nathwani, M., & Doering, L. C. (2010). Fragile X astrocytes induce developmental delays in dendrite maturation and synaptic protein expression. *BMC Neuroscience, 11*, 132.

Jaisser, F. (2000). Inducible gene expression and gene modification in transgenic mice. *Journal of the American Society of Nephrology: JASN, 11*(Suppl 16), S95–S100.

Jamain, S., Quach, H., Betancur, C., Rastam, M., Colineaux, C., Gillberg, I. C., et al. (2003). Mutations of the X-linked genes encoding neuroligins NLGN3 and NLGN4 are associated with autism. *Nature Genetics, 34*, 27–29.

Jay, V., Becker, L. E., Chan, F. W., & Perry, T. L., Sr. (1991). Puppet-like syndrome of Angelman: A pathologic and neurochemical study. *Neurology, 41*, 416–422.

Jover, M., Ayoun, C., Berton, C., & Carlier, M. (2014). Development of motor planning for dexterity tasks in trisomy 21. *Research in Developmental Disabilities, 35*, 1562–1570.

Kalueff, A. V., Gebhardt, M., Stewart, A. M., Cachat, J. M., Brimmer, M., Chawla, J. S., et al. (2013). Towards a comprehensive catalog of zebrafish behavior 1.0 and beyond. *Zebrafish, 10*, 70–86.

Kane, M. T. (1992). Argument-based approach to validation. *Psychological Bulletin, 112*(3), 527–535.

Kane, M. (2013). The argument-based approach to validation. *School Psychology Review, 42*(4), 448–457.

Karmiloff-Smith, A., D'Souza, D., Dekker, T. M., Van Herwegen, J., Xu, F., Rodic, M., et al. (2012). Genetic and environmental vulnerabilities in children with neurodevelopmental disorders. *Proceedings of the National Academy of Sciences of the United States of America, 109*(Suppl 2), 17261–17265.

Kato, T. A., Yamauchi, Y., Horikawa, H., Monji, A., Mizoguchi, Y., Seki, Y., et al. (2013). Neurotransmitters, psychotropic drugs and microglia: Clinical implications for psychiatry. *Current Medicinal Chemistry, 20*, 331–344.

Kelley, K., Chang, S. J., & Lin, S. L. (2012). Mechanism of repeat-associated microRNAs in fragile X syndrome. *Neural Plasticity, 2012*, 104796.

Konopka, W., Kiryk, A., Novak, M., Herwerth, M., Parkitna, J. R., Wawrzyniak, M., et al. (2010). MicroRNA loss enhances learning and memory in mice. *The Journal of Neuroscience: The Official Journal of the Society for Neuroscience, 30*, 14835–14842.

Koopman, P., Munsterberg, A., Capel, B., Vivian, N., & Lovell-Badge, R. (1990). Expression of a candidate sex-determining gene during mouse testis differentiation. *Nature, 348*, 450–452.

Kuehn, M. R., Bradley, A., Robertson, E. J., & Evans, M. J. (1987). A potential animal model for Lesch-Nyhan syndrome through introduction of HPRT mutations into mice. *Nature, 326*, 295–298.

Laumonnier, F., Bonnet-Brilhault, F., Gomot, M., Blanc, R., David, A., Moizard, M. P., et al. (2004). X-linked mental retardation and autism are associated with a mutation in the NLGN4 gene, a member of the neuroligin family. *American Journal of Human Genetics, 74*, 552–557.

Leonard, H., & Wen, X. (2002). The epidemiology of mental retardation: Challenges and opportunities in the new millennium. *Mental Retardation and Developmental Disabilities Research Reviews, 8*, 117–134.

Leroy, C., Landais, E., Briault, S., David, A., Tassy, O., Gruchy, N., et al. (2013). The 2q37-deletion syndrome: An update of the clinical spectrum including overweight, brachydactyly and behavioural features in 14 new patients. *European Journal of Human Genetics: EJHG, 21*, 602–612.

Lévy, N., & Roubertoux, P. L. (2015). Organism models: Choosing the right model. In P. L. Roubertoux (Ed.), *Organism models of autism spectrum disorders* (pp. 3–28). New York: Springer.

Li, W., Cressy, M., Qin, H., Fulga, T., Van Vactor, D., & Dubnau, J. (2013). MicroRNA-276a functions in ellipsoid body and mushroom body neurons for naive and conditioned olfactory

avoidance in Drosophila. *The Journal of Neuroscience: The Official Journal of the Society for Neuroscience, 33*, 5821–5833.
Li, Y., Tian, C., Yang, Y., Yan, Y., Ni, Y., Wei, Y., et al. (2011). An inducible transgenic Cre mouse line for the study of hippocampal development and adult neurogenesis. *Genesis, 49*, 919–926.
Lo-Castro, A., Galasso, C., Cerminara, C., El-Malhany, N., Benedetti, S., Nardone, A. M., et al. (2009). Association of syndromic mental retardation and autism with 22q11.2 duplication. *Neuropediatrics, 40*, 137–140.
Lu, Y., Wang, F., Li, Y., Ferris, J., Lee, J. A., & Gao, F. B. (2009). The Drosophila homologue of the Angelman syndrome ubiquitin ligase regulates the formation of terminal dendritic branches. *Human Molecular Genetics, 18*, 454–462.
Madisen, L., Mao, T., Koch, H., Zhuo, J. M., Berenyi, A., Fujisawa, S., et al. (2012). A toolbox of Cre-dependent optogenetic transgenic mice for light-induced activation and silencing. *Nature Neuroscience, 15*, 793–802.
Madisen, L., Zwingman, T. A., Sunkin, S. M., Oh, S. W., Zariwala, H. A., Gu, H., et al. (2010). A robust and high-throughput Cre reporting and characterization system for the whole mouse brain. *Nature Neuroscience, 13*, 133–140.
Madrigal, I., Rodriguez-Revenga, L., Xuncla, M., & Mila, M. (2012). 15q11.2 microdeletion and FMR1 premutation in a family with intellectual disabilities and autism. *Gene, 508*, 92–95.
Marchetto, M. C., Carromeu, C., Acab, A., Yu, D., Yeo, G. W., Mu, Y., et al. (2010a). A model for neural development and treatment of Rett syndrome using human induced pluripotent stem cells. *Cell, 143*, 527–539.
Marchetto, M. C., Winner, B., & Gage, F. H. (2010b). Pluripotent stem cells in neurodegenerative and neurodevelopmental diseases. *Human Molecular Genetics, 19*, R71–R76.
Mason, R. P., & Giorgini, F. (2011). Modeling Huntington disease in yeast: Perspectives and future directions. *Prion, 5*, 269–276.
Matagne, V., Budden, S., Ojeda, S. R., & Raber, J. (2013). Correcting deregulated Fxyd1 expression ameliorates a behavioral impairment in a mouse model of Rett syndrome. *Brain Research, 1496*, 104–114.
Maulik, P. K., Mascarenhas, M. N., Mathers, C. D., Dua, T., & Saxena, S. (2011). Prevalence of intellectual disability: A meta-analysis of population-based studies. *Research in Developmental Disabilities, 32*, 419–436.
Mazzone, L., Ruta, L., & Reale, L. (2012a). Psychiatric comorbidities in Asperger syndrome and high functioning autism: Diagnostic challenges. *Annals of General Psychiatry, 11*, 16.
Mazzone, L., Vassena, L., Ruta, L., Mugno, D., Galesi, O., & Fichera, M. (2012b). Brief report: Peculiar evolution of autistic behaviors in two unrelated children with brachidactyly-mental retardation syndrome. *Journal of Autism and Developmental Disorders, 42*, 2202–2207.
McDermott, S., Durkin, M. S., Schupf, N., & Stein, Z. A. (2007). Epidemiology and etiology of mental retardation. In J. W. Jacobson, J. A. Mulick, & J. Rojahn (Eds.), *Handbook of intellectual and developmental disabilities* (pp. 3–40). New York: Springer.
McKinney, W. T. (2001). Overview of the past contributions of animal models and their changing place in psychiatry. *Seminars in Clinical Neuropsychiatry, 6*, 68–78.
McMichael, G., Girirajan, S., Moreno-De-Luca, A., Gecz, J., Shard, C., Nguyen, L. S., et al. (2014). Rare copy number variation in cerebral palsy. *European Journal of Human Genetics: EJHG, 22*, 40–45.
Mertz, L. G., Christensen, R., Vogel, I., Hertz, J. M., Nielsen, K. B., Gronskov, K., et al. (2013). Angelman syndrome in Denmark. Birth incidence, genetic findings, and age at diagnosis. *American Journal of Medical Genetics Part A, 161A*, 2197–2203.
Mervis, C. B., & John, A. E. (2008). Vocabulary abilities of children with Williams syndrome: Strengths, weaknesses, and relation to visuospatial construction ability. *Journal of Speech, Language, and Hearing Research: JSLHR, 51*, 967–982.
Mervis, C. B., & John, A. E. (2010). Cognitive and behavioral characteristics of children with Williams syndrome: Implications for intervention approaches. *American Journal of Medical Genetics Part C: Seminars in Medical Genetics, 154C*, 229–248.

Mouse Genome Sequencing, C., Waterston, R. H., Lindblad-Toh, K., Birney, E., Rogers, J., Abril, J. F., et al. (2002). Initial sequencing and comparative analysis of the mouse genome. *Nature, 420*, 520–562.

Mukaddes, N. M., & Herguner, S. (2007). Autistic disorder and 22q11.2 duplication. *The World Journal of Biological Psychiatry: The Official Journal of the World Federation of Societies of Biological Psychiatry, 8*, 127–130.

Muly, E. C., Nairn, A. C., Greengard, P., & Rainnie, D. G. (2008). Subcellular distribution of the Rho-GEF Lfc in primate prefrontal cortex: Effect of neuronal activation. *The Journal of Comparative Neurology, 508*, 927–939.

Ng, M. C., Yang, Y. L., & Lu, K. T. (2013). Behavioral and synaptic circuit features in a zebrafish model of fragile X syndrome. *PLoS One, 8*, e51456.

Nowaczyk, M. J., & Irons, M. B. (2012). Smith-Lemli-Opitz syndrome: Phenotype, natural history, and epidemiology. *American Journal of Medical Genetics Part C: Seminars in Medical Genetics, 160C*, 250–262.

Oeseburg, B., Dijkstra, G. J., Groothoff, J. W., Reijneveld, S. A., & Jansen, D. E. (2011). Prevalence of chronic health conditions in children with intellectual disability: A systematic literature review. *Intellectual and Developmental Disabilities, 49*, 59–85.

Osorio, F. G., Navarro, C. L., Cadinanos, J., Lopez-Mejia, I. C., Quiros, P. M., Bartoli, C., et al. (2011). Splicing-directed therapy in a new mouse model of human accelerated aging. *Science Translational Medicine, 3*, 106ra107.

Paterson, S. J., Brown, J. H., Gsodl, M. K., Johnson, M. H., & Karmiloff-Smith, A. (1999). Cognitive modularity and genetic disorders. *Science, 286*, 2355–2358.

Phelan, M. C. (2008). Deletion 22q13.3 syndrome. *Orphanet Journal of Rare Diseases, 3*, 14.

Philippe, A., Boddaert, N., Vaivre-Douret, L., Robel, L., Danon-Boileau, L., Malan, V., et al. (2008). Neurobehavioral profile and brain imaging study of the 22q13.3 deletion syndrome in childhood. *Pediatrics, 122*, e376–e382.

Provost, P. (2010a). Interpretation and applicability of microRNA data to the context of Alzheimer's and age-related diseases. *Aging, 2*, 166–169.

Provost, P. (2010b). MicroRNAs as a molecular basis for mental retardation, Alzheimer's and prion diseases. *Brain Research, 1338*, 58–66.

Qin, M., Entezam, A., Usdin, K., Huang, T., Liu, Z. H., Hoffman, G. E., et al. (2011). A mouse model of the fragile X premutation: Effects on behavior, dendrite morphology, and regional rates of cerebral protein synthesis. *Neurobiology of Disease, 42*, 85–98.

Quintana-Murci, L., & Fellous, M. (2001). The human Y chromosome: The biological role of a "Functional Wasteland". *Journal of Biomedicine & Biotechnology, 1*, 18–24.

Reep, R. (1984). Relationship between prefrontal and limbic cortex: A comparative anatomical review. *Brain, Behavior and Evolution, 25*, 5–80.

Reep, R. L., Corwin, J. V., Hashimoto, A., & Watson, R. T. (1984). Afferent connections of medial precentral cortex in the rat. *Neuroscience Letters, 44*, 247–252.

Roubertoux, P. L. (Ed.). (2015). *Organism models of autism spectrum disorders* (pp. ix–xiii). New York: Springer.

Roubertoux, P. L., Bichler, Z., Pinoteau, W., Jamon, M., Seregaza, Z., Smith, D. J., et al. (2006). Pre-weaning sensorial and motor development in mice transpolygenic for the critical region of trisomy 21. *Behavior Genetics, 36*, 377–386.

Roubertoux, P. L., Bichler, Z., Pinoteau, W., Seregaza, Z., Fortes, S., Jamon, M., et al. (2005). Functional analysis of genes implicated in Down syndrome: 2. Laterality and corpus callosum size in mice transpolygenic for Down syndrome chromosomal region-1 (DCR-1). *Behavior Genetics, 35*, 333–341.

Roubertoux, P. L., & Carlier, M. (2007). From DNA to mind. The decline of causality as a general rule for living matter. *EMBO Reports, 8 Spec No*, S7–S11.

Roubertoux, P. L. C., & Carlier, M. (2010). Mouse models of cognitive disabilities in trisomy 21 (Down syndrome). *American Journal of Medical Genetics Part C, Seminars in Medical Genetics, 154C*, 400–416.

Roubertoux, P. L., & Kerdelhue, B. (2006). Trisomy 21: From chromosomes to mental retardation. *Behavior Genetics, 36*, 346–354.

Ruis-Rubio, M., Calahorro, F., & Gámez-del-Estal, M. M. (2015). Invertebrate models of synaptic transmission in autism spectrum disorders. In P. L. Roubertoux (Ed.), *Organism models of autism spectrum disorders* (pp. 157–182). New York: Springer.

Ryan, S. J., Ehrlich, D. E., Jasnow, A. M., Daftary, S., Madsen, T. E., & Rainnie, D. G. (2012). Spike-timing precision and neuronal synchrony are enhanced by an interaction between synaptic inhibition and membrane oscillations in the amygdala. *PLoS One, 7*, e35320.

Sakamoto, M., Imayoshi, I., Ohtsuka, T., Yamaguchi, M., Mori, K., & Kageyama, R. (2011). Continuous neurogenesis in the adult forebrain is required for innate olfactory responses. *Proceedings of the National Academy of Sciences of the United States of America, 108*, 8479–8484.

Sanders, S. J., Ercan-Sencicek, A. G., Hus, V., Luo, R., Murtha, M. T., Moreno-De-Luca, D., et al. (2011). Multiple recurrent de novo CNVs, including duplications of the 7q11.23 Williams syndrome region, are strongly associated with autism. *Neuron, 70*, 863–885.

Sato, M., & Stryker, M. P. (2010). Genomic imprinting of experience-dependent cortical plasticity by the ubiquitin ligase gene Ube3a. *Proceedings of the National Academy of Sciences of the United States of America, 107*, 5611–5616.

Seregaza, Z., Roubertoux, P. L., Jamon, M., & Soumireu-Mourat, B. (2006). Mouse models of cognitive disorders in trisomy 21: A review. *Behavior Genetics, 36*, 387–404.

Serfontein, J., Nisbet, R. E., Howe, C. J., & de Vries, P. J. (2011). Conservation of structural and functional elements of TSC1 and TSC2: A bioinformatic comparison across animal models. *Behavior Genetics, 41*, 349–356.

Simmons, J. P., Nelson, L. D., & Simonsohn, U. (2011). False-positive psychology: Undisclosed flexibility in data collection and analysis allows presenting anything as significant. *Psychological Science, 2*, 359–366.

Sinclair, A. H., Berta, P., Palmer, M. S., Hawkins, J. R., Griffiths, B. L., Smith, M. J., et al. (1990). A gene from the human sex-determining region encodes a protein with homology to a conserved DNA-binding motif. *Nature, 346*, 240–244.

Smith, D. J., Stevens, M. E., Sudanagunta, S. P., Bronson, R. T., Makhinson, M., Watabe, A. M., et al. (1997). Functional screening of 2 Mb of human chromosome 21q22.2 in transgenic mice implicates minibrain in learning defects associated with Down syndrome. *Nature Genetics, 16*, 28–36.

Smith, D. J., Zhu, Y., Zhang, J., Cheng, J. F., & Rubin, E. M. (1995). Construction of a panel of transgenic mice containing a contiguous 2-Mb set of YAC/P1 clones from human chromosome 21q22.2. *Genomics, 27*, 425–434.

Song, G., Li, Q., Long, Y., Hackett, P. B., & Cui, Z. (2012). Effective expression-independent gene trapping and mutagenesis mediated by Sleeping Beauty transposon. *Journal of genetics and genomics, 39*, 503–520.

Staal, W. G., de Krom, M., & de Jonge, M. V. (2012). Brief report: The dopamine-3-receptor gene (DRD3) is associated with specific repetitive behavior in autism spectrum disorder (ASD). *Journal of Autism and Developmental Disorders, 42*, 885–888.

Stevens, H. E. (2010). Prefrontal cortex: Disorders and development. *Journal of the American Academy of Child and Adolescent Psychiatry, 49*, 203–204.

Stevens, H. E., Smith, K. M., Maragnoli, M. E., Fagel, D., Borok, E., Shanabrough, M., et al. (2010). Fgfr2 is required for the development of the medial prefrontal cortex and its connections with limbic circuits. *The Journal of Neuroscience: The Official Journal of the Society for Neuroscience, 30*, 5590–5602.

Straub, C., Granger, A. J., Saulnier, J. L., & Sabatini, B. L. (2014). CRISPR/Cas9-mediated gene knock-down in post-mitotic neurons. *PLoS One, 9*, e105584.

Suen, H. K., & French, J. L. (2003). A history of the development of psychological and educational testing. In C. R. Reynolds & R. W. Kamphaus (Eds.), *Handbook of psychological and educational assessment of children* (pp. 3–23). New York: Guilford.

Summerton, J., Stein, D., Huang, S. B., Matthews, P., Weller, D., & Partridge, M. (1997). Morpholino and phosphorothioate antisense oligomers compared in cell-free and in-cell systems. *Antisense & Nucleic Acid Drug Development, 7*, 63–70.

Summerton, J., & Weller, D. (1997). Morpholino antisense oligomers: Design, preparation, and properties. *Antisense & Nucleic Acid Drug Development, 7*, 187–195.

Swiech, L., Heidenreich, M., Banerjee, A., Habib, N., Li, Y., Trombetta, J., et al. (2015). In vivo interrogation of gene function in the mammalian brain using CRISPR-Cas9. *Nature Biotechnology, 33*, 102–106.

Szczypka, M. S., Kwok, K., Brot, M. D., Marck, B. T., Matsumoto, A. M., Donahue, B. A., et al. (2001). Dopamine production in the caudate putamen restores feeding in dopamine-deficient mice. *Neuron, 30*, 819–828.

Taniguchi, H., He, M., Wu, P., Kim, S., Paik, R., Sugino, K., et al. (2011). A resource of Cre driver lines for genetic targeting of GABAergic neurons in cerebral cortex. *Neuron, 71*, 995–1013.

Tarkar, A., Loges, N. T., Slagle, C. E., Francis, R., Dougherty, G. W., Tamayo, J. V., et al. (2013). DYX1C1 is required for axonemal dynein assembly and ciliary motility. *Nature Genetics, 45*, 995–1003.

Tasse, M. J., Luckasson, R., & Nygren, M. (2013). AAIDD proposed recommendations for ICD-11 and the condition previously known as mental retardation. *Intellectual and Developmental Disabilities, 51*, 127–131.

Tauber, E., Miller-Fleming, L., Mason, R. P., Kwan, W., Clapp, J., Butler, N. J., et al. (2011). Functional gene expression profiling in yeast implicates translational dysfunction in mutant huntingtin toxicity. *The Journal of Biological Chemistry, 286*, 410–419.

Teipel, S. J., & Hampel, H. (2006). Neuroanatomy of Down syndrome in vivo: A model of preclinical Alzheimer's disease. *Behavior Genetics, 36*, 405–415.

Tordjman, S., Drapier, D., Bonnot, O., Graignic, R., Fortes, S., Cohen, D., et al. (2007). Animal models relevant to schizophrenia and autism: Validity and limitations. *Behavior Genetics, 37*, 61–78.

Urraca, N., Cleary, J., Brewer, V., Pivnick, E. K., McVicar, K., Thibert, R. L., et al. (2013). The interstitial duplication 15q11.2-q13 syndrome includes autism, mild facial anomalies and a characteristic EEG signature. *Autism Research: Official Journal of the International Society for Autism Research, 6*, 268–279.

van Spronsen, M., van Battum, E. Y., Kuijpers, M., Vangoor, V. R., Rietman, M. L., Pothof, J., et al. (2013). Developmental and activity-dependent miRNA expression profiling in primary hippocampal neuron cultures. *PLoS One, 8*, e74907.

Vicari, S. (2006). Motor development and neuropsychological patterns in persons with Down syndrome. *Behavior Genetics, 36*, 355–364.

Vicari, S., & Carlesimo, G. A. (2006). Short-term memory deficits are not uniform in Down and Williams syndromes. *Neuropsychology Review, 16*, 87–94.

Villard, L. (2007). MECP2 mutations in males. *Journal of Medical Genetics, 44*, 417–423.

Wang, G. X., Smith, S. J., & Mourrain, P. (2014). Fmr1 KO and fenobam treatment differentially impact distinct synapse populations of mouse neocortex. *Neuron, 84*, 1273–1286.

Weber, P., Metzger, D., & Chambon, P. (2001). Temporally controlled targeted somatic mutagenesis in the mouse brain. *The European Journal of Neuroscience, 14*, 1777–1783.

Willemsen, M. H., Valles, A., Kirkels, L. A., Mastebroek, M., Olde Loohuis, N., Kos, A., et al. (2011). Chromosome 1p21.3 microdeletions comprising DPYD and MIR137 are associated with intellectual disability. *Journal of Medical Genetics, 48*, 810–818.

Witsch-Baumgartner, M., Schwentner, I., Gruber, M., Benlian, P., Bertranpetit, J., Bieth, E., et al. (2008). Age and origin of major Smith-Lemli-Opitz syndrome (SLOS) mutations in European populations. *Journal of Medical Genetics, 45*, 200–209.

Wu, C. L., & Melton, D. W. (1993). Production of a model for Lesch-Nyhan syndrome in hypoxanthine phosphoribosyltransferase-deficient mice. *Nature Genetics, 3*, 235–240.

Yang, M., Bozdagi, O., Scattoni, M. L., Wohr, M., Roullet, F. I., Katz, A. M., et al. (2012). Reduced excitatory neurotransmission and mild autism-relevant phenotypes in adolescent Shank3 null mutant mice. *The Journal of Neuroscience: The Official Journal of the Society for Neuroscience, 32*, 6525–6541.

Zeng, H., Shen, E. H., Hohmann, J. G., Oh, S. W., Bernard, A., Royall, J. J., et al. (2012). Large-scale cellular-resolution gene profiling in human neocortex reveals species-specific molecular signatures. *Cell, 149*, 483–496.

Zhuang, X., Masson, J., Gingrich, J. A., Rayport, S., & Hen, R. (2005). Targeted gene expression in dopamine and serotonin neurons of the mouse brain. *Journal of Neuroscience Methods, 143*, 27–32.

Zufferey, F., Sherr, E. H., Beckmann, N. D., Hanson, E., Maillard, A. M., Hippolyte, L., et al. (2012). A 600 kb deletion syndrome at 16p11.2 leads to energy imbalance and neuropsychiatric disorders. *Journal of Medical Genetics, 49*, 660–668.

Chapter 6
Neurodegenerative Diseases and Dementia

Christopher Janus and Hans Welzl

Introduction

"A man is but what he knoweth" wrote Sir Francis Bacon (Dubos 1968). This phrase defines one of our traits we value most in being human. In this respect, neurodegenerative diseases are devastating age-associated disorders that compromise and eventually destroy our memory and ability to learn, "de-humanizing" (dementia) our lives.

Cognition is a loosely defined term with divergent meanings in different disciplines. In psychology, human cognition is equivalent to "mind" or "higher mental functions," and cognitive abilities comprise perception, attention, learning and memory, thinking, planning, decision-making, and language. In more general terms, cognition includes the ability that helps to collect, process, and store information. Cognition in humans can only be studied indirectly by creating a suitable experimental environment, collecting behavioral data or writing down introspective reports. Brain activities can also be simultaneously recorded; but they are by themselves still inconclusive of the exact content of ongoing higher mental functions. Despite its inaccessibility as a research object and intrinsic complexity, an increasing number of psychologists, neuroscientist, and geneticists study cognitive abilities in humans and animals (Cromwell and Panksepp 2011). One of the hopes is that

C. Janus, Ph.D. (✉)
Department of Neuroscience, Center for Translational Research in Neurodegenerative Disease, University of Florida, 1275 Center Drive, BMS J-495, P. O. Box 100159, Gainesville, FL 32610, USA
e-mail: janus.christopher@gmail.com

H. Welzl, Ph.D.
Division of Neuroanatomy and Behavior, Institute of Anatomy, University of Zurich, Winterthurerstrasse 190, Zurich CH-8057, Switzerland
e-mail: h.welzl@bluewin.ch

J.C. Gewirtz, Y.-K. Kim (eds.), *Animal Models of Behavior Genetics*, Advances in Behavior Genetics, DOI 10.1007/978-1-4939-3777-6_6

such combined efforts will increase our knowledge of neural mechanisms underlying cognitive faculties and why they decline in many disease states. In this respect the goal of translational research is to develop animal models, which will adequately represent cognitive profiles relevant to homologous behavioral systems in humans (Sharbaugh et al. 2003). Models with such propensities would likely be most promising in the preclinical studies targeting potential therapeutics mediating dementia in humans. Consequently, in this chapter we will use the term "cognition" as a broad term encompassing mnemonic functions which might be successfully investigated in animals and which might translate reasonably well to cognitive disorders in neurodegenerative diseases.

Defining Cognition

Behavioral genetic studies have provided ample evidence that variability in cognition is, like almost all other types of behavior to a lesser or larger part, genetically determined (Plomin et al. 2001). Knowing the genetic contribution to behavior is also essential to determine the extent of environmental influences on behavioral variability. However, investigating a genotype–behavior relationship is difficult since few genes impact a single type of behavior, and many genes show pleiotropic effects affecting activity of several different genes and consequently modulating behavior (Denenberg 2000).

Research investigating the genetic contribution to cognition in rodents has followed three major lines. In the first line of research, a few laboratories use a quantitative genetic approach to look for genetic influences on general cognitive ability in rats and mice. General cognitive ability in humans (g) (Spearman 1904) determines performance in a battery of cognitive tests. The goal of the equivalent animal research is to develop a battery of cognitive tasks that best reflects their 'g' (Anderson 2000). Such a battery will then help in the search for a linkage between performance levels in cognitive tasks and different alleles. The second line of research attempts to unravel the contribution of specific genes to cognitive abilities using mainly rodent species. To that end, researchers manipulate genes and record the resultant changes in learning and memory. Based on the results they construct models of molecular and cellular mechanisms crucial for learning and memory. The third line of research attempts to model human mental retardation, including senile and pathological dementias in mice and rats (Cohen et al. 2013; Do Carmo and Cuello 2013; Kastner et al. 2012; Kelly et al. 2009). Deficits in cognitive abilities during early development are the hallmark of mental retardation, and decline in cognitive function with age characterizes senile dementias. For both syndromes genetic causes or predispositions have been detected in humans. Rodent models carrying similar gene defects or transgenically over-expressing mutated human genes implicated in particular diseases are tools to investigate the etiology and neuropathological changes responsible for reduced cognitive function.

Quantitative Genetic Approaches to General Cognitive Ability

The concept of general cognitive ability assumes a *hierarchical organization* of cognitive function. At the bottom level are different cognitive tasks. On a higher level, specific cognitive abilities such as spatial orientation or working memory each affect performance in several cognitive tasks. Only general cognitive ability at the top level affects performance in all cognitive tasks (Fig. 6.1). This psychometric simple model implies that to measure 'g' subjects should be submitted to a battery of tasks probing several different specific cognitive abilities. A performance index averaged across these tasks can then be taken as a value for 'g'. Whereas non-cognitive factors such as activity, stress, anxiety, or motivation might affect performance in some tasks, only 'g' is assumed to affect performance in all cognitive tasks. Quantitative genetic studies in humans strongly support a genetic influence on general cognitive ability (Plomin et al. 2001). In the case of mice and rats, quantitative genetic research of 'g' uses selectively bred animals (outbred strains, F2 of inbred strains) and submits them to a battery of tasks to evaluate their general cognitive ability. If general cognitive ability considerably influences performance in all cognitive tasks then an individual should maintain a similar rank score within a group over all tasks. It is important to stress that cognitive behavior is always prone to be affected by non-cognitive aspects of a task. To control for such confounding effects, cognitive tasks are selected that vary with respect to the stress and motivation involved, and further, tests evaluating predominantly non-mnemonic behavior should be included in the global phenotyping evaluation. Psychometric analysis should reveal a factor 'g' that loads positively with performance in each cognitive task but not with performance in non-cognitive tasks. Thus, to obtain a score of 'g': (1) a test battery of diverse measures has to be established, (2) the measures must be reliable at the level of the individual, and (3) large samples of animals have to be employed (Plomin 2001). A battery of cognitive tasks that measures 'g' specifically in mice has yet to be established.

These theoretical assumptions are demanding and, as the available evidence indicates, difficult to meet in the experimental laboratory studies. During the last decade a few laboratories attempted to establish a battery of tasks measuring 'g' in mice (Galsworthy et al. 2002, 2005; Locurto et al. 2003; Matzel et al. 2003). These studies used different batteries of cognitive tasks with varying types of motivation and stress levels, and all of them found evidence for 'g' in mice with 'g' accounting for approximately 30–60 % of the variance in performance. More complex tasks had a higher loading for 'g'. However, different measures of performance in one task did not load equally well with 'g'. Other tasks were strongly influenced by non-cognitive factors that overshadowed the influence of 'g'. Such tasks are less suitable for testing cognitive abilities since the variability in test scores is largely due to variability in non-cognitive behavior. Furthermore, the variability in 'g' can be correlated with allele variability in a top-down approach. Defined outbred lines of mice (see, e.g., Galsworthy et al. 2005) and the F2 generation of inbred strains can be used in studies looking for a linkage between a high, or low, score of 'g' and specific genetic markers.

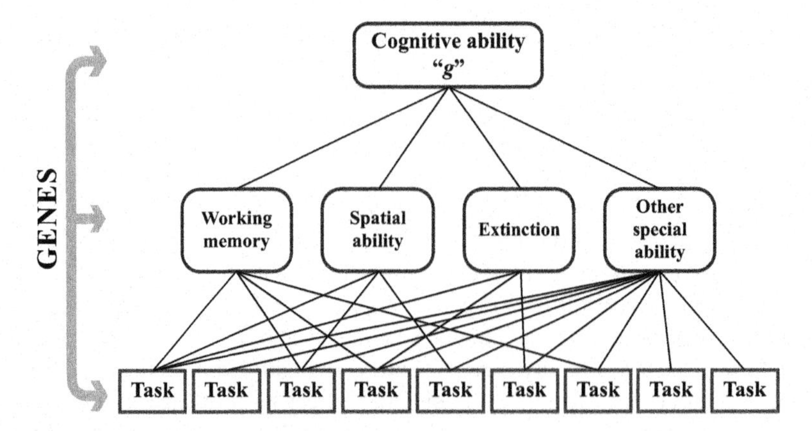

Fig. 6.1 Performance in cognitive tasks (*bottom level*) is affected by specific cognitive abilities (*middle level*), and all these specific cognitive abilities are influenced by a general cognitive ability (*top level*). Genes have an effect on performance by modulating abilities on all three levels (adapted from Plomin et al. 2001)

This seemingly straightforward approach might be complicated by the fact that a large number of genes likely influence 'g'. Also, such studies would require a large number of mice subjected to several tasks in order to detect individual gene influences, which would be a time-consuming and costly process, without guarantee that genes contributing only modestly to 'g' might be detected. While support for a general cognitive ability index in mice accumulates, the question arises as to what might be its neural basis. Research in humans suggests that variability in brain structure or changes in brain structure during development might be related to 'g' (for review see Shaw 2007; Toga and Thompson 2005). Tentative data for a similar relationship between brain size or fiber architecture and performance in cognitive tasks also exist (Anderson 1993; Chiang et al. 2009; Rushton and Ankney 2009). Variability in processes involved in neural development and plasticity is even more likely to influence 'g'. This is especially relevant with reference to genetically engineered mouse models of neurodegenerative diseases, in which the expression of mutated human genes might considerably change the developmental trajectory and affect the expression of other genes due to the insertion or linkage effects (Gerlai 1996), or compensatory developmental processes. Although the quantitative approach to cognition in rodents has not yet isolated specific genes that quantitatively contribute to the variability in 'g', the concept of what 'g' is and how it could be measured in a test battery is highly relevant for all other studies looking for genetic influences on cognition. The search for batteries of cognitive tasks that measure 'g' made us aware of the limitations of different learning and memory tasks. A low score of genetically manipulated mice in learning and memory tests could be due to compromised cognitive behavior or due to changes in non-mnemonic processes. Continued efforts to improve batteries of cognitive tasks should reveal the extent to which mnemonic as well as non-mnemonic processes influence individual tasks, hopefully paving the way to the development of a representative battery of tests that will provide an index of cognition in mice, which will be task-independent and comparable with 'g' in humans. Only then might the preclinical studies focusing on the mechanisms underlying dementia be aligned closely with the clinical efforts aiming to prevent or treat dementia in humans.

Dementia in Neurodegenerative Diseases

Dementia with neurodegeneration is characterized by a progressive decline in mental function that results from loss of the underlying neuronal architecture. The regional brain specificity and pathological hallmarks of each disease manifest behaviorally in patients as distinct, although sometimes partially overlapping clinical entities (Lee et al. 2001). Alzheimer's disease (AD) is the most prominent of these disorders, and is characterized by progressive memory loss leading to overt dementia and patient's death. More than 100 years ago, Alois Alzheimer published a seminal paper in which he documented the behavioral symptoms of his patient, Auguste Deter, who suffered from a mental illness (Alzheimer 1907; see Stelzmann et al. 1995 for English translation of the original paper). His main observation was that "*... she developed a rapid loss of memory. She was disoriented in her home ... She is completely disoriented in time and space. Her memory is seriously impaired. If objects are shown to her, she names them correctly, but almost immediately afterwards she has forgotten everything.*" In the final description of the patient's state Dr. Alzheimer wrote: "*As the illness progressed ... the imbecility of the patient increased in general. Her death occurred after four and a half years of illness. At the end, the patient was lying in bed in a foetal position completely pathetic, incontinent.*" An autopsy performed after death revealed dense deposits outside and around the nerve cells in her brain as well as twisted strands of fiber inside dead neurons. AD is one of the most devastating age-associated neurodegenerative disorders that compromises and eventually destroys our memory and ability to learn. Therefore, the understanding of the etiology, risk factors, and mechanisms of cognitive impairment caused by AD and other neurodegenerative diseases is pivotal for the development of effective therapeutics that might prevent or ameliorate symptoms of these diseases which may reach pandemic proportions by the end of the century.

Neurodegenerative Diseases: Alzheimer's Disease and Frontotemporal Dementia

The identification of gene mutations implicated in familial forms of AD (FAD) opened a new research field for disease modeling. Although no mutations directly associated with tau neurofibrillary tangles (NFTs) have been identified in AD, the accumulation of hyperphosphorylated forms of the mircotubule associated protein tau (MAPT) in NFTs is the most obvious pathological hallmark of AD and other neurodegenerative dementias collectively called neurodegenerative tauopathies (Lee et al. 2001). NFTs are ubiquitously present in a variety of other dementias like frontotemporal dementia with Parkinsonism linked to chromosome 17 (FTDP-17), Pick's disease, progressive supranuclear palsy (PSP), Argyrophilic grain dementia, Creutzfeldt–Jakob disease, Down's syndrome, Gerstmann–Sträussler–Scheinker disease, Hallervorden–Spatz disease, Tangle-only dementia, and other tauopathies (Lee et al. 2001). The discovery of FAD and over 30 mutations in *MAPT* made these genes powerful candidates for generation of transgenic models of neurodegeneration in a mouse.

Genes Implicated in AD and FTD

The first mutation identified as implicated in familial AD was a missense mutation in the gene encoding amyloid-β precursor protein (*APP*) on chromosome 21 (Chartier-Harlin et al. 1991a, b; Goate et al. 1991), followed by mutations in presenilin 1 (*PS1*) on chromosome 14 (Campion et al. 1995; Sherrington et al. 1995) and presenilin 2 (*PS2*) on chromosome 1, identified in Volga German and Italian families (Levy-Lahad et al. 1995; Rogaev et al. 1995). Although initially it was unclear how mutations in *APP* and *PS* could result in the AD phenotype, it soon became apparent that mutations in these genes had a direct effect on amyloid beta (Aβ) levels. Studies using assays of Aβ levels in plasma and cultured fibroblasts from patients harboring these mutations revealed a selective enhancement in the production of Aβ, which was considered to be a hallmark of the disease (Scheuner et al. 1996). At present, significant evidence supports a hypothesis that elevated production of Aβ plays a pivotal role in the pathogenesis of AD (Golde 2003; Hardy and Higgins 1992; Selkoe 2002) (Fig. 6.2).

A fourth FAD locus is believed to be located on chromosome 12, with the *α-2 macroglobulin* (*A2M*) and its receptor, the low-density lipoprotein receptor-related protein (LRP1) being under consideration (Bertram and Tanzi 2001; Blacker et al. 1998; Depboylu et al. 2006; Zappia et al. 2002). Despite extensive genetic screens, investigators have identified that mutations in *APP, PS1*, and *PS2* genes remain rare in the disease. To date, 33 mutations have been reported for *APP*, 185 for *PS1*, and only 13 for *PS2*; Alzheimer Disease and Frontotemporal Dementia Mutation Database source (http://www.molgen.vib-ua.be/ADMutations) (Cruts et al. 2012).

Epidemiological studies also revealed that 40–50 % of all AD patients are carriers of the ε4-allele of the *apolipoprotein E* gene (*ApoE*, located on chromosome 19), in contrast to only 15 % in the normal, healthy population (Strittmatter et al. 1993). Since no direct causal links or inheritance patterns are evident, the presence of *ApoE4* identifies an allele-specific risk factor of late-onset AD (see Chap. 7 in this volume). The presence of both *ApoE-ε4* alleles leads to an earlier onset of AD in comparison to patients with one or no *ApoE-ε4* alleles (Corder et al. 1993). Interestingly, there is evidence suggesting that its effect might vary amongst different

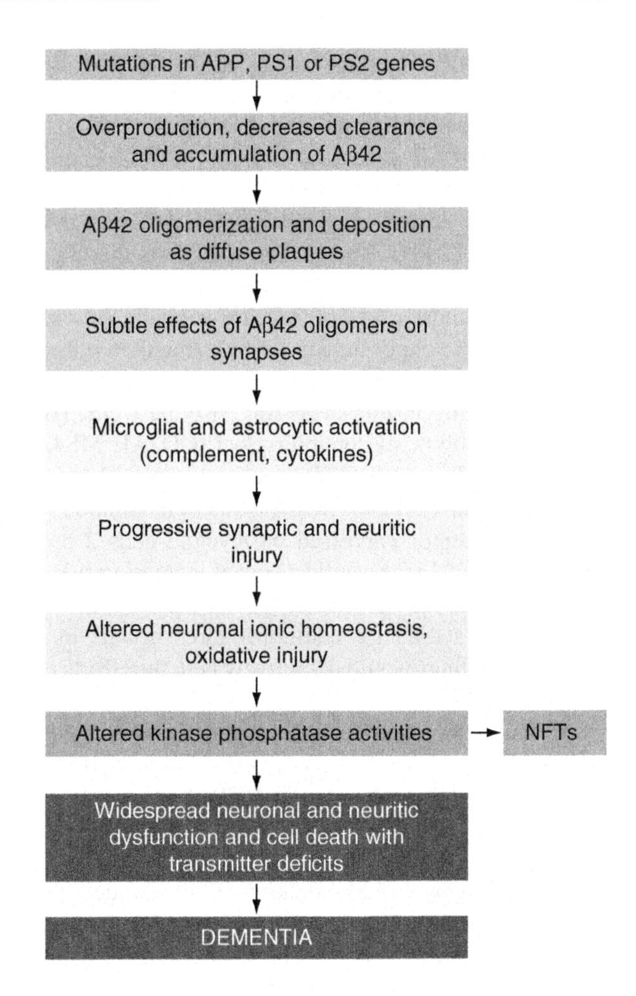

Fig. 6.2 The amyloid cascade hypothesis. The cascade is initiated by overproduction of Aβ42 as a result of pathogenic mutations, which trigger a sequence of pathological events, including downstream formation of neurofibrillary tau tangles (NFT), with neuronal loss and dementia as final endpoints

races (Evans et al. 2003). The presence of *ApoE-ε4* was found to increase the risk of developing AD amongst white Americans, but have no significant effect amongst African Americans or Hispanics (Evans et al. 2003; Tang et al. 1998). The reasons for this differential risk factor remain unknown, but linkage disequilibrium between *ApoE* and another allele or an unknown environmental or genetic factor present in some races and not others was suggested as one possibility (Evans et al. 2003; Kim et al. 2001). Moreover, a study based on patients in metropolitan New York confirmed the results for carriers of the *ApoE-ε4* allele, but also showed that African Americans and Hispanics without the presence of *ApoE-ε4* were more likely than white Americans to develop AD by the age of 90 years (Tang et al. 1998), indicating that more studies are needed to elucidate further the complex interplay between various genetic risk factors of AD. Together, these observations indicate that both environmental and genetic factors trigger a chain of events, which converge onto a final pathogenic pathway and lead to the stereotypic neuropathology (Hardy and Higgins 1992). The inheritance of AD follows a dichotomous pattern. Rare mutations in

APP, *PS1*, and *PS2* guarantee early-onset (<60 year) FAD, which is characterized by Mendelian inheritance and represents about 5 % of AD (Tanzi 2012). The *ApoE-ε4* allele represents a risk-modifying factor for late-onset AD. Together, *APP*, *PS1*, *PS2*, and *ApoE-ε4* genes account for 30–50 % of genetic variance in AD (Tanzi 2012; Tanzi and Bertram 2001). In addition to the *ApoE* locus, two other loci at *CLU* and *CR1* were identified as potentially increasing risk of AD (Harold et al. 2009; Lambert et al. 2009). The evidence suggests that the genes at these loci, together with *ApoE*, are involved in Aβ clearance.

However, most cases of AD are sporadic (idiopathic) of unknown etiology, which can be strongly influenced by variants in other genes and/or environmental factors. Recent Genome-Wide Associations Studies (GWAS) identified variants of a gene encoding insulin-degrading enzyme (*IDE*) (Bertram et al. 2000; Ertekin-Taner et al. 2000), susceptibility loci (CD2AP, MS4A6A/MS4A4E, EPHA1, and ABCA7) implicated in ATP regulation and influx of lipids (Hollingworth et al. 2011; Naj et al. 2011), and loss-of-function mutation in *TREM2* gene, encoding the *t*riggering *r*eceptor *e*xpressed on *m*yeloid cells *2* protein (Guerreiro et al. 2012; Jonsson et al. 2013) as candidate genes increasing risk of late-onset of AD. Although all these identified susceptibility variants confer only small effects of risk, it is important to search for additional rare variants in other genes that predispose AD. Such findings would definitively help the efforts to develop novel strategies for preventing AD.

Although no mutations that are directly associated with NFTs have been identified in AD, the accumulation of various hyperphosphorylated forms of the mircotubule associated protein tau in NFTs is the most obvious pathological hallmark of AD. Many pathological and functional studies pointed to disturbances in the neuronal cytoskeleton involving tau processing as a likely cause leading to cell loss (Mandelkow and Mandelkow 1998; Mandelkow et al. 2003; Mattson 2004; Rizzu et al. 2000). The discovery of over 30 mutations in *MAPT* in patients with FTDP-17, an autosomal dominant inherited tauopathy (Clark et al. 1998; Hutton et al. 1998; Spillantini et al. 1998), and evidence that excessive intraneuronal deposition of tau protein is also a key feature of dying neurons during normal aging, occurring most often without amyloid deposits (Braak and Braak 1997), suggests that tau dysfunction may be sufficient to cause neuronal death, thus making this gene a particularly interesting candidate for models of neurodegeneration.

Pathology

Pathological diagnosis of AD is based on the presence of extracellular senile plaques and intracellular neurofibrillary tangles found throughout specific brain regions (Price et al. 1991). Senile plaques are found in large numbers in the limbic and associative cortices of Alzheimer's disease patients and are composed principally of extracellular Aβ protein in fibrillar form (see Chap. 7 in this volume). The bulk of the fibrillar Aβ consists of a 42 amino acid peptide (Aβ42), cleaved from amyloid precursor protein (APP) by the γ-secretase complex (Hardy and Selkoe 2002). Although this species is generated by γ-secretase less frequently than the shorter 40 amino acid (Aβ40) species, Aβ42 is considerably more hydrophobic and is prone to aggregation and is thought to be pivotal in the pathogenesis of AD (Hardy and Higgins 1992; Selkoe 2002) (Fig. 6.2). Both Aβ40 and Aβ42 typically co-localize in amyloid plaques (Price et al. 1991). NFTs are composed of various hyperphosphorylated forms of tau filaments wound into paired helical filaments (PHF) (Ksiezak-Reding and Wall 2005; Selkoe 2001; von Bergen et al. 2005). In AD and other tauopathies, many neurons in the affected brain regions—including the frontal, temporal, parietal and occipital associative cortices, the entorhinal cortex, hippocampus, parahippocampal gyrus and amygdala—contain large bundles of perinuclear tau fibers that fill most of the cytoplasmic compartment (Ittner et al. 2011).

The regional age-progressing formations of extra- and/or intracellular deposits comprised of misfolded and aggregated proteins, include: extracellular amyloid plaques and intracellular tau deposits in AD (Cummings et al. 1996; Davies et al. 2005; Hardy and Selkoe 2002; Kosik and Shimura 2005; Ksiezak-Reding and Wall 2005; Price et al. 1991); intracellular NFTs in frontotemporal dementia with parkinsonism linked to chromosome 17 (FTDP-17) (Bugiani et al. 1999; Goedert et al. 1998; Hutton et al. 1998; Mirra et al. 1999; Spillantini et al. 1998); intracellular deposits of α-synuclein contained in Lewy bodies in Parkinson disease (Chartier-Harlin et al. 2004; Clayton and George 1999; Conway et al. 2000; Cummings 2004). These formations lead to severe loss of neurons and dysfunction of neuronal circuits, which is a prominent feature of all tauopathies (Lee et al. 2001).

Behavioral Hallmarks of Tauopathies

The major clinical hallmark that signals the onset of AD and other tauopathies is a progressive decline in cognitive faculties with compromised learning and memory, and reduced problem-solving speed (Albert 1996). Another phenotypic feature, often concurrent to progressing dementia, encompasses a disturbance in other behavioral systems, with manifestations of delusions, depression, agitation, or aggressive behavior (Burns et al. 1990a, b, c, d; Victoroff et al. 1996). Such behavioral disturbances may not be directly related to the decline in memory per se, but may profoundly affect patient life and the results of cognitive evaluation, thus potentially biasing the severity of diagnosed dementia. The most prominent facet of neurodegeneration is progressing with age dementia. Aging is not only one of the major risk factors in neurodegenerative disorders but, in conjunction with genetic, epigenetic and environmental factors, often results in the deterioration of cognitive functions and coinciding neuronal loss in healthy individuals.

In summary, the main hallmarks of tauopathies are (1) age-progressing dementia, (2) disturbance in behavioral systems not related directly to cognitive function,

(3) coinciding neuronal loss in disease-relevant brain regions and cytoarchitecture, and (4) intra- and/or extracellular deposition of misfolded proteins that are specific to a given disorder. An animal model, which would recapitulate all or at least the main key hallmarks of the clinical phenotype, would present a powerful tool helping to delineate the causes underlying cognitive impairment in tauopathies. Here, the identification of gene mutations implicated in familial forms of tauopathies has opened a new research field for disease modeling. Recently generated in vitro cellular or in vivo transgenic animal models present unique and important experimental systems for the analysis of clinical phenomena.

Modeling Dementia with Neurodegeneration in a Mouse

From a historical perspective molecular genetics, using predominantly inbred mice, has been vigorously investigating whether specific genes are essential for or modify learning and memory. This forward-genetics or bottom-up approach involves the manipulation of specific genes and the comparison of the behavior of such genetically engineered mice with the behavior of their wild-type littermates (King and Takahashi 1996; Takahashi et al. 1994). Gene targeting or transgenesis can disable, reduce, or increase the expression of genes (Muller 1999). While so-called constitutive mutant mice carry the mutation throughout their life, in conditional models the mutation is activated only during an experimentally restricted time in development. Furthermore, knock-in methods allow mutated forms of a gene to replace the mouse's endogenous genes, thus further humanizing the models. Although the use of mutant mice has expanded our knowledge about mechanisms of synaptic plasticity implicated in almost all forms of learning and memory processes (see Grant et al. 1992; Silva et al. 1992 for the pioneering work), several cautionary aspects of research related to the use of mutant mouse models should be emphasized. (1) Mutant mice submitted to a battery of behavioral tests often revealed unexpected pleiotropic effects of genes; manipulating a gene may also change, in addition to learning and memory, other behaviors including anxiety, locomotor activity, or aggression. Such effects have appeared even when gene manipulation has been restricted to certain brain areas and restricted in time. (2) Consequently, manipulating genes may up- or downregulate expression of other genes, and it may lead to an adjustment of the equilibrium in various systems not directly targeted by the genetic manipulation (Blendy et al. 1996; Parish et al. 2005). (3) Mouse strain genetic background may exacerbate or compensate for a deleted gene, for example. Even when careful breeding schedules are employed and mutant and wild-type littermates are used for testing, a mutation-induced defect may show up with one genetic background but not with another. Mouse strains greatly differ in mnemonic (Brooks et al. 2005) as well as in non-mnemonic behaviors (Kim et al. 2005). (4) Lines of mice with the same targeted gene but created in different laboratories sometimes differ in their behavioral phenotype (see Brambilla et al. 1997; versus Giese et al. 2001). (5) Performance in learning and memory tasks cannot be uncritically equated with cognition. As quantitative genetic studies have convincingly demonstrated, each individual task depends on non-mnemonic factors, as well as on specific and general cognitive ability, as discussed in section "Quantitative Genetic Approaches to General Cognitive Ability."

From this perspective, the identification of gene mutations implicated in neurodegenerative diseases, at a the time when forward genetic studies have aggressively focused on mutant mouse models targeting the mechanisms of memory formation, has created unique opportunities to generate mouse models that over-express mutated human genes implicated in AD, allowing for investigation of the role of these genes in the mediation of cognitive function. The first models focused on reproducing specific aspects of neuropathology, like neuronal damage to cholinergic depletion (Beeri et al. 1995; Crutcher et al. 1993; Cummings and Masterman 1998; Lin et al. 1998; Whishaw 1985). While informative, these models suffered from being too restricted, from being acute (rather then gradually progressive), and from a focus on often one specific cognitive task (Alvarez et al. 1997; Baxter and Gallagher 1996; Beeri et al. 1995; Dickson and Vanderwolf 1990). It is not surprising that the advent of animal models began with the identification of gene mutations associated with neurodegenerative diseases.

Among many of available models, cellular models have been especially powerful in helping to elucidate the intracellular pathways leading to neuronal death and perform biochemical analyses leading to high-throughput screening of potential therapeutic targets. Genetic models based on invertebrate species like the fruit fly *Drosophila* sp. (Ali et al. 2011; Bonner and Boulianne 2011; Jahn et al. 2011) and nematode *Caenorhabditis elegans* (Calahorro and Ruiz-Rubio 2011; Ewald and Li 2010; Morcos and Hutter 2009) have indisputable economic advantages, including the possibility of maintaining whole populations in the laboratory. The comprehensive knowledge of complete cell lineages of the species provided new opportunities to develop genetic screens for identification of modifier or suppressor genes implicated in a disease. The use of a plethora of genetic approaches in concert with genomic technologies make these genetic models a powerful tool for understanding basic cellular processes and exploring the molecular basis of human neurodegenerative diseases (Tickoo and Russell 2002). Finally, transgenic, knock-in (KI), and conditional mouse and rat models of neurodegenerative diseases made a significant contribution to our understanding of the intracellular cascades leading to neuronal dysfunction and coinciding behavioral deficits associated with the expression of mutated or dysfunctional proteins (Gama Sosa et al. 2012). The first successful mouse model replicating major hallmarks of AD was characterized about two decades ago by Games and colleagues (Games et al. 1995). The ability to study the effect of expression of human mutated genes linked to AD and other tauopathies in mouse models proved extremely informative and has been chronicled in a number of scholarly reviews (Ashe 2001, 2005; Ashe and Zahs 2010; Dodart et al. 2002b; Eriksen and Janus 2007; Greenberg et al. 1996; Higgins and Jacobsen 2003; Janus and Westaway 2001; Price et al. 1998; Seabrook and Rosahl 1999; Spires and Hyman 2005; van Leuven 2000; Wong et al. 2002; Zahs and Ashe 2010). The work using mouse models of AD also led to major advances in the development and testing of new therapeutic strategies (Dodart et al. 2002a; Janus et al. 2000b; Jensen et al. 2005; Kotilinek et al. 2002; Oddo et al. 2004; Schenk et al. 1999), which has led to clinical trials (for reviews and detailed reports see Delrieu et al. 2012; Elder et al. 2010; Hock et al. 2003; Huang and Mucke 2012; Nicoll et al. 2003; Orgogozo et al. 2003; Schenk 2002). An additional advantage of using rodent models is that the preponderance of our understanding of the biological changes during aging comes from in vivo studies in rodents, whose similarities to humans in terms of physiology and cell biology of aging makes them a valuable model for studying the effects of neurodegeneration.

Mouse Models: Criteria

At a juncture of about two decades of using mouse models for AD, it seems useful to revise some of the criteria and expectations of a robust and good model. Since the cognitive decline and region specific neuronal loss are central to neurodegenerative diseases, the model of a disease should recapitulate accurately these facets of the clinical phenotype. Furthermore, a credible model should exhibit age-progressive neuropathology and cognitive deficits, which should be evident, to various extents, in paradigms addressing different memory systems. The extent of age-related behavioral impairment may eventually encompass non-cognitive systems due to significantly progressing pathology. Although this may raise operational complexities associated with the interpretation of the results confounded by the late emergence of impairment in non-cognitive systems, which we discussed earlier, the use of such models during screens of potential therapeutics may reveal which behavioral deficiencies can be ameliorated by a treatment at a given stage of pathology. The oft-suggested independent confirmation and replication of the results in several laboratories (Janus et al. 2000a) is important, however, as recent advances in the field of behavioral genetics showed, may not be so easily obtainable, even when experiments are conducted in well-established behavioral laboratories under supervision of experienced researchers (Crabbe et al. 1999). Idiosyncratic, difficult to control, and almost impossible to standardize differences existing between various labs and animal colonies, including expertise of technical personnel and differences in handling methods (Wahlsten et al. 2003), make the replication of subtle differences rather difficult. A pragmatic approach would dictate that robust phenotypes obtained in less liable tests (in which data collection is based on motor or strong sensory inputs) should be replicable within tolerable margins, while more labile phenotypes based on emotional or social behaviors may be strongly affected by differences in laboratories (Wahlsten et al. 2006), especially in poorly managed animal colonies (Janus and Welzl 2010).

Cognitive Impairment in Mouse Models of Neurodegeneration

Given the disparities between species, it can be challenging to draw definitive conclusions about the association of cognitive function between humans and transgenic mouse models. In order to draw appropriate comparisons, tests of memory in rodent models of neurodegeneration should use cognitive systems that are found and conserved across species and have a clearly delineated function and a well-defined neuroanatomy. To this end, spatial navigation, being highly conserved in mammals (Squire 1992) and dependent on the hippocampus meet the above assumption. The neuroanatomical structure of the hippocampus, together with changes in its synaptic plasticity during memory formation (Barnes et al. 1996; Bliss and Collingridge 1993; Collingridge et al. 1983; Eichenbaum 1996; Fazeli et al. 1988; Malenka and Nicoll 1993; Morris 1989, 1990), presents a well-defined model of memory that has been frequently employed in studies using rodent species (Eichenbaum 1996; Morris 1989; Morris et al. 1990; O'Keefe and Nadel 1978; Olton et al. 1979). The involvement of the hippocampus in spatial memory in humans has also been confirmed in clinical observations that revealed that humans with temporal lobe damage have a severe impairment in learning and memory, including the recall of spatial locations and solving spatial maze tasks (Milner 1965; Milner and Scoville 1957; Smith and Milner 1981). AD patients show significantly increased atrophy of the hippocampus (Elgh et al. 2006; Rodriguez et al. 2000) and have inferior performance when subjected to spatial orientation tests in laboratory settings (Carlesimo et al. 1998; Ghilardi et al. 1999; Kavcic and Duffy 2003; Monacelli et al. 2003; Pai and Jacobs 2004; Rizzo et al. 2000). One potential caveat regarding mouse models is that the performance in behavioral tests may not only differ due to inherent differences in learning abilities between strains of mice, but also may be seriously compromised by the presence of recessive mutations compromising their behavior. Retinal degeneration (rd), caused by an autosomal recessive mutation ($Pde6b^{rd1}$) that causes rapid age-progressing degeneration of rods and cones (Jimenez et al. 1996; Ogilvie and Speck 2002), is one of the most infamous. About 20 % of all inbred mouse strains carry the *rd* gene (Sidman and Green 1965), and it is fixed in such strains like C3H, FVB, SJL which are often present in hybrid strains of mouse models of neurodegeneration. Depending on a genetic strain background, breeding scheme of transgenic lines, or generation of multiple transgenic lines, the presence of rd mutation in homozygous state may create a serious confound affecting mostly visually dependent spatial tasks (Garcia et al. 2004).

Transgenic Mouse Models of Amyloid Deposition

Consequently, the overexpression of *APP* and *PS* genes was used in the first transgenic mouse models to replicate amyloid pathology. Although these first-generation transgenic mice did not develop robust amyloid deposition (Hsiao et al. 1995), a second generation of transgenic lines employing mutant forms of these genes was generated in the mid-1990s. Regardless of the particular genetic background used, the models expressed high levels of APP and Aβ peptide, showing robust Aβ deposition. The temporal and spatial expression patterns of Aβ deposition depended upon the transgene promoter. The most common promoters used in APP mouse models include: APP promoter (Lamb et al. 1993), brain enriched prion protein promoter (Chishti et al. 2001; Hsiao et al. 1995,

1996), the platelet-derived growth factor B-chain (PDGF-B) promoter (Games et al. 1995) (both PrP and PDGF promoters resulting in a transgene expression outside of the CNS), and neuronal specific Thy-1 promoter (Sturchler-Pierrat et al. 1997).

It is commonly thought that the ratio of Aβ42 to Aβ40 strongly influences the degree and location of amyloid deposition. However, early experiments using APP and APP/PS1 transgenic mouse models were unable to fully differentiate the Aβ species that initiated seeding and growth of amyloid plaques in the parenchyma from those species that initiated deposits in the vasculature. To address this question, lines of mice expressing human Aβ42 or Aβ40 without overexpression of human APP were generated (McGowan et al. 2005). This was achieved by fusing Aβ40 or Aβ42 peptide sequences to C-terminal end of BRI protein. Mutations in *BRI* gene result in a form of cerebral amyloidosis that is associated with familial British and Danish dementias (Vidal et al. 1999, 2000). In the normal, non-mutated form, BRI protein is cut by furin or furin-like proteases near the COOH-terminus to release a soluble, 23 amino acid peptide. The generated BRI2-Aβ fusion constructs took advantage of this cleavage activity. Transgenic mice were generated containing a construct of Aβ40 or Aβ42 sequence fused to the C-terminal cleavage site of BRI protein. The resultant cleavage activity allowed the release and efficient secretion of specific Aβ species into the lumen and extracellular space. The obtained results were somewhat surprising. Although Aβ42 and Aβ40 are often associated in senile plaques, mice expressing only Aβ40 did not develop any form of amyloid pathology, despite producing higher Aβ levels than the Aβ42-expressing mice. Mice expressing Aβ42 developed overt amyloid deposits in the form of parenchymal plaques and in the walls of brain vasculature, providing evidence that only Aβ42 is necessary for the formation of amyloid plaques in vivo. When BRI2-Aβ42 mice were crossed with the Tg2576 line, the bigenic mice showed a substantial and synergetic increase in the amyloid deposition in the brain, supporting the hypothesis that the increase in Aβ42 to Aβ40 ratio mediates the rate of amyloid deposition.

These findings confirm that Aβ42 is the initiating molecule in the pathogenesis in AD (Hardy and Selkoe 2002; Younkin 1998). However, our recent behavioral research, using these novel BRI2-Aβ mouse models that express Aβ40, Aβ42, or both Aβ40/Aβ42 peptides in the secretory pathway, revealed no decline in established cognitive tests that demonstrated mnemonic impairment in APP transgenic mice, despite the presence of overt Aβ42 plaques in BRI2-Aβ42 and BRI2-Aβ40/Aβ42 mice (Kim et al. 2013). These results were confirmed independently in the conditional APP mouse model. The conditional suppression of human *APP* transgene at the stage of florid Aβ pathology restored cognitive function in this model, despite the abundant presence of Aβ plaques (Melnikova et al. 2013). In conclusion, the new mouse models of AD-like amyloidosis provide evidence that Aβ species alone might not be ultimately responsible for downstream neuronal loss and decline in the cognitive function.

Aβ Oligomers and Cognitive Decline

While there is limited evidence that Aβ is associated with widespread neuronal death in transgenic mouse lines of AD-like pathology, many studies show that Aβ can exert strong effects on synaptic maintenance and function, and this effect may be responsible for the cognitive deficits seen in these lines. For instance, 21–25-month-old Tg2576 mice show a loss in synaptophysin immunoreactivity around senile plaques, accompanied by abnormalities in synaptic functions (as determined by electrophysiological studies). Confocal microscopy studies showed that neurons that are proximal to dense cored plaques in Tg2576 and PDAPP models are associated with abnormal morphology, including loss of dendritic spines and reduction in the number of dendrites surrounding plaques. These functional data strongly suggest that Aβ and/or APP fragments may play a role in cognitive dysfunction. Also, immunization studies further demonstrated that removal of Aβ could resolve some of these morphological abnormalities. Nevertheless, these findings cannot completely account for the behavioral impairments seen in APP and APP/PS1 mice, since many of these decrements develop before the onset of widespread plaque deposition.

Recent studies have suggested that soluble Aβ assemblies, rather than deposited insoluble Aβ, may underlie the memory deficits (Hsia et al. 1999; Klein et al. 2001; Walsh et al. 2002a; Westerman et al. 2002). Studies utilizing immunization against Aβ substantiated this hypothesis by showing that reductions of largely soluble Aβ species in the APP and APP/PS1 mice could significantly improve their cognitive performance, despite modest reduction of preexisting amyloid pathology (Dodart et al. 2002a; Janus et al. 2000b; Kotilinek et al. 2002; Morgan et al. 2000; Schroeter et al. 2008; Spires-Jones et al. 2009). The initial support for the hypothesis that Aβ assemblies with higher molecular weight induced cognitive dysfunction came from work that demonstrated that Aβ oligomeric species, produced from cell culture, decreased long-term potentiation in rats in vivo (Walsh et al. 2002b), and induced significant memory deficits when injected into the brain of young rats (Cleary et al. 2005). More recently, investigators identified a 56-kDa species of Aβ (Aβ*56, a dodecamer aggregate) in the brains of 6-month-old Tg2576 mice that strongly correlated with impairments in spatial memory (Lesne et al. 2006). Furthermore, the direct injection of Aβ*56 species impaired the cognitive performance in rats. The discovery that specific oligomeric Aβ species may be responsible for cognitive deficits was replicated in some other transgenic mouse models (Cheng et al. 2007; Lefterov et al. 2009; Tomiyama et al. 2010).

Transgenic Mouse Models of Amyloid Deposition and Tau Tangles

Since AD involves the development of plaques and tangles, various groups have generated mouse models to address the functional interaction between *MAPT*, *APP*, and *PS* genes. Crossing APP Tg2576 mice with the mutant P301L tau (JNPL3) mice generated the bigenic TAPP line, which developed both plaques and tangles, and had increased NFT pathology as compared to JNPL3 littermates (Lewis et al. 2001), indicating that Aβ production enhanced NFT formation. TAPP mice also developed motor disturbances comprable to JNPL3, with similiar age of onset comparably to JNPL3 mice motor disturbances with similar age of onset. Work by Gotz and colleagues complemented these findings, showing that injection of synthetic Aβ42 into cortex and hippocampus of MAP *P301L* tau mice caused a fivefold increase in NFT numbers in the amygdala which receives projections from both injected areas (Gotz et al. 2001b; but see also Gotz et al. 2010). However, the injection of Aβ42 into mice expressing wild-type tau did not result in NFT formation (Gotz et al. 2001b).

The interaction between Aβ and mutated tau was independently confirmed in another bigenic model generated by crossing Tg2576 with VLW mice expressing mutant human *MAPT* genes (Ribé et al. 2005). VLM mice over-express in cortex and hippocampus the human 4R tau that contains a triple mutation (*G272V*, *P301L*, *R406W*). The brain pathology included dystrophic neurites, pretangles, and increased lysosomal bodies (Lim et al. 2001). The bigenic Tg2576/VLW mice showed a widely spread and accelerated Aβ deposition, as compared to Tg2576 littermates, and neuronal loss in brain limbic areas (Ribé et al. 2005).

A triple transgenic mouse model (3×Tg-AD), expressing APP (*Swedish*), PS1 (*M146V*), and tau (*P301L*) mutations, enabled investigation of the relationship between amyloid and tau pathology (Oddo et al. 2003b). Although the pathological development of these mice has not been completely characterized, investigators reported that development of NFT clearly followed after Aβ pathology. Plaques developed by 3 months of age, whereas tangles appeared first in the hippocampus around 12 months and later spread to the cortex (Oddo et al. 2003a). The 3×Tg-AD mice also develop age-dependent synaptic dysfunction, including impaired LTP, which preceded plaque and tangle formation (Oddo et al. 2003a), and age-progressing memory impairment that correlated with the accumulation of intraneuronal Aβ (Billings et al. 2005). Studies using these mice showed that Aβ species directly promoted MAPT pathology. An injection of anti-Aβ antibodies into the brains of 3×Tg-AD mice rapidly cleared Aβ deposits, followed by a reduction in tau pathology (Oddo et al. 2004, 2006). This time-dependent clearance of Aβ and tau pathology due to immunization was reversed, with Aβ deposits re-emerging prior to tau pathology after clearance of the injected antibody (Oddo et al. 2004). The regional and temporal pattern of pathology development observed in the 3×Tg-AD mice is reminiscent of the development of AD pathology. This sequential set of events supports amyloid cascade hypothesis of AD pathogenesis, where Aβ drives the formation of MAPT, leading to neurodegeneration. These rather exciting characteristics of 3×Tg-AD model were recently carefully scrutinized by Virginia Lee and colleagues, who showed in a series of well-controlled experiments that the positive intraneuronal Aβ immunoreactivity in the 3×Tg-AD mice did not reflect Aβ peptides but rather Aβ sequences within APP fragments (Winton et al. 2011). Furthermore, it was also demonstrated that the onset and progression of tau pathology was independent from the generation of any specific Aβ species in this unique model (Winton et al. 2011). Despite the absence of the intracellular Aβ, the 3×Tg-AD mouse model presents a useful tool that allows the simultaneous examination of both Aβ and tau pathologies.

Clinical Trials of Anti-Amyloid Approach

Although the mouse models provided a powerful tool for investigating Aβ-centric therapeutics, the successes of active and passive anti-Aβ immunization in preventing and clearing parenchymal amyloid in these models (Janus 2003) were not, however, replicated in clinical trials that had to be prematurely halted when a subset of patients developed meningoencephalitis (Orgogozo et al. 2003). Furthermore, recent phase-3 trials testing two more refined anti-Aβ antibodies: bapineuzmab (the humanized IgG1antibody of the mouse monoclonal antibody 3D6) and solanezumab (the humanized IgG1antibody of the mouse monoclonal antibody 266) reported negative results (Doody et al. 2014; Salloway et al. 2014). Also, attempts focusing on the lowering of the production of toxic Aβ42 species through inhibition of γ-secretase were unsuccessful. The phase 3 trial evaluating efficacy of semagacestat, a small-molecule γ-secretase inhibitor had to be stopped following the results that indicated that patients receiving the drug showed worse performance than patients receiving placebo (Doody et al. 2013). These negative results likely indicate that either the treatment was administered too late in the course of the disease, or Aβ alone is the wrong target for an effective treatment of AD. Both hypothetical explanations have some support from recent mouse studies. First, the significance of cognitive decline in a mouse model, based on limited number of tests without reporting the amount of variance accounted for by genotype and/or treatment effects, is bound to seriously bias the translational application of such results to clinical trials. The cognitive impairment manifested in the models of amyloidosis is usually mild, likely reflecting stages of disease that precede the diagnosis of mild cognitive impairment. It might be argued that the lifespan of the mouse is too short for the disease to progress in a manner analogous to its clinical stages. If such an assumption is correct, the translation of results obtained from mouse models to symptomatic AD patients might present high experimental risk.

Neuronal Loss and Transgenic Models of MAPT Pathology

The disturbance in the neuronal cytoskeleton found in most types of tauopathies suggests that this event can be a major contributing factor leading to cell loss (Barghorn and Mandelkow 2002; Delacourte et al. 2002; Dermaut et al. 2005; Mandelkow and Mandelkow 1998; Mattson 2004). It has been hypothesized that the disruption of normal tau function due to its abnormal phosphorylation or other biochemical perturbations represents a common mechanism in clinically diverse neurodegenerative diseases in which neuronal death is one of the main hallmarks (Mandelkow et al. 2003; Stamer et al. 2002). It is intriguing, therefore, that this most prominent feature of neurodegeneration has not been robustly replicated in most mouse models expressing *APP* and/or *PS1* genes (reviewed by Eriksen and Janus 2007; Janus et al. 2000a; McGowan et al. 2006). Although dystrophic neurites and loss of synaptic densities were reported in some APP single transgenic mouse lines (Calhoun et al. 1998; Games et al. 1995; Irizarry et al. 1997; Masliah et al. 2001; Masliah et al. 1996), and loss of large pyramidal neurons in the brain was reported in transgenic mice carrying 5×FAD mutation (Oakley et al. 2006), the rate and intensity of neuronal loss in those models did not replicate the magnitude of loss observed in human patients. Here, an exception is APP/PS1 KI double transgenic mice which exhibit a robust brain axonal degeneration and hippocampal CA1 neuron loss starting at 6 months of age, with concomitant spinal cord axonal degeneration, which adversely affects their motor function at that age (Wirths et al. 2006, 2008), thus preventing their unbiased evaluation in most cognitive tests.

In this light, recently generated transgenic mouse models that express either the wild-type human or mutant *tau* gene linked to FTDP-17 and show age-dependent accumulation of NFT accompanied by significantly increased neuronal loss without compromised motor function were notable long-awaited models. Although these mice showed NFT-like structures or tau-positive inclusions in the brain and spinal cord (Gotz et al. 2001a; Lewis et al. 2000; Tanemura et al. 2002), they did not capture the full spectrum of pathological features of FTDP-17 or other tauopathies, largely due to the distribution of pathology in the spinal cord or the lack of profound cognitive deficits. The spinal cord atrophy limited the life expectancy of these mice and their ability to develop full pathologies with definite age-related onset (Lewis et al. 2000). Later transgenic models, mostly free from the motor deficit confound, showed age-progressing accumulation of NFTs and neuronal degeneration (see Eriksen and Janus 2007 for a review). However, two recently published lines of work shed an interesting new light on the issue of the onset of neuronal death in tauopathies. First, a line expressed wild-type human tau (Htau) on a null mouse tau background (Andorfer et al. 2003). These mice developed a fronto-temporal dementia like pathology, with NFTs and cell death at the age of 15 months. Pathological analysis of these mice revealed that the number of NFTs did not correlate directly with observed cell death, strongly suggesting that cell death could occur independently of the observed accumulation of hyperphosphorylated tau (Andorfer et al. 2005). These studies suggested that it was unlikely that the insoluble accumulations of abnormally phosphorylated tau were the major cause of toxicity underlying neuronal death. It is more likely that, similar to recently refined amyloid cascade hypothesis (Hardy and Selkoe 2002; Lesne et al. 2006), intermediate toxic forms of tau of molecular weights between soluble forms of phosphorylated tau and aggregated insoluble tau might be responsible for neuronal death. The support of this hypothesis comes from the second recently published work, which described a mouse model denoted rTg4510, in which the expression of the mutated P301L *tau* gene could be conditionally switched off (SantaCruz et al. 2005). This study provided substantial insight into the relationship between neurofibrillary pathology, cell death, and memory decline. The study carried out by Karen Ashe and colleagues clearly demonstrated that the recovery of cognitive function was dissociated, at least in this mouse model, from the progressing NFT pathology, and that the overt presence of NFTs did not significantly affect cognitive function (SantaCruz et al. 2005). The findings also implied dissociation between the pathological processes leading to the formation of NFTs and the processes underlying cognitive dysfunction, indicating that the presence of NFTs was not a sine qua non factor leading to memory impairment. The relatively rapid recovery of memory function observed after suppression of the *tau* transgene suggests reversible neuronal dysfunction, which may involve a certain level of neuronal remodeling including improved synaptic function and/or significantly decreased synaptic loss observed in tauopathies (Yoshiyama et al. 2007). Also, the fact that the memory recovery was possible despite irreversible structural degeneration, including substantial neuronal loss, is encouraging, especially from the therapeutic point of view. The results indicated that not only could preventive therapy at an early stage of pathology be efficacious, but also therapy applied at more advanced stages of pathology.

Conclusions and Future Directions

Finding a cure that will prevent or dramatically slow down the progressive cognitive decline that characterizes AD and other tauopathies is a major unmet medical need. Mouse models expressing FAD-linked APP and/or MAPT mutations provide a powerful research tool to study cerebral pathology and the effects of Aβ or tau accumulation on the central nervous system function. Despite the large body of available evidence there are still major gaps in our knowledge regarding how relevant behavioral impairments seen in the transgenic mouse models are to human dementia. Though it is generally acknowledged that these models exhibit a variety of behavioral alterations, including deficits in learning and memory, the temporal and quantitative relationship of these deficits to underlying pathologies is not consistent from model to model. Furthermore, there is limited evidence that treatments that rapidly reverse cognitive deficits in mice exhibit the same efficacy when tested in the clinic. There are also disconnects between the lack of a robust neurodegenerative phenotype and neurofibrillary pathology in the mouse models, and the strong correlation of these disease features with cognitive impairments in humans. Perhaps this is best observed by the massive loss of the cholinergic neurotransmitter system in human AD, and the relative sparing of this system in mice. There is no doubt that presently available models have limited utility in replicating advanced stages of dementia, where both tangle pathology and neuronal loss almost certainly contribute to severe cognitive impairment. However, despite unquestionable progress in our understanding of pathologies in tauopathies, there are still many unanswered questions focusing, among others, on (1) the possible interaction between amyloid-β and APP derivatives to mediate cognitive function, (2) the existence of specific toxic species of Aβ in AD or tau in FTD that trigger the decline in cognitive function, or (3) the efficacy of potential therapeutics in mediating cognitive recovery. The answers to these important fundamental questions can only be satisfactorily obtained in experimental studies using animal models, and this knowledge may be pivotal for better understanding of the mechanisms underlying dementia. Given the huge costs of conducting human trials, especially prevention or early intervention, there is a need to understand both the scientific validity of the available mouse models and behavioral tests, as well as the limitations of translating the preclinical results observed in the current mouse models to cognitive decline in humans. Such understanding will aid us in designing the most appropriate studies evaluating potential therapeutics addressing specific stages of pathology and dementia in humans.

References

Albert, M. S. (1996). Cognitive and neurobiologic markers of early Alzheimer's disease. *Proceedings of the National Academy of Sciences of the United States of America, 93*, 13547–13551.

Ali, Y. O., Escala, W., Ruan, K., & Zhai, R. G. (2011). Assaying locomotor, learning, and memory deficits in Drosophila models of neurodegeneration. *Journal of Visualized Experiments*, (49).

Alvarez, X. A., Miguel-Hidalgo, J. J., Fernandez-Novoa, L., & Cacabelos, R. (1997). Intrahippocampal injections of the beta-amyloid 1–28 fragment induces behavioral deficits in rats. *Methods and Findings in Experimental and Clinical Pharmacology, 19*(7), 471–479.

Alzheimer, A. (1907). Über eine eigenartige Erkankung der Hirnrinde. *Allgemeine Zeitschrift fur Psychiatrie und Psychisch-gerichtliche Medizin, 64*, 146–148.

Anderson, B. (1993). Evidence from the rat for a general factor that underlies cognitive performance and that relates to brain size: Intelligence? *Neuroscience Letters, 153*(1), 98–102.

Anderson, B. (2000). The g factor in non-human animals. *Novartis Foundation Symposium, 233*, 79–90; discussion 90–75.

Andorfer, C., Acker, C. M., Kress, Y., Hof, P. R., Duff, K., & Davies, P. (2005). Cell-cycle reentry and cell death in transgenic mice expressing nonmutant human tau isoforms. *The Journal of Neuroscience, 25*(22), 5446–5454.

Andorfer, C., Kress, Y., Espinoza, M., de Silva, R., Tucker, K. L., Barde, Y. A., et al. (2003). Hyperphosphorylation and aggregation of tau in mice expressing normal human tau isoforms. *Journal of Neurochemistry, 86*(3), 582–590.

Ashe, K. (2001). Learning and memory in transgenic mice modelling Alzhiemer's disease. *Learning & Memory, 8*, 301–308.

Ashe, K. H. (2005). Mechanisms of memory loss in Abeta and tau mouse models. *Biochemical Society Transactions, 33*(Pt 4), 591–594.

Ashe, K. H., & Zahs, K. R. (2010). Probing the biology of Alzheimer's disease in mice. *Neuron, 66*(5), 631–645.

Barghorn, S., & Mandelkow, E. (2002). Toward a unified scheme for the aggregation of tau into Alzheimer paired helical filaments. *Biochemistry, 41*(50), 14885–14896.

Barnes, C. A., Rao, G., & McNaughton, B. L. (1996). Functional integrity of NMDA-dependent LTP induction mechanisms across the lifespan of F-344 rats. *Learning & Memory, 3*, 124–137.

Baxter, M. G., & Gallagher, M. (1996). Intact spatial learning in both young and aged rats following selective removal of hippocampal cholinergic input. *Behavioral Neuroscience, 110*(3), 460–467.

Beeri, R., Andres, C., Lev-Lehman, E., Timberg, R., Huberman, T., Shani, M., et al. (1995). Transgenic expression of human acetylcholinesterase induces progressive cognitive deterioration in mice. *Current Biology, 5*(9), 1063–1071.

Bertram, L., Blacker, D., Crystal, A., Mullin, K., Keeney, D., Jones, J., et al. (2000). Candidate genes showing no evidence for association or linkage with Alzheimer's disease using family-based methodologies. *Experimental Gerontology, 35*(9–10), 1353–1361.

Bertram, L., & Tanzi, R. E. (2001). Of replications and refutations: The status of Alzheimer's disease genetic research. *Current Neurology and Neuroscience Reports, 1*(5), 442–450.

Billings, L. M., Oddo, S., Green, K. N., McGaugh, J. L., & LaFerla, F. M. (2005). Intraneuronal Abeta causes the onset of early Alzheimer's disease-related cognitive deficits in transgenic mice. *Neuron, 45*(5), 675–688.

Blacker, D., Wilcox, M. A., Laird, N. M., Rodes, L., Horvath, S. M., Go, R. C., et al. (1998). Alpha-2 macroglobulin is genetically associated with Alzheimer disease. *Nature Genetics, 19*(4), 357–360.

Blendy, J. A., Kaestner, K. H., Schmid, W., Gass, P., & Schutz, G. (1996). Targeting of the CREB gene leads to up-regulation of a novel CREB mRNA isoform. *EMBO Journal, 15*(5), 1098–1106.

Bliss, T. V. P., & Collingridge, G. L. (1993). A synaptic model of memory: Long-term potentiation in the hippocampus. *Nature, 361*, 31–39.

Bonner, J. M., & Boulianne, G. L. (2011). Drosophila as a model to study age-related neurodegenerative disorders: Alzheimer's disease. *Experimental Gerontology, 46*(5), 335–339.

Braak, H., & Braak, E. (1997). Frequency of stages of Alzheimer-related lesions in different age categories. *Neurobiology of Aging, 18*(4), 351–357.

Brambilla, R., Gnesutta, N., Minichiello, L., White, G., Roylance, A. J., Herron, C. E., et al. (1997). A role for the Ras signalling pathway in synaptic transmission and long-term memory. *Nature, 390*(6657), 281–286.

Brooks, S. P., Pask, T., Jones, L., & Dunnett, S. B. (2005). Behavioural profiles of inbred mouse strains used as transgenic backgrounds. II: Cognitive tests. *Genes, Brain and Behavior, 4*(5), 307–317.

Bugiani, O., Murrell, J. R., Giaccone, G., Hasegawa, M., Ghigo, G., Tabaton, M., et al. (1999). Frontotemporal dementia and corticobasal degeneration in a family with a P301S mutation in tau. *Journal of Neuropathology and Experimental Neurology, 58*(6), 667–677.

Burns, A., Jacoby, R., & Levy, R. (1990a). Psychiatric phenomena in Alzheimer's disease. I: Disorders of thought content. *The British Journal of Psychiatry, 157*, 72–76, 92–74.

Burns, A., Jacoby, R., & Levy, R. (1990b). Psychiatric phenomena in Alzheimer's disease. II: Disorders of perception. *The British Journal of Psychiatry, 157*, 76–81, 92–74.

Burns, A., Jacoby, R., & Levy, R. (1990c). Psychiatric phenomena in Alzheimer's disease. III: Disorders of mood. *The British Journal of Psychiatry, 157*, 81–86, 92–84.

Burns, A., Jacoby, R., & Levy, R. (1990d). Psychiatric phenomena in Alzheimer's disease. IV: Disorders of behaviour. *The British Journal of Psychiatry, 157*, 86–94.

Calahorro, F., & Ruiz-Rubio, M. (2011). Caenorhabditis elegans as an experimental tool for the study of complex neurological diseases: Parkinson's disease, Alzheimer's disease and autism spectrum disorder. *Invertebrate Neuroscience, 11*(2), 73–83.

Calhoun, M. E., Wiederhold, K. H., Abramowski, D., Phinney, A. L., Probst, A., Sturchler-Pierrat, C., et al. (1998). Neuron loss in APP transgenic mice. *Nature, 395*(6704), 755–756.

Campion, D., Flaman, J. M., Brice, A., Hannequin, D., Dubois, B., Martin, C., et al. (1995). Mutations of the presenilin-1 gene in families with early-onset alzheimer's disease. *Human Molecular Genetics, 4*, 2373–2377.

Carlesimo, G. A., Mauri, M., Graceffa, A. M., Fadda, L., Loasses, A., Lorusso, S., et al. (1998). Memory performances in young, elderly, and very old healthy individuals versus patients with Alzheimer's disease: Evidence for discontinuity between normal and pathological aging. *Journal of Clinical and Experimental Neuropsychology, 20*(1), 14–29.

Chartier-Harlin, M. C., Crawford, F., Hamandi, K., Mullan, M., Goate, A., Hardy, J., et al. (1991a). Screening for the beta-amyloid precursor protein mutation (APP717: Val----Ile) in extended pedigrees with early onset alzheimer's disease. *Neuroscience Letters, 129*(1), 134–135.

Chartier-Harlin, M.-C., Crawford, F., Houlden, H., Warren, A., Hughes, D., Fidani, L., et al. (1991b). Early-onset alzheimer's Disease caused by mutations at codon 717 of the ß-amyloid precursor protein gene. *Nature, 353*, 844–846.

Chartier-Harlin, M. C., Kachergus, J., Roumier, C., Mouroux, V., Douay, X., Lincoln, S., et al. (2004). Alpha-synuclein locus duplication as a cause of familial Parkinson's disease. *Lancet, 364*(9440), 1167–1169.

Cheng, I. H., Scearce-Levie, K., Legleiter, J., Palop, J. J., Gerstein, H., Bien-Ly, N., et al. (2007). Accelerating amyloid-beta fibrillization reduces oligomer levels and functional deficits in Alzheimer disease mouse models. *The Journal of Biological Chemistry, 282*(33), 23818–23828.
Chiang, M. C., Barysheva, M., Shattuck, D. W., Lee, A. D., Madsen, S. K., Avedissian, C., et al. (2009). Genetics of brain fiber architecture and intellectual performance. *The Journal of Neuroscience, 29*(7), 2212–2224.
Chishti, M. A., Yang, D. S., Janus, C., Phinney, A. L., Horne, P., Pearson, J., et al. (2001). Early-onset amyloid deposition and cognitive deficits in transgenic mice expressing a double mutant form of amyloid precursor protein 695. *Journal of Biological Chemistry, 276*(24), 21562–21570.
Clark, L. N., Poorkaj, P., Wszolek, Z., Geschwind, D. H., Nasreddine, Z. S., Miller, B., et al. (1998). Pathogenic implications of mutations in the tau gene in pallido-ponto-nigral degeneration and related neurodegenerative disorders linked to chromosome 17. *Proceedings of the National Academy of Sciences of the United States of America, 95*(22), 13103–13107.
Clayton, D. F., & George, J. M. (1999). Synucleins in synaptic plasticity and neurodegenerative disorders. *Journal of Neuroscience Research, 58*(1), 120–129.
Cleary, J. P., Walsh, D. M., Hofmeister, J. J., Shankar, G. M., Kuskowski, M. A., Selkoe, D. J., et al. (2005). Natural oligomers of the amyloid-beta protein specifically disrupt cognitive function. *Nature Neuroscience, 8*(1), 79–84.
Cohen, R. M., Rezai-Zadeh, K., Weitz, T. M., Rentsendorj, A., Gate, D., Spivak, I., et al. (2013). A transgenic Alzheimer rat with plaques, tau pathology, behavioral impairment, oligomeric abeta, and frank neuronal loss. *The Journal of Neuroscience, 33*(15), 6245–6256.
Collingridge, G. L., Kehl, S. J., & McLennan, H. (1983). Excitatory amino acids in synaptic transmission in the Schaffer collateral-commissural pathway of the rat hippocampus. *Journal of Physiology (London), 334*, 33–46.
Conway, K. A., Harper, J. D., & Lansbury, P. T., Jr. (2000). Fibrils formed in vitro from alpha-synuclein and two mutant forms linked to Parkinson's disease are typical amyloid. *Biochemistry, 39*(10), 2552–2563.
Corder, E. H., Saunders, A. M., Strittmatter, W. J., Schmechel, D. E., Gaskell, P. C., Small, G. W., et al. (1993). Gene dose of apolipoprotein E type 4 allele and the risk of Alzheimer's disease in late onset families. *Science, 261*(5123), 921–923.
Crabbe, J. C., Wahlsten, D., & Dudek, B. C. (1999). Genetics of mouse behavior: Interactions with laboratory environment. *Science, 284*(5420), 1670–1672.
Cromwell, H. C., & Panksepp, J. (2011). Rethinking the cognitive revolution from a neural perspective: How overuse/misuse of the term 'cognition' and the neglect of affective controls in behavioral neuroscience could be delaying progress in understanding the BrainMind. *Neuroscience and Biobehavioral Reviews, 35*(9), 2026–2035.
Crutcher, K. A., Anderton, B. H., Barger, S. W., Ohm, T. G., & Snow, A. D. (1993). Cellular and molecular pathology in Alzheimer's disease. *Hippocampus, 3*(Spec No), 271–287.
Cruts, M., Theuns, J., & Van Broeckhoven, C. (2012). Locus-specific mutation databases for neurodegenerative brain diseases. *Human Mutation, 33*(9), 1340–1344.
Cummings, J. L. (2004). Dementia with lewy bodies: Molecular pathogenesis and implications for classification. *Journal of Geriatric Psychiatry and Neurology, 17*(3), 112–119.
Cummings, J. L., & Masterman, D. L. (1998). Assessment of treatment-associated changes in behavior and cholinergic therapy of neuropsychiatric symptoms in Alzheimer's disease. *The Journal of Clinical Psychiatry, 59*(Suppl 13), 23–30.
Cummings, B. J., Pike, C. J., Shankle, R., & Cotman, C. W. (1996). Beta-amyloid deposition and other measures of neuropathology predict cognitive status in Alzheimer's disease. *Neurobiology of Aging, 17*(6), 921–933.
Davies, R. R., Hodges, J. R., Kril, J. J., Patterson, K., Halliday, G. M., & Xuereb, J. H. (2005). The pathological basis of semantic dementia. *Brain, 128*(Pt 9), 1984–1995.
Delacourte, A., Sergeant, N., Wattez, A., Maurage, C. A., Lebert, F., Pasquier, F., et al. (2002). Tau aggregation in the hippocampal formation: An ageing or a pathological process? *Experimental Gerontology, 37*(10–11), 1291–1296.
Delrieu, J., Ousset, P. J., Caillaud, C., & Vellas, B. (2012). 'Clinical trials in Alzheimer's disease': Immunotherapy approaches. *Journal of Neurochemistry, 120*(Suppl 1), 186–193.

Denenberg, V. H. (2000). Evolution proposes and ontogeny disposes. *Brain and Language, 73*(2), 274–296.

Depboylu, C., Lohmuller, F., Du, Y., Riemenschneider, M., Kurz, A., Gasser, T., et al. (2006). Alpha2-macroglobulin, lipoprotein receptor-related protein and lipoprotein receptor-associated protein and the genetic risk for developing Alzheimer's disease. *Neuroscience Letters, 400*(3), 187–190.

Dermaut, B., Kumar-Singh, S., Rademakers, R., Theuns, J., Cruts, M., & Van Broeckhoven, C. (2005). Tau is central in the genetic Alzheimer-frontotemporal dementia spectrum. *Trends in Genetics, 21*(12), 664–672.

Dickson, C. T., & Vanderwolf, C. H. (1990). Animal-models of human amnesia and dementia—Hippocampal and amygdala ablation compared with serotonergic and cholinergic blockade in the rat. *Behavioural Brain Research, 41*, 215–227.

Do Carmo, S., & Cuello, A. C. (2013). Modeling Alzheimer's disease in transgenic rats. *Molecular Neurodegeneration, 8*, 37.

Dodart, J. C., Bales, K. R., Gannon, K. S., Greene, S. J., DeMattos, R. B., Mathis, C., et al. (2002a). Immunization reverses memory deficits without reducing brain Abeta burden in Alzheimer's disease model. *Nature Neuroscience, 5*(5), 452–457.

Dodart, J. C., Mathis, C., Bales, K. R., & Paul, S. M. (2002b). Does my mouse have Alzheimer's disease? *Genes, Brain, and Behavior, 1*(3), 142–155.

Doody, R. S., Raman, R., Farlow, M., Iwatsubo, T., Vellas, B., Joffe, S., et al. (2013). A phase 3 trial of semagacestat for treatment of Alzheimer's disease. *The New England Journal of Medicine, 369*(4), 341–350.

Doody, R. S., Thomas, R. G., Farlow, M., Iwatsubo, T., Vellas, B., Joffe, S., et al. (2014). Phase 3 trials of solanezumab for mild-to-moderate Alzheimer's disease. *The New England Journal of Medicine, 370*(4), 311–321.

Dubos, R. (1968). *So human an animal*. New York: Charles Scribner's Sons.

Eichenbaum, H. (1996). Learning from LTP: A comment on recent attempts to identify cellular and molecular mechanisms of memory. *Learning & Memory, 3*, 61–73.

Elder, G. A., Gama Sosa, M. A., & De Gasperi, R. (2010). Transgenic mouse models of Alzheimer's disease. *The Mount Sinai Journal of Medicine, 77*(1), 69–81.

Elgh, E., Lindqvist Astot, A., Fagerlund, M., Eriksson, S., Olsson, T., & Nasman, B. (2006). Cognitive dysfunction, hippocampal atrophy and glucocorticoid feedback in Alzheimer's disease. *Biological Psychiatry, 59*(2), 155–161.

Eriksen, J. L., & Janus, C. G. (2007). Plaques, tangles, and memory loss in mouse models of neurodegeneration. *Behavioural Genetics, 37*(1), 79–100.

Ertekin-Taner, N., Graff-Radford, N., Younkin, L. H., Eckman, C., Baker, M., Adamson, J., et al. (2000). Linkage of plasma Abeta42 to a quantitative locus on chromosome 10 in late-onset alzheimer's disease pedigrees. *Science, 290*(5500), 2303–2304.

Evans, D. A., Bennett, D. A., Wilson, R. S., Bienias, J. L., Morris, M. C., Scherr, P. A., et al. (2003). Incidence of Alzheimer disease in a biracial urban community: Relation to apolipoprotein E allele status. *Archives of Neurology, 60*(2), 185–189.

Ewald, C. Y., & Li, C. (2010). Understanding the molecular basis of Alzheimer's disease using a Caenorhabditis elegans model system. *Brain Structure and Function, 214*(2–3), 263–283.

Fazeli, M. S., Errington, M. L., Dolphin, A. C., & Bliss, T. V. P. (1988). Long–term potentiation in the dentate gyrus of the anaesthetized rat is accompanied by an increase in protein efflux into push–pull cannula perfusates. *Brain Research, 473*, 51–59.

Galsworthy, M. J., Paya-Cano, J. L., Liu, L., Monleon, S., Gregoryan, G., Fernandes, C., et al. (2005). Assessing reliability, heritability and general cognitive ability in a battery of cognitive tasks for laboratory mice. *Behavior Genetics, 35*(5), 675–692.

Galsworthy, M. J., Paya-Cano, J. L., Monleon, S., & Plomin, R. (2002). Evidence for general cognitive ability (g) in heterogeneous stock mice and an analysis of potential confounds. *Genes, Brain and Behavior, 1*(2), 88–95.

Gama Sosa, M. A., De Gasperi, R., & Elder, G. A. (2012). Modeling human neurodegenerative diseases in transgenic systems. *Human Genetics, 131*(4), 535–563.

Games, D., Adams, D., Alessandrini, R., Barbour, R., Berthelette, P., Blackwell, C., et al. (1995). Alzheimer-type neuropathology in transgenic mice overexpressing V717F beta-amyloid precursor protein. *Nature, 373*(6514), 523–527.

Garcia, M. F., Gordon, M. N., Hutton, M., Lewis, J., McGowan, E., Dickey, C. A., et al. (2004). The retinal degeneration (rd) gene seriously impairs spatial cognitive performance in normal and Alzheimer's transgenic mice. *Neuroreport, 15*(1), 73–77.

Gerlai, R. (1996). Gene-targeting studies of mammalian behavior: Is it the mutation or the background genotype? *Trends in Neurosciences, 19*(5), 177–181.

Ghilardi, M. F., Alberoni, M., Marelli, S., Rossi, M., Franceschi, M., Ghez, C., et al. (1999). Impaired movement control in Alzheimer's disease. *Neuroscience Letters, 260*(1), 45–48.

Giese, K. P., Friedman, E., Telliez, J. B., Fedorov, N. B., Wines, M., Feig, L. A., et al. (2001). Hippocampus-dependent learning and memory is impaired in mice lacking the Ras-guanine-nucleotide releasing factor 1 (Ras-GRF1). *Neuropharmacology, 41*(6), 791–800.

Goate, A., Chartier-Harlin, M. C., Mullan, M., Brown, J., Crawford, F., Fidani, L., et al. (1991). Segregation of a missense mutation in the amyloid precursor protein gene with familial Alzheimer's disease. *Nature, 349*(6311), 704–706.

Goedert, M., Crowther, R. A., & Spillantini, M. G. (1998). Tau mutations cause frontotemporal dementias. *Neuron, 21*(5), 955–958.

Golde, T. E. (2003). Alzheimer disease therapy: Can the amyloid cascade be halted? *The Journal of Clinical Investigation, 111*(1), 11–18.

Gotz, J., Chen, F., Barmettler, R., & Nitsch, R. M. (2001a). Tau filament formation in transgenic mice expressing P301L tau. *Journal of Biological Chemistry, 276*(1), 529–534.

Gotz, J., Chen, F., van Dorpe, J., & Nitsch, R. M. (2001b). Formation of neurofibrillary tangles in P301l tau transgenic mice induced by Abeta 42 fibrils. *Science, 293*(5534), 1491–1495.

Gotz, J., Lim, Y. A., Ke, Y. D., Eckert, A., & Ittner, L. M. (2010). Dissecting toxicity of tau and beta-amyloid. *Neurodegenerative Diseases, 7*(1–3), 10–12.

Grant, S. G., O'Dell, T. J., Karl, K. A., Stein, P. L., Soriano, P., & Kandel, E. R. (1992). Impaired long-term potentiation, spatial learning, and hippocampal development in fyn mutant mice. *Science, 258*(5090), 1903–1910.

Greenberg, B. D., Savage, M. J., Howland, D. S., Ali, S. M., Siedlak, S. L., Perry, G., et al. (1996). APP transgenesis: Approaches toward the development of animal models for Alzheimer disease neuropathology. *Neurobiology of Aging, 17*(2), 153–171.

Guerreiro, R., Wojtas, A., Bras, J., Carrasquillo, M., Rogaeva, E., Majounie, E., et al. (2012). TREM2 variants in Alzheimer's disease. *The New England Journal of Medicine, 368*(2), 117–127.

Hardy, J. A., & Higgins, G. A. (1992). Alzheimer's disease: The amyloid cascade hypothesis. *Science, 256*(5054), 184–185.

Hardy, J., & Selkoe, D. J. (2002). The amyloid hypothesis of Alzheimer's disease: Progress and problems on the road to therapeutics. *Science, 297*(5580), 353–356.

Harold, D., Abraham, R., Hollingworth, P., Sims, R., Gerrish, A., Hamshere, M. L., et al. (2009). Genome-wide association study identifies variants at CLU and PICALM associated with Alzheimer's disease. *Nature Genetics, 41*(10), 1088–1093.

Higgins, G. A., & Jacobsen, H. (2003). Transgenic mouse models of Alzheimer's disease: Phenotype and application. *Behavioural Pharmacology, 14*(5–6), 419–438.

Hock, C., Konietzko, U., Streffer, J. R., Tracy, J., Signorell, A., Muller-Tillmanns, B., et al. (2003). Antibodies against beta-amyloid slow cognitive decline in Alzheimer's disease. *Neuron, 38*(4), 547–554.

Hollingworth, P., Harold, D., Sims, R., Gerrish, A., Lambert, J. C., Carrasquillo, M. M., et al. (2011). Common variants at ABCA7, MS4A6A/MS4A4E, EPHA1, CD33 and CD2AP are associated with Alzheimer's disease. *Nature Genetics, 43*(5), 429–435.

Hsia, A. Y., Masliah, E., McConlogue, L., Yu, G. Q., Tatsuno, G., Hu, K., et al. (1999). Plaque-independent disruption of neural circuits in Alzheimer's disease mouse models. *Proceedings of the National Academy of Sciences of the United States of America, 96*(6), 3228–3233.

Hsiao, K. K., Borchelt, D. R., Olson, K., Johannsdottir, R., Kitt, C., Yunis, W., et al. (1995). Age-related CNS disorder and early death in transgenic FVB/N mice overexpressing Alzheimer amyloid precursor proteins. *Neuron, 15*(5), 1203–1218.

Hsiao, K., Chapman, P., Nilsen, S., Eckman, C., Harigaya, Y., Younkin, S., et al. (1996). Correlative memory deficits, Abeta elevation, and amyloid plaques in transgenic mice. *Science, 274*(5284), 99–102.

Huang, Y., & Mucke, L. (2012). Alzheimer mechanisms and therapeutic strategies. *Cell, 148*(6), 1204–1222.

Hutton, M., Lendon, C. L., Rizzu, P., Baker, M., Froelich, S., Houlden, H., et al. (1998). Association of missense and 5′-splice-site mutations in tau with the inherited dementia FTDP-17. *Nature, 393*(6686), 702–705.

Irizarry, M. C., McNamara, M., Fedorchak, K., Hsiao, K., & Hyman, B. T. (1997). APPSw transgenic mice develop age-related A beta deposits and neuropil abnormalities, but no neuronal loss in CA1. *Journal of Neuropathology and Experimental Neurology, 56*(9), 965–973.

Ittner, A., Ke, Y. D., van Eersel, J., Gladbach, A., Gotz, J., & Ittner, L. M. (2011). Brief update on different roles of tau in neurodegeneration. *IUBMB Life, 63*(7), 495–502.

Jahn, T. R., Kohlhoff, K. J., Scott, M., Tartaglia, G. G., Lomas, D. A., Dobson, C. M., et al. (2011). Detection of early locomotor abnormalities in a Drosophila model of Alzheimer's disease. *Journal of Neuroscience Methods, 197*(1), 186–189.

Janus, C. (2003). Vaccines for Alzheimer's disease: How close are we? *CNS Drugs, 17*(7), 457–474.

Janus, C., Chishti, M. A., & Westaway, D. (2000a). Transgenic mouse models of Alzheimer's disease. *Biochimica et Biophysica Acta, 1502*(1), 63–75.

Janus, C., Pearson, J., McLaurin, J., Mathews, P. M., Jiang, Y., Schmidt, S. D., et al. (2000b). A beta peptide immunization reduces behavioural impairment and plaques in a model of Alzheimer's disease. *Nature, 408*(6815), 979–982.

Janus, C., & Welzl, H. (2010). Mouse models of neurodegenerative diseases: Criteria and general methodology. *Methods in Molecular Biology, 602*, 323–345.

Janus, C., & Westaway, D. (2001). Transgenic mouse models of Alzheimer's disease. *Physiology and Behavior, 73*(5), 873–886.

Jensen, M. T., Mottin, M. D., Cracchiolo, J. R., Leighty, R. E., & Arendash, G. W. (2005). Lifelong immunization with human beta-amyloid (1–42) protects Alzheimer's transgenic mice against cognitive impairment throughout aging. *Neuroscience, 130*(3), 667–684.

Jimenez, A. J., Garcia-Fernandez, J. M., Gonzalez, B., & Foster, R. G. (1996). The spatio-temporal pattern of photoreceptor degeneration in the aged rd/rd mouse retina. *Cell and Tissue Research, 284*(2), 193–202.

Jonsson, T., Stefansson, H., Ph, D. S., Jonsdottir, I., Jonsson, P. V., Snaedal, J., et al. (2013). Variant of TREM2 associated with the risk of Alzheimer's disease. *The New England Journal of Medicine, 368*, 107–116.

Kastner, A., Grube, S., El-Kordi, A., Stepniak, B., Friedrichs, H., Sargin, D., et al. (2012). Common variants of the genes encoding erythropoietin and its receptor modulate cognitive performance in schizophrenia. *Molecular Medicine, 18*, 1029–1040.

Kavcic, V., & Duffy, C. J. (2003). Attentional dynamics and visual perception: Mechanisms of spatial disorientation in Alzheimer's disease. *Brain, 126*(Pt 5), 1173–1181.

Kelly, M. P., Stein, J. M., Vecsey, C. G., Favilla, C., Yang, X., Bizily, S. F., et al. (2009). Developmental etiology for neuroanatomical and cognitive deficits in mice overexpressing Galphas, a G-protein subunit genetically linked to schizophrenia. *Molecular Psychiatry, 14*(4), 398–415. 347.

Kim, J., Chakrabarty, P., Hanna, A., March, A., Dickson, D. W., Borchelt, D. R., et al. (2013). Normal cognition in transgenic BRI2-Abeta mice. *Molecular Neurodegeneration, 8*, 15.

Kim, H. C., Kim, D. K., Choi, I. J., Kang, K. H., Yi, S. D., Park, J., et al. (2001). Relation of apolipoprotein E polymorphism to clinically diagnosed Alzheimer's disease in the Korean population. *Psychiatry and Clinical Neurosciences, 55*(2), 115–120.

Kim, J. U., Lee, H. J., Kang, H. H., Shin, J. W., Ku, S. W., Ahn, J. H., et al. (2005). Protective effect of isoflurane anesthesia on noise-induced hearing loss in mice. *The Laryngoscope, 115*(11), 1996–1999.

King, D. P., & Takahashi, J. S. (1996). Forward genetic approaches to circadian clocks in mice. *Cold Spring Harbor Symposia on Quantitative Biology, 61*, 295–302.

Klein, W. L., Krafft, G. A., & Finch, C. E. (2001). Targeting small Abeta oligomers: The solution to an Alzheimer's disease conundrum? *Trends in Neurosciences, 24*(4), 219–224.

Kosik, K. S., & Shimura, H. (2005). Phosphorylated tau and the neurodegenerative foldopathies. *Biochimica et Biophysica Acta, 1739*(2–3), 298–310.

Kotilinek, L. A., Bacskai, B., Westerman, M., Kawarabayashi, T., Younkin, L., Hyman, B. T., et al. (2002). Reversible memory loss in a mouse transgenic model of Alzheimer's disease. *The Journal of Neuroscience, 22*(15), 6331–6335.

Ksiezak-Reding, H., & Wall, J. S. (2005). Characterization of paired helical filaments by scanning transmission electron microscopy. *Microscopy Research and Technique, 67*(3–4), 126–140.

Lamb, B. T., Sisodia, S. S., Lawler, A. M., Slunt, H. H., Kitt, C. A., Kearns, W. G., et al. (1993). Introduction and expression of the 400 kilobase amyloid precursor protein gene in transgenic mice. *Nature Genetics, 5*(1), 22–30.

Lambert, J. C., Heath, S., Even, G., Campion, D., Sleegers, K., Hiltunen, M., et al. (2009). Genome-wide association study identifies variants at CLU and CR1 associated with Alzheimer's disease. *Nature Genetics, 41*(10), 1094–1099.

Lee, V. M., Goedert, M., & Trojanowski, J. Q. (2001). Neurodegenerative tauopathies. *Annual Review of Neuroscience, 24*, 1121–1159.

Lefterov, I., Fitz, N. F., Cronican, A., Lefterov, P., Staufenbiel, M., & Koldamova, R. (2009). Memory deficits in APP23/Abca1+/– mice correlate with the level of Abeta oligomers. *ASN Neuro, 1*(2).

Lesne, S., Koh, M. T., Kotilinek, L., Kayed, R., Glabe, C. G., Yang, A., et al. (2006). A specific amyloid-beta protein assembly in the brain impairs memory. *Nature, 440*(7082), 352–357.

Levy-Lahad, E., Wasco, W., Poorkaj, P., Romano, D. M., Oshima, J., Pettingell, W. H., et al. (1995). Candidate gene for the chromosome 1 familial Alzheimer's disease locus. *Science, 269*(5226), 973–977.

Lewis, J., Dickson, D. W., Lin, W. L., Chisholm, L., Corral, A., Jones, G., et al. (2001). Enhanced neurofibrillary degeneration in transgenic mice expressing mutant tau and APP. *Science, 293*(5534), 1487–1491.

Lewis, J., McGowan, E., Rockwood, J., Melrose, H., Nacharaju, P., Van Slegtenhorst, M., et al. (2000). Neurofibrillary tangles, amyotrophy and progressive motor disturbance in mice expressing mutant (P301L) tau protein. *Nature Genetics, 25*(4), 402–405.

Lim, F., Hernandez, F., Lucas, J. J., Gomez-Ramos, P., Moran, M. A., & Avila, J. (2001). FTDP-17 mutations in tau transgenic mice provoke lysosomal abnormalities and Tau filaments in forebrain. *Molecular and Cellular Neurosciences, 18*(6), 702–714.

Lin, L., LeBlanc, C. J., Deacon, T. W., & Isacson, O. (1998). Chronic cognitive deficits and amyloid precursor protein elevation after selective immunotoxin lesions of the basal forebrain cholinergic system. *Neuroreport, 9*(3), 547–552.

Locurto, C., Fortin, E., & Sullivan, R. (2003). The structure of individual differences in heterogeneous stock mice across problem types and motivational systems. *Genes, Brain and Behavior, 2*(1), 40–55.

Malenka, R. C., & Nicoll, R. A. (1993). NMDA-receptor-dependent synaptic plasticity: Multiple forms and mechanisms. *Trends in Neurosciences, 16*(12), 521–527.

Mandelkow, E. M., & Mandelkow, E. (1998). Tau in Alzheimer's disease. *Trends in Cell Biology, 8*(11), 425–427.

Mandelkow, E. M., Stamer, K., Vogel, R., Thies, E., & Mandelkow, E. (2003). Clogging of axons by tau, inhibition of axonal traffic and starvation of synapses. *Neurobiology of Aging, 24*(8), 1079–1085.

Masliah, E., Sisk, A., Mallory, M., Mucke, L., Schenk, D., & Games, D. (1996). Comparison of neurodegenerative pathology in transgenic mice overexpressing V717F beta-amyloid precursor protein and Alzheimer's disease. *The Journal of Neuroscience, 16*(18), 5795–5811.

Masliah, E., Sisk, A., Mallory, M., & Games, D. (2001). Neurofibrillary pathology in transgenic mice overexpressing V717F beta-amyloid precursor protein. *Journal of Neuropathology and Experimental Neurology, 60*(4), 357–368.

Mattson, M. P. (2004). Pathways towards and away from Alzheimer's disease. *Nature, 430*(7000), 631–639.

Matzel, L. D., Han, Y. R., Grossman, H., Karnik, M. S., Patel, D., Scott, N., et al. (2003). Individual differences in the expression of a "general" learning ability in mice. *The Journal of Neuroscience, 23*(16), 6423–6433.

McGowan, E., Eriksen, J., & Hutton, M. (2006). A decade of modeling Alzheimer's disease in transgenic mice. *Trends in Genetics, 22*(5), 281–289.

McGowan, E., Pickford, F., Kim, J., Onstead, L., Eriksen, J., Yu, C., et al. (2005). Abeta42 is essential for parenchymal and vascular amyloid deposition in mice. *Neuron, 47*(2), 191–199.

Melnikova, T., Fromholt, S., Kim, H., Lee, D., Xu, G., Price, A., et al. (2013). Reversible pathologic and cognitive phenotypes in an inducible model of Alzheimer-amyloidosis. *The Journal of Neuroscience, 33*(9), 3765–3779.

Milner, B. (1965). Visually-guided maze-learning in man: Effects of bilateral hippocampal, bilateral frontal hippocampal lesions. *Neuropsychologia, 3*, 317–338.

Milner, B., & Scoville, W. B. (1957). Loss of recent memory after bilateral hippocampal lesions. *Journal of Neurology, Neurosurgery, & Psychiatry, 20*, 11–21.

Mirra, S. S., Murrell, J. R., Gearing, M., Spillantini, M. G., Goedert, M., Crowther, R. A., et al. (1999). Tau pathology in a family with dementia and a P301L mutation in tau. *Journal of Neuropathology and Experimental Neurology, 58*(4), 335–345.

Monacelli, A. M., Cushman, L. A., Kavcic, V., & Duffy, C. J. (2003). Spatial disorientation in Alzheimer's disease: The remembrance of things passed. *Neurology, 61*(11), 1491–1497.

Morcos, M., & Hutter, H. (2009). The model Caenorhabditis elegans in diabetes mellitus and Alzheimer's disease. *Journal of Alzheimer's Disease, 16*(4), 897–908.

Morgan, D., Diamond, D. M., Gottschall, P. E., Ugen, K. E., Dickey, C., Hardy, J., et al. (2000). A beta peptide vaccination prevents memory loss in an animal model of Alzheimer's disease. *Nature, 408*(6815), 982–985.

Morris, R. G. (1989). Synaptic plasticity and learning: Selective impairment of learning rats and blockade of long-term potentiation in vivo by the N-methyl-D-aspartate receptor antagonist AP5. *The Journal of Neuroscience, 9*(9), 3040–3057.

Morris, R. G. M. (1990). Toward a representational hypothesis of the role of hippocampal synaptic plasticity in spatial and other forms of learning. *Cold Spring Harbor Symposia on Quantitative Biology, 55*, 161–173.

Morris, R. G., Davis, S., & Butcher, S. P. (1990). Hippocampal synaptic plasticity and NMDA receptors: A role in information storage? *Philosophical Transactions of the Royal Society of London. Series B, Biological Sciences, 329*(1253), 187–204.

Muller, U. (1999). Ten years of gene targeting: Targeted mouse mutants, from vector design to phenotype analysis. *Mechanisms of Development, 82*(1–2), 3–21.

Naj, A. C., Jun, G., Beecham, G. W., Wang, L. S., Vardarajan, B. N., Buros, J., et al. (2011). Common variants at MS4A4/MS4A6E, CD2AP, CD33 and EPHA1 are associated with late-onset alzheimer's disease. *Nature Genetics, 43*(5), 436–441.

Nicoll, J. A. R., Wilkinson, D., Holmes, C., Steart, P., Markham, H., & Weller, R. O. (2003). Neuropathology of human Alzheimer disease after immunization with amyloid-ß peptide: A case report. *Nature Medicine, 9*, 448–452.

O'Keefe, J., & Nadel, L. (1978). *The hippocampus as a cognitive map*. Oxford: Oxford University Press.

Oakley, H., Cole, S. L., Logan, S., Maus, E., Shao, P., Craft, J., et al. (2006). Intraneuronal beta-amyloid aggregates, neurodegeneration, and neuron loss in transgenic mice with five familial Alzheimer's disease mutations: Potential factors in amyloid plaque formation. *The Journal of Neuroscience, 26*(40), 10129–10140.

Oddo, S., Billings, L., Kesslak, J. P., Cribbs, D. H., & LaFerla, F. M. (2004). Abeta immunotherapy leads to clearance of early, but not late, hyperphosphorylated tau aggregates via the proteasome. *Neuron, 43*(3), 321–332.

Oddo, S., Caccamo, A., Kitazawa, M., Tseng, B. P., & LaFerla, F. M. (2003a). Amyloid deposition precedes tangle formation in a triple transgenic model of Alzheimer's disease. *Neurobiology of Aging, 24*(8), 1063–1070.

Oddo, S., Caccamo, A., Shepherd, J. D., Murphy, M. P., Golde, T. E., Kayed, R., et al. (2003b). Triple-transgenic model of Alzheimer's disease with plaques and tangles: Intracellular Abeta and synaptic dysfunction. *Neuron, 39*(3), 409–421.

Oddo, S., Caccamo, A., Tran, L., Lambert, M. P., Glabe, C. G., Klein, W. L., et al. (2006). Temporal profile of amyloid-beta (Abeta) oligomerization in an in vivo model of Alzheimer disease. A link between Abeta and tau pathology. *Journal of Biological Chemistry, 281*(3), 1599–1604.

Ogilvie, J. M., & Speck, J. D. (2002). Dopamine has a critical role in photoreceptor degeneration in the rd mouse. *Neurobiology of Disease, 10*(1), 33–40.

Olton, D. S., Becker, J. T., & Handelman, G. E. (1979). Hippocampus space and memory. *Behavioral and Brain Sciences, 2*, 313–365.

Orgogozo, J. M., Gilman, S., Dartigues, J. F., Laurent, B., Puel, M., Kirby, L. C., et al. (2003). Subacute meningoencephalitis in a subset of patients with AD after Abeta42 immunization. *Neurology, 61*(1), 46–54.

Pai, M. C., & Jacobs, W. J. (2004). Topographical disorientation in community-residing patients with Alzheimer's disease. *International Journal of Geriatric Psychiatry, 19*(3), 250–255.

Parish, C. L., Nunan, J., Finkelstein, D. I., McNamara, F. N., Wong, J. Y., Waddington, J. L., et al. (2005). Mice lacking the alpha4 nicotinic receptor subunit fail to modulate dopaminergic neuronal arbors and possess impaired dopamine transporter function. *Molecular Pharmacology, 68*(5), 1376–1386.

Plomin, R. (2001). The genetics of g in human and mouse. *Nature Reviews Neuroscience, 2*(2), 136–141.

Plomin, R., DeFries, J. C., McClearn, G. E., & McGuffin, P. (2001). *Behavioral genetics* (4th ed.). New York: Worth Publishers.

Price, J. L., Davis, P. B., Morris, J. C., & White, D. L. (1991). The distribution of tangles, plaques and related immunohistochemical markers in healthy aging and Alzheimer's disease. *Neurobiology of Aging, 12*(4), 295–312.

Price, D. L., Tanzi, R. E., Borchelt, D. R., & Sisodia, S. S. (1998). Alzheimer's disease: Genetic studies and transgenic models. *Annual Review of Genetics, 32*, 461–493.

Ribé, E. M., Perez, M., Puig, B., Gich, I., Lim, F., Cuadrado, M., et al. (2005). Accelerated amyloid deposition, neurofibrillary degeneration and neuronal loss in double mutant APP/tau transgenic mice. *Neurobiology of Disease, 20*(3), 814–822.

Rizzo, M., Anderson, S. W., Dawson, J., & Nawrot, M. (2000). Vision and cognition in Alzheimer's disease. *Neuropsychologia, 38*(8), 1157–1169.

Rizzu, P., Joosse, M., Ravid, R., Hoogeveen, A., Kamphorst, W., van Swieten, J. C., et al. (2000). Mutation-dependent aggregation of tau protein and its selective depletion from the soluble fraction in brain of P301L FTDP-17 patients. *Human Molecular Genetics, 9*(20), 3075–3082.

Rodriguez, G., Vitali, P., Calvini, P., Bordoni, C., Girtler, N., Taddei, G., et al. (2000). Hippocampal perfusion in mild Alzheimer's disease. *Psychiatry Research, 100*(2), 65–74.

Rogaev, E. I., Sherrington, R., Rogaeva, E. A., Levesque, G., Ikeda, M., Liang, Y., et al. (1995). Familial Alzheimer's disease in kindreds with missense mutations in a gene on chromosome 1 related to the Alzheimer's disease type 3 gene. *Nature, 376*(6543), 775–778.

Rushton, J. P., & Ankney, C. D. (2009). Whole brain size and general mental ability: A review. *The International Journal of Neuroscience, 119*(5), 691–731.

Salloway, S., Sperling, R., Fox, N. C., Blennow, K., Klunk, W., Raskind, M., et al. (2014). Two phase 3 trials of bapineuzumab in mild-to-moderate Alzheimer's disease. *The New England Journal of Medicine, 370*(4), 322–333.

SantaCruz, K., Lewis, J., Spires, T., Paulson, J., Kotilinek, L., Ingelsson, M., et al. (2005). Tau suppression in a neurodegenerative mouse model improves memory function. *Science, 309*(5733), 476–481.

Schenk, D. (2002). Opinion: Amyloid-ß immunotherapy for Alzheimer's disease: The end of the beginning. *Nature Reviews Neuroscience, 3*(10), 824–828.

Schenk, D., Barbour, R., Dunn, W., Gordon, G., Grajeda, H., Guido, T., et al. (1999). Immunization with amyloid-beta attenuates Alzheimer-disease-like pathology in the PDAPP mouse. *Nature, 400*(6740), 173–177.
Scheuner, D., Eckman, C., Jensen, M., Song, X., Citron, M., Suzuki, N., et al. (1996). Secreted amyloid beta-protein similar to that in the senile plaques of Alzheimer's disease is increased in vivo by the presenilin 1 and 2 and APP mutations linked to familial Alzheimer's disease. *Nature Medicine, 2*(8), 864–870.
Schroeter, S., Khan, K., Barbour, R., Doan, M., Chen, M., Guido, T., et al. (2008). Immunotherapy reduces vascular amyloid-beta in PDAPP mice. *The Journal of Neuroscience: The Official Journal of the Society for Neuroscience, 28*(27), 6787–6793.
Seabrook, G. R., & Rosahl, T. W. (1999). Transgenic animals relevant to Alzheimer's disease. *Neuropharmacology, 38*(1), 1–17.
Selkoe, D. J. (2001). Alzheimer's disease results from the cerebral accumulation and cytotoxicity of amyloid beta-protein. *Journal of Alzheimer's Disease, 3*(1), 75–80.
Selkoe, D. J. (2002). Deciphering the genesis and fate of amyloid beta-protein yields novel therapies for Alzheimer disease. *The Journal of Clinical Investigation, 110*(10), 1375–1381.
Sharbaugh, C., Viet, S. M., Fraser, A., & McMaster, S. B. (2003). Comparable measures of cognitive function in human infants and laboratory animals to identify environmental health risks to children. *Environmental Health Perspectives, 111*(13), 1630–1639.
Shaw, P. (2007). Intelligence and the developing human brain. *Bioessays, 29*(10), 962–973.
Sherrington, R., Rogaev, E. I., Liang, Y., Rogaeva, E. A., Levesque, G., Ikeda, M., et al. (1995). Cloning of a gene bearing missense mutations in early-onset familial Alzheimer's disease. *Nature, 375*(6534), 754–760.
Sidman, R. L., & Green, M. C. (1965). Retinal degeneration in the mouse: Location of the Rd locus in linkage group Xvii. *The Journal of Heredity, 56*, 23–29.
Silva, A. J., Stevens, C. F., Tonegawa, S., & Wang, Y. (1992). Deficient hippocampal long-term potentiation in alpha-calcium-calmodulin kinase II mutant mice. *Science, 257*(5067), 201–206.
Smith, M. L., & Milner, B. (1981). The role of the right hippocampus in the recall of spatial location. *Neuropsychologia, 19*(6), 781–793.
Spearman, C. (1904). "General Intelligence," objectively determined and measured. *The American Journal of Psychology, 15*(2), 201–292.
Spillantini, M. G., Murrell, J. R., Goedert, M., Farlow, M. R., Klug, A., & Ghetti, B. (1998). Mutation in the tau gene in familial multiple system tauopathy with presenile dementia. *Proceedings of the National Academy of Sciences of the United States of America, 95*(13), 7737–7741.
Spires, T. L., & Hyman, B. T. (2005). Transgenic models of Alzheimer's disease: Learning from animals. *NeuroRx, 2*(3), 423–437.
Spires-Jones, T. L., Mielke, M. L., Rozkalne, A., Meyer-Luehmann, M., de Calignon, A., Bacskai, B. J., et al. (2009). Passive immunotherapy rapidly increases structural plasticity in a mouse model of Alzheimer disease. *Neurobiology of Disease, 33*(2), 213–220.
Squire, L. R. (1992). Memory and the hippocampus: A synthesis from findings with rats, monkeys, and humans. *Psychological Review, 99*(2), 195–231.
Stamer, K., Vogel, R., Thies, E., Mandelkow, E., & Mandelkow, E. M. (2002). Tau blocks traffic of organelles, neurofilaments, and APP vesicles in neurons and enhances oxidative stress. *The Journal of Cell Biology, 156*(6), 1051–1063.
Stelzmann, R. A., Schnitzlein, H. N., & Murtagh, F. R. (1995). An English translation of Alzheimer's 1907 paper, "Uber eine eigenartige Erkankung der Hirnrinde". *Clinical Anatomy, 8*(6), 429–431.
Strittmatter, W. J., Saunders, A. M., Schmechel, D., Pericak-Vance, M., Enghild, J., Salvesen, G. S., et al. (1993). Apolipoprotein E: High-avidity binding to beta-amyloid and increased frequency of type 4 allele in late-onset familial Alzheimer disease. *Proceedings of the National Academy of Sciences of the United States of America, 90*(5), 1977–1981.
Sturchler-Pierrat, C., Abramowski, D., Duke, M., Wiederhold, K. H., Mistl, C., Rothacher, S., et al. (1997). Two amyloid precursor protein transgenic mouse models with Alzheimer disease-

like pathology. *Proceedings of the National Academy of Sciences of the United States of America, 94*(24), 13287–13292.
Takahashi, J. S., Pinto, L. H., & Vitaterna, M. H. (1994). Forward and reverse genetic approaches to behavior in the mouse. *Science, 264*(5166), 1724–1733.
Tanemura, K., Murayama, M., Akagi, T., Hashikawa, T., Tominaga, T., Ichikawa, M., et al. (2002). Neurodegeneration with tau accumulation in a transgenic mouse expressing V337M human tau. *The Journal of Neuroscience, 22*(1), 133–141.
Tang, M. X., Stern, Y., Marder, K., Bell, K., Gurland, B., Lantigua, R., et al. (1998). The APOE-epsilon4 allele and the risk of Alzheimer disease among African Americans, whites, and Hispanics. *JAMA, 279*(10), 751–755.
Tanzi, R. E. (2012). The genetics of Alzheimer disease. *Cold Spring Harbor Perspectives in Medicine, 2*(10), a006296.
Tanzi, R. E., & Bertram, L. (2001). New frontiers in Alzheimer's disease genetics. *Neuron, 32*(2), 181–184.
Tickoo, S., & Russell, S. (2002). Drosophila melanogaster as a model system for drug discovery and pathway screening. *Current Opinion in Pharmacology, 2*(5), 555–560.
Toga, A. W., & Thompson, P. M. (2005). Genetics of brain structure and intelligence. *Annual Review of Neuroscience, 28*, 1–23.
Tomiyama, T., Matsuyama, S., Iso, H., Umeda, T., Takuma, H., Ohnishi, K., et al. (2010). A mouse model of amyloid {beta} oligomers: Their contribution to synaptic alteration, abnormal tau phosphorylation, glial activation, and neuronal loss in vivo. *The Journal of Neuroscience, 30*(14), 4845–4856.
van Leuven, F. (2000). Single and multiple transgenic mice as models for Alzheimer's disease. *Progress in Neurobiology, 61*(3), 305–312.
Victoroff, J., Zarow, C., Mack, W. J., Hsu, E., & Chui, H. C. (1996). Physical aggression is associated with preservation of substantia nigra pars compacta in Alzheimer disease. *Archives of Neurology, 53*(5), 428–434.
Vidal, R., Frangione, B., Rostagno, A., Mead, S., Revesz, T., Plant, G., et al. (1999). A stop-codon mutation in the BRI gene associated with familial British dementia. *Nature, 399*(6738), 776–781.
Vidal, R., Revesz, T., Rostagno, A., Kim, E., Holton, J. L., Bek, T., et al. (2000). A decamer duplication in the 3′ region of the BRI gene originates an amyloid peptide that is associated with dementia in a Danish kindred. *Proceedings of the National Academy of Sciences of the United States of America, 97*(9), 4920–4925.
von Bergen, M., Barghorn, S., Biernat, J., Mandelkow, E. M., & Mandelkow, E. (2005). Tau aggregation is driven by a transition from random coil to beta sheet structure. *Biochimica et Biophysica Acta, 1739*(2–3), 158–166.
Wahlsten, D., Bachmanov, A., Finn, D. A., & Crabbe, J. C. (2006). Stability of inbred mouse strain differences in behavior and brain size between laboratories and across decades. *Proceedings of the National Academy of Sciences of the United States of America, 103*(44), 16364–16369.
Wahlsten, D., Metten, P., Phillips, T. J., Boehm, S. L., 2nd, Burkhart-Kasch, S., Dorow, J., et al. (2003). Different data from different labs: Lessons from studies of gene-environment interaction. *Journal of Neurobiology, 54*(1), 283–311.
Walsh, D. M., Klyubin, I., Fadeeva, J. V., Cullen, W. K., Anwyl, R., Wolfe, M. S., et al. (2002a). Naturally secreted oligomers of amyloid beta protein potently inhibit hippocampal long-term potentiation in vivo. *Nature, 416*(6880), 535–539.
Walsh, D. M., Klyubin, I., Fadeeva, J. V., Rowan, M. J., & Selkoe, D. J. (2002b). Amyloid-beta oligomers: Their production, toxicity and therapeutic inhibition. *Biochemical Society Transactions, 30*(4), 552–557.
Westerman, M. A., Cooper-Blacketer, D., Mariash, A., Kotilinek, L., Kawarabayashi, T., Younkin, L. H., et al. (2002). The relationship between Abeta and memory in the Tg2576 mouse model of Alzheimer's disease. *The Journal of Neuroscience, 22*(5), 1858–1867.

Whishaw, I. Q. (1985). Cholinergic receptor blockade in the rat impairs locale but not taxon strategies for place navigation in a swimming pool. *Behavioral Neuroscience, 99*(5), 979–1005.

Winton, M. J., Lee, E. B., Sun, E., Wong, M. M., Leight, S., Zhang, B., et al. (2011). Intraneuronal APP, Not Free A{beta} Peptides in 3xTg-AD Mice: Implications for Tau versus A{beta}-Mediated Alzheimer Neurodegeneration. *The Journal of Neuroscience: The Official Journal of the Society for Neuroscience, 31*(21), 7691–7699.

Wirths, O., Breyhan, H., Schafer, S., Roth, C., & Bayer, T. A. (2008). Deficits in working memory and motor performance in the APP/PS1ki mouse model for Alzheimer's disease. *Neurobiology of Aging, 29*(6), 891–901.

Wirths, O., Weis, J., Szczygielski, J., Multhaup, G., & Bayer, T. A. (2006). Axonopathy in an APP/PS1 transgenic mouse model of Alzheimer's disease. *Acta Neuropathologica, 111*(4), 312–319.

Wong, P. C., Cai, H., Borchelt, D. R., & Price, D. L. (2002). Genetically engineered mouse models of neurodegenerative diseases. *Nature Neuroscience, 5*(7), 633–639.

Yoshiyama, Y., Higuchi, M., Zhang, B., Huang, S. M., Iwata, N., Saido, T. C., et al. (2007). Synapse loss and microglial activation precede tangles in a P301S tauopathy mouse model. *Neuron, 53*(3), 337–351.

Younkin, S. G. (1998). The role of A beta 42 in Alzheimer's disease. *Journal of Physiology, Paris, 92*(3–4), 289–292.

Zahs, K. R., & Ashe, K. H. (2010). 'Too much good news'—Are Alzheimer mouse models trying to tell us how to prevent, not cure, Alzheimer's disease? *Trends in Neurosciences, 33*(8), 381–389.

Zappia, M., Cittadella, R., Manna, I., Nicoletti, G., Andreoli, V., Bonavita, S., et al. (2002). Genetic association of alpha2-macroglobulin polymorphisms with AD in southern Italy. *Neurology, 59*(5), 756–758.

Chapter 7
Genetic Models of Alzheimer's Disease: The Influence of Apolipoprotein E allele Isoforms on Behaviour in Laboratory Animals

Matthew A. Albrecht and Jonathan K. Foster

Introduction

Alzheimer's disease (AD) is a neuro degenerative disorder that initially presents as a neurocognitive impairment focused on impaired episodic memory. The clinical course and severity of the cognitive symptoms of AD follows the observed post-mortem pathology of the disease and generally occurs in distinct stages (Braak and Braak 1991; Thal et al. 2006). The initial impairment in episodic memory has been associated with neuropathology originating within the hippocampus and the entorhinal cortex. AD brain pathology then typically radiates from the hippocampus and entorhinal cortex to association cortices and subcortical structures (including the amygdala and nucleus basalis of Meynert) leading to a gradual erosion of other cognitive domains (including executive functioning, speed of perceptuo-motor processing and visuospatial skills; Arnold et al. 1991).

In this chapter, we will first discuss the major pathological characteristics of AD, including beta amyloid plaques and neurofibrillary tangles. In the next section of this chapter, we will discuss genetic factors which are relevant in AD, with a specific focus on variants of the apolipoprotein E gene and how these polymorphisms influence the pathophysiological mechanisms that have been previously outlined. Once these topics have been introduced, the last section of this chapter will detail

M.A. Albrecht
School of Psychology and Speech Pathology, Curtin University, Perth, WA, Australia

MPRC, 55 Wade Avenue, Catonsville, MD 21228, USA
e-mail: malbrecht@mprc.umaryland.edu

J.K. Foster (✉)
School of Psychology and Speech Pathology, Curtin University, Perth, WA, Australia

Neurosciences Unit, Health Department of WA, GPO Box U1987, Perth, WA 6844, Australia
e-mail: j.foster@curtin.edu.au; Jonathan.Foster@health.wa.gov.au

J.C. Gewirtz, Y.-K. Kim (eds.), *Animal Models of Behavior Genetics*,
Advances in Behavior Genetics, DOI 10.1007/978-1-4939-3777-6_7

current research that has used apolipoprotein E genetic mouse models to investigate cognitive outcomes that are relevant for AD. This last section of the chapter will focus on behavioural models that target episodic memory function as—in a clinical context—this is the cognitive capacity of primary interest in the early stages of 'conversion' from healthy ageing to AD.

Pathology of Alzheimer's Disease

The main histopathological features seen in post-mortem AD brain tissue are (1) protein accumulations underlying the formation of extracellular neuritic/senile plaques containing aggregates of amyloid beta (Aβ) and (2) intraneuronal fibrillary tangles (NFT) composed of hyperphosphorylated tau. These characteristic features of the disease are accompanied by substantial neuronal cell body loss and marked reductions in synaptic connections (Gandy 2005; Katzman and Saitoh 1991; Terry et al. 1991). The neural atrophy occurring in AD can also be corroborated through in vivo neuroimaging studies of AD patients who have identified neural atrophy in the cerebral cortex and the hippocampus (Apostolova and Thompson 2008; De Leon et al. 1997).

Amyloid Beta

The Aβ peptide was first isolated from AD blood vessels by Glenner and Wong (1984). This group of peptides are between 39 and 43 amino acid chains in length and are derived from the amyloid precursor protein (APP). The suggested role of APP in the non-diseased state is to promote neuronal cell survival by influencing adhesive interactions, neurite outgrowth, synaptogenesis and synaptic plasticity (Mattson 1997). APP is enzymatically cleaved via three major enzymes: the alpha-, beta- and gamma-secretases (Mills and Reiner 1999; Roßner et al. 1998; Selkoe 1998). Alpha-secretase cleaves APP within the Aβ sequence, yielding the N-terminal soluble alpha-APP(s) fragment that is considered to be non-amyloidogenic (i.e. this cleavage precludes the formation of Aβ; Esch et al. 1990; Sisodia 1992; Sisodia et al. 1990). By contrast, an alternative pathway is facilitated by beta-secretase, which cleaves APP in a potentially amyloidogenic manner at the N-terminal of the Aβ sequence yielding an N-terminal soluble fragment beta-APP(s). In the final stage of Aβ synthesis, this fragment can then undergo cleavage by gamma-secretase to expose the C-terminal of Aβ, thereby generating the amyloidogenic insoluble Aβ fragment which has been centrally implicated in the pathophysiology of AD (Busciglio et al. 1993; Haass et al. 1992; Strooper and Annaert 2000).

Aβ pathology in AD progresses in well-documented phases, as detailed below (see also Thal et al. 2000). Aβ initially deposits exclusively within the neocortex. In the next stages of AD pathology, Aβ deposits spread to allocortical areas (entorhinal and subiculum/CA1 regions) [stage 2], to diencephalic nuclei, the striatum, cholinergic nuclei (nucleus basalis of Meynert, basal ganglia, thalamus and hypothalamus) [stage 3] and then to brainstem nuclei (midbrain and medulla oblongata) [stage 4]. In the fifth and final stage of the disease, Aβ deposition is eventually found within cerebellar regions. Of note with respect to the clinical manifestation of the illness, clinical AD cases have been identified as being in stages 3, 4 or 5 of Aβ pathology, while prodromal individuals have been identified as being in the first three stages (Thal et al. 2000, 2002, 2006).

The Aβ peptides that are found in senile plaques exhibit N- and C-terminal variations (Seubert et al. 1992, 1993). Of major relevance for AD, processing at the C-terminal end of the Aβ peptide by gamma-secretase forms the two major species of full length Aβ: $A\beta_{1-40}$ and $A\beta_{1-42}$. These two species account for approximately 90 % and 10 % of secreted Aβ, respectively (Haass et al. 1992; Seubert et al. 1993). However, the $A\beta_{1-42}$ peptide is the predominant species of Aβ that is aggregated and deposited in plaques found throughout much of the cerebral cortex in AD, whereas $A\beta_{1-40}$ is less readily deposited and is more localised to the cored plaques of the cerebral cortex and cerebellum (Fukumoto et al. 1996; Gravina et al. 1995; Iwatsubo et al. 1994; Mak et al. 1994; Mann et al. 1998).

The amyloid beta (Aβ) hypothesis of AD initially stressed the importance of Aβ accumulation and deposition as being centrally causal in the pathophysiology of the disease (Gandy 2005; Glenner and Wong 1984; Hardy and Selkoe 2002; Masters et al. 1985). However, subsequent investigations have shown that non-demented healthy individuals can sustain considerable brain deposition of amyloid while remaining behaviourally symptom-free (Mintun et al. 2006; Price and Morris 1999; Rowe et al. 2007). Furthermore, Aβ peptides are not only associated with pathological changes in the brain, but are also present in the CSF of non-demented people (Haass et al. 1992; Tamaoka et al. 1997). Indeed, it has been demonstrated that in some contexts and in appropriate concentrations Aβ possesses some neuroprotective effects. In particular, depletion of endogenous Aβ production via enzymatic inhibition or immunodepletion has been shown to cause neuronal cell death that is reversed by the addition of physiological levels of Aβ, with $A\beta_{1-40}$ being the most effective restorative species (Plant et al. 2003). The neuroprotective effects of Aβ are potentially mediated via modifying synaptic activity in order to down-regulate excessive glutamate release, thereby preventing glutamate-mediated excitotoxicity (Pearson and Peers 2006).

By contrast, there are several mechanisms by which the Aβ peptide can initiate neurotoxicity. First, the addition of Aβ to neuronal cells has been demonstrated to initiate oxidative and free-radical stress mechanisms via the production of H_2O_2 (Behl 1994; Behl et al. 1992; Markesbery 1997). Secondly, Aβ has been shown to destabilise Ca^{2+} homeostasis in a manner that renders neurons vulnerable to excitotoxicity (Koh et al. 1990; Mattson et al. 1992); this action of Aβ can be blocked by Ca^{2+} channel blockers (Weiss et al. 1994). Thirdly, Aβ is able to induce apoptotic cell death in a mouse model (LaFerla et al. 1995; Loo et al. 1993). Fourthly, Aβ can induce microglia to produce proteolytic enzymes, cytokines, free radicals and nitric oxide which can all induce neurotoxicity (Giulian et al. 1996; Meda et al. 1995).

The effect of Aβ on neuroprotection or neurological degeneration is likely to be dose-dependent (Pearson and Peers 2006).

In addition to its potentially neurotoxic properties in isolation, Aβ has also been shown to interact with tau protein in a manner that may be additive or synergistic with respect to the pathophysiological processes underlying AD-related neurodegeneration. Through one interactive mechanism, Aβ oligomers may trigger the phosphorylation of tau protein. The hyperphosphorylation of tau protein is considered to be a core mechanism for inducing tau-associated neurodegeneration (Takashima 2008). Injection of Aβ into mouse brain increases tau hyperphosphorylation (Götz et al. 2001). In vivo human evidence for the interaction of Aβ with tau has been provided by early-onset familial AD cases, whereby mutations to the presenilin or APP genes that cause increased levels of Aβ precede and trigger tauopathy (Jaworski et al. 2010). There may also be a two-way interaction between Aβ and tau, whereby reducing levels of tau in amyloid transgenic mice can reduce behavioural deficits in these animals (Roberson et al. 2007).

Neurofibrillary Tangles: Tauopathy

The main competitor theory to the Aβ hypothesis concerning the central causal pathophysiological factors underlying the symptomatology of AD concerns the tau protein (Jaworski et al. 2010; Olgiati et al. 2011). Specifically, intracellular neurofibrillary lesions containing tau appear to correlate more strongly with cognitive impairment in AD than the occurrence of amyloid plaques (Arriagada et al. 1992; Gómez-Isla et al. 1997; Neve and Robakis 1998).

The tau protein is located within the axon. It functions to assemble and stabilise microtubules that maintain the structure of the cell and form the neuron's infrastructure for axonal transport. The primary tauopathies found in AD are neurofibrillary tangles (NFT), neuropil threads and dystrophic neurites. Tau-rich dystrophic neurites are common components of plaques in AD, including small amyloid deposits, diffuse plaques and perivascular plaques in the hippocampal formation of Alzheimer brain. Neurofibrillary tangles are composed of hyperphosphorylated, fibrillar aggregates of tau located within the cytoplasm of neuron cell-bodies and proximal dendrites. Neuropil threads are filamentous accumulations of tau located within dendrites. Histologically, the result of the tauopathy induced by these intracellular components is a literal twisting of the neuronal cytoskeleton.

Researchers have noted that the hallmark tauopathy associated with AD displays remarkably little inter-individual variation. This has resulted in the identification of six stages of tau-related pathophysiology in AD: the transentorhinal stages I and II (which represents a 'silent' stage of the disease with no noticeable cognitive deficits); the limbic stages III and IV; and, finally, the near complete destruction of neocortical association areas which occurs in stages V and VI (Braak and Braak 1991).

Similar to the Aβ hypothesis, the relationship between tau pathology and clinical features of dementia has undergone some recent revisions (Jaworski et al. 2010). These revisions have relegated NFT and fibril pathologies from a primary, causal, neurotoxic role to instead representing a putative final end product of neurotoxicity. This is consistent with research demonstrating that while NFTs are correlated with neuronal loss in AD, neuronal loss in the disease is disproportionately greater than can be accounted for by NFT formation (Gómez-Isla et al. 1997; Kril et al. 2002). In addition, there is evidence to suggest that CA1 hippocampal neurons (which are thought to be critically important in the formation of new memories) can survive for approximately 20 years despite the presence of NFTs (Morsch et al. 1999).

Consequently, the focus of interest in tau-associated neurodegeneration has shifted towards alterations in the function of the tau protein. Specifically, Jaworski et al. (2010) suggest that tau-related neurotoxicity occurs via a hypothesised tau species they have termed 'Tau-P'. This species is phosphorylated to a greater extent than regular tau, but less so than hyper-phosphorylated fibrils (Jaworski et al. 2010). According to this model, it is the rapid accumulation of Tau-P that leads to cell death, whereas gradual accumulation forms tau aggregates (which prevent cell death, as a means of detoxification). Rapid accumulation of this Tau-P species in neurons results in the destabilisation and hindrance of the motor proteins responsible for neuronal transport. By comparison, hyperphosphorylated tau can cause destabilisation by inducing detachment from microtubules, and can cause hindrance of motor proteins by saturating microtubules with abnormal levels of tau (Iqbal et al. 2010; Jaworski et al. 2010). These actions impede transport of integral cellular components and energy to distal neuronal locations, leading to a progressive degeneration of the neuron from synapses to axons, progressing eventually in a retrograde manner to the cell soma (Braak and Braak 1997; Jaworski et al. 2010).

Current Issues

Having reviewed the principal histopathological features of AD, we now turn to a consideration of genetic factors, focusing on the APOE gene and the behavioural features of the different isoforms (in particular, the ε4 variant which has been linked to increased risk of AD).

Genetic Factors in AD

The literature classifies several subtypes of AD, depending upon familial association and the genetic mutation observed. Despite the differences in the age of onset and genetic mutation involved, the pathology between different forms of AD is very similar. Familial AD presents in early- and late-onset varieties and is associated with approximately 25 % of AD cases, although the early-onset form of familial AD is much rarer than the late-onset form and accounts for less than 2 % of familial cases (Bird 1998). Early-onset AD exhibits the strongest genetic linkages. Specifically, the early-onset familial form is linked to three distinct genetic mutations: the amyloid precursor protein gene (APP), the presenilin 1 gene (PSEN1) and the presenilin 2 gene (PSEN2) (Bird 1999). Each of these individual genetic mutations results in the abnormal processing of APP, causing Aβ overproduction (Hardy 1997). The presenilins are components of the gamma-secretase complex that (together with beta-secretase) generate Aβ fragments from APP (Jankowsky et al. 2004; Larner and Doran 2005).

In contrast, the later onset familial form of AD and the remaining 75 % of AD cases that represent non-familial AD have a complex etiology, with many risk factors cited in the literature (e.g. genetic, lifestyle and environmental influences). More than 600 genes have been investigated with respect to their potential role (Olgiati et al. 2011). So far, the most consistent and strongest association concerns the finding that different isoforms of the apolipoprotein E gene (APOE) lead to differential risk with respect to diagnosis of these forms of AD. Other genes have also been associated with AD, but these relationships are much weaker; for example, CLU encoding clusterin, PICALM (encoding phosphatidylinositol-binding clathrin assembly protein) and CR1 (the complement component (3b/4b) receptor 1) (Harold et al. 2009; Lambert et al. 2009; Olgiati et al. 2011). We here focus on the APOE gene, because it is the most important risk factor associated with the largest number of AD cases, and it has been most extensively investigated in non-human animal models.

The APOE gene is located at chromosome 19q13.2. There are three isoforms of the APOE gene, which have been labelled ε2 (cys112, cys158), ε3 (cys112, arg158) and ε4 (arg112, arg158). These polymorphisms code for three different species of apolipoprotein E, namely: apoE2, apoE3 and apoE4 (Mahley and Rall 2000; Mahley et al. 2006). These genetic isoforms are present in frequencies of 6 %, 78 % and 15 %, respectively. However, these frequencies are highly variable depending on the geographical location of the population (Eisenberg et al. 2010). The ε4 allele shows a dose–response risk relationship between the number of copies of the ε4 allele (0, 1, or 2) and the risk of developing AD: one copy of the ε4 allele is associated with a 2–5 times increased risk of developing AD, whereas two copies are associated with an 8–20 times increased risk (Corder et al. 1993; Saunders et al. 1993; Strittmatter et al. 1993).

Apolipoprotein

The human APOE gene codes for the 34 kDa, 299 amino acid glycoprotein, apolipoprotein E (apoE), which is the most abundant apolipoprotein in the body (Mahley 1988; Mahley et al. 2008). This glycoprotein was discovered in the 1970s as a component in extracted triglyceride-rich lipoproteins (Shore and Shore 1973).

Functionally, it is responsible for aiding reverse cholesterol transport, cholesterol/lipid delivery and clearance of cholesterol and triglyceride-rich lipoproteins (Mahley and Rall 2000). ApoE protein is expressed by several cell types. Most apoE production occurs in the liver, which accounts for more than 75 % of total apoE concentration, while CNS production of apoE accounts for most of the remaining proportion of apoE in humans (Elshourbagy et al. 1985). ApoE protein that is found within the brain is expressed predominantly by astrocytes (Boyles et al. 1985; Grehan et al. 2001). Neurons have also been shown to be able to synthesise apoE under conditions of excitotoxicity or when stressed (Xu et al. 2006, 2008).

ApoE4 in AD

Individuals who possess the ε4 allele have been identified as having an increased amyloid plaque and NFT load *post-mortem* compared to individuals who possess the ε3 allele or the ε2 allele (Ghebremedhin et al. 1998; Nagy et al. 1995; Ohm et al. 1995; Schmechel et al. 1993). This is likely to reflect the increased risk and earlier onset of AD for individuals with the ε4 allele (Ohm et al. 1995). Indeed, when differences in the duration or severity of dementia are controlled for, there appears to be no difference on the burden of pathophysiology (as defined by NFTs) between individuals who possess at least one copy of the ε4 allele ($\varepsilon4^+$) and individuals who do not possess a copy of the ε4 allele ($\varepsilon4^-$) (Landén et al. 1996). Together with several other lines of evidence, this suggests that the ε4 allele contributes to the likelihood, age of onset and perhaps time course of AD (Cosentino et al. 2008; Kanai et al. 1999; Packard et al. 2007), but it may not influence all aspects of the disease (Berg et al. 1998; Gomez-Isla et al. 1996).

There are several potential mechanisms whereby the ε4 allele may increase AD associated pathology and neurodegeneration. Much of this research has focused on Aβ mechanisms (due to the aforementioned focus on Aβ and plaque formation). For example:

- apoE4 has been shown to promote the fibrillization of Aβ peptides to a greater extent than apoE3, and this can result in a higher accumulation of Aβ (Kim et al. 2009; Ma et al. 1994)
- apoE4 can stimulate APP recycling to a greater extent than apoE3, increasing the potential for Aβ production (Ye et al. 2005)
- apoE4 can induce lysosomal leakage and apoptosis in neurons cultured with $A\beta_{1-42}$, whereas apoE3 appears to offer some protection from $A\beta_{1-42}$ toxicity (Ji et al. 2002, 2006)
- apoE4 binds with lesser affinity and forms less stable complexes with Aβ compared to apoE3, while apoE3 is more effective at binding and clearing Aβ via the apoE receptor and thereby aiding in the prevention of potential toxicity (Castellano et al. 2011; Deane et al. 2008; Kim et al. 2009; LaDu et al. 1994; Tokuda et al. 2000)

- the reduced rate of clearance of Aβ in APOE ε4 mice mirrors the extent of Aβ deposition in older mice (Castellano et al. 2011); when the interaction between Aβ and apoE is antagonised there is less Aβ pathology evident in mice (Sadowski et al. 2006)

Relationships have also been proposed between different isoforms of apoE and their influence on tau-associated neurodegeneration. Although these links have not been investigated as extensively as the relationship between apoE and Aβ pathology (Kim et al. 2009), it appears that presence of apoE4 can increase the phosphorylation of tau species (Tesseur et al. 2000). In particular, the proteolytically cleaved carboxyl-terminal fragments of apoE4 have been identified as being the most significant contributors to apoE4 induced hyperphosphorylation of tau (Brecht et al. 2004; Harris et al. 2003; Huang et al. 2001). Increased apoE4-related phosphorylation of tau may be mediated via second messenger signalling pathways induced by apoE (e.g. Erk activation) rather than via a direct effect of apoE4 on tau phosphorylation (Harris et al. 2004).

In addition to the effects of apoE variants on tau and Aβ-mediated pathology, several other potential mechanisms potentially related to neurodegeneration have also emerged from the literature. For example: (1) apoE4 possesses the least amount of antioxidant activity across the three apoE species (Miyata and Smith 1996); (2) apoE4 fails to protect neurons from kainic acid-induced excitotoxicity, whereas apoE3 is protective in transgenic mice (Buttini et al. 1999); (3) exogenously applied apoE4 increases production of pro-inflammatory cytokines more than apoE3 in glial cells (Guo et al. 2004); (4) lipopolysaccharide administration induces greater TNF-α and IL-6 in mice expressing apoE4 compared with mice expressing apoE3 (Lynch et al. 2003).

Therefore, there are many potential pathways through which possession of the ε4 allele may elicit toxicity in the brain, with commensurate increased risk of AD in APOE ε4 individuals. Furthermore, a considerable proportion of the research aimed at deducing possible mechanisms underlying increased neurotoxicity and neurodegeneration influenced by the ε4 allele has been conducted in organisms expressing human variants of the apoE protein. Having outlined the relevant pathophysiological mechanisms underlying AD and the influence of APOE status on these mechanisms, the remainder of this chapter will focus on investigations of the behavioural effects of APOE polymorphisms in laboratory animals (Fig. 7.1).

Transgenic APOE Animal Models of AD

As noted earlier, AD in humans is characterized by deficits in a range of cognitive domains as the disease progresses, suggesting a general decline in cognitive ability (see Chapter XX in this volume). However, the most prominent early loss in AD occurs in the domain of episodic memory. Many different behavioural tests have been used to evaluate episodic memory loss in genetic models of AD. The majority

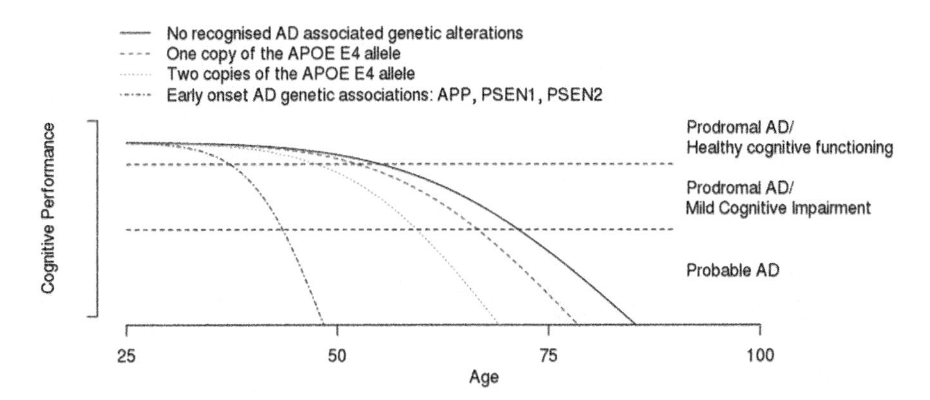

Fig. 7.1 Progression of cognitive changes in Alzheimer's disease according to APOE ε4 status

of APOE transgenic behavioural experiments have used mice. The principal episodic memory tasks that have been used to investigate the relationship between APOE and cognitive performance are the Morris Water Maze (MWM), novel object learning, novel location learning and the passive avoidance test. Memory-related tasks that attempt to measure endogenous affective behaviours (especially anxiety-related behaviours), including the elevated plus and zero mazes, have also been reported in the literature. These tasks are also relevant to AD research in humans, as AD patients often report increased levels of anxiety compared to healthy elderly people.

As previously noted, the major APOE polymorphisms that occur in humans are the ε2, ε3 and ε4 alleles; it is the ε4 allele in particular which confers an increased risk of developing AD in older adults. However, these APOE polymorphisms do not occur naturally in mice, which possess their own murine APOE gene. In order to isolate the influence of human apoE on animal behaviour, the influence of murine apoE has first to be negated. Two of the earlier variants of human APOE mice were generated by using pronuclear injection of human DNA into mice and then crossing these mice with $APOE^{-/-}$ mice to control for the potential influence of endogenous murine apoE (Bellosta et al. 1995; Raber et al. 1998, 2002; Xu et al. 1996). These mice were then administered a promoter (either the neuron-specific enolase [NSE] promoter or the glial fibrillary acidic protein [GFAP] promoter) to induce expression of human apoE3 or apoE4. (While the NSE promoter induces the expression of apoE in neurons, the GFAP promoter induces its expression in astrocytes.) As such, these earlier 'human APOE mice' had undergone both genetic transmission and genetic deletion, in conjunction with artificial promotion.

The use of NSE and GFAP mice confers a number of advantages, including the ability to investigate the effect of neuronal-specific, astrocytic-specific and brain-specific expression of apoE. In addition, NSE promoters offer significant control over the amount of apoE that is expressed by the animal, allowing more consistent levels of apoE to be expressed regardless of the specific human isoform. However, these earlier models also incurred a number of problems that may impede the validity of this approach. NSE and GFAP mice manifest varying levels of transgene expression due to differences in chromosomal location and copy number, and they do not express a widespread physiological/anatomical distribution of apoE due to the use of these artificial promoters (Sullivan et al. 1997, 2004). This raises a potentially important issue, because apoE that is naturally expressed outside of the CNS may have important physiological effects within the CNS. For example, apoE has been shown to interact with adrenal and sex hormone steroidogenesis in humans and animals (Reyland et al. 1991; Travert et al. 2000; Zofkova et al. 2002) and manipulations of these hormones can significantly affect learning and memory in humans and animals, particularly when hormonal manipulations of sex steroids occur during key developmental periods (Janowsky et al. 2000; Luine et al. 1998; Van Haaren et al. 1990). Sullivan et al. (2004) also note that APOE manipulations can have significant effects on cholesterol and systemic inflammation that may lead to "unknown isoform-specific consequences in the CNS". Recently, a postulated role for neuroinflammation has been postulated in many psychiatric disturbances (Dantzer et al. 2008; Harrison et al. 2009; Miller et al. 2009). It is therefore possible that there may be further significant indirect effects of APOE manipulation on psychological and emotional processes.

More recently developed genetic lines of APOE transgenic mice have overcome some of these problematic issues. Sullivan et al. (1997, 2004) developed human apoE targeted-replacement (TR) mice that express apoE via the endogenous mouse apoE promoter, i.e. in a manner that is more physiologically comparable (both spatially and temporally) to wild-type (WT) mice (with apoE expressed both in the brain and periphery; Sullivan et al. 2004). These latter types of mice may therefore offer a more valid model of the cognitive sequelae of APOE status. Nevertheless, the human apoE that is induced in these animal models is also non-murine, and there may be potentially significant physiological differences between human and murine apoE. For example, one potentially relevant issue raised by Sullivan et al. (1997) concerns the possibility that apoE3 may have a lower affinity for the mouse LDL receptor compared to the endogenous murine apoE. This consideration may have significant behavioural consequences via differential moderation of synaptogenesis; this issue could be investigated further using mice expressing the human LDL receptor.

It should therefore be borne in mind that some of the conceptual and methodological issues surrounding artificial genetic manipulations discussed above could limit interpretation of the findings that will be discussed in the following sections of this chapter. We suggest that it is more meaningful in evaluating the role of any particular human allele on cognitive performance to contrast mice that have undergone similar genetic manipulations. The following sections of this chapter will therefore focus on comparisons between mice with the ε4 allele and mice with the ε3 allele or the ε2 allele. Further, we consider that comparisons between different types of *human* APOE genetic mice (i.e. groups of mice in which different variants of human APOE have been inserted) are more likely to confer the greatest validity for an AD model of the human illness.

Behavioural Paradigms

Morris Water Maze

Performance on the Morris Water Maze (MWM) has been shown to be dependent on the integrity of the hippocampus, which is one of the first brain regions to show reliable AD pathology in humans. The MWM is considered to be one of the best currently available behavioural assays of episodic memory capacity in laboratory animals. The MWM consists of a 1.5–2 m diameter water-filled pool with a submerged, invisible 'escape' platform (which is located just below the water's surface) 10–15 cm in diameter. The water in the MWM is rendered opaque using milk or chalk in order to conceal the escape platform. This is important, because hiding the platform ensures that navigation to the escape platform is achieved by the animal using distinctive spatial features located outside the maze rather than using the direct sight of the platform. Mice quickly learn to swim to the location of the submerged platform. Performance is often measured simply by the duration that it takes the animal to reach the platform or (in the case where motor activity is suspected of either being impaired or enhanced due to the experimental manipulation) by evaluating the trajectory of the animal's route to the platform (instead of duration). Researchers can also use a probe trial, whereby the platform is removed and the path that the animal takes is analyzed. The probe trial is considered to be the most sensitive measure of episodic memory performance for the MWM.

MWM in NSE and GFAP APOE Mice

The behavioural analysis of NSE and GFAP mice that has been undertaken to date in the MWM indicates a gender- and age-dependent effect on spatial learning and memory. In the first study to investigate the cognitive/behavioural profile of human apoE expressing mice, Raber et al. (1998) found learning impairments on the MWM in NSE apoE4 expressing mice compared to apoE3 expressing mice, as indicated by increased swim latency to the platform on hidden trials ($N=6$–8 per group). These impairments were more pronounced in female apoE4 expressing mice compared to male apoE4 expressing mice. There were limited performance differences between the female mice strains at 3 months of age. However, there were significant performance reductions (i.e. longer latency to reach the platform) in the 6-month-old female apoE4 expressing mice compared to the male and female apoE3 and the male apoE4 expressing mice of the same age (Raber et al. 1998). In follow-up studies, Raber et al. (2000, 2002) replicated this result, finding that the performance of

6–8-month-old female apoE4 expressing mice under both the NSE and GFAP promoters worsened on the learning trials of the MWM relative to female apoE3 expressing mice (*N*=5–11 per group). It was also found that, on probe trials, 6-month-old and 18-month-old female apoE4 expressing mice spent less time in the target quadrant. Performance deficits were also reported by Raber et al. (2000) on the probe trial in male 6-month-old apoE4 expressing mice. However, in another study there were no performance deficits identified in male apoE4 expressing mice compared to male apoE3 expressing mice (Raber et al. 2002). This finding has been confirmed by two other research groups that found no genotype differences between ε4 and ε3 10–13-month-old GFAP (Hartman et al. 2001) (*N*=8–9 per group) and 6–7-month-old NSE (Pfankuch et al. 2005) male mice on learning or probe trials in the MWM (*N*=5–8 per group).

However, in contrast to their earlier results (e.g. Raber et al. 1998, 2000, 2002), Van Meer et al. (2007) found no differences between different APOE genotypes on learning trials of the MWM in 6-month-old GFAP female mice, although they did find performance deficits in the apoE4 expressing female mice on probe trials (*N*=8–9 per group). There appears to be no clear explanation given for differences in outcome found by the Raber et al. group with respect to the hidden platform present trials (see preceding paragraph) versus the probe trials observed. It may be that differences in housing interacted with genotype. Specifically, the mice in the Raber et al. (2002) study were housed singly for 1 week before behavioural testing began, whereas the mice reported in Van Meer et al. (2007) were housed singly for 48 h before behavioural testing. These differences in the duration of social isolation may induce different levels of stress in the animals (Matsumoto et al. 2005). Levels of stress hormones have also been shown to be influenced by APOE genetic manipulations (Zhou et al. 1998). It may also be the case that some reported studies are statistically underpowered to detect consistent effects. Finally, differences in the promoter used (i.e. GFAP used in the Van Meer et al. (2007) study versus the NSE promoter used in the earlier studies) may at least partially account for these differences, i.e. with respect to by differential expression of apoE in neurons or in astrocytes.

MWM in TR Mice

Research investigating the expression of human apoE in TR mice (which are more physiologically comparable to wild-type mice, as previously discussed) has also found inconsistent genotype effects. The first study to explore the cognitive-behavioural profile of these mice was conducted by Grootendorst et al. (2005). These researchers investigated 4–5-month-old male and female mice and did not find any differences between ε3 and ε4 mice on performance on the hidden platform training trials of the MWM across the 4 days of testing (*N*=8–9 per group). On the probe trials, Grootendorst et al. (2005) stated that ε4 mice showed the lowest retention performance on the MWM. However, there was no direct comparison of ε3 versus ε4 mice reported. In addition, despite female aopE3 mice failing to demonstrate significant learning across the hidden platform trials, this group exhibited the best performance on the probe trial. Indeed, male ε3, male ε4 and female ε4 mice all demonstrated similar levels of performance.

In a follow-up study by the same group (conducted on an older group of mice aged approximately 15 months old), Bour et al. (2008) found that ε3 female mice displayed a similar level of learning on the hidden platform trials when compared with mice of other genotypes (N=8–9 per group). Interestingly, both male and female ε4 mice performed slightly better than ε3 mice across all learning trials, but on the probe trial the female ε4 mice were recorded as spending the least amount of time swimming in the target zone. However, this group was not directly compared with ε3 female mice, and so it cannot be determined whether there was or was not a statistically significant difference between the two genotypes. The combined results from these two studies suggest that, if an APOE-associated deficit is present on the MWM, it is likely to be subtle and only apparent in older female mice.

Adding to the inconsistent findings, further studies by the Raber et al. group have found that female ε4 mice across the lifespan learned the MWM better (during both visible and hidden trials), compared to age matched female ε3 and ε2 mice (Siegel et al. 2012) (N=6–19 per group). However, there were no differences in learning performance between the genotypes when 5-month-old female mice (Villasana et al. 2006) and 15-month-old male and female mice (Villasana et al. 2011) were evaluated in isolation (N=6–10 per group). Similarly, on the probe trials there were no differences in performance between the genotypes reported by Siegel et al. (2012) in 5-month-old female mice (Villasana et al. 2006) or in 15-month-old male mice (Villasana et al. 2011). On the probe trial, 15-month-old ε3 female mice did spend more time in the target quadrant than 15-month-old female ε4 mice in this study (Villasana et al. 2011); however, this difference was not statistically significant.

Other research groups have also failed to observe consistent effects of the ε4 allele on MWM performance. Andrews-Zwilling et al. (2010) found that, while 12-month-old ε3 and ε4 female mice performed equally well on learning and probe trials of the MWM, 16- and 21-month-old ε4 mice show deficits in learning/acquisition trials (N=10–12 per group). The findings of this study suggested a genotype-dependent deficit in learning trials, in addition to age-related effects. Despite worse performance during the learning trials, 16-month-old apoE4 expressing mice performed at a similar level to apoE3 expressing mice during the 24-h and 72-h probe trials. However, the 16-month-old apoE4 expressing mice spent less time over the target zone at the 120-h probe. In contrast, Reverte et al. (2012) found MWM learning deficits in 5-month-old ε4 and ε2 mice compared to ε3 mice (N=18–22 per group, pooled across sexes), i.e. at a younger age than that reported by Andrews-Zwilling et al. (2010). In addition, Salomon-Zimri et al. (2014) demonstrated a similar ε4 effect in 4-month-old male mice (N=8–10 per group). However, despite showing apparent learning deficits, the ε4 mice described in Reverte et al. (2012)

still performed adequately on the 72 h probe trials. Indeed, male ε4 mice performed better than male ε3 mice and female ε4 mice in this study. Furthermore, while male ε4 mice performed worse than ε3 mice, the difference between the female ε3 and ε4 groups was not statistically significant and there were no sex by genotype interactions observed.

MWM: Summary

Taken together, it appears from findings on the MWM that there may be subtle learning and retention deficits in older female ε4 mice compared to mice of other 'knock in' genotypes, sex and age. One potential mechanism proposed by Raber et al. (2002) concerns the possibility that effects that have been observed in older apoE4 expressing female mice may be mediated via reduced androgen receptor levels (i.e. compared to the other apoE expressing mice). These researchers found that androgen administration to NSE and GFAP apoE4 expressing female mice (note: these experiments were not conducted in TR mice) reduced performance deficits on the MWM, while androgen receptor antagonists worsened performance in male apoE4 expressing mice (Raber et al. 2002). However, the MWM results discussed throughout this section have not yet shown robust effects across studies, and further replications are necessary to demonstrate consistent age- and sex-associated differences in spatial working memory across different genetic strains of APOE mice. There are also considerable differences in the MWM training protocols between studies, which may represent another important factor underlying in the lack of reliable effects on spatial episodic memory across mice expressing human apoE variants. Furthermore, it might be expected that younger APOE strains may not provide the most consistent results with respect to cognitive deficits, as variants of the APOE gene are associated with vulnerability in older age in humans. Therefore, the findings in older mice (i.e. >15 months) presented here perhaps best reflect the relationship between polymorphisms in the APOE gene and cognitive features of early Alzheimer's disease.

Passive Avoidance Test

While the MWM evaluates spatial learning, memory and navigation, the passive avoidance test measures memory in the context of shock avoidance. The test takes place in a chamber with two compartments separated by a gate; there is one light compartment and one dark compartment. Mice are placed firstly into the light compartment. After a short period of acclimatization the gate opens, allowing entry into the dark compartment. Mice are more phobic of lighter environments and tend to move quickly into the dark compartment. After entry into the dark compartment, they are given a brief foot shock and removed from the apparatus. After a short period, the mouse is returned to the light compartment and then timed to see how long it takes for the mouse to venture back into the dark compartment. The testing can either stop at this point or continue until the mouse reaches a criterion of three consecutive trials without entering the dark compartment (after which it is typically given a probe trial 24 h later).

The influence of the APOE gene on the single learning trial version of the passive avoidance test has also been somewhat inconsistent. Early trials by Raber et al. (1998) found no APOE-related differences between 6-month-old male or female NSE mice on a single learning trial version of the passive avoidance test. Similarly, 5- and 15-month-old male and female TR mice also showed no performance differences on the single learning trial passive avoidance test (Grootendorst et al. 2005; Bour et al. 2008).

On the version of the passive avoidance test where mice are trained to a criterion before undergoing a probe trial, Villasana et al. (2006) and Siegel et al. (2012) demonstrated that young male and female ε4 TR mice (5–8 months old) and older female ε4 TR mice (14–22 months old) required fewer trials to reach criterion than ε3 mice, but middle aged female ε4 mice (10–13 months) performed at a level that was consistent with ε3 mice and reached criterion faster than ε2 mice. Furthermore, while there were no differences between ε3 and ε4 mice on the retention probe trial, ε2 mice had a shorter latency to re-enter the dark shock compartment than both ε3 and ε4 mice (Siegel et al. 2012).

In contrast to the results in the TR mice trained to criterion, Pfankuch et al. (2005) found that 6-month-old male NSE ε4 mice required more trials to reach criterion than 6-month-old ε3 mice, while Van Meer et al. (2007) failed to find a genotype effect in 6-month-old female GFAP mice. Both Pfankuch et al. (2005) and Van Meer et al. (2007) also failed to find any genotype-related differences on probe trials in the passive avoidance test.

Taken together, the findings reported here on the passive avoidance test suggest that, somewhat paradoxically, there may be enhanced learning effects in ε4 mice. This could be because, as some authors have noted, there appears to be increased levels of anxiety-type behaviours in ε4 mice compared to ε3 mice (e.g. Siegel et al. 2012; Villasana et al. 2006, 2011; see also the following sections of this chapter discussing anxiety-related issues). Therefore, increased learning on this task is perhaps a consequence of an enhanced fear response in ε4 mice. However, these ε4-related results seem to be specific to the TR mice which were used by Villasana et al. (2006) and Siegel et al. (2012) and for the version of the passive avoidance test that trains mice to a set criterion before testing their subsequent performance via a probe trial. Moreover, with little direct replication of findings across studies, it is difficult to derive a clear conclusion concerning passive avoidance behaviour in different strains of APOE transgenic mice.

Contextual Fear Conditioning

The previous task tested an animal's ability to remember and avoid a punishing stimulus. In contextual fear conditioning, animals are conditioned to associate a specific environment with an unavoidable shock, leading to a freeze response when they are placed in that environment. The primary measure of interest in this task is extinction learning. Extinction learning is tested by repeatedly placing the animal in the same environment associated with shock, but without presenting the aversive shock stimulus. Gradually, the animal learns that the environment is no longer associated with shock and begins to show less freezing behaviour, i.e. extinction of the conditioned freeze response occurs.

Olsen et al. (2012) investigated the contextual fear conditioning response in 3–6-month-old apoE2, apoE3 and apoE4 TR mice ($N = 16–19$ per genotype). On the initial freeze response to the shock stimulus, they found that apoE3 mice froze for a longer period compared to apoE2 mice, while apoE4 mice showed an intermediate level of freezing. After several shocks, the apoE2 mice displayed the lowest freeze response compared to the apoE3 and apoE4 mice, with the latter displaying a similar level of freeze response to each other. All apoE groups showed acquisition of conditioned fear. However, apoE2 mice did not show a significant reduction in freezing behaviour, indicating a failure to extinguish the conditioned fear response. By comparison, apoE3 and apoE4 mice did show a significant extinction response. One interpretation of these findings is that apoE2 mice show a failure of fear extinction learning. However, this finding might be confounded by a floor effect, as the apoE2 mice demonstrated the least amount of conditioned freezing compared to apoE3 and apoE4 mice, with the latter groups manifesting a higher level of freezing behaviour prior to the extinction learning phase.

Novel Object/Place Recognition

A number of studies of APOE mice have incorporated novel object or novel location recognition tasks in their behavioural assessment battery. Both the novel location and novel object tests use a similar experimental design. In the novel location test, mice are placed into an environment where several objects are located in a set position. The mouse is put into the environment for a certain length of time, and explores the objects before habituating to the overall environment. A test phase occurs either the same day or 24 h later. During the test phase of the novel location test, an object is moved to a new location in the environment, and the mouse is observed to determine if it spends longer exploring the re-located objected compared to the objects that remain in the same position. For the test phase of the novel object test, a new object is placed into the environment in the place of an existing object (i.e. instead of moving an object), and the time spent exploring the novel object is recorded and compared to the length of time exploring habituated objects. These tests take advantage of a mouse's natural behavioural inclination to explore changes in its habituated environment.

The findings obtained from using these two tests have been inconsistent. Results from Raber et al. (2002; assessing NSE and GFAP mice) and Grootendorst et al. (2005; assessing TR mice) indicated that young female ε4 mice spent less time exploring the replaced and displaced objects than ε3 female mice and male mice of both genotypes. By comparison, Benice and Raber (2008; N=12–16 per group) and Salomon-Zimri et al. (2014) tested TR mice and found that young ε4 male mice demonstrated reduced novel location and object behaviour compared to ε3 male mice. Other studies have failed to find an effect of APOE in young male TR mice (Belinson et al. 2008; Villasana et al. 2006) (N=5 per group and N=7–10 per group), female TR mice (Villasana et al. 2006) and young male NSE mice (Pfankuch et al. 2005). It is difficult to identify a satisfactory coherent explanation for these divergent findings, because these studies include evidence obtained across strains of genetic mice (TR, NSE or GFAP) using novel object and novel location tests under heterogeneous training and testing methods. Moreover, in addition to the divergent findings in younger mice, it has been shown that older female ε3 TR mice (10–15 months) demonstrate less exploration of a displaced object in the novel location test compared to older ε3 male TR mice and older ε4 male and female TR mice (Bour et al. 2008). This finding contradicts the hypothesis that ε4 mice should perform more poorly than ε3 mice on memory-associated measures (due to the association of ε4 with increased risk of AD in humans). However, Nichol et al. (2009) did find that male and female 10–12-month ε3 TR mice performed better than male and female ε4 mice (N=15–16 per group) on a novel location task but not on a novel object task.

Y-Maze Active Avoidance

The Y-maze active avoidance task is designed to test spatial discrimination. The apparatus consists of a walled maze constructed in the shape of a "Y". A mouse is placed in the starting arm of the maze and within a short period of time it is given a foot-shock that punishes the mouse for staying in the starting arm. The mouse thereby learns to move out of the training arm and enter one of the two other arms of the Y-maze. One of these arms is the target arm; no shock is given to the mouse if it enters this arm. Entry into the non-target arm is punished with another shock. This task yields two types of errors: the first is an avoidance error that arises as a consequence of failing to leave the start arm within the specified period, while the second is a discrimination error that arises from choosing the wrong arm to enter once the mouse has left the starting arm.

So far, only Grootendorst et al. (2005) and the follow-up study undertaken by Bour et al. (2008) in older mice have incorporated the Y maze into their behavioural test batteries. The earlier study in younger TR mice reported by Grootendorst et al. (2005) found that male ε3 mice made more avoidance errors during both acquisition and retention trials than female ε3 mice, while ε4 mice of both sexes performed at a level that was intermediate between male and female ε3 mice. No differences in the number of discrimination errors were apparent across sex and genotype. In contrast, the follow-up study reported by Bour et al. (2008) in older TR animals found that ε4 female mice made more avoidance and discrimination errors compared to ε3 female mice, ε3 male mice and ε4 male mice. These results suggest an age- and gender-dependent effect of the ε4 allele. However, only two studies have evaluated the Y-maze so far in APOE mice; therefore, drawing definitive conclusions from these data must at this stage be considered premature.

Elevated Plus Maze and Elevated Zero Maze

The elevated plus maze and the elevated zero maze are used to measure anxiety-like behaviour in animals. The apparatus for the plus maze consists of an elevated platform in the shape of a plus sign. Two of the arms have ~15 cm high walls (termed the closed arms), while the remaining two arms are unwalled and open to the laboratory environment. Similarly, the zero maze contains two closed areas and two open areas located on a narrow circular platform. The total time spent (or the ratio of time spent) in the open arms or open areas of the maze offers a measure of the animal's anxiety (i.e. with respect to the preference of rodents to stay within confined spaces and away from spaces that are in the open).

The evidence so far seems to suggest that ε4 mice demonstrate increased avoidance of the open arms or areas of the maze, perhaps indicating increased anxiety levels. For example, Siegel et al. (2012) showed that both male and female ε4 TR mice across all age groups (range: 6–20 months) spent less time in open sections of the elevated plus maze and the zero maze compared to ε3 and ε2 mice. Several studies have replicated this finding using both the elevated plus maze and the zero maze in ε4 mice compared to ε3 mice (a finding replicated in both TR and GFAP mice), suggesting differential levels of fear/avoidance behaviour across genotypes (Hartman et al. 2001; Reverte et al. 2012; Robertson et al. 2005; Villasana et al. 2006, 2011).

The behaviour demonstrated in the elevated plus and elevated zero maze appears consistent with the previous suggestion that performance in the passive avoidance test may be a direct consequence of increased anxiety/fear in ε4 mice.

Open Field

The open field test is perhaps the simplest behavioural test of those discussed here. Rodents are placed into an open area surrounded by walls that are high enough to stop them from escaping. In the case of mice, the animal will usually explore the novel environment, enabling the researcher to measure behaviours such as total distance travelled, time spent in the centre, rearing, sniffing, licking, scratching, grooming and freezing (depending on the paradigm being tested). The two behaviours that will be the focus here are total locomotor activity (which gives an indication of psychomotor integrity) and the time spent in the middle sections of the open field (sometimes referred to as thigmotaxis) which can provide an indication of the animal's level of anxiety.

The genetic behavioural profiles observed on these tests have, once again, not been consistent. On the measure of total locomotor activity, some studies have shown reduced activity in ε4 mice compared to ε3 mice (Villasana et al. 2006, 2011; Bour et al. 2008; Siegel et al. 2012), while other studies have failed to find a significant effect of genotype (Olsen et al. 2012; Pfankuch et al. 2005; Raber et al. 1998; Reverte et al. 2012). However, Raber et al. (1998) found increased rearing in NSE/GFAP apoE4 expressing mice compared to apoE3 mice. These differences are not readily explained by the influences of sex or genotype, as both sets of studies tested both male and female Sullivan TR mice. However, there may be age-dependent effects, as studies that have failed to find reduced locomotor activity in ε4 mice have generally tested younger mice, whereas studies that have found a significant ε4 effect have generally tested older mice.

Similar to the total locomotor activity results, the thigmotaxis results have also shown some inconsistencies. Reverte et al. (2012), Villasana et al. (2006) and Olsen et al. (2012) demonstrated a reduced amount of activity in the centre of the field in ε4 TR mice compared to ε3 mice, while Villasana et al. (2011) failed to find any differences in time spent in the central zone across different genotypes. The study conducted by Villasana et al. (2011) that failed to observe a genotype effect tested older mice, compared to the studies that did find such an effect (Reverte et al. 2012; Villasana et al. 2006; Olsen et al. 2012).

Conclusions

The majority of the behavioural assessments of human apoE mice cited in this chapter have reported conflicting findings within and across behavioural paradigms with respect to the impact of the AD-related higher risk allele ε4 when compared with the lower risk alleles ε3 and ε2. We next consider some potential explanations for these divergent findings.

Given rodents' sensitivity to their environmental conditions and potential stress responses induced by laboratory testing, one possibility is that the mice tested in the reported studies may have been influenced by different laboratory conditions across studies. Indeed, there have been discussions in the literature regarding potential difficulties in interpreting experimental outcomes in animal models due to differences in conditions across studies, with suggestions that some of these difficulties are inherent in laboratory animal research (Akhtar et al. 2008). However, these considerations do not appear to offer a likely explanation in the current context, as many of the reported studies were undertaken within the same research group (in which handling procedures were conserved across experiments—according to the methodological details reported). We instead suggest that the most likely reason for inconsistent findings across studies is due to a lack of statistical power in investigating genetic effects. Most studies reviewed here (indeed, animal behavioural studies more generally) have used an N of between 6 and 10 animals per cell. This sample size may be adequate for the identification of large effect sizes, but it is insufficient for identifying genetic associations that, in humans, can often require hundreds if not thousands of participants to detect. Another plausible explanation for the conflicting set of findings across animal genetic studies concerns the established finding that the presence of the ε4 allele represents only a risk factor for late-onset AD and is not deterministic with respect to cognitive impairment and/or incidence of AD. Indeed, many older individuals possessing the ε4 allele are cognitively healthy (Bondi et al. 1999; Bunce et al. 2004; Foster et al. 2013; Smith et al. 1998), and in some cases may be cognitively more healthy (Carrión-Baralt et al. 2009; Negash et al. 2009) than non-ε4 individuals.

Nevertheless, there appears to be a potential age- and sex-dependent effect of the ε4 allele on spatial 'working memory' in APOE transgenic mice. This mirrors, to some extent, the demographic susceptibilities of neurodegenerative cognitive decline in humans, i.e. the ε4 allele is a significant contributor to late-onset AD and particularly increases susceptibility for older females (Corder et al. 2004; Farrer et al. 1997). Furthermore, one of the earliest clinical signs of AD (i.e. decline in episodic memory) may be characterized as the functional equivalent in humans of spatial 'working memory' in mice.

Future Directions

From the literature reviewed here, it is clear that more work needs to be undertaken in order to identify consistent effects of APOE polymorphisms on behaviour. Larger sample sizes are needed (i.e. estimated >20 mice per group) using more mature animals, in order to gain a clearer picture of the effects of the human forms of apoE on animal behaviour across the lifespan. The use of older animals is particularly important given that the ε4 polymorphism of the APOE gene represents a significant risk factor in older humans, such that the use of younger animals may produce misleading information concerning the impact of APOE isoforms on behavioural outcomes.

Aside from this issue, there is significant potential for manipulating gene–gene interactions in animal models to substantially increase our understanding of the mechanisms underlying the pathophysiology of AD. As mentioned previously, mice possess their own murine apoE and there is considerable methodological effort required in order for mice to express human versions of apoE (as has been outlined in this chapter). It may be that interactions between human forms of apoE and other human-specific factors are responsible for conferring an increased risk of cognitive decline in humans. Specifically, investigations of potential gene–gene interactions in APOE mice were reported in Raber et al. (2000) and more recently by Kornecook et al. (2010). In their study, Raber et al. (2000) used doubly transgenic APOE/APP 6-month-old male and female mice, while Kornecook et al. (2010) tested combined APOE/APP 6–7-month-old female mice. Kornecook et al. (2010) found that the combination of human apoE4 and human APP expressing mice demonstrated reduced locomotor activity, reduced exploration of objects in novel locations, poorer visual discrimination performance and slower reaction times on a discrimination test. Unfortunately, Kornecook et al. (2010) did not examine performance on the Morris Water Maze. However, the study by Raber et al. (2000) did use the Morris Water Maze; these researchers observed significant deficits in combined human apoE4 and APP compared to apoE3 and APP mice. Work using combined transgenic strains of mice offers a promising avenue for future investigations, as these approaches may offer more realistic animal models of AD-related deficits in humans.

Another potential direction for future research relates to the interaction between the APOE gene and environmental factors and/or accumulative neural insults on cognitive decline. For example, the cognitive performance of individuals carrying the ε4 isoform has been shown to improve in response to physical activity comparatively more than that of individuals without the ε4 isoform (Lautenschlager et al. 2008). Similar findings in apoE4 expressing mice have been demonstrated by Nichol et al. (2009) and Chaudhari et al. (2014), who found that several weeks of exercise significantly improved performance in 10–12-month-old and 5–6-month-old apoE4 expressing mice compared to sedentary apoE4 mice on the radial arm water maze and on an active avoidance task. Other environmental factors such as diet, cognitive enrichment and history of head trauma may also significantly interact with APOE status and age to influence cognitive functioning, and would be useful to evaluate in transgenic animal models.

References

Akhtar, A. Z., Pippin, J. J., & Sandusky, C. B. (2008). Animal models in spinal cord injury: A review. *Reviews in the Neurosciences, 19*, 47–60. doi:10.1515/REVNEURO.2008.19.1.47.

Andrews-Zwilling, Y., Bien-Ly, N., Xu, Q., Li, G., Bernardo, A., Yoon, S. Y., et al. (2010). Apolipoprotein E4 causes age- and tau-dependent impairment of GABAergic interneurons, leading to learning and memory deficits in mice. *The Journal of Neuroscience, 30*(41), 13707–13717. doi:10.1523/JNEUROSCI.4040-10.2010.

Apostolova, L. G., & Thompson, P. M. (2008). Mapping progressive brain structural changes in early Alzheimer's disease and mild cognitive impairment. *Neuropsychologia, 46*(6), 1597–1612. doi:10.1016/j.neuropsychologia.2007.10.026.

Arnold, S. E., Hyman, B. T., Flory, J., Damasio, A. R., & Hoesen, G. W. V. (1991). The topographical and neuroanatomical distribution of neurofibrillary tangles and neuritic plaques in the cere-

bral cortex of patients with Alzheimer's disease. *Cerebral Cortex, 1*(1), 103–116. doi:10.1093/cercor/1.1.103.
Arriagada, P. V., Growdon, J. H., Hedley-Whyte, E. T., & Hyman, B. T. (1992). Neurofibrillary tangles but not senile plaques parallel duration and severity of Alzheimer's disease. *Neurology, 42*(3), 631–639. doi:10.1212/WNL.42.3.631.
Behl, C. (1994). Hydrogen peroxide mediates amyloid β protein toxicity. *Cell, 77*(6), 817–827. doi:10.1016/0092-8674(94)90131-7.
Behl, C., Davis, J., Cole, G. M., & Schubert, D. (1992). Vitamin E protects nerve cells from amyloid β-protein toxicity. *Biochemical and Biophysical Research Communications, 186*(2), 944–950. doi:10.1016/0006-291X(92)90837-B.
Belinson, H., Lev, D., Masliah, E., & Michaelson, D. M. (2008). Activation of the amyloid cascade in apolipoprotein E4 transgenic mice induces lysosomal activation and neurodegeneration resulting in marked cognitive deficits. *The Journal of Neuroscience, 28*(18), 4690–4701. doi:10.1523/JNEUROSCI.5633-07.2008.
Bellosta, S., Nathan, B. P., Orth, M., Dong, L.-M., Mahley, R. W., & Pitas, R. E. (1995). Stable expression and secretion of apolipoproteins E3 and E4 in mouse neuroblastoma cells produces differential effects on neurite outgrowth. *Journal of Biological Chemistry, 270*(45), 27063–27071. doi:10.1074/jbc.270.45.27063.
Benice, T. S., & Raber, J. (2008). Object recognition analysis in mice using nose-point digital video tracking. *Journal of Neuroscience Methods, 168*(2), 422–430. doi:10.1016/j.jneumeth.2007.11.002.
Berg, L., McKeel, D. W., Jr., Miller, J. P., Storandt, M., Rubin, E. H., Morris, J. C., et al. (1998). Clinicopathologic studies in cognitively healthy aging and Alzheimer disease: Relation of histologic markers to dementia severity, age, sex, and apolipoprotein E genotype. *Archives of Neurology, 55*(3), 326.
Bird, T. (1998). *Alzheimer disease overview (Updated 2012).* Seattle, WA: University of Washington. Retrieved from http://www.ncbi.nlm.nih.gov/books/NBK1161/.
Bird, T. (1999). *Early-onset familial Alzheimer disease (Updated 2012).* Seattle, WA: University of Washington. Retrieved from http://www.ncbi.nlm.nih.gov/books/NBK1236/.
Bondi, M. W., Salmon, D. P., Galasko, D., Thomas, R. G., & Thal, L. J. (1999). Neuropsychological function and apolipoprotein E genotype in the preclinical detection of Alzheimer's disease. *Psychology and Aging, 14*(2), 295.
Bour, A., Grootendorst, J., Vogel, E., Kelche, C., Dodart, J.-C., Bales, K., et al. (2008). Middle-aged human apoE4 targeted-replacement mice show retention deficits on a wide range of spatial memory tasks. *Behavioural Brain Research, 193*(2), 174–182. doi:10.1016/j.bbr.2008.05.008.
Boyles, J. K., Pitas, R. E., Wilson, E., Mahley, R. W., & Taylor, J. M. (1985). Apolipoprotein E associated with astrocytic glia of the central nervous system and with nonmyelinating glia of the peripheral nervous system. *Journal of Clinical Investigation, 76*(4), 1501–1513.
Braak, H., & Braak, E. (1991). Neuropathological stageing of Alzheimer-related changes. *Acta Neuropathologica, 82*(4), 239–259.
Braak, E., & Braak, H. (1997). Alzheimer's disease: Transiently developing dendritic changes in pyramidal cells of sector CA1 of the Ammon's horn. *Acta Neuropathologica, 93*(4), 323–325. doi:10.1007/s004010050622.
Brecht, W. J., Harris, F. M., Chang, S., Tesseur, I., Yu, G.-Q., Xu, Q., et al. (2004). Neuron-specific apolipoprotein E4 proteolysis is associated with increased tau phosphorylation in brains of transgenic mice. *The Journal of Neuroscience, 24*(10), 2527–2534. doi:10.1523/JNEUROSCI.4315-03.2004.
Bunce, D., Fratiglioni, L., Small, B. J., Winblad, B., & Bäckman, L. (2004). APOE and cognitive decline in preclinical Alzheimer disease and non-demented aging. *Neurology, 63*(5), 816–821.

Busciglio, J., Gabuzda, D. H., Matsudaira, P., & Yankner, B. A. (1993). Generation of beta-amyloid in the secretory pathway in neuronal and nonneuronal cells. *Proceedings of the National Academy of Sciences, 90*(5), 2092–2096.

Buttini, M., Orth, M., Bellosta, S., Akeefe, H., Pitas, R. E., Wyss-Coray, T., et al. (1999). Expression of human apolipoprotein E3 or E4 in the brains of Apoe−/− mice: Isoform-specific effects on neurodegeneration. *The Journal of Neuroscience, 19*(12), 4867–4880.

Carrión-Baralt, J. R., Meléndez-Cabrero, J., Rodríguez-Ubiñas, H., Schmeidler, J., Beeri, M. S., Angelo, G., et al. (2009). Impact of APOE ε4 on the cognitive performance of a sample of non-demented Puerto Rican nonagenarians. *Journal of Alzheimer's Disease, 18*(3), 533–540. doi:10.3233/JAD-2009-1160.

Castellano, J. M., Kim, J., Stewart, F. R., Jiang, H., DeMattos, R. B., Patterson, B. W., et al. (2011). Human apoE isoforms differentially regulate brain amyloid-peptide clearance. *Science Translational Medicine, 3*(89), 89ra57. doi:10.1126/scitranslmed.3002156.

Chaudhari, K., Wong, J. M., Vann, P. H., & Sumien, N. (2014). Exercise training and antioxidant supplementation independently improve cognitive function in adult male and female GFAP-APOE mice. *Journal of Sport and Health Science, 3*(3), 196–205.

Corder, E. H., Ghebremedhin, E., Taylor, M. G., Thal, D. R., Ohm, T. G., & Braak, H. (2004). The biphasic relationship between regional brain senile plaque and neurofibrillary tangle distributions: Modification by age, sex, and APOE polymorphism. *Annals of the New York Academy of Sciences, 1019*(1), 24–28. doi:10.1196/annals.1297.005.

Corder, E. H., Saunders, A. M., Strittmatter, W. J., Schmechel, D. E., Gaskell, P. C., Small, G. W., et al. (1993). Gene dose of apolipoprotein E type 4 allele and the risk of Alzheimer's disease in late onset families. *Science, 261*(5123), 921–923. doi:10.1126/science.8346443.

Cosentino, S., Scarmeas, N., Helzner, E., Glymour, M. M., Brandt, J., Albert, M., et al. (2008). APOE ε4 allele predicts faster cognitive decline in mild Alzheimer's disease. *Neurology, 70*(19 Pt 2), 1842–1849. doi:10.1212/01.wnl.0000304038.37421.cc.

Dantzer, R., Capuron, L., Irwin, M. R., Miller, A. H., Ollat, H., Hugh Perry, V., et al. (2008). Identification and treatment of symptoms associated with inflammation in medically ill patients. *Psychoneuroendocrinology, 33*(1), 18–29. doi:10.1016/j.psyneuen.2007.10.008.

De Leon, M. J., Convit, A., DeSanti, S., Bobinski, M., George, A. E., Wisniewski, H. M., et al. (1997). Contribution of structural neuroimaging to the early diagnosis of Alzheimer's disease. *International Psychogeriatrics, 9*(Suppl. S1), 183–190. doi:10.1017/S1041610297004900.

Deane, R., Sagare, A., Hamm, K., Parisi, M., Lane, S., Finn, M. B., et al. (2008). apoE isoform–specific disruption of amyloid β peptide clearance from mouse brain. *The Journal of Clinical Investigation, 118*(12), 4002–4013. doi:10.1172/JCI36663.

Eisenberg, D. T. A., Kuzawa, C. W., & Hayes, M. G. (2010). Worldwide allele frequencies of the human apolipoprotein E gene: Climate, local adaptations, and evolutionary history. *American Journal of Physical Anthropology, 143*(1), 100–111. doi:10.1002/ajpa.21298.

Elshourbagy, N. A., Liao, W. S., Mahley, R. W., & Taylor, J. M. (1985). Apolipoprotein E mRNA is abundant in the brain and adrenals, as well as in the liver, and is present in other peripheral tissues of rats and marmosets. *Proceedings of the National Academy of Sciences, 82*(1), 203–207.

Esch, F. S., Keim, P. S., Beattie, E. C., Blacher, R. W., Culwell, A. R., Oltersdorf, T., et al. (1990). Cleavage of amyloid beta peptide during constitutive processing of its precursor. *Science, 248*(4959), 1122–1124. doi:10.1126/science.2111583.

Farrer, L. A., Cupples, L. A., Haines, J. L., Hyman, B., Kukull, W. A., Mayeux, R., et al. (1997). Effects of age, sex, and ethnicity on the association between apolipoprotein E genotype and Alzheimer disease. *JAMA, the Journal of the American Medical Association, 278*(16), 1349–1356.

Foster, J. K., Albrecht, M. A., Savage, G., Lautenschlager, N. T., Ellis, K. A., Maruff, P., et al. (2013). Lack of reliable evidence for a distinctive ε4–related cognitive phenotype that is independent from clinical diagnostic status: Findings from the Australian Imaging. Biomarkers and Lifestyle Study. *Brain, 136*(Pt 7), 2201–2216. doi:10.1093/brain/awt127.

Fukumoto, H., Asami-Odaka, A., Suzuki, N., Shimada, H., Ihara, Y., & Iwatsubo, T. (1996). Amyloid beta protein deposition in normal aging has the same characteristics as that in Alzheimer's disease. Predominance of A beta 42(43) and association of A beta 40 with cored plaques. *The American Journal of Pathology, 148*(1), 259–265.

Gandy, S. (2005). The role of cerebral amyloid β accumulation in common forms of Alzheimer disease. *Journal of Clinical Investigation, 115*(5), 1121–1129. doi:10.1172/JCI200525100.

Ghebremedhin, E., Schultz, C., Braak, E., & Braak, H. (1998). High frequency of apolipoprotein E ϵ4 allele in young individuals with very mild Alzheimer's disease-related neurofibrillary changes. *Experimental Neurology, 153*(1), 152–155. doi:10.1006/exnr.1998.6860.

Giulian, D., Haverkamp, L. J., Yu, J. H., Karshin, W., Tom, D., Li, J., et al. (1996). Specific domains of β-amyloid from Alzheimer plaque elicit neuron killing in human microglia. *The Journal of Neuroscience, 16*(19), 6021–6037.

Glenner, G. G., & Wong, C. W. (1984). Alzheimer's disease: Initial report of the purification and characterization of a novel cerebrovascular amyloid protein. *Biochemical and Biophysical Research Communications, 120*(3), 885–890.

Gómez-Isla, T., Hollister, R., West, H., Mui, S., Growdon, J. H., Petersen, R. C., et al. (1997). Neuronal loss correlates with but exceeds neurofibrillary tangles in Alzheimer's disease. *Annals of Neurology, 41*(1), 17–24. doi:10.1002/ana.410410106.

Gomez-Isla, T., West, H. L., Rebeck, G. W., Harr, S. D., Growdon, J. H., Locascio, J. J., et al. (1996). Clinical and pathological correlates of apolipoprotein E ε4 in Alzheimer's disease. *Annals of Neurology, 39*(1), 62–70. doi:10.1002/ana.410390110.

Götz, J., Chen, F., van Dorpe, J., & Nitsch, R. M. (2001). Formation of neurofibrillary tangles in P301L tau transgenic mice induced by Aβ42 fibrils. *Science, 293*(5534), 1491–1495. doi:10.1126/science.1062097.

Gravina, S. A., Ho, L., Eckman, C. B., Long, K. E., Otvos, L., Younkin, L. H., et al. (1995). Amyloid β protein (Aβ) in Alzheimeri's disease brain. *Journal of Biological Chemistry, 270*(13), 7013–7016. doi:10.1074/jbc.270.13.7013.

Grehan, S., Tse, E., & Taylor, J. M. (2001). Two distal downstream enhancers direct expression of the human apolipoprotein E gene to astrocytes in the brain. *The Journal of Neuroscience, 21*(3), 812–822.

Grootendorst, J., Bour, A., Vogel, E., Kelche, C., Sullivan, P. M., Dodart, J.-C., et al. (2005). Human apoE targeted replacement mouse lines: h-apoE4 and h-apoE3 mice differ on spatial memory performance and avoidance behavior. *Behavioural Brain Research, 159*(1), 1–14. doi:10.1016/j.bbr.2004.09.019.

Guo, L., LaDu, M., & Van Eldik, L. (2004). A dual role for apolipoprotein E in neuroinflammation. *Journal of Molecular Neuroscience, 23*(3), 205–212. doi:10.1385/JMN:23:3:205.

Haass, C., Schlossmacher, M. G., Hung, A. Y., Vigo-Pelfrey, C., Mellon, A., Ostaszewski, B. L., et al. (1992). Amyloid β-peptide is produced by cultured cells during normal metabolism. *Nature, 359*(6393), 322–325. doi:10.1038/359322a0.

Hardy, J. (1997). Amyloid, the presenilins and Alzheimer's disease. *Trends in Neurosciences, 20*(4), 154–159.

Hardy, J., & Selkoe, D. J. (2002). The amyloid hypothesis of Alzheimer's disease: Progress and problems on the road to therapeutics. *Science, 297*(5580), 353–356. doi:10.1126/science.1072994.

Harold, D., Abraham, R., Hollingworth, P., Sims, R., Gerrish, A., Hamshere, M., et al. (2009). Genome-wide association study identifies variants at CLU and PICALM associated with Alzheimer's disease, and shows evidence for additional susceptibility genes. *Nature Genetics, 41*(10), 1088–1093. doi:10.1038/ng.440.

Harris, F. M., Brecht, W. J., Xu, Q., Mahley, R. W., & Huang, Y. (2004). Increased tau phosphorylation in apolipoprotein E4 transgenic mice is associated with activation of extracellular signal-regulated kinase modulation by Zinc. *Journal of Biological Chemistry, 279*(43), 44795–44801. doi:10.1074/jbc.M408127200.

Harris, F. M., Brecht, W. J., Xu, Q., Tesseur, I., Kekonius, L., Wyss-Coray, T., et al. (2003). Carboxyl-terminal-truncated apolipoprotein E4 causes Alzheimer's disease-like neurodegeneration and behavioral deficits in transgenic mice. *Proceedings of the National Academy of Sciences, 100*(19), 10966–10971. doi:10.1073/pnas.1434398100.

Harrison, N. A., Brydon, L., Walker, C., Gray, M. A., Steptoe, A., & Critchley, H. D. (2009). Inflammation causes mood changes through alterations in subgenual cingulate activity and mesolimbic connectivity. *Biological Psychiatry, 66*(5), 407–414. doi:10.1016/j.biopsych.2009.03.015.

Hartman, R. E., Wozniak, D. F., Nardi, A., Olney, J. W., Sartorius, L., & Holtzman, D. M. (2001). Behavioral phenotyping of GFAP-ApoE3 and -ApoE4 transgenic mice: ApoE4 mice show profound working memory impairments in the absence of Alzheimer's-like neuropathology. *Experimental Neurology, 170*(2), 326–344. doi:10.1006/exnr.2001.7715.

Huang, Y., Liu, X. Q., Wyss-Coray, T., Brecht, W. J., Sanan, D. A., & Mahley, R. W. (2001). Apolipoprotein E fragments present in Alzheimer's disease brains induce neurofibrillary tangle-like intracellular inclusions in neurons. *Proceedings of the National Academy of Sciences, 98*(15), 8838–8843. doi:10.1073/pnas.151254698.

Iqbal, K., Liu, F., Gong, C.-X., & Grundke-Iqbal, I. (2010). Tau in Alzheimer disease and related tauopathies. *Current Alzheimer Research, 7*(8), 656–664.

Iwatsubo, T., Odaka, A., Suzuki, N., Mizusawa, H., Nukina, N., & Ihara, Y. (1994). Visualization of Aβ42(43) and Aβ40 in senile plaques with end-specific Aβ monoclonals: Evidence that an initially deposited species is Aβ42(43). *Neuron, 13*(1), 45–53. doi:10.1016/0896-6273(94)90458-8.

Jankowsky, J. L., Fadale, D. J., Anderson, J., Xu, G. M., Gonzales, V., Jenkins, N. A., et al. (2004). Mutant presenilins specifically elevate the levels of the 42 residue β-amyloid peptide in vivo: Evidence for augmentation of a 42-specific γ secretase. *Human Molecular Genetics, 13*(2), 159–170. doi:10.1093/hmg/ddh019.

Janowsky, J. S., Chavez, B., & Orwoll, E. (2000). Sex steroids modify working memory. *Journal of Cognitive Neuroscience, 12*(3), 407–414. doi:10.1162/089892900562228.

Jaworski, T., Kügler, S., & Van Leuven, F. (2010). Modeling of tau-mediated synaptic and neuronal degeneration in Alzheimer's disease. *International Journal of Alzheimer's Disease, 2010*, 1–10. doi:10.4061/2010/573138.

Ji, Z.-S., Miranda, R. D., Newhouse, Y. M., Weisgraber, K. H., Huang, Y., & Mahley, R. W. (2002). Apolipoprotein E4 potentiates amyloid β peptide-induced lysosomal leakage and apoptosis in neuronal cells. *Journal of Biological Chemistry, 277*(24), 21821–21828. doi:10.1074/jbc.M112109200.

Ji, Z.-S., Müllendorff, K., Cheng, I. H., Miranda, R. D., Huang, Y., & Mahley, R. W. (2006). Reactivity of apolipoprotein E4 and amyloid β peptide lysosomal stability and neurodegeneration. *Journal of Biological Chemistry, 281*(5), 2683–2692. doi:10.1074/jbc.M506646200.

Kanai, M., Shizuka, M., Urakami, K., Matsubara, E., Harigaya, Y., Okamoto, K., et al. (1999). Apolipoprotein E4 accelerates dementia and increases cerebrospinal fluid tau levels in Alzheimer's disease. *Neuroscience Letters, 267*(1), 65–68.

Katzman, R., & Saitoh, T. (1991). Advances in Alzheimer's disease. *The FASEB Journal, 5*(3), 278–286.

Kim, J., Basak, J. M., & Holtzman, D. M. (2009). The role of apolipoprotein E in Alzheimer's disease. *Neuron, 63*(3), 287–303. doi:10.1016/j.neuron.2009.06.026.

Koh, J., Yang, L. L., & Cotman, C. W. (1990). β-Amyloid protein increases the vulnerability of cultured cortical neurons to excitotoxic damage. *Brain Research, 533*(2), 315–320. doi:10.1016/0006-8993(90)91355-K.

Kornecook, T. J., McKinney, A. P., Ferguson, M. T., & Dodart, J.-C. (2010). Isoform-specific effects of apolipoprotein E on cognitive performance in targeted-replacement mice overexpressing human APP. *Genes, Brain and Behavior, 9*(2), 182–192. doi:10.1111/j.1601-183X.2009.00545.x.

Kril, J., Patel, S., Harding, A., & Halliday, G. (2002). Neuron loss from the hippocampus of Alzheimer's disease exceeds extracellular neurofibrillary tangle formation. *Acta Neuropathologica, 103*(4), 370–376. doi:10.1007/s00401-001-0477-5.

LaDu, M. J., Falduto, M. T., Manelli, A. M., Reardon, C. A., Getz, G. S., & Frail, D. E. (1994). Isoform-specific binding of apolipoprotein E to beta-amyloid. *Journal of Biological Chemistry, 269*(38), 23403–23406.

LaFerla, F. M., Tinkle, B. T., Bieberich, C. J., Haudenschild, C. C., & Jay, G. (1995). The Alzheimer's Aβ peptide induces neurodegeneration and apoptotic cell death in transgenic mice. *Nature Genetics, 9*(1), 21–30. doi:10.1038/ng0195-21.

Lambert, J.-C., Heath, S., Even, G., Campion, D., Sleegers, K., Hiltunen, M., et al. (2009). Genome-wide association study identifies variants at CLU and CR1 associated with Alzheimer's disease. *Nature Genetics, 41*(10), 1094–1099. doi:10.1038/ng.439.

Landén, M., Thorsell, A., Wallin, A., & Blennow, K. (1996). The apolipoprotein E allele epsilon 4 does not correlate with the number of senile plaques or neurofibrillary tangles in patients with Alzheimer's disease. *Journal of Neurology, Neurosurgery, and Psychiatry, 61*(4), 352–356.

Larner, A. J., & Doran, M. (2005). Clinical phenotypic heterogeneity of Alzheimer's disease associated with mutations of the presenilin-1 gene. *Journal of Neurology, 253*(2), 139–158. doi:10.1007/s00415-005-0019-5.

Lautenschlager, N. T., Cox, K. L., Flicker, L., Foster, J. K., van Bockxmeer, F. M., Xiao, J., et al. (2008). Effect of physical activity on cognitive function in older adults at risk for Alzheimer disease—A randomized trial. *JAMA, the Journal of the American Medical Association, 300*(9), 1027–1037. doi:10.1001/jama.300.9.1027.

Loo, D. T., Copani, A., Pike, C. J., Whittemore, E. R., Walencewicz, A. J., & Cotman, C. W. (1993). Apoptosis is induced by beta-amyloid in cultured central nervous system neurons. *Proceedings of the National Academy of Sciences, 90*(17), 7951–7955.

Luine, V. N., Richards, S. T., Wu, V. Y., & Beck, K. D. (1998). Estradiol enhances learning and memory in a spatial memory task and effects levels of monoaminergic neurotransmitters. *Hormones and Behavior, 34*(2), 149–162. doi:10.1006/hbeh.1998.1473.

Lynch, J. R., Tang, W., Wang, H., Vitek, M. P., Bennett, E. R., Sullivan, P. M., et al. (2003). APOE genotype and an ApoE-mimetic peptide modify the systemic and central nervous system inflammatory response. *Journal of Biological Chemistry, 278*(49), 48529–48533. doi:10.1074/jbc.M306923200.

Ma, J., Yee, A., Brewer, H. B., Das, S., & Potter, H. (1994). Amyloid-associated proteins α1-antichymotrypsin and apolipoprotein E promote assembly of Alzheimer β-protein into filaments. *Nature, 372*(6501), 92–94. doi:10.1038/372092a0.

Mahley, R. W. (1988). Apolipoprotein E: Cholesterol transport protein with expanding role in cell biology. *Science, 240*(4852), 622–630. doi:10.1126/science.3283935.

Mahley, R. W., & Rall, S. C., Jr. (2000). Apolipoprotein E: Far more than a lipid transport protein. *Annual Review of Genomics and Human Genetics, 1*(1), 507–537.

Mahley, R. W., Weisgraber, K. H., & Huang, Y. (2006). Apolipoprotein E4: A causative factor and therapeutic target in neuropathology, including Alzheimer's disease. *Proceedings of the National Academy of Sciences of the United States of America, 103*(15), 5644–5651. doi:10.1073/pnas.0600549103.

Mahley, R. W., Weisgraber, K. H., & Huang, Y. (2008). Apolipoprotein E: Structure determines function, from atherosclerosis to Alzheimer's disease to AIDS. *The Journal of Lipid Research, 50*(Suppl.), S183–S188. doi:10.1194/jlr.R800069-JLR200.

Mak, K., Yang, F., Vinters, H. V., Frautschy, S. A., & Cole, G. M. (1994). Polyclonals to β-amyloid(1–42) identify most plaque and vascular deposits in Alzheimer cortex, but not striatum. *Brain Research, 667*(1), 138–142. doi:10.1016/0006-8993(94)91725-6.

Mann, D. M., Brown, S. M., Owen, F., Baba, M., & Iwatsubo, T. (1998). Amyloid β protein (Aβ) deposition in dementia with Lewy bodies: Predominance of Aβ42(43) and paucity of Aβ40 compared with sporadic Alzheimer's disease. *Neuropathology and Applied Neurobiology, 24*(3), 187–194. doi:10.1046/j.1365-2990.1998.00112.x.

Markesbery, W. R. (1997). Oxidative stress hypothesis in Alzheimer's disease. *Free Radical Biology and Medicine, 23*(1), 134–147. doi:10.1016/S0891-5849(96)00629-6.

Masters, C. L., Simms, G., Weinman, N. A., Multhaup, G., McDonald, B. L., & Beyreuther, K. (1985). Amyloid plaque core protein in Alzheimer disease and down syndrome. *Proceedings of the National Academy of Sciences, 82*(12), 4245–4249.

Matsumoto, K., Pinna, G., Puia, G., Guidotti, A., & Costa, E. (2005). Social isolation stress-induced aggression in mice: A model to study the pharmacology of neurosteroidogenesis. *Stress, 8*(2), 85–93. doi:10.1080/10253890500159022.

Mattson, M. P. (1997). Cellular actions of beta-amyloid precursor protein and its soluble and fibrillogenic derivatives. *Physiological Reviews, 77*(4), 1081–1132.

Mattson, M. P., Cheng, B., Davis, D., Bryant, K., Lieberburg, I., & Rydel, R. E. (1992). beta-Amyloid peptides destabilize calcium homeostasis and render human cortical neurons vulnerable to excitotoxicity. *The Journal of Neuroscience, 12*(2), 376–389.

Meda, L., Cassatella, M. A., Szendrei, G. I., Otvos, L., Baron, P., Villalba, M., et al. (1995). Activation of microglial cells by β-amyloid protein and interferon-γ. *Nature, 374*(6523), 647–650. doi:10.1038/374647a0.

Miller, A. H., Maletic, V., & Raison, C. L. (2009). Inflammation and its discontents: The role of cytokines in the pathophysiology of major depression. *Biological Psychiatry, 65*(9), 732–741. doi:10.1016/j.biopsych.2008.11.029.

Mills, J., & Reiner, P. B. (1999). Regulation of amyloid precursor protein cleavage. *Journal of Neurochemistry, 72*(2), 443–460. doi:10.1046/j.1471-4159.1999.0720443.x.

Mintun, M. A., LaRossa, G. N., Sheline, Y. I., Dence, C. S., Lee, S. Y., Mach, R. H., et al. (2006). [11C]PIB in a nondemented population. Potential antecedent marker of Alzheimer disease. *Neurology, 67*(3), 446–452. doi:10.1212/01.wnl.0000228230.26044.a4.

Miyata, M., & Smith, J. D. (1996). Apolipoprotein E allele–specific antioxidant activity and effects on cytotoxicity by oxidative insults and β–amyloid peptides. *Nature Genetics, 14*(1), 55–61. doi:10.1038/ng0996-55.

Morsch, R., Simon, W., & Coleman, P. D. (1999). Neurons may live for decades with neurofibrillary tangles. *Journal of Neuropathology and Experimental Neurology, 58*(2), 188–197.

Nagy, Z. S., Esiri, M. M., Jobst, K. A., Johnston, C., Litchfield, S., Sim, E., et al. (1995). Influence of the apolipoprotein E genotype on amyloid deposition and neurofibrillary tangle formation in Alzheimer's disease. *Neuroscience, 69*(3), 757–761. doi:10.1016/0306-4522(95)00331-C.

Negash, S., Greenwood, P. M., Sunderland, T., Parasuraman, R., Geda, Y. E., Knopman, D. S., et al. (2009). The influence of Apolipoprotein E genotype on visuospatial attention dissipates after age 80. *Neuropsychology, 23*(1), 81–89. doi:10.1037/a0014014.

Neve, R. L., & Robakis, N. K. (1998). Alzheimer's disease: A re-examination of the amyloid hypothesis. *Trends in Neurosciences, 21*(1), 15–19. doi:10.1016/S0166-2236(97)01168-5.

Nichol, K., Deeny, S. P., Seif, J., Camaclang, K., & Cotman, C. W. (2009). Exercise improves cognition and hippocampal plasticity in APOE ε4 mice. *Alzheimer's and Dementia, 5*(4), 287–294. doi:10.1016/j.jalz.2009.02.006.

Ohm, T. G., Kirca, M., Bohl, J., Scharnagl, H., Groβ, W., & März, W. (1995). Apolipoprotein E polymorphism influences not only cerebral senile plaque load but also Alzheimer-type neurofibrillary tangle formation. *Neuroscience, 66*(3), 583–587. doi:10.1016/0306-4522(94)00596-W.

Olgiati, P., Politis, A. M., Papadimitriou, G. N., De Ronchi, D., & Serretti, A. (2011). Genetics of late-onset alzheimer's disease: Update from the Alzgene database and analysis of shared pathways. *International Journal of Alzheimer's Disease, 2011*, 832379. doi:10.4061/2011/832379.

Olsen, R. H. J., Agam, M., Davis, M. J., & Raber, J. (2012). ApoE isoform-dependent deficits in extinction of contextual fear conditioning. *Genes, Brain and Behavior, 11*(7), 806–812. doi:10.1111/j.1601-183X.2012.00833.x.

Packard, C. J., Westendorp, R. G. J., Stott, D. J., Caslake, M. J., Murray, H. M., Shepherd, J., et al. (2007). Association between apolipoprotein E4 and cognitive decline in elderly adults. *Journal of the American Geriatrics Society, 55*(11), 1777–1785. doi:10.1111/j.1532-5415.2007.01415.x.

Pearson, H. A., & Peers, C. (2006). Physiological roles for amyloid β peptides. *The Journal of Physiology, 575*(1), 5–10. doi:10.1113/jphysiol.2006.111203.

Pfankuch, T., Rizk, A., Olsen, R., Poage, C., & Raber, J. (2005). Role of circulating androgen levels in effects of apoE4 on cognitive function. *Brain Research, 1053*(1–2), 88–96. doi:10.1016/j.brainres.2005.06.028.

Plant, L. D., Boyle, J. P., Smith, I. F., Peers, C., & Pearson, H. A. (2003). The production of amyloid β peptide is a critical requirement for the viability of central neurons. *The Journal of Neuroscience, 23*(13), 5531–5535.

Price, J. L., & Morris, J. C. (1999). Tangles and plaques in nondemented aging and "preclinical" Alzheimer's disease. *Annals of Neurology, 45*(3), 358–368.

Raber, J., Bongers, G., LeFevour, A., Buttini, M., & Mucke, L. (2002). Androgens protect against apolipoprotein E4-induced cognitive deficits. *The Journal of Neuroscience, 22*(12), 5204–5209.

Raber, J., Wong, D., Buttini, M., Orth, M., Bellosta, S., Pitas, R. E., et al. (1998). Isoform-specific effects of human apolipoprotein E on brain function revealed in ApoE knockout mice: Increased susceptibility of females. *Proceedings of the National Academy of Sciences, 95*(18), 10914–10919.

Raber, J., Wong, D., Yu, G.-Q., Buttini, M., Mahley, R. W., Pitas, R. E., et al. (2000). Alzheimer's disease: Apolipoprotein E and cognitive performance. *Nature, 404*(6776), 352–354. doi:10.1038/35006165.

Reverte, I., Klein, A. B., Ratner, C., Domingo, J. L., & Colomina, M. T. (2012). Behavioral phenotype and BDNF differences related to apoE isoforms and sex in young transgenic mice. *Experimental Neurology, 237*(1), 116–125. doi:10.1016/j.expneurol.2012.06.015.

Reyland, M. E., Gwynne, J. T., Forgez, P., Prack, M. M., & Williams, D. L. (1991). Expression of the human apolipoprotein E gene suppresses steroidogenesis in mouse Y1 adrenal cells. *Proceedings of the National Academy of Sciences, 88*(6), 2375–2379.

Roberson, E. D., Scearce-Levie, K., Palop, J. J., Yan, F., Cheng, I. H., Wu, T., et al. (2007). Reducing endogenous tau ameliorates amyloid ß-induced deficits in an Alzheimer's disease mouse model. *Science, 316*(5825), 750–754. doi:10.1126/science.1141736.

Robertson, J., Curley, J., Kaye, J., Quinn, J., Pfankuch, T., & Raber, J. (2005). apoE isoforms and measures of anxiety in probable AD patients and Apoe$^{-/-}$ mice. *Neurobiology of Aging, 26*(5), 637–643. doi:10.1016/j.neurobiolaging.2004.06.003.

Roßner, S., Ueberham, U., Schliebs, R., Regino Perez-Polo, J., & Bigl, V. (1998). The regulation of amyloid precursor protein metabolism by cholinergic mechanisms and neurotrophin receptor signaling. *Progress in Neurobiology, 56*(5), 541–569. doi:10.1016/S0301-0082(98)00044-6.

Rowe, C. C., Ng, S., Ackermann, U., Gong, S. J., Pike, K., Savage, G., et al. (2007). Imaging beta-amyloid burden in aging and dementia. *Neurology, 68*(20), 1718–1725.

Sadowski, M. J., Pankiewicz, J., Scholtzova, H., Mehta, P. D., Prelli, F., Quartermain, D., et al. (2006). Blocking the apolipoprotein E/amyloid-β interaction as a potential therapeutic approach for Alzheimer's disease. *Proceedings of the National Academy of Sciences, 103*(49), 18787–18792. doi:10.1073/pnas.0604011103.

Salomon-Zimri, S., Boehm-Cagan, A., Liraz, O., & Michaelson, D. M. (2014). Hippocampus-related cognitive impairments in young apoE4 targeted replacement mice. *Neurodegenerative Diseases, 13*, 86–92. doi:10.1159/000354777.

Saunders, A. M., Strittmatter, W. J., Schmechel, D., St. George-Hyslop, P. H., Pericak-Vance, M. A., Joo, S. H., et al. (1993). Association of apolipoprotein E allele E4 with late-onset familial and sporadic Alzheimer's disease. *Neurology, 43*(8), 1467–1472.

Schmechel, D. E., Saunders, A. M., Strittmatter, W. J., Crain, B. J., Hulette, C. M., Joo, S. H., et al. (1993). Increased amyloid beta-peptide deposition in cerebral cortex as a consequence of apolipoprotein E genotype in late-onset alzheimer disease. *Proceedings of the National Academy of Sciences, 90*(20), 9649–9653.

Selkoe, D. J. (1998). The cell biology of β-amyloid precursor protein and presenilin in Alzheimer's disease. *Trends in Cell Biology, 8*(11), 447–453. doi:10.1016/S0962-8924(98)01363-4.

Seubert, P., Oltersdorf, T., Lee, M. G., Barbour, R., Blomquist, C., Davis, D. L., et al. (1993). Secretion of β-amyloid precursor protein cleaved at the amino terminus of the β-amyloid peptide. *Nature, 361*(6409), 260–263. doi:10.1038/361260a0.

Seubert, P., Vigo-Pelfrey, C., Esch, F., Lee, M., Dovey, H., Davis, D., et al. (1992). Isolation and quantification of soluble Alzheimer's β-peptide from biological fluids. *Nature, 359*(6393), 325–327. doi:10.1038/359325a0.

Shore, V. G., & Shore, B. (1973). Heterogeneity of human plasma very low density lipoproteins. Separation of species differing in protein components. *Biochemistry, 12*(3), 502–507. doi:10.1021/bi00727a022.

Siegel, J. A., Haley, G. E., & Raber, J. (2012). Apolipoprotein E isoform-dependent effects on anxiety and cognition in female TR mice. *Neurobiology of Aging, 33*(2), 345–358. doi:10.1016/j.neurobiolaging.2010.03.002.

Sisodia, S. S. (1992). Beta-amyloid precursor protein cleavage by a membrane-bound protease. *Proceedings of the National Academy of Sciences, 89*(13), 6075–6079.

Sisodia, S. S., Koo, E. H., Beyreuther, K., Unterbeck, A., & Price, D. L. (1990). Evidence that beta-amyloid protein in Alzheimer's disease is not derived by normal processing. *Science, 248*(4954), 492–495. doi:10.1126/science.1691865.

Smith, G. E., Bohac, D., Waring, S., Kokmen, E., Tangalos, E., Ivnik, R., et al. (1998). Apolipoprotein E genotype influences cognitive "phenotype" in patients with Alzheimer's disease but not in healthy control subjects. *Neurology, 50*(2), 355–362.

Strittmatter, W. J., Saunders, A. M., Schmechel, D., Pericak-Vance, M., Enghild, J., Salvesen, G. S., et al. (1993). Apolipoprotein E: High-avidity binding to beta-amyloid and increased frequency of type 4 allele in late-onset familial Alzheimer disease. *Proceedings of the National Academy of Sciences, 90*(5), 1977–1981.

Strooper, B. D., & Annaert, W. (2000). Proteolytic processing and cell biological functions of the amyloid precursor protein. *Journal of Cell Science, 113*(11), 1857–1870.

Sullivan, P., Mace, B., Maeda, N., & Schmechel, D. (2004). Marked regional differences of brain human apolipoprotein e expression in targeted replacement mice. *Neuroscience, 124*(4), 725–733. doi:10.1016/j.neuroscience.2003.10.011.

Sullivan, P. M., Mezdour, H., Aratani, Y., Knouff, C., Najib, J., Reddick, R. L., et al. (1997). Targeted replacement of the mouse apolipoprotein E gene with the common human APOE3 allele enhances diet-induced hypercholesterolemia and atherosclerosis. *Journal of Biological Chemistry, 272*(29), 17972–17980. doi:10.1074/jbc.272.29.17972.

Takashima, A. (2008). Hyperphosphorylated tau is a cause of neuronal dysfunction in tauopathy. *Journal of Alzheimer's Disease, 14*(4), 371–375.

Tamaoka, A., Sawamura, N., Fukushima, T., Shoji, S., Matsubara, E., Shoji, M., et al. (1997). Amyloid β protein 42(43) in cerebrospinal fluid of patients with Alzheimer's disease. *Journal of the Neurological Sciences, 148*(1), 41–45. doi:10.1016/S0022-510X(96)00314-0.

Terry, R. D., Masliah, E., Salmon, D. P., Butters, N., DeTeresa, R., Hill, R., et al. (1991). Physical basis of cognitive alterations in Alzheimer's disease: Synapse loss is the major correlate of cognitive impairment. *Annals of Neurology, 30*(4), 572–580. doi:10.1002/ana.410300410.

Tesseur, I., Van Dorpe, J., Spittaels, K., Van den Haute, C., Moechars, D., & Van Leuven, F. (2000). Expression of human apolipoprotein E4 in neurons causes hyperphosphorylation of protein tau in the brains of transgenic mice. *The American Journal of Pathology, 156*(3), 951–964.

Thal, D. R., Capetillo-Zarate, E., Del Tredici, K., & Braak, H. (2006). The development of amyloid beta protein deposits in the aged brain. *Science of Aging Knowledge Environment, 2006*(6), re1. doi:10.1126/sageke.2006.6.re1.

Thal, D. R., Rüb, U., Orantes, M., & Braak, H. (2002). Phases of Aβ-deposition in the human brain and its relevance for the development of AD. *Neurology, 58*(12), 1791–1800. doi:10.1212/WNL.58.12.1791.

Thal, D. R., Rub, U., Schultz, C., Sassin, I., Ghebremedhin, E., Tredici, K. D., et al. (2000). Sequence of (A(beta))-protein deposition in the human medial temporal lobe. *Journal of Neuropathology and Experimental Neurology, 59*(8), 733–748.

Tokuda, T., Calero, M., Matsubara, E., Vidal, R., Kumar, A., Permanne, B., et al. (2000). Lipidation of apolipoprotein E influences its isoform-specific interaction with Alzheimer's amyloid beta peptides. *Biochemical Journal, 348*(Pt 2), 359–365.

Travert, C., Forfana, M., Carreau, S., & Le Goff, D. (2000). Rat Leydig cells use apolipoprotein E depleted high density lipoprotein to regulate testosterone production. *Molecular and Cellular Biochemistry, 213*(1), 51–59.

Van Haaren, F., van Hest, A., & Heinsbroek, R. P. W. (1990). Behavioral differences between male and female rats: Effects of gonadal hormones on learning and memory. *Neuroscience & Biobehavioral Reviews, 14*(1), 23–33. doi:10.1016/S0149-7634(05)80157-5.

Van Meer, P., Acevedo, S., & Raber, J. (2007). Impairments in spatial memory retention of GFAP-apoE4 female mice. *Behavioural Brain Research, 176*(2), 372–375. doi:10.1016/j.bbr.2006.10.024.

Villasana, L., Acevedo, S., Poage, C., & Raber, J. (2006). Sex-and APOE isoform-dependent effects of radiation on cognitive function. *Radiation Research, 166*(6), 883–891.

Villasana, L. E., Benice, T. S., & Raber, J. (2011). Long-term effects of 56Fe irradiation on spatial memory of mice: Role of sex and apolipoprotein E isoform. *International Journal of Radiation Oncology, Biology, Physics, 80*(2), 567–573. doi:10.1016/j.ijrobp.2010.12.034.

Weiss, J. H., Pike, C. J., & Cotman, C. W. (1994). Rapid communication: Ca2+ channel blockers attenuate β-amyloid peptide toxicity to cortical neurons in culture. *Journal of Neurochemistry, 62*(1), 372–375. doi:10.1046/j.1471-4159.1994.62010372.x.

Xu, Q., Bernardo, A., Walker, D., Kanegawa, T., Mahley, R. W., & Huang, Y. (2006). Profile and regulation of apolipoprotein E (ApoE) expression in the CNS in mice with targeting of green fluorescent protein gene to the ApoE locus. *The Journal of Neuroscience, 26*(19), 4985–4994. doi:10.1523/JNEUROSCI.5476-05.2006.

Xu, P.-T., Schmechel, D., Rothrock-Christian, T., Burkhart, D. S., Qiu, H.-L., Popko, B., et al. (1996). Human apolipoprotein E2, E3, and E4 isoform-specific transgenic mice: Human-like pattern of glial and neuronal immunoreactivity in central nervous system not observed in wild-type mice. *Neurobiology of Disease, 3*(3), 229–245. doi:10.1006/nbdi.1996.0023.

Xu, Q., Walker, D., Bernardo, A., Brodbeck, J., Balestra, M. E., & Huang, Y. (2008). Intron-3 retention/splicing controls neuronal expression of apolipoprotein E in the CNS. *The Journal of Neuroscience, 28*(6), 1452–1459. doi:10.1523/JNEUROSCI.3253-07.2008.

Ye, S., Huang, Y., Müllendorff, K., Dong, L., Giedt, G., Meng, E. C., et al. (2005). Apolipoprotein (apo) E4 enhances amyloid β peptide production in cultured neuronal cells: ApoE structure as a potential therapeutic target. *Proceedings of the National Academy of Sciences of the United States of America, 102*(51), 18700–18705. doi:10.1073/pnas.0508693102.

Zhou, Y., Elkins, P. D., Howell, L. A., Ryan, D. H., & Harris, R. B. S. (1998). Apolipoprotein-E deficiency results in an altered stress responsiveness in addition to an impaired spatial memory in young mice. *Brain Research, 788*(1–2), 151–159. doi:10.1016/S0006-8993(97)01533-3.

Zofkova, I., Zajickova, K., Hill, M., & Horinek, A. (2002). Apolipoprotein E gene determines serum testosterone and dehydroepiandrosterone levels in postmenopausal women. *European Journal of Endocrinology, 147*(4), 503–506. doi:10.1530/eje.0.1470503.

Part III
Animal Models of Psychopathology

Chapter 8
Social Endophenotypes in Mouse Models of Psychiatric Disease

Marc T. Pisansky, Irving I. Gottesman, and Jonathan C. Gewirtz

"...the difference in mind between man and the higher animals, great as it is, certainly is one of degree and not of kind. We have seen that the senses and intuitions, the various emotions and faculties, such as love, memory, attention, curiosity, imitation, reason, etc., of which man boasts, may be found in an incipient, or even sometimes in a well-developed condition, in the lower animals." Charles Darwin, The Descent of Man, page 86 (1871).

Introduction

Humans are a highly affiliative species. From the onset of parental care to the formation and dissolution of friendships or partnerships, much of life is spent in dynamic social relationships. These relationships greatly influence our health and livelihood (Seeman 1996). Hence, diseases that affect social functioning have drastic impacts on our well-being. Indeed, disruptions of socio-communicative behaviors represent pervasive features across several psychiatric diseases. These features include the social interaction and communication impairments of autism spectrum disorders (ASDs) and negative symptoms of schizophrenia, and the multifarious social impairments of mood and personality disorders. Similar impairments in socio-communicative behaviors across psychiatric diseases hint at common genetic and neurobiological

M.T. Pisansky (✉)
Graduate Program in Neuroscience, University of Minnesota—Twin Cities,
N90 Elliott Hall, 75 East River Parkway, Minneapolis, MN 55455, USA
e-mail: pisansky@umn.edu

I.I. Gottesman • J.C. Gewirtz
Department of Psychology, University of Minnesota—Twin Cities,
75 East River Parkway, Minneapolis, MN, USA
e-mail: gotte003@umn.edu; jgewirtz@umn.edu

J.C. Gewirtz, Y.-K. Kim (eds.), *Animal Models of Behavior Genetics*,
Advances in Behavior Genetics, DOI 10.1007/978-1-4939-3777-6_8

sources. In two of these diseases, ASDs and schizophrenia, socio-communicative impairments have been put forth as candidate endophenotypes, or components of disease that reside between genes and symptoms. Being linked to underlying genetic sources, endophenotypes are tractable features for study in animal models, particularly mice. Although mice possess a much more rudimentary behavioral repertoire than humans, under close examination they exhibit a range of complex social behaviors. Furthermore, with current transgenic technologies, mouse models embodying targeted gene mutations can inform our understanding of genetic sources, and therefore endophenotypes, of psychiatric disease.

In this chapter, characteristic socio-communicative deficits of ASDs and schizophrenia will be discussed in the context of the endophenotype concept and mapped onto contemporary behavioral assays employed in mouse models of psychiatric disease (Fig. 8.1). First, the genetic etiology of ASDs and schizophrenia and the application of the endophenotype concept will be overviewed. Then, candidate socio-communicative endophenotypes in these disorders will be discussed. Against this background, the application of various social interaction and communication paradigms in mouse models of psychiatric disease will be reviewed. The overarching aim of this chapter is to highlight the growing capabilities of studying socio-communication behavior in mouse models so that associated endophenotypes in psychiatric diseases can be better understood.

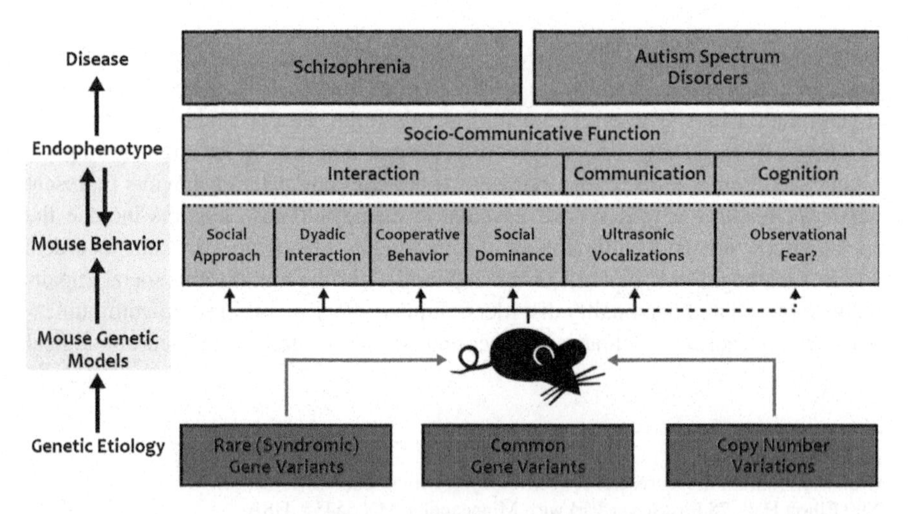

Fig. 8.1 The mouse model approach to studying socio-communicative endophenotypes of psychiatric disease. The mouse model approach (*yellow box*) aims to connect genetic etiologies with endophenotypes of disease. Genetic sources of disease (*red boxes*) are recapitulated in mice (*red lines*), which are then characterized for social and communicative behaviors comparable to socio-communicative endophenotypes identified in schizophrenia and ASDs (*gray boxes*). Mouse behavioral paradigms relevant to social cognition endophenotypes are less established, and therefore a significant field of development

Socio-Communicative Dysfunction in Psychiatric Disease

Autism Spectrum Disorders

Perhaps the most prototypical psychiatric disease states reflecting socio-communicative abnormalities are ASDs, a set of neurodevelopmental disabilities exemplified by impairments in social interaction and communication, as well as restricted interests or stereotypic behavior (APA 2013). The prevalence of ASDs has been estimated at 1 % of the general population (Pedersen et al. 2014; CDC 2012). While some epidemiological reports suggest a dramatic increase in prevalence over recent years (e.g., (Blumberg et al. 2013)), these findings have not been without opposition (Gernsbacher et al. 2005; Mandell and Lecavalier 2014). Indeed, one of the largest epidemiological studies to date reported no change in diagnosis rates worldwide (Baxter et al. 2015). Earliest diagnosis of ASDs typically occurs at 18–36 months with the identification of symptoms that include atypical eye contact and visual attention/fixation, as well as delayed expressive/receptive language abilities (Zwaigenbaum et al. 2005). Remarkably, deficits in eye contact at 2–6 months have been shown to be predictive of later ASD diagnosis, implying that socio-communicative deficits may arise much earlier than previously thought (Jones and Klin 2013). One hypothesis for these characteristic socio-communicative symptoms suggests that they arise from a generalized failure to orient towards social stimuli in early life, which in turn results in impaired development of associated neural systems (Dawson 2008). ASDs are also thought to involve deficits in social cognition, notably those related to "theory of mind," which may contribute to more recognizable socio-communication deficits (Baron-Cohen et al. 1985; Loveland et al. 1997) as well as to general psychopathology (Brüne 2005).

Schizophrenia

Schizophrenia represents another psychiatric disease in which socio-communicative deficits are characteristic features. Diagnostic symptoms of schizophrenia can be broadly classified into three categories. Positive symptoms include false perceptions (hallucinations) or beliefs (delusions); negative symptoms include social withdrawal, lack of motivation (avolition), or abnormal social interactions; and cognitive symptoms include deficits in attention, thought, working memory, etc. With a lifetime risk commensurate to ASDs, schizophrenia is present in roughly 1 % of the general population. In schizophrenia, symptoms relevant to social behavior are sources of significant impairment, and are present throughout the course of disease presentation (Addington and Addington 2000). Indeed, social impairments can be identified before the onset of psychosis (Häfner et al. 1999), and serve as strong predictors of outcome (Carrión et al. 2013; Nieman et al. 2013; Strauss and Carpenter 1974). Impairments in communication are also evident in schizophrenia, frequently arising during the prodromal stage of disease (Häfner et al. 1999). Schizophrenia patients exhibiting flat affect use normal word patterns to express emotions, yet convey this information using less prosodic inflection (Alpert et al. 2000;

Dickinson et al. 2007). Furthermore, schizophrenia patients are impaired in the recognition and repetition of affective prosody in others, indicating—much like ASDs—defective social cognition abilities (Leentjens et al. 1998).

ASDs and Schizophrenia as "Socially Diametric" Diseases

In light of similar socio-communicative dysfunction between ASDs and schizophrenia, attention has been directed towards understanding common neuropathological underpinnings. In fact, ASDs and schizophrenia have been termed diseases of the "social brain" (Baron-Cohen and Belmonte 2005; Burns 2004, 2006). This view proposes that these diseases are socially diametric—ASD patients possess a socially underdeveloped brain, whereas schizophrenia patients possess a highly developed social brain (Crespi and Badcock 2008). Albeit a simplified perspective of these diseases, one line of evidence supporting this view has been the discovery of hyper- and hypo-connectivity across neural networks in ASDs and schizophrenia, respectively (Belmonte et al. 2004; White and Gottesman 2012). Both schizophrenic and ASD patients also display decrements across various measures of social cognition (Couture et al. 2010), which are reflected in functional differences in associated brain regions (Pinkham et al. 2008). Lastly, the socially diametric view of ASDs and schizophrenia has been supported by various genetic findings, including reciprocal chromosomal abnormalities (to be discussed in section "Genetic Commonalities Between ASDs and Schizophrenia"). This evidence is particularly persuasive, given the known etiological role of genetics in both diseases.

The Genetic Origins of Psychiatric Disease: ASDs and Schizophrenia

Genetic Heterogeneity

Twin and family studies of psychiatric disease have historically inferred a complex etiology, conceptualized as a combination of genetic, epigenetic, environmental, and stochastic factors. While these factors impact the initiation and/or progression of disease processes, ASDs and schizophrenia exhibit the highest heritability of any psychiatric disease, thus are thought to involve significant genetic sources (Bailey et al. 1995; Sandin et al. 2014; Sullivan et al. 2012a). These genetic sources are highly heterogeneous, incompletely understood, and confer only a portion of disease risk. Variants identified by modern whole-genome sequencing and association methods account for a paltry fraction of common (nonsyndromic) forms of these diseases (Lee et al. 2012; Purcell et al. 2009; Šestan et al. 2013). Likewise, exome-wide sequencing studies, which focus selectively on coding regions of the genome, have only identified deleterious mutations in select cases (e.g., 4–8 % of ASDs), many of which have failed to replicate across studies (Gratten et al. 2013; Neale et al. 2012; O'Roak et al. 2012; Sanders et al. 2012). With the current literature pointing towards a complex and multifactorial genetic contribution, contemporary theories now posit that many psychiatric diseases arise from several "common" allelic variants (Lohmueller et al. 2003; Manolio et al. 2009). This "common disease-common variant" (CDCV) model has gained significant traction in psychiatric disease research (Klei et al. 2012; Stefansson et al. 2009). For ASDs and schizophrenia, the collection of recognized variants is now becoming categorized into functional biological pathways relevant to disease etiology. This reductionist approach has converged on processes related to synapse function/development and transcriptional regulation (Fromer et al. 2014; Gilman et al. 2012; Parikshak et al. 2013; Pinto et al. 2014; Purcell et al. 2014).

Syndromic Forms of Psychiatric Disease

While the majority of psychiatric cases likely result from numerous common variants, a portion of cases arise from a single, rare gene variant. These syndromic forms of ASDs and schizophrenia manifest characteristic phenotypic features making them clinically more identifiable and diagnosable. The most common rare variant associated with ASD diagnosis involves the FMR1 gene. Mutations in FMR1, while giving rise to fragile X syndrome and intellectual disability, also explain approximately 2% of ASD cases (Kielinen et al. 2004). Other ASD-associated single-gene mutations include tuberous sclerosis (TSC1/2), Angelman syndrome (UBE3A), Rett syndrome (MECP2), and CHARGE syndrome (CHD7) (Betancur and Coleman 2013). Although classification of these syndromes as ASDs remains imperfect, comorbidity of diagnoses and overlapping symptomology support common pathogenic sources. Several genes involved with synaptic function have also been identified in syndromic forms of both ASDs and schizophrenia (Fromer et al. 2014; Heck and Lu 2012; Singh and Eroglu 2013), including NLG3/4 (Singh and Eroglu 2013), NRXN1/3 (Gai et al. 2011), and SHANK1/2/3 (Berkel et al. 2010; Gauthier et al. 2009, 2010; Sato et al. 2012b). The highly penetrant nature of these syndromes facilitates a relatively straightforward means of study. As will be discussed shortly (section "Types of Mouse Models"), these examples are particularly suitable for modeling in mice.

Copy Number Variations

The CDCV model generally presumes that most disease-related variants are single nucleotide polymorphisms (SNPs) or mutations within a single gene. In fact, a growing body of research has highlighted the pathogenic significance of copy number variations (CNVs), which comprise large multi-gene structural variations in the form of duplications, deletions, or inversions. Recently, human genome-wide association studies have highlighted the striking contribution of CNVs to both ASDs and schizophrenia, which account for an estimated 10–15 % of all cases (Ku et al. 2010; Marshall et al. 2008; Merikangas et al. 2009; Pinto et al. 2010; Sanders et al. 2012; Sebat et al. 2007). These large structural mutations occur either de novo (i.e., via nonallelic homologous recombination within the germline) or by transmission from an unaffected parent, and often represent highly penetrant risk factors for disease. CNVs at several genomic loci have been identified to underlie either ASDs or schizophrenia, including 1q21.1, 15q11-13, 15q13.3, 16p11.2, and 22q11.2. As well as being replicated in many disease cohorts, many of these loci have been investigated in mouse models (see section "Types of Mouse Models"), an approach that offers a unique vantage point for understanding the underlying neuropathology in these forms of psychiatric disease.

Genetic Commonalities Between ASDs and Schizophrenia

Along with the similarity between socio-communicative impairments in ASDs and schizophrenia, several lines of evidence also suggest a genetic overlap. The first line of evidence comes from CNVs that are either of the same (e.g., deletion of 15q13.3 (Ben-Shachar and Lanpher 2009; Shinawi et al. 2009)) or opposite gene dosage (e.g., 1q21.1—duplicated in ASDs (Szatmari et al. 2007); deleted in schizophrenia (International Schizophrenia Consortium 2008; Stefansson et al. 2008)). These examples raise questions as to the relationship of gene dosage to both schizophrenia and ASDs, and provide clues as to the genes that have causal roles in pathogenesis (Crespi et al. 2010; King and Lord 2011). For instance, CNVs that exhibit opposite dosage effects (e.g., 1q21.1) further the hypothesis that these diseases are "socially diametric" (see section "ASDs and Schizophrenia as 'Socially Diametric' Diseases"). Several functionally similar gene families have also been linked to both disorders, namely those involving synapse development, epigenetic regulation, and excitation/inhibition balance (Di Cristo 2007; Fromer et al. 2014; Huguet et al. 2013; Marín 2012). These ontologically similar gene families point to commonly dysregulated biological pathways in both ASDs and schizophrenia. In addition to these examples, family studies have identified an increased risk of ASDs in individuals with family members diagnosed with schizophrenia (Daniels et al. 2008; Larsson et al. 2005; Sullivan et al. 2012b). This particular evidence implies an intriguing, yet to be identified, genetic commonality to these diseases.

The Endophenotype Concept in Psychiatric Disease

The Endophenotype Concept

The genetic contribution to psychiatric disease is indisputably complex. Accordingly, connecting the array of genetic variants associated with schizophrenia and ASDs to symptomatology represents a long-standing and arduous quest in the field of psychiatry. One strategy to simplify this endeavor has been to link genetic evidence with intermediary features, termed endophenotypes, that more closely bridge the genotype–phenotype relationship. The endophenotype concept was first applied to the field of psychiatry by Gottesman and Shields in 1973 (Gottesman and Shields 1973). At that time the endophenotype was conceptualized as an "internal phenotype," or component of the disease process residing between proximal, disease-causing genes and more distal symptoms. The rationale behind the endophenotype concept supposes that genetic variants coding for a related set of aberrant protein products subsequently give rise to detectable and causal markers of the disease. The methods used to identify these markers, which range from biochemical, endocrinological, neuroanatomical, cognitive, and physiological, are often (but not always completely) independent of clinical symptoms, and use primarily quantifiable measures best obtained in a laboratory environment. At present, the core criteria of an endophenotype can be enumerated as follows (see (Gottesman and McGue 2015; Gottesman et al. 2003) for detailed discussion of criteria):

1. The endophenotype is associated with illness in the population.
2. The endophenotype is heritable.
3. The endophenotype is primarily state-independent (i.e., manifests if the illness is active or not, but may require an appropriate challenge).
4. The endophenotype and illness co-segregate within families.
5. The endophenotype found in an individual is found in non-affected family members at a higher rate than in the general population (see (Leboyer et al. 1998)).
6. The endophenotype is a trait measured reliably, and ideally is more strongly associated with the disease of interest than with other psychiatric conditions.

Endophenotypes in Psychiatric Disease

The endophenotype concept was well received at its introduction into the field of psychiatry. However, not until a perspective article in the *American Journal of Psychiatry* (Gottesman et al. 2003) did the concept gain traction in widespread psychiatric research. Now nearing the end of the "Decade of the Endophenotype" (Miller and Rockstroh 2013), the endophenotype concept has emerged as a valuable strategy for understanding complex disease. Specifically, the endophenotype concept holds potential for uncovering pathogenic commonalities between diseases, thereby pointing to shared underlying genetic variants. Identification of endophenotypes in psychiatric disease also stands to bolster both diagnostic and prognostic avenues of clinical practice and treatment (Ritsner and Gottesman 2009). Two psychiatric diseases that are particularly amenable to the endophenotype concept are ASDs and schizophrenia. In both diseases, several candidate endophenotypes have been identified. While outside the scope of this review, a comprehensive listing of candidate endophenotypes can be found elsewhere (Persico and Sacco 2014; Ritsner and Gottesman 2009).

Endophenotype Considerations

Issues surrounding endophenotypes are twofold: identification and validation. Not only must the endophenotype be differentiated in affected or (to a lesser extent) related, undiagnosed individuals, but it must largely adhere to core criteria (see section "The Endophenotype Concept"). Importantly, endophenotypes are not readily discernable by clinical observation, but rather require highly sensitive measurement. Therefore, in addition to endophenotypes heretofore identified in ASDs and schizophrenia, it remains likely that other, yet to be discovered, endophenotypes exist. Such latent endophenotypes could include prenatal/early-life indicators of disease or features that are otherwise difficult to disidentify with current methodology.

At present, endophenotypes stand to enhance our understanding of psychiatric diseases. Diagnosis currently involves identifying core phenotypic symptoms as outlined in the Diagnostic and Statistical Manual of Mental Disorders (DSM). These symptoms are used in light of any coherent link between manifested behavior and underlying neuropathology. Endophenotypes aim to provide concrete markers of disease processes that can advance diagnostic practice, and ultimately improve treatment. In disorders such as ASDs, which hold similar diagnoses but may arise from diverse sources, this approach may provide a more valid diagnosis schema. The endophenotype concept also stands to complement the Research Domain Criteria (RDoC) strategy (Simmons and Quinn 2014), which aims to reduce the complexity of clinical diagnostic criteria into observable behaviors and neurobiological measures. Together, the endophenotype and RDoC strategies aim to provide more quantitative and neurobiologically relevant measures for understanding diseases such as ASDs and schizophrenia.

The Socio-Communicative Endophenotype

Historical Perspective on Endophenotypes in Psychiatric Disease

Endophenotypes conceivably related to socio-communicative functioning are less studied, yet have been hinted at since the first known cases. In fact, among some of the first documented cases of ASDs and schizophrenia were subtle indications of social abnormalities in undiagnosed relatives. Leo Kanner, the original documenter of autism, was also the first to note in relatives strong preoccupations with "abstractions of a scientific, literary, or artistic nature" and "limited genuine interest in people" (Kanner 1943). Likewise, observations of social withdrawal and eccentricities among family members of schizophrenia patients can be found among the earliest descriptions by Kraepelin and Bleuler—both of whom hypothesize the existence of a "latent schizophrenia" syndrome among relatives (Kendler 1985). These initial observations in ASDs and schizophrenia have since been substantiated by a field of research that continues to uncover a range of socio-communicative endophenotypes in both disorders. These behavioral endophenotypes are by definition not diagnostic symptoms nor clinical observations, but rather more covert manifestations of the disease and therefore only measured by close experimental analysis. Moreover, unlike symptoms, endophenotypes are integrally grounded in genetic underpinnings, and therefore closer to the true sources of disease.

Socio-Communicative Endophenotypes in ASDs

Following Kanner's initial suggestion of a broader phenotype, the first evidence for latent indicators of ASDs was proposed by Folstein and colleagues in two papers documenting the genetic etiology and predisposition of undiagnosed relatives to the disease (Folstein and Rutter 1977a, b). This work has since been expanded upon in several family and twin studies (Bolton et al. 1994; Le Couteur et al. 1996). It is now estimated that nearly 25 % of first-degree relatives of individuals with an ASD exhibit at least one of the core symptoms of the disorder (criteria 2, 5) (Goussé et al. 2002). For instance, language difficulties are elevated among undiagnosed monozygotic twins compared to dizygotic twins (Bailey et al. 1995; Folstein and Rutter 1977a) or other close relatives (Bartak et al. 1975; Piven et al. 1997), with the severity of these impairments ranging drastically by relatedness (criteria 2, 4) (Bailey et al. 1998). Infants later diagnosed with an ASD exhibit a delayed capacity or inability to read and/or mimic facial expressions (criteria 3) (McIntosh et al. 2006; Oberman et al. 2009), a deficit that has also been observed in undiagnosed siblings (Oerlemans et al. 2014) and related adult family members (Sucksmith et al. 2013). Similarly, social cognition impairments have been observed in undiagnosed identical co-twins (Bailey et al. 1995; Le Couteur et al. 1996) as well as among the first-degree (Losh et al. 2009; Oerlemans et al. 2013; Piven et al. 1997) or more distant relatives (Pickles et al. 2000) of patients. Together, these findings have contributed to the notion that socio-communicative impairments in ASDs are indeed endophenotypes and conform accordingly to the criteria explicated in section "The Endophenotype Concept."

Socio-Communicative Endophenotypes in Schizophrenia

Evidence to support the existence of latent indicators or endophenotypes of schizophrenia has accumulated in recent years. Some of the first supporting evidence appeared from family studies showing a higher than expected prevalence of schizophrenia-like traits in reared-apart relatives of patients (criteria 2) (Alanen 1966; Kety et al. 1976; Kety et al. 1968). Since these pioneering observations, several other schizophrenia-like features have been reported in relatives (Faraone et al. 1995).

Most compelling have been observations of disruptions of socio-communicative functioning. For instance, social schizotypal features are elevated among undiagnosed relatives, whereas cognitive abilities remain nearly intact (criteria 6) (Tarbox and Pogue-Geile 2011). This implies that social impairments serve as more promising candidate endophenotypes than cognitive impairments. Individuals with high familial risk of schizophrenia also show deficits in social cognition, and these impairments are strongly associated with positive symptoms and overall prodromal psychopathology (Eack et al. 2010). In concordance with these findings, several other groups have found that undiagnosed relatives show similar, albeit more mild, patterns of general social cognition impairments such as face recognition and theory of mind (criteria 5) (de Achával et al. 2010; Irani et al. 2006). A range of other deficits in communication have also been documented in undiagnosed relatives (Docherty et al. 1999); however, there remains an overall dearth of studies investigating candidate endophenotypes related to communicative function. Akin to ASDs, these findings in schizophrenia implicate socio-communicative impairments as endophenotypes of the disease.

Mouse Models of Socio-Communicative Endophenotypes

Overview of Mouse Models

The common mouse (*mus musculus*) serves as an indispensable tool for connecting human genetic findings to neuropathological sequelae. Current transgenic technology, paired with the high fecundity and shorter lifespan of mice (compared to humans and nonhuman primates), allows for artificial transposition or mutation of candidate gene variants in the mouse genome. These mouse models can then be readily examined for features matching those in human psychiatric disease.

While a widely exploited animal system, the study of psychiatric disease in mouse models is fraught with obvious issues of validity: mice cannot be said to truly express the hallucinations/delusions of schizophrenia, restricted interests of ASDs, or other quintessentially human cognitive impairments. Even so, many of the brain regions and circuits thought to be affected by psychiatric diseases are also evolutionarily conserved and functionally similar across mammalian species. Subcortical and brainstem structures have similar connectivity in both humans and mice, and these neural circuits give rise to analogous behavioral output. Social behavior is no exception to this, with associated neurobiological substrates remaining relatively conserved across species (Stanley and Adolphs 2013). For instance, oxytocin and vasopressin, neuropeptides recently implicated in affiliative behavior, are synthesized and act on homologous neuroanatomical sites in both mice and humans (Domes et al. 2007; Ferguson et al. 2001; Gur et al. 2014; Hurlemann et al. 2010). Thus, several indicators of these diseases, in the form of quantifiable endophenotypes, have tangible antecedents in mice.

The ideal mouse model should recapitulate the pathogenic source of the disease (construct validity), while also presenting with several parallel features (face validity) and responding in a similar manner to therapeutics (predictive validity). In light of the inadequacy of models based largely on face validity (e.g., characterizing positive symptoms of schizophrenia), recent efforts have shifted to identifying endophenotypes in mice that mirror those of humans (Gould and Gottesman 2006). Indeed, connecting genetic alterations with molecular, neurobiological, or behavioral endophenotypes is relatively straightforward in mice compared to humans. However, given the genetic complexity of psychiatric disease, it is unreasonable to presume that one mouse model may recapitulate all endophenotypes manifest in humans. Instead, mouse models that exhibit particular endophenotypes provide powerful tools for simplifying and elucidating the inherent complexity of psychiatric disease.

Types of Mouse Models

Mouse models of psychiatric disease can be generally classified into two categories: environmental or genetic. Environmental models involve exposure of mice (oftentimes in utero or perinatally) to teratogens or depleted social/nutritional conditions. Numerous studies have been published on environmental models, suggesting that they mimic certain features of these diseases (Jones et al. 2011). For instance, one environmental model involves lesioning the ventral hippocampus in early life, a manipulation that leads to features of psychiatric disease later in life (Tseng et al. 2009). Interestingly, social isolation in rodents results in several behavioral and neurochemical endophenotypes of psychiatric disease (Fone and Porkess 2008; Reser 2014), indicating an influence of social environment in disease pathogenesis. While there has been no shortage of research into environmental models, genetic mouse models serve as congruent and robust tools for understanding the genetic source of disease, and are therefore more relevant models for studying endophenotypes of psychiatric disease.

Genetic models are generated by modification of the mouse genome. In this approach, candidate genes in human psychiatric cases are either deleted (knockout model) or inserted (knock-in model) into the mouse genome using standard techniques with mouse embryonic stem cells. Genetic mouse models have been generated that mimic a range of genetic variants identified in human cohorts. These genetic variants fall into functional categories, of which several are shared between schizophrenia and ASDs. Perhaps the most widely studied category of genetically modified mice investigates genes involved in synapse function. Numerous mouse models have been generated by deleting genes encoding SH3 and multiple ankyrin repeat domains (SHANKs) (Peça et al. 2011; Wang et al. 2011; Yang et al. 2012), neuroligins/neurexins (Blundell et al. 2010; Etherton et al. 2009; Jamain et al. 2008), neural cell adhesion molecules (NCAMs) (Albrecht and Stork 2012), contactin-associated proteins (CNTNAPs) (Peñagarikano et al. 2011), or other proteins enriched at synapses. Many of these genes have also been investigated for their known involvement in syndromic forms of these diseases (Ching et al. 2010; Gauthier et al. 2009, 2011; Gregor et al. 2011; Jamain et al. 2003; Kim et al. 2008; Kirov et al. 2009; Sato et al. 2012b). Another category of genetically modified mouse models are those mimicking epigenetically relevant sources of disease. These include models having mutations of MeCP2 or FMR1(Chen et al. 2001; Guy et al. 2001; Mineur et al. 2006; Picker et al. 2006; Santos et al. 2007), genes implicated in ASDs and schizophrenia (Amir et al. 1999; Purcell et al. 2014). Although mouse models that conform to these categories are some of the most widely studied, a wide range of models focusing on other, disparate gene families exist. See Jones et al. (2011) and Provenzano et al. (2012) for a more extensive listing of these models.

Beyond more canonical knockout/knock-in genetic mouse models, mice can also be generated to harbor CNVs found in human psychiatric cases. Modeling CNVs in mice can be accomplished using chromosomal engineering, a technique allowing large, mega-base pair rearrangements in the mouse genome (Mills and Bradley 2001; Ramirez-Solis et al. 1995). The feasibility of this approach for modeling human-specific CNVs in mice relies on the evolutionary conservation of genes and their chromosomal locations between mouse and man. Not only do 96 % of human genes have mouse orthologs, but there exists significant synteny (i.e., consistent ordering of genes on chromosomes) between species. To date, a select few groups have utilized chromosomal engineering to generate CNV-specific mouse models of ASD and schizophrenia, including loci at 22q11, 7q11.23, 15q11-13, 15q13.3, and 16p11.2 (Fejgin et al. 2014; Horev et al. 2011; Li et al. 2009; Nakatani et al. 2009; Stark et al. 2008).

Behavioral Endophenotyping Methods in Mouse Models

The armamentarium of behavioral assays for use in mouse models is extensive. Here, established assays that focus on social and communicative behavior will be discussed. The overarching goal of developing and implementing behavioral tests for evaluating endophenotypes is to classify a pattern of behavioral differences in a mouse model that mimic endophenotypes in patient populations. Because social endophenotypes in ASDs and schizophrenia are often not independent of diagnostic symptomology, these behavioral assays have also been used extensively to characterize symptom-like behaviors in mouse models. However, employing these assays in an endophenotyping approach is more fitting, given the genetic basis of psychiatric disease modeling and inability of mice to exhibit the full range of quintessentially human symptoms.

Social Interaction

Mice are a gregarious species that exhibit a range of social behaviors. In the laboratory, social behavior can be observed after birth as the close, nurturing relationship between the infant (pup) mouse and its mother (dam). As the pup develops, it forms social bonds first with its siblings and then—after weaning—with other conspecifics. Social interactions at the juvenile stage are primarily of a playful quality (Panksepp and Lahvis 2007; Spinka et al. 2001). At the onset of sexual maturity, mouse social behavior transitions to serve reproductive purposes (i.e., mating with opposite-sex conspecifics) in addition to both affiliative and antagonistic behavior with same-sex conspecifics. Throughout these developmental phases there is a fundamental predilection for social behavior, a phenomenon that has remained consistent with classical theories of motivated behavior (Glickman and Schiff 1967; Insel 2003; Young 1959).

Social behavior as related to psychiatric disease endophenotypes has been applied to numerous mouse models, in different contexts, and at various developmental periods. Below are a select set of behavioral endophenotyping methods used to characterize mouse models of psychiatric disease. Along with methodological information, each section includes reference to mouse models of either ASDs or schizophrenia that have characterized by these assays.

Social Approach

The standard test for assessing social approach behavior in mice is the three-chamber social approach task developed by Jackie Crawley and colleagues (Fig. 8.2a). The task is situated within an arena partitioned into three chambers. Mice are individually tested for their predilection to investigate an enclosed conspecific (object) mouse over an empty enclosure. Typically, the test is composed of three phases. In the first phase, the test mouse is habituated to the three-chamber arena without inclusion of an object mouse. In the second phase, an unfamiliar object mouse is introduced to one of chambers, having been placed within a small enclosure that allows transmission of visual, auditory, and olfactory cues. In the third phase, a second novel object mouse is placed in the opposite chamber and the interaction of the test mouse is assessed. The final phase investigates preference for social novelty—that is, the preference for the novel mouse over the now familiar mouse from the second phase. Each phase is typically 10 min in duration, and the test animal is confined to the center chamber during phase transitions. Sociability is defined as the amount of time that the test mouse spends in the chamber containing the familiar (phase 2) or novel (phase 3) conspecific. Time investigating either cup can also be quantified and used as a metric of sociability. Because object mice are contained within small enclosures at all times social approach is initiated by the test mouse only. Variations in this task include incorporating object mice from the same/different home cages, same/opposite sex, same/different strain, various stages of development, or following isolate housing.

This task has been utilized extensively for modeling social impairments in mouse models of ASDs (Bader et al. 2011; Jamain et al. 2008; Molina et al. 2008; Moy et al. 2009). It currently represents a common means of measuring social behavior in candidate mouse models of ASDs, and is therefore recommended as a standard task to be incorporated into any social behavior array.

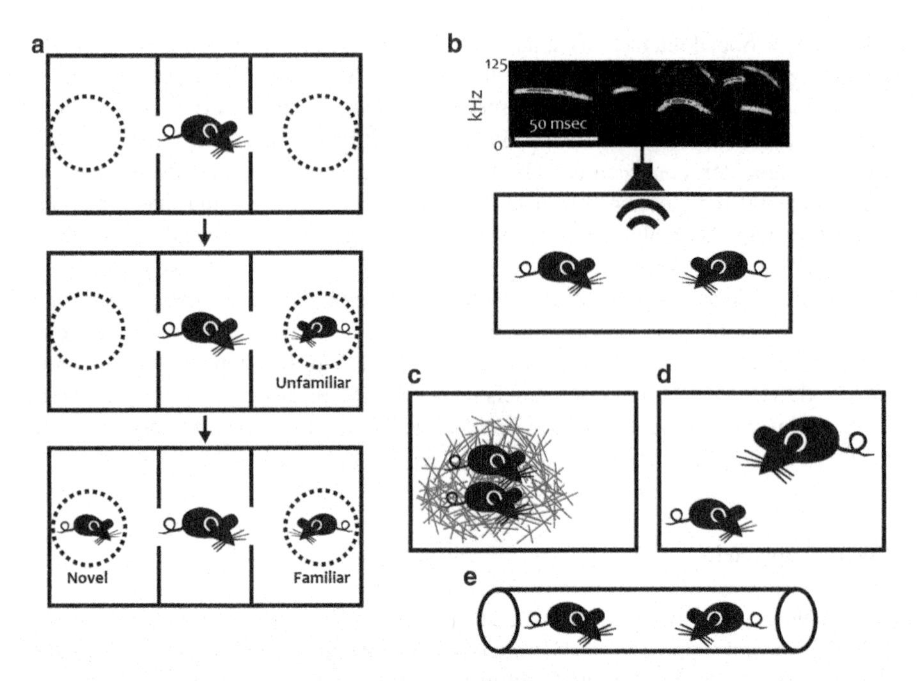

Fig. 8.2 Social interaction tasks in mice. (**a**) Social approach using the three chamber interaction task. In the first phase, the test mouse is habituated to the arena; in the second phase, an unfamiliar conspecific is introduced to a small enclosure in one of the chambers; finally, in the third phase, a second, more novel conspecific is introduced to a small enclose in the opposite chamber. (**b**) The dyadic interaction task, including ultrasonic vocalization recording. (**c**) Cooperative behavior assessed by social nest building. Dominance behavior using the (**d**) resident-intruder test and (**e**) tube test

Dyadic Interaction

A second task applied to mouse models of psychiatric disease is the dyadic interaction test (Fig. 8.2b). In this task, two age-matched conspecifics are placed into an open arena and allowed to freely interact for a set amount of time. Interactions between animals include oral-to-oral and oral-to-anogenital sniffing, following, chasing, mounting, and wrestling (Bolivar et al. 2007; McFarlane et al. 2008; Silverman et al. 2010b). Various experimental parameters can be modified to suit the research aims, including session duration, time of day, and mouse age, sex, strain, or prior social isolation.

Albeit less popular than the three-chamber interaction task, dyadic interaction has been used for characterizing the social behaviors of several mouse models of psychiatric disease (Defensor and Pearson 2011; Duncan et al. 2004; Jamain et al. 2008; Sato et al. 2012a). Dyadic interactions can be recorded over prolonged periods of time, including overnight, and behavior assessed using manual or automated methods. One modification of this task entails bisecting the arena with a transparent partition and quantifying the time and distance spent by each test mouse from the separation (Moretti et al. 2005; Spencer et al. 2008).

In our experience, it is also feasible to record ultrasonic vocalizations (USVs) between juvenile pairings during the dyadic interaction task (see also section on "Communication" below). While not a commonly adopted practice, this coincident measurement of social interaction and vocal communication provides a unique vantage for assaying socio-communicative functioning in mouse models. Importantly, one requirement for this variation to the dyadic interaction task is the use of adolescent or opposite-sex mice, as older, same-sex mice (particularly males) do not produce spontaneous USVs in this context. Even though this approach provides information as to socio-communicative functioning of mouse models, one limitation remains the distinction of USVs emitted from each mouse within dyadic pairs. One recent solution to this issue has involved stereophonic recording with post-hoc sound source analysis (Neunuebel et al. 2015). In our experience, this analysis is tedious, but is possible if experimentally necessary.

Cooperative Social Behavior

Nest Building

Nest construction represents a socially relevant construct across a wide variety of animals. In standard laboratory environments, mice are housed in small groups (2–4 animals) within home cages. Cotton bedding material is introduced into these home cages for nesting. One basic form of cooperative social behavior entails the formation of a common nest from this material (Fig. 8.2c). Animals can either be placed individually or in small groups, and the nesting pattern can be assessed at a predetermined time point (e.g., 1 h) (Koh et al. 2008). A standard rating scale has been formulated (Deacon 2006) in which the amount and use of a provided housing material is quantified (Albrecht and Stork 2012). Use of nest building as a social interaction behavior has been limited. Even so, a few mouse models of schizophrenia (Albrecht and Stork 2012; Koh et al. 2008) and ASD (Goorden et al. 2007) have shown gross impairments in nest formation.

Barbering

Another cooperative social behavior exhibited by mice is barbering, or allo-grooming-associated hair or whisker trimming. Barbering is thought to reflect social hierarchy or dominance (Kalueff et al. 2006; Sarna et al. 2000) in that mice with higher social status have increased exposure to conspecifics resulting in increased trimming and shorter whisker lengths. Mice may also show differential barbering behavior as a result of strain backgrounds or targeted gene mutations. Two genetic mouse models of schizophrenia—lacking either genes Dvl1 or PLCβ1—displayed longer whiskers, a feature indicative of limited interaction with other conspecifics (Koh et al. 2008; Lijam et al. 1997). A standardizing scale of barbering has been devised and includes four primary grooming behaviors—holding, grasping, plucking, and manipulating—that can be used to quantifiably score mouse behavior (Bergner et al. 2009).

Social Dominance

Resident-Intruder Test

The resident-intruder test assesses the territorial or social dominance behavior of mice (Thurmond 1975), and is particularly useful for investigating aggressive behavior (Duncan et al. 2004; Irie et al. 2012; Moretti et al. 2005). In this test, mice are housed individually for one to several days after which time an unfamiliar, age- and sex-matched conspecific is introduced into the home cage of the resident mouse (Fig. 8.2d). The resident mouse typically exhibits attack behavior within 5 min of introduction of the intruder (Thurmond 1975). The primary measure is attack behavior; however, various other social behaviors can be recorded, including sniffing, digging, huddling, and mounting. Aggressive behaviors can be potentiated in the resident mouse by prolonging pretest isolation or employing repeated testing (Moretti et al. 2005). USVs can also be recorded simultaneously during this task, particularly between female test animals (Scattoni et al. 2008).

Tube Test

The tube test represents another assay designed to quantify social dominance. In this test, two age-matched mice are placed at opposing ends of a long, transparent tube and allowed to approach each another (Fig. 8.2e). A winner mouse is designated when the opposing mouse retreats from the tube. Typical bouts do not exceed 5 min. Mice should be habituated to the tube prior to testing, and pairs can be tested multiple times. Several mouse models of ASDs have exhibited increased dominance behavior in this task (Molina et al. 2008; Shahbazian et al. 2002), whereas models of schizophrenia have generally found the opposite (Irie et al. 2012; Koh et al. 2008).

Experimental Considerations

In parallel with any task aimed to assess social behavior should be control experiments that rule out the influence of other confounding factors. At the most basic level, novel mouse models should be tested in standard assays for body morphology/health and gross sensory/motoric function. Most relevant for social behavior is assessment of olfactory functioning, the sensory modality with the most influence on sociability in rodents (Jamain et al. 2008). Standardized assays have been developed to assess olfactory function (Yang and Crawley 2009). Impaired motoric function in genetically modified mice may also confound measurement of social behavior. For example, in an assay such as the three-chamber interaction paradigm, mice with compromised motor faculties may show less movement between chambers. Given the emergence of symptoms in ASDs and schizophrenia during adolescence and early adulthood, it is also exceedingly important to characterize these basic behavioral attributes across analogous developmental stages in mice. In our experience, social interaction paradigms should ideally be pursued at one time point during adolescence ((postnatal day) PND 21–35) and one time point in adulthood (PND >35).

One design consideration should be the sequential ordering of assays. The same cohort of mice may undergo several behavioral tests in succession. However, generally less stressful paradigms (e.g., open field exploration, home cage behavior) ought to be completed prior to more stressful paradigms (e.g., fear conditioning, resident-intruder social interaction) in order to control for the persistent effects of prior stress and for generalization of anxiety- or fear-like behavior into other contexts. It also remains important to test animals using multiple assays that target similar socio-communicative functioning and corroborate one another. For example, it has become common practice to conduct at least two social interaction paradigms, such as both the three-chamber and dyadic interaction tests (Crawley 2004).

One technical consideration in mouse social behavior testing is the application of automated equipment for data collection and/or analysis. For the most part, quantification of social behavior has been completed by visual observation and manual scoring. This has historically been a reliable means of assessing effects in various social behavioral paradigms. However, with the development of computerized laboratory equipment has been the advent of automated software to improve the throughput and objectivity of behavioral assessment. More long-term assessment of social behavior can be facilitated using radio frequency identification device technology (e.g., telemetry), which allows an ethologically valid approach to monitoring behavior while in the home cage with or without conspecifics (Howerton et al. 2012). With the three-chamber social approach assay, it is also possible to use a photocell-equipped apparatus with infrared beams to record time spent by the subject mouse in each chamber (Nadler et al. 2004; Yang et al. 2007).

Communication

Communication is an essential component of social behavior. Whether vocal or otherwise, communication serves to convey emotion, intent, or information regarding the environment to a conspecific. Much like social interaction, vocal communication is markedly abnormal in psychiatric disease. In ASDs, inklings of communication impairments arise in infants (later diagnosed with ASDs) as a reduction or absence of cooing/babbling, and instead the presence of grunting, humming, or extended periods of inconsolable crying (Johnson 2008; Nadig et al. 2007; Osterling et al. 2002; Zwaigenbaum et al. 2005). Later, autistic individuals may stress the wrong syllables of words (McCann and Peppé 2003), squeal stereotypically, laugh inappropriately, and have difficulty modulating the pitch and volume of their speech (Johnson 2008; Shriberg et al. 2001; Zwaigenbaum et al. 2005). While not typically observed in early life, mild disturbances in communication have also been reported in schizophrenia (Docherty et al. 1996; Langdon et al. 2002).

Overview of Ultrasonic Vocalizations

A wide variety of animals communicate vocally. In mammals, information encoded by vocal communication can include social status or relationship, environmental features, or affect (Knutson et al. 2002; Portfors 2007). Much of the subcortical architecture underlying vocalization is conserved across mammalian species (Newman 2007). In particular, the periaqueductal gray area and various brainstem nuclei comprise integral hubs for vocal motor production. Most small mammals, including laboratory rats and mice, emit context- and state-dependent vocalizations that exceed the audible range of humans (>20 kHz). These USVs can be parsed into "syllables," representing units of sound separated by variable silent periods, then further classified by their intensity, duration, frequency modulation, or harmonics (Fig. 8.3). Much like human speech, rodent USVs convey information through changes in frequency and amplitude, both features of prosody (Lahvis et al. 2011). While this chapter focuses on mouse USVs, rats are known to have a rich repertoire of USVs that have been well characterized (Portfors 2007). In contrast to rats, mice have been presumed to be non-learning vocalizers—that is, unable to modify their

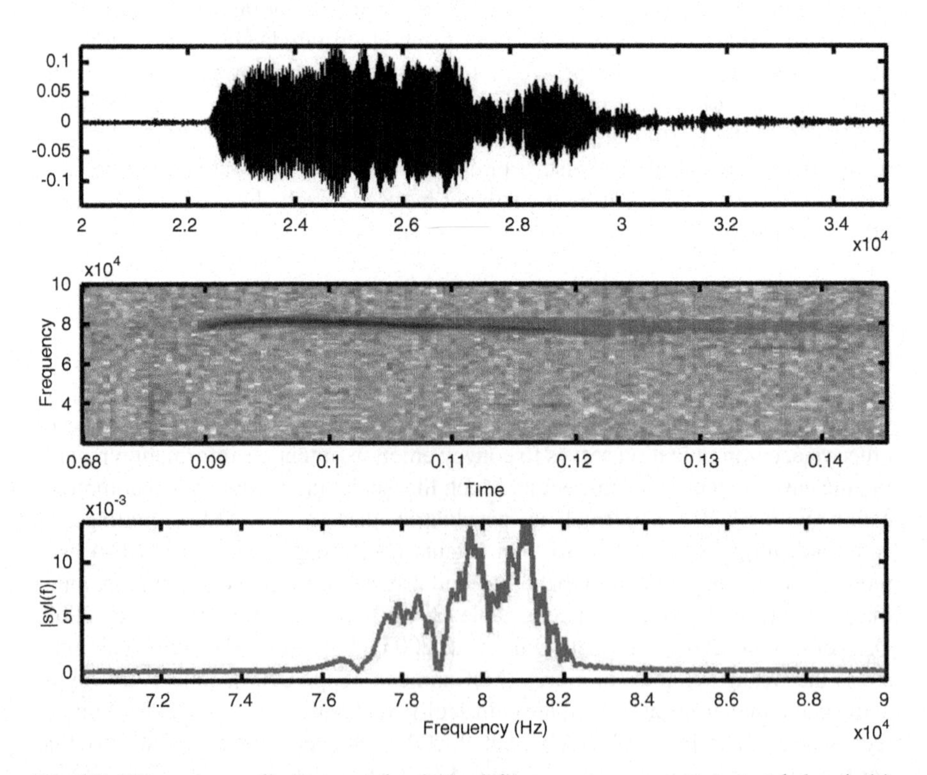

Fig. 8.3 Ultrasonic vocalization sample obtained from a mouse pup at postnatal day 4. (**a**) Oscillogram; (**b**) Spectrogram; (**c**) Frequency Spectrum

innate USV repertoire in response to external factors (Mahrt et al. 2013). However, some features of mouse USVs have been shown to change over development (Arriaga et al. 2012).

USVs have been proposed recently as useful for validating communication functioning in mouse models of psychiatric disease (Wöhr and Schwarting 2010). Aberrations in USV features provide modest face validity for the communication deficits of ASDs (Crawley 2004) and less so schizophrenia (Scattoni et al. 2009). The following sections provide descriptions of age-specific USVs along with example mouse models that corroborate the use of USVs as measures of communication impairments in psychiatric disease.

Pup USVs

Mice are born blind and deaf, have limited motor abilities, lack fur and subcutaneous fat, and cool rapidly when displaced from the nest (Fox 1965). Following removal from the nest, pups emit stereotypic separation calls (30–100 kHz) reminiscent of infant crying in humans and other mammals (Branchi et al. 2004; Newman 2007). These separation calls in mice are elicited at ages (PNDs 2–14) corresponding to late gestational to infancy periods in humans. While one theory suggests that these calls are merely a physiological response to the thermal challenge of leaving the nest (Blumberg and Alberts 1990), the most ethologically accepted theory posits that calls elicit orientation and retrieval from the dam and are therefore important for mother–infant social communication, conveyance of emotional state, and pro-survival behavior (Noirot et al. 1972; Scattoni et al. 2009). One piece of evidence favoring this view has been observation of pup head raising immediately preceding USVs, indicative of orientation towards the mother (Branchi et al. 2004). Another piece of evidence has been findings that maternal responsiveness "tunes" the rate of USVs in pups (Amato et al. 2005).

The extensive methodology for eliciting pup separation-induced USVs has been described previously (Hofer et al. 2001; Winslow 2009). Briefly, a single pup (PNDs 2–14) is removed from the litter and transported to an isolated test area where USVs are collected through an ultrasonic microphone. A 10–15 min period between separation and testing is standard practice in order to promote steady-state conditions. Typical USV numbers approximate 100 calls/min in younger pups, diminishing to 10 calls/min in older pups. Collected USVs are either transduced into an audible frequency range for manual scoring or processed using any number of software packages. Emission of USVs can be potentiated by briefly exposing the pups to the dam, then allowing another period of isolation. Conversely, emission of USVs can be markedly reduced by presenting pups with the scent of an unfamiliar adult male, a phenomenon known as "predator-induced USV suppression." Detailed explanation of these manipulations is discussed elsewhere (Hofer et al. 2001).

Even though pup USVs have been studied ethologically over the past several decades, only recently has their measurement become a tool for refining endophenotypes of psychiatric disease. For instance, numerous mouse models of ASDs have shown variations in pup USVs (Jamain et al. 2008; Long et al. 2004; Picker et al. 2006; Scattoni et al. 2008), findings that are reminiscent of reports of abnormal patterns of vocalizations in autistic infants prior to diagnosis (Zwaigenbaum et al. 2005). A mouse model of Timothy Syndrome, an ASD resulting from mutation of an L-type calcium channel ($Ca_V1.2$), displayed reduced duration of calls across several early PNDs (Bader et al. 2011). Disruption of Foxp2, a gene putatively linked to language impairments in ASDs, resulted in reduced USV number at PND 10 (Shu et al. 2005). Interestingly, this mouse model also produced fewer "clicks," a transient, broadband type of USV with unknown function (Liu et al. 2003).

Mouse models of CNVs have also shown abnormalities in USVs. One model harboring duplication of human 15q11-13, a commonly observed CNV in ASDs, displayed increased USV numbers at PNDs 7 and 14 (Nakatani et al. 2009). Deletion at the maternal allele of the same chromosomal locus in the mouse (a model of Angelman syndrome) resulted in increased USVs at PND 10 and 12 (Jiang et al. 2010). An experimental variation introduced in this study was the use of bedding from either the mother or a stranger female to elicit USVs in pups. In mutant pups, only the mother's bedding resulted in an increased USV number. Another study has extended the complexity of the pup–dam relationship. A mouse model of tuberous sclerosis, a disorder often comorbid with ASDs, showed increased USV rate but only when separated from a heterozygous dam (Young et al. 2010). This effect was independent of maternal retrieval behavior, suggesting that the pups' response to the dam was a contributing factor.

Consistent with endophenotypic evidence in humans, fewer mouse models of schizophrenia have utilized pup USVs as a measure of communication impairments (see Scearce-Levie et al. (2008)). Even so, this measure holds significant potential for validating mouse models of schizophrenia.

Juvenile/Adult USVs

In contrast to rats, adult mice do not vocalize spontaneously but rather primarily in response to social cues (e.g., odors from opposite sex conspecifics) or innately adverse stimuli (e.g., foot shock). Furthermore, unlike rats, there exists little evidence that USVs encode affective state. Adult male mice will vocalize in the presence of females or to urine from a female in estrous, but not during affiliative encounters with other males. While the restricted expression of USVs in mice suggests a limited role in communication, the stereotypic nature of USVs in social settings makes them characteristic of song and therefore more complex than traditionally thought (Holy and Guo 2005). Compared to pups, adult USVs differ in frequency and temporal characteristics (Grimsley et al. 2011; Liu et al. 2003). Adult females also vocalize, and do so particularly during initial same-sex social investigation or during pup separation (Gourbal et al. 2004). Environment can also greatly influence the repertoire of vocalizations. For instance, more naturalistic conditions have been reported to give rise to more complex vocalizations

(Holy and Guo 2005; Portfors 2007). Mice raised in social isolation also exhibit a reduced number of USVs, an effect that correlates with the duration of isolation (Chabout et al. 2012).

A limited number of mouse models have incorporated adult USV production as a measure of communicative impairments analogous to schizophrenia or ASDs. Male mice with deletion of the gene encoding an enzyme essential for synthesis of heparan sulfate, an extracellular polysaccharide involved with cell adhesion and synaptic function, demonstrated a reduced number, duration, and amplitude of USVs in response to female odor (Irie et al. 2012). Female knockout mice of the vasopressin 1b receptor gene (Avpr1b) emitted USVs that were lower in number and average frequency during a resident-intruder test (Scattoni et al. 2008).

Although less studied, USVs have been identified in juvenile mice during play behavior (Panksepp and Lahvis 2007). Unlike adult USVs, juveniles exhibit more whistle-like calls. In males, USVs diminish after adolescence and into sexual maturity, whereas females continue to vocalize in social situations (Wöhr and Schwarting 2010).

Experimental Considerations

The methodology for obtaining pup USVs, including the various conditions and parameters, has been reviewed more completely elsewhere (Hahn and Schanz 2002). Recording USVs from pups is by its nature a stressful event that involves maternal separation and repeated handling. Because of this, measures should be taken to minimize stress wherever possible. Foremost, it is important to maintain the core temperature of pups during USV collection, as moderate changes (>3–4C) can greatly influence call rate. Handling should remain consistent across pups, avoiding grasping the skin on the back of the pup. In addition, repeated removal of pups across early postnatal days may have considerable effects on the dam, nest, and the home cage environment. Minimizing the displacement of nesting materials and the duration of maternal separation should be considered.

Analysis of USV data can be completed by several different means. The most common method is manual scoring. This approach is accomplished by converting the raw audio data to audible frequency levels or by generating visual spectrograms. While this method is most reliable, it is also the most time-intensive. In response to this issue, several labs have developed software to process USV data automatically, and various proprietary software licenses are available (e.g., SASLabPro, Avisoft Bioacoustics, Inc.) and commonly utilized (Panksepp et al. 2007; Scattoni et al. 2008). Other labs, including our own, use MATLAB, R, or other coding environments for the same purpose (Holy and Guo 2005; Liu et al. 2006). While much less time-intensive, such automated methods may produce spurious results (e.g., syllables may be combined thereby reducing total number identified), and should be validated with manual scoring for each cohort of animals tested.

Conclusion and Future Directions

Endophenotypes comprise components of disease that are fundamentally grounded in genetic sources and require detailed experimental investigation. These characteristics make endophenotypes highly suitable for study in mouse models, for which robust genetic technologies and an expanding repertoire of behavioral and neurobiological assays are available. In the field of psychiatric disease, these facets of the endophenotype approach contribute to the growing movement away from symptom-based characterization of mouse models. This shift has already been successful for diseases such as schizophrenia, for which endophenotypes such as prepulse inhibition have been connected to underlying genetic sources using genetically modified mice (Gould and Gottesman 2006). Even so, other behavioral endophenotypes, including those related to socio-communicative dysfunction, remain relatively unexplored in mouse models.

Application of the endophenotype concept in mouse models, while an improvement from traditional symptom-based phenotyping approaches, requires refinement. In this endeavor, emphasis ought to be placed on better translating the core endophenotype criteria from humans to mice. For instance, one criterion states that the endophenotype must be present to a higher degree in undiagnosed family members compared to the general population. In mice, this criterion could be met by quantifying the co-occurrence of putative markers of disease in selectively bred animals, especially as a function of the degree of relatedness. Modern genomic sequencing techniques could then feasibly identify genomic loci that co-segregate with the observed endophenotype.

In measuring socio-communicative endophenotypes in mouse models of psychiatric disease, several areas of development remain. For one, assays that characterize social cognition-like behavior in mice hold significant potential for informing our understanding of analogous endophenotypes in ASDs and schizophrenia. To this end, a select few studies have reported observational fear behavior in rodents (Chen et al. 2009; Jeon et al. 2010; Panksepp 2013), a measure of emotional contagion and putative form of social cognition. These methods for understanding social cognition-like behavior have yet to be applied to mouse models of psychiatric disease or used to screen novel pharmacological agents. Moreover, these assays have largely focused on behavioral measures, whereas neurobiological correlates of social cognition-like behavior have not been investigated. This is a particularly important direction, given the accessibility of mouse neurobiology and use of mice as preclinical models for drug development.

Another area of development is the design of experimental paradigms that incorporate both social and communicative behavior of mice. Similar to the dyadic interaction and resident-intruder tasks that allow simultaneous recording of social interaction and USVs, novel tests of socio-communicative functioning could enhance characterization of mouse models. Lastly, more precise methods for collecting and analyzing USVs are required to increase experimental throughput and delineate USVs among multiple test animals.

Characterization of socio-communicative endophenotypes in mouse models of psychiatric disease may also provide a foundation for developing and screening pharmacological agents. To date, numerous candidate drugs have been shown to reduce neuropathological or behavioral endophenotypes of disease, including: mGlu5 antagonists (Silverman et al. 2010a; Yan et al. 2005), rapamycin (Ehninger et al. 2008), BDNF (Ogier et al. 2007), and oxytocin (Ferguson et al. 2001). Of these drugs, oxytocin has emerged as arguably the most promising, having been shown to enhance social behavior in mice (Huang et al. 2013), as well as normal humans (Bartz et al. 2010; Hurlemann et al. 2010) and individuals with ASDs (Andari et al. 2010; Guastella et al. 2010; Watanabe et al. 2014) or schizophrenia (Pedersen et al. 2011). Mouse models continue to serve as ideal systems through which novel pharmacological agents can be validated. However, in order to accurately judge the therapeutic effects of drugs on mouse models, an endophenotype approach is paramount.

Ultimately, our understanding and treatment of socio-communicative impairments in psychiatric disease may be furthered through endophenotyping approaches in mice. These approaches both simplify the genotype–phenotype link and facilitate the translational utility of mouse models to human psychiatric disease.

References

Addington, J., & Addington, D. (2000). Neurocognitive and social functioning in schizophrenia: A 2.5 year follow-up study. *Schizophrenia Research, 44*, 47–56.

Alanen, Y. O. (1966). The family in the pathogenesis of schizophrenic and neurotic disorders. *Acta Psychiatrica Scandinavica. Supplementum, 189*, 1–654.

Albrecht, A., & Stork, O. (2012). Are NCAM deficient mice an animal model for schizophrenia? *Frontiers in Behavioral Neuroscience, 6*, 43.

Alpert, M., Rosenberg, S. D., Pouget, E. R., & Shaw, R. J. (2000). Prosody and lexical accuracy in flat affect schizophrenia. *Psychiatry Research, 97*, 107–118.

Amato, F. R. D., Scalera, E., Sarli, C., Moles, A., & D'Amato, F. R. (2005). Pups call, mothers rush: Does maternal responsiveness affect the amount of ultrasonic vocalizations in mouse pups? *Behavior Genetics, 35*, 103–112.

American Psychiatric Association Publishing. (2013). *Diagnostic and statistical manual of mental disorders* (5th ed.). Arlington, VA: American Psychiatric Association Publishing.

Amir, R. E., Van den Veyver, I. B., Wan, M., Tran, C. Q., Francke, U., & Zoghbi, H. Y. (1999). Rett syndrome is caused by mutations in X-linked MECP2, encoding methyl-CpG-binding protein 2. *Nature Genetics, 23*, 185–188.

Andari, E., Duhamel, J.-R., Zalla, T., Herbrecht, E., Leboyer, M., & Sirigu, A. (2010). Promoting social behavior with oxytocin in high-functioning autism spectrum disorders. *Proceedings of the National Academy of Sciences of the United States of America, 107*, 4389–4394.

Arriaga, G., Zhou, E. P., & Jarvis, E. D. (2012). Of mice, birds, and men: The mouse ultrasonic song system has some features similar to humans and song-learning birds. *PLoS One, 7*, e46610.

Bader, P., Faizi, M., & Kim, L. (2011). Mouse model of Timothy syndrome recapitulates triad of autistic traits. *Proceedings of the National Academy of Sciences, 108*, 15432–15437.

Bailey, A., Le Couteur, A., Gottesman, I., Bolton, P., Simonoff, E., Yuzda, E., et al. (1995). Autism as a strongly genetic disorder: Evidence from a British twin study. *Psychological Medicine, 25*, 63–77.

Bailey, A., Palferman, S., Heavey, L., & Le Couteur, A. (1998). Autism: The phenotype in relatives. *Journal of Autism and Developmental Disorders, 28*, 369–392.

Baron-Cohen, S., & Belmonte, M. K. (2005). Autism: A window onto the development of the social and the analytic brain. *Annual Review of Neuroscience, 28*, 109–126.

Baron-Cohen, S., Leslie, A. M., & Frith, U. (1985). Does the autistic child have a "theory of mind"? *Cognition, 21*, 37–46.

Bartak, L., Rutter, M., & Cox, A. (1975). A comparative study of infantile autism and specific developmental receptive language disorder: I. The children. *The British Journal of Psychiatry, 126*, 127–145.

Bartz, J. A., Zaki, J., Bolger, N., Hollander, E., Ludwig, N. N., Kolevzon, A., et al. (2010). Oxytocin selectively improves empathic accuracy. *Psychological Science, 21*, 1426–1428.

Baxter, A. J., Brugha, T. S., Erskine, H. E., Scheurer, R. W., Vos, T., & Scott, J. G. (2015). The epidemiology and global burden of autism spectrum disorders. *Psychological Medicine, 45*(3), 601–613.

Belmonte, M. K., Allen, G., Beckel-Mitchener, A., Boulanger, L. M., Carper, R. A., & Webb, S. J. (2004). Autism and abnormal development of brain connectivity. *The Journal of Neuroscience: The Official Journal of the Society for Neuroscience, 24*, 9228–9231.

Ben-Shachar, S., & Lanpher, B. (2009). Microdeletion 15q13. 3: A locus with incomplete penetrance for autism, mental retardation, and psychiatric disorders. *Journal of Medical Genetics, 46*, 382–388.

Bergner, C., Smolinsky, A., Dufour, B., Laporte, J., Hart, P., Egan, R., et al. (2009). Phenotyping and genetics of rodent grooming and barbering: Utility for experimental neuroscience research. In A. V. Kalueff, J. L. LaPorte, & C. L. Bergner (Eds.), *Neurobiology of grooming behavior* (pp. 46–65). Cambridge: Cambridge University Press.

Berkel, S., Marshall, C. R., Weiss, B., Howe, J., Roeth, R., Moog, U., et al. (2010). Mutations in the SHANK2 synaptic scaffolding gene in autism spectrum disorder and mental retardation. *Nature Genetics, 42*, 489–491.

Betancur, C., & Coleman, M. (2013). Etiological heterogeneity in autism spectrum disorders: Role of rare variants. In J. D. Buxbaum & P. R. Hof (Eds.), *The neuroscience of autism spectrum disorders*. Chicago: Elsevier.

Blumberg, M. S., & Alberts, J. R. (1990). Ultrasonic vocalizations by rat pups in the cold: An acoustic by-product of laryngeal braking? *Behavioral Neuroscience, 104*, 808–817.

Blumberg, S. J., Bramlett, M. D., Kogan, M. D., Schieve, L. A, & Jones, J. R. (2013). Changes in prevalence of parent-reported autism spectrum disorder in school-aged U.S. children: 2007 to 2011–2012. *National Health Statistics Reports*, (65), 1–11.

Blundell, J., Blaiss, C. A., Etherton, M. R., Espinosa, F., Tabuchi, K., Walz, C., et al. (2010). Neuroligin-1 deletion results in impaired spatial memory and increased repetitive behavior. *The Journal of Neuroscience, 30*, 2115–2129.

Bolivar, V., Walters, S., & Phoenix, J. (2007). Assessing autism-like behavior in mice: Variations in social interactions among inbred strains. *Behavioural Brain Research, 176*, 21–26.

Bolton, P., Macdonald, H., Pickles, A., Rios, P., Goode, S., Crowson, M., et al. (1994). A case-control family history study of autism. *Journal of Child Psychology and Psychiatry, and Allied Disciplines, 35*, 877–900.

Branchi, I., Santucci, D., Puopolo, M., & Alleva, E. (2004). Neonatal behaviors associated with ultrasonic vocalizations in mice (mus musculus): A slow-motion analysis. *Developmental Psychobiology, 44*, 37–44.

Brüne, M. (2005). Emotion recognition, "theory of mind", and social behavior in schizophrenia. *Psychiatry Research, 133*, 135–147.

Burns, J. K. (2004). An evolutionary theory of schizophrenia: Cortical connectivity, metarepresentation, and the social brain. *The Behavioral and Brain Sciences, 27*, 831–855.

Burns, J. (2006). The social brain hypothesis of schizophrenia. *Psychiatria Danubina, 18*, 225–229.

Carrión, R. E., McLaughlin, D., Goldberg, T. E., Auther, A. M., Olsen, R. H., Olvet, D. M., et al. (2013). Prediction of functional outcome in individuals at clinical high risk for psychosis. *JAMA Psychiatry, 70*, 1133–1142.

Centers for Disease Control and Prevention. (2012). Prevalence of autism spectrum disorders—Autism and Developmental Disabilities Monitoring Network, 14 sites, United States, 2008. *Morbidity and Mortality Weekly Report Surveillance Summaries, 61*, 1–19.
Chabout, J., Serreau, P., Ey, E., Bellier, L., Aubin, T., Bourgeron, T., et al. (2012). Adult male mice emit context-specific ultrasonic vocalizations that are modulated by prior isolation or group rearing environment. *PLoS One, 7*, e29401.
Chen, R. Z., Akbarian, S., Tudor, M., & Jaenisch, R. (2001). Deficiency of methyl-CpG binding protein-2 in CNS neurons results in a Rett-like phenotype in mice. *Nature Genetics, 27*, 327–331.
Chen, Q., Panksepp, J. B., & Lahvis, G. P. (2009). Empathy is moderated by genetic background in mice. *PLoS One, 4*, e4387.
Ching, M. S. L., Shen, Y., Tan, W.-H., Jeste, S. S., Morrow, E. M., Chen, X., et al. (2010). Deletions of NRXN1 (neurexin-1) predispose to a wide spectrum of developmental disorders. *American Journal of Medical Genetics, 153B*, 937–947.
Couture, S. M., Penn, D. L., Losh, M., Adolphs, R., Hurley, R., & Piven, J. (2010). Comparison of social cognitive functioning in schizophrenia and high functioning autism: More convergence than divergence. *Psychological Medicine, 40*, 569–579.
Crawley, J. N. (2004). Designing mouse behavioral tasks relevant to autistic-like behaviors. *Mental Retardation and Developmental Disabilities Research Reviews, 10*, 248–258.
Crespi, B., & Badcock, C. (2008). Psychosis and autism as diametrical disorders of the social brain. *The Behavioral and Brain Sciences, 31*, 241–261.
Crespi, B., Stead, P., & Elliot, M. (2010). Evolution in health and medicine Sackler colloquium: Comparative genomics of autism and schizophrenia. *Proceedings of the National Academy of Sciences of the United States of America, 107*, 1736–1741.
Daniels, J. L., Forssen, U., Hultman, C. M., Cnattingius, S., Savitz, D. A., Feychting, M., et al. (2008). Parental psychiatric disorders associated with autism spectrum disorders in the offspring. *Pediatrics, 121*, e1357–e1362.
Dawson, G. (2008). Early behavioral intervention, brain plasticity, and the prevention of autism spectrum disorder. *Development and Psychopathology, 20*, 775–803.
de Achával, D., Costanzo, E. Y., Villarreal, M., Jáuregui, I. O., Chiodi, A., Castro, M. N., et al. (2010). Emotion processing and theory of mind in schizophrenia patients and their unaffected first-degree relatives. *Neuropsychologia, 48*, 1209–1215.
Deacon, R. M. J. (2006). Assessing nest building in mice. *Nature Protocols, 1*, 1117–1119.
Defensor, E., & Pearson, B. (2011). A novel social proximity test suggests patterns of social avoidance and gaze aversion-like behavior in BTBR T+ tf/J mice. *Behavioural Brain Research, 31*, 1169–1182.
Di Cristo, G. (2007). Development of cortical GABAergic circuits and its implications for neurodevelopmental disorders. *Clinical Genetics, 72*, 1–8.
Dickinson, D., Bellack, A. S., & Gold, J. M. (2007). Social/communication skills, cognition, and vocational functioning in schizophrenia. *Schizophrenia Bulletin, 33*, 1213–1220.
Docherty, N. M., Gordinier, S. W., Hall, M. J., & Cutting, L. P. (1999). Communication disturbances in relatives beyond the age of risk for schizophrenia and their associations with symptoms in patients. *Schizophrenia Bulletin, 25*, 851–862.
Docherty, N. M., Hawkins, K. A., Hoffman, R. E., Quinlan, D. M., Rakfeldt, J., & Sledge, W. H. (1996). Working memory, attention, and communication disturbances in schizophrenia. *Journal of Abnormal Psychology, 105*, 212–219.
Domes, G., Heinrichs, M., Gläscher, J., Büchel, C., Braus, D. F., & Herpertz, S. C. (2007). Oxytocin attenuates amygdala responses to emotional faces regardless of valence. *Biological Psychiatry, 62*, 1187–1190.
Duncan, G., Moy, S., & Perez, A. (2004). Deficits in sensorimotor gating and tests of social behavior in a genetic model of reduced NMDA receptor function. *Behavioural Brain Research, 153*, 507–519.
Eack, S. M., Mermon, D. E., Montrose, D. M., Miewald, J., Gur, R. E., Gur, R. C., et al. (2010). Social cognition deficits among individuals at familial high risk for schizophrenia. *Schizophrenia Bulletin, 36*, 1081–1088.

Ehninger, D., Han, S., Shilyansky, C., & Zhou, Y. (2008). Reversal of learning deficits in a Tsc2+/– mouse model of tuberous sclerosis. *Nature Medicine, 14*, 843–848.

Etherton, M. R., Blaiss, C. A., Powell, C. M., Etherton, M. R., Blaiss, C. A., Powell, C. M., et al. (2009). Mouse neurexin-1alpha deletion causes correlated electrophysiological and behavioral changes consistent with cognitive impairments. *Proceedings of the National Academy of Sciences, 106*, 17998–18003.

Faraone, S. V., Seidman, L. J., Kremen, W. S., Pepple, J. R., Lyons, M. J., & Tsuang, M. T. (1995). Neuropsychological functioning among the nonpsychotic relatives of schizophrenic patients: A diagnostic efficiency analysis. *Journal of Abnormal Psychology, 104*, 286–304.

Fejgin, K., Nielsen, J., Birknow, M. R., Bastlund, J. F., Nielsen, V., Lauridsen, J. B., et al. (2014). A Mouse model that recapitulates cardinal features of the 15q13.3 microdeletion syndrome including schizophrenia- and epilepsy-related alterations. *Biological Psychiatry, 76*, 128–137.

Ferguson, J. N., Aldag, J. M., Insel, T. R., & Young, L. J. (2001). Oxytocin in the medial amygdala is essential for social recognition in the mouse. *The Journal of Neuroscience, 21*, 8278–8285.

Folstein, S. E., & Rutter, M. L. (1977a). Infantile autism: A genetic study of 21 twin pairs. *Journal of Child Psychology and Psychiatry, 18*, 297–321.

Folstein, S., & Rutter, M. (1977b). Genetic influences and infantile autism. *Nature, 265*, 726–728.

Fone, K. C. F., & Porkess, M. V. (2008). Behavioural and neurochemical effects of post-weaning social isolation in rodents-relevance to developmental neuropsychiatric disorders. *Neuroscience and Biobehavioral Reviews, 32*, 1087–1102.

Fox, W. (1965). Reflex-ontogeny and behavioural development of the mouse. *Animal Behaviour, 13*, 234–241.

Fromer, M., Pocklington, A. J., Kavanagh, D. H., Williams, H. J., Dwyer, S., Gormley, P., et al. (2014). De novo mutations in schizophrenia implicate synaptic networks. *Nature, 506*, 179–184.

Gai, X., Xie, H. M., Perin, J. C., Takahashi, N., Murphy, K., Wenocur, A. S., et al. (2011). Rare structural variation of synapse and neurotransmission genes in autism. *Molecular Psychiatry, 17*, 402–411.

Gauthier, J., Champagne, N., Lafrenière, R. G., Xiong, L., Spiegelman, D., & Brustein, E. (2010). De novo mutations in the gene encoding the synaptic scaffolding protein SHANK3 in patients ascertained for schizophrenia. *Proceedings of the National Academy of Sciences of the United States of America, 107*, 7863–7868.

Gauthier, J., Siddiqui, T. J., Huashan, P., Yokomaku, D., Hamdan, F. F., Champagne, N., et al. (2011). Truncating mutations in NRXN2 and NRXN1 in autism spectrum disorders and schizophrenia. *Human Genetics, 130*, 563–573.

Gauthier, J., Spiegelman, D., Piton, A., Lafrenière, R. G., Laurent, S., St-Onge, J., et al. (2009). Novel de novo SHANK3 mutation in autistic patients. *American Journal of Medical Genetics, 150B*, 421–424.

Gernsbacher, M. A., Dawson, M., & Hill Goldsmith, H. (2005). Three reasons not to believe in an autism epidemic. *Current Directions in Psychological Science, 14*, 55–58.

Gilman, S. R., Chang, J., Xu, B., Bawa, T. S., Gogos, J. A., Karayiorgou, M., et al. (2012). Diverse types of genetic variation converge on functional gene networks involved in schizophrenia. *Nature Neuroscience, 15*, 1723–1728.

Glickman, S. E., & Schiff, B. B. (1967). A biological theory of reinforcement. *Psychological Review, 74*, 81–109.

Goorden, S. M. I., van Woerden, G. M., van der Weerd, L., Cheadle, J. P., & Elgersma, Y. (2007). Cognitive deficits in Tsc1+/– mice in the absence of cerebral lesions and seizures. *Annals of Neurology, 62*, 648–655.

Gottesman, I. I., & McGue, M. (2015). Endophenotype. In R. L. Cautin & S. O. Lilienfeld (Eds.), *Encyclopedia of clinical psychology*. New York: Wiley-Blackwell.

Gottesman, I. I., Ph, D., & Gould, T. D. (2003). The endophenotype concept in psychiatry: Etymology and strategic intentions. *The American Journal of Psychiatry, 160*, 636–645.

Gottesman, I. I., & Shields, J. (1973). Genetic theorizing and schizophrenia. *The British Journal of Psychiatry, 122*, 15–30.
Gould, T. D., & Gottesman, I. I. (2006). Psychiatric endophenotypes and the development of valid animal models. *Genes, Brain, and Behavior, 5*, 113–119.
Gourbal, B. E. F., Barthelemy, M., Petit, G., & Gabrion, C. (2004). Spectrographic analysis of the ultrasonic vocalisations of adult male and female BALB/c mice. *Die Naturwissenschaften, 91*, 381–385.
Goussé, V., Plumet, M. H., Chabane, N., Mouren-Siméoni, M.-C., Ferradian, N., & Leboyer, M. (2002). Fringe phenotypes in autism: A review of clinical, biochemical and cognitive studies. *European Psychiatry, 17*, 120–128.
Gratten, J., Visscher, P. M., Mowry, B. J., & Wray, N. R. (2013). Interpreting the role of de novo protein-coding mutations in neuropsychiatric disease. *Nature Genetics, 45*, 234–238.
Gregor, A., Albrecht, B., Bader, I., Bijlsma, E. K., Ekici, A. B., Engels, H., et al. (2011). Expanding the clinical spectrum associated with defects in CNTNAP2 and NRXN1. *BMC Medical Genetics, 12*, 106.
Grimsley, J. M. S., Monaghan, J. J. M., & Wenstrup, J. J. (2011). Development of social vocalizations in mice. *PLoS One, 6*, e17460.
Guastella, A. J., Einfeld, S. L., Gray, K. M., Rinehart, N. J., Tonge, B. J., Lambert, T. J., et al. (2010). Intranasal oxytocin improves emotion recognition for youth with autism spectrum disorders. *Biological Psychiatry, 67*, 692–694.
Gur, R., Tendler, A., & Wagner, S. (2014). Long-term social recognition memory is mediated by oxytocin-dependent synaptic plasticity in the medial amygdala. *Biological Psychiatry, 76*, 377–386.
Guy, J., Hendrich, B., Holmes, M., Martin, J. E., & Bird, A. (2001). A mouse Mecp2-null mutation causes neurological symptoms that mimic Rett syndrome. *Nature Genetics, 27*, 322–326.
Häfner, H., Löffler, W., Maurer, K., Hambrecht, M., & van der Heiden, W. (1999). Depression, negative symptoms, social stagnation and social decline in the early course of schizophrenia. *Acta Psychiatrica Scandinavica, 100*, 105–118.
Hahn, M. E., & Schanz, N. (2002). The effects of cold, rotation, and genotype on the production of ultrasonic calls in infant mice. *Behavior Genetics, 32*, 267–273.
Heck, D. H., & Lu, L. (2012). The social life of neurons: Synaptic communication deficits as a common denominator of autism, schizophrenia, and other cognitive disorders. *Biological Psychiatry, 72*, 173–174.
Hofer, M., Shair, H., & Brunelli, S. (2001). Ultrasonic vocalizations in rat and mouse pups. *Current Protocols in Neuroscience, 8*, 1–16.
Holy, T. E., & Guo, Z. (2005). Ultrasonic songs of male mice. *PLoS Biology, 3*, e386.
Horev, G., Ellegood, J., Lerch, J. P., Son, Y. E., Muthuswamy, L., & Vogel, H. (2011). Dosage-dependent phenotypes in models of 16p11.2 lesions found in autism. *PNAS, 108*, 17076–17081.
Howerton, C. L., Garner, J. P., & Mench, J. a. (2012). A system utilizing radio frequency identification (RFID) technology to monitor individual rodent behavior in complex social settings. *Journal of Neuroscience Methods, 209*, 74–78.
Huang, H., Michetti, C., Busnelli, M., Managò, F., Sannino, S., Scheggia, D., et al. (2013). Chronic and acute intranasal oxytocin produce divergent social effects in mice. *Neuropsychopharmacology, 39*(5), 1102–1114.
Huguet, G., Ey, E., & Bourgeron, T. (2013). The genetic landscapes of autism spectrum disorders. *Annual Review of Genomics and Human Genetics, 14*, 191–213.
Hurlemann, R., Patin, A., Onur, O. A., Cohen, M. X., Baumgartner, T., Metzler, S., et al. (2010). Oxytocin enhances amygdala-dependent, socially reinforced learning and emotional empathy in humans. *The Journal of Neuroscience, 30*, 4999–5007.
Insel, T. R. (2003). Is social attachment an addictive disorder? *Physiology & Behavior, 79*, 351–357.
International Schizophrenia Consortium. (2008). Rare chromosomal deletions and duplications increase risk of schizophrenia. *Nature, 455*, 237–241.

Irani, F., Platek, S. M., Panyavin, I. S., Calkins, M. E., Kohler, C., Siegel, S. J., et al. (2006). Self-face recognition and theory of mind in patients with schizophrenia and first-degree relatives. *Schizophrenia Research, 88*, 151–160.

Irie, F., Badie-Mahdavi, H., & Yamaguchi, Y. (2012). Autism-like socio-communicative deficits and stereotypies in mice lacking heparan sulfate. *Proceedings of the National Academy of Sciences of the United States of America, 2012*, 5052–5056.

Jamain, S., Quach, H., Betancur, C., Råstam, M., Colineaux, C., Gillberg, I. C., et al. (2003). Mutations of the X-linked genes encoding neuroligins NLGN3 and NLGN4 are associated with autism. *Nature Genetics, 34*, 27–29.

Jamain, S., Radyushkin, K., Hammerschmidt, K., Granon, S., Boretius, S., Varoqueaux, F., et al. (2008). Reduced social interaction and ultrasonic communication in a mouse model of monogenic heritable autism. *Proceedings of the National Academy of Sciences of the United States of America, 105*, 1710–1715.

Jeon, D., Kim, S., Chetana, M., Jo, D., Ruley, H. E., Lin, S.-Y., et al. (2010). Observational fear learning involves affective pain system and Cav1.2 Ca2+ channels in ACC. *Nature Neuroscience, 13*, 482–488.

Jiang, Y.-H., Pan, Y., Zhu, L., Landa, L., Yoo, J., Spencer, C., et al. (2010). Altered ultrasonic vocalization and impaired learning and memory in Angelman syndrome mouse model with a large maternal deletion from Ube3a to Gabrb3. *PLoS One, 5*, e12278.

Johnson, C. P. (2008). Recognition of autism before age 2 years. *Pediatrics in Review/American Academy of Pediatrics, 29*, 86–96.

Jones, W., & Klin, A. (2013). Attention to eyes is present but in decline in 2–6-month-old infants later diagnosed with autism. *Nature, 504*, 427–431.

Jones, C. A., Watson, D. J. G., & Fone, K. C. F. (2011). Animal models of schizophrenia. *British Journal of Pharmacology, 164*, 1162–1194.

Kalueff, A. V., Minasyan, A., Keisala, T., Shah, Z. H., & Tuohimaa, P. (2006). Hair barbering in mice: Implications for neurobehavioural research. *Behavioural Processes, 71*, 8–15.

Kanner, L. (1943). Autistic disturbances of affective contact. *Nervous Child, 2*, 217–250.

Kendler, K. S. (1985). Diagnostic approaches to schizotypal personality disorder: A historical perspective. *Schizophrenia Bulletin, 11*, 538–553.

Kety, S. S., Rosenthal, D., Wender, P. H., Schulsinger, F., & Jacobsen, B. (1976). Mental illness in the biological and adoptive families of adopted individuals who have become schizophrenic. *Behavior Genetics, 6*, 219–225.

Kety, S. S., Rosenthal, D., Wender, P. H., & Schulsinger, F. (1968). The types and prevalence of mental illness in the biological and adoptive families of adopted schizophrenics. *Journal of Psychiatric Research, 6*, 345–362.

Kielinen, M., Rantala, H., Timonen, E., Linna, S.-L., & Moilanen, I. (2004). Associated medical disorders and disabilities in children with autistic disorder: A population-based study. *Autism, 8*, 49–60.

Kim, H. G., Kishikawa, S., Higgins, A. W., Seong, I. S., Donovan, D. J., Shen, Y., et al. (2008). Disruption of neurexin 1 associated with autism spectrum disorder. *American Journal of Human Genetics, 82*, 199–207.

King, B. H., & Lord, C. (2011). Is schizophrenia on the autism spectrum? *Brain Research, 1380*, 34–41.

Kirov, G., Rujescu, D., Ingason, A., Collier, D. A., O'Donovan, M. C., & Owen, M. J. (2009). Neurexin 1 (NRXN1) deletions in schizophrenia. *Schizophrenia Bulletin, 35*, 851–854.

Klei, L., Sanders, S. J., Murtha, M. T., Hus, V., Lowe, J. K., Willsey, A. J., et al. (2012). Common genetic variants, acting additively, are a major source of risk for autism. *Molecular Autism, 3*, 9.

Knutson, B., Burgdorf, J., & Panksepp, J. (2002). Ultrasonic vocalizations as indices of affective states in rats. *Psychological Bulletin, 128*, 961–977.

Koh, H.-Y., Kim, D., Lee, J., Lee, S., & Shin, H.-S. (2008). Deficits in social behavior and sensorimotor gating in mice lacking phospholipase Cbeta1. *Genes, Brain, and Behavior, 7*, 120–128.

Ku, C. S., Loy, E. Y., Salim, A., Pawitan, Y., & Chia, K. S. (2010). The discovery of human genetic variations and their use as disease markers: Past, present and future. *Journal of Human Genetics, 55*, 403–415.

Lahvis, G., Alleva, E., & Scattoni, M. (2011). Translating mouse vocalizations: Prosody and frequency modulation. *Genes, Brain, and Behavior, 10*, 4–16.

Langdon, R., Coltheart, M., Ward, P. B., & Catts, S. V. (2002). Disturbed communication in schizophrenia: The role of poor pragmatics and poor mind-reading. *Psychological Medicine, 32*, 1273–1284.

Larsson, H. J., Eaton, W. W., Madsen, K. M., Vestergaard, M., Olesen, A. V., Agerbo, E., et al. (2005). Risk factors for autism: Perinatal factors, parental psychiatric history, and socioeconomic status. *American Journal of Epidemiology, 161*, 916–925.

Le Couteur, A., Bailey, A., Goode, S., Pickles, A., Gottesman, I., Robertson, S., et al. (1996). A broader phenotype of autism: The clinical spectrum in twins. *Journal of Child Psychology and Psychiatry, and Allied Disciplines, 37*, 785–801.

Leboyer, M., Bellivier, F., Nosten-Bertrand, M., Jouvent, R., Pauls, D., & Mallet, J. (1998). Psychiatric genetics: Search for phenotypes. *Trends in Neurosciences, 21*, 102–105.

Lee, S. H., DeCandia, T. R., Ripke, S., Yang, J., Sullivan, P. F., Goddard, M. E., et al. (2012). Estimating the proportion of variation in susceptibility to schizophrenia captured by common SNPs. *Nature Genetics, 44*, 247–250.

Leentjens, A. F., Wielaert, S. M., van Harskamp, F., & Wilmink, F. W. (1998). Disturbances of affective prosody in patients with schizophrenia: A cross sectional study. *Journal of Neurology, Neurosurgery, and Psychiatry, 64*, 375–378.

Li, H. H., Roy, M., Kuscuoglu, U., Spencer, C. M., Halm, B., Harrison, K. C., et al. (2009). Induced chromosome deletions cause hypersociability and other features of Williams-Beuren syndrome in mice. *EMBO Molecular Medicine, 1*, 50–65.

Lijam, N., Paylor, R., McDonald, M. P., Crawley, J. N., Deng, C. X., Herrup, K., et al. (1997). Social interaction and sensorimotor gating abnormalities in mice lacking Dvl1. *Cell, 90*, 895–905.

Liu, R. C., Linden, J. F., & Schreiner, C. E. (2006). Improved cortical entrainment to infant communication calls in mothers compared with virgin mice. *The European Journal of Neuroscience, 23*, 3087–3097.

Liu, R. C., Miller, K. D., Merzenich, M. M., & Schreiner, C. E. (2003). Acoustic variability and distinguishability among mouse ultrasound vocalizations. *The Journal of the Acoustical Society of America, 114*, 3412.

Lohmueller, K. E., Pearce, C. L., Pike, M., Lander, E. S., & Hirschhorn, J. N. (2003). Meta-analysis of genetic association studies supports a contribution of common variants to susceptibility to common disease. *Nature Genetics, 33*, 177–182.

Long, J. M., Laporte, P., Paylor, R., & Wynshaw-Boris, A. (2004). Expanded characterization of the social interaction abnormalities in mice lacking Dvl1. *Genes, Brain and Behavior, 3*(1), 51–62.

Losh, M., Adolphs, R., Poe, M. D., Couture, S., Penn, D., Baranek, G. T., et al. (2009). Neuropsychological profile of autism and the broad autism phenotype. *Archives of General Psychiatry, 66*, 518–526.

Loveland, K. A., Tunali-Kotoski, B., Chen, Y. R., Ortegon, J., Pearson, D. A., Brelsford, K. A., et al. (1997). Emotion recognition in autism: Verbal and nonverbal information. *Development and Psychopathology, 9*, 579–593.

Mahrt, E. J., Perkel, D. J., Tong, L., Rubel, E. W., & Portfors, C. V. (2013). Engineered deafness reveals that mouse courtship vocalizations do not require auditory experience. *The Journal of Neuroscience, 33*, 5573–5583.

Mandell, D., & Lecavalier, L. (2014). Should we believe the centers for disease control and prevention's autism spectrum disorder prevalence estimates? *Autism, 18*, 482–484.

Manolio, T. A., Collins, F. S., Cox, N. J., Goldstein, D. B., Hindorff, L. A., Hunter, D. J., et al. (2009). Finding the missing heritability of complex diseases. *Nature, 461*, 747–753.

Marín, O. (2012). Interneuron dysfunction in psychiatric disorders. *Nature Reviews. Neuroscience, 13*, 107–120.
Marshall, C. R., Noor, A., Vincent, J. B., Lionel, A. C., Feuk, L., Skaug, J., et al. (2008). Structural variation of chromosomes in autism spectrum disorder. *Journal of Human Genetics, 82*(2), 477–488.
McCann, J., & Peppé, S. (2003). Prosody in autism spectrum disorders: A critical review. *International Journal of Language & Communication Disorders/Royal College of Speech & Language Therapists, 38*, 325–350.
McFarlane, H. G., Kusek, G. K., Yang, M., Phoenix, J. L., Bolivar, V. J., & Crawley, J. N. (2008). Autism-like behavioral phenotypes in BTBR T+tf/J mice. *Genes, Brain, and Behavior, 7*, 152–163.
McIntosh, D. N., Reichmann-Decker, A., Winkielman, P., & Wilbarger, J. L. (2006). When the social mirror breaks: Deficits in automatic, but not voluntary, mimicry of emotional facial expressions in autism. *Developmental Science, 9*, 295–302.
Merikangas, A. K., Corvin, A. P., & Gallagher, L. (2009). Copy-number variants in neurodevelopmental disorders: Promises and challenges. *Trends in Genetics, 25*, 536–544.
Miller, G. A., & Rockstroh, B. (2013). Endophenotypes in psychopathology research: Where do we stand? *Annual Review of Clinical Psychology, 9*, 177–213.
Mills, A. A., & Bradley, A. (2001). From mouse to man: Generating megabase chromosome rearrangements. *Trends in Genetics, 17*, 331–339.
Mineur, Y., Huynh, L., & Crusio, W. (2006). Social behavior deficits in the Fmr 1 mutant mouse. *Behavioural Brain Research, 168*(1), 172e5. doi:10.1016/j.bbr.2005.11.004.
Molina, J., Carmona-Mora, P., Chrast, J., Krall, P. M., Canales, C. P., Lupski, J. R., et al. (2008). Abnormal social behaviors and altered gene expression rates in a mouse model for Potocki-Lupski syndrome. *Human Molecular Genetics, 17*, 2486–2495.
Moretti, P., Bouwknecht, J. A., Teague, R., Paylor, R., & Zoghbi, H. Y. (2005). Abnormalities of social interactions and home-cage behavior in a mouse model of Rett syndrome. *Human Molecular Genetics, 14*, 205–220.
Moy, S. S., Nadler, J. J., Young, N. B., Nonneman, R. J., Grossman, A. W., Murphy, D. L., et al. (2009). Social approach in genetically engineered mouse lines relevant to autism. *Genes, Brain, and Behavior, 8*, 129–142.
Nadig, A. S., Ozonoff, S., Young, G. S., Rozga, A., Sigman, M., & Rogers, S. J. (2007). A prospective study of response to name in infants at risk for autism. *Archives of Pediatrics & Adolescent Medicine, 161*, 378–383.
Nadler, J. J., Moy, S. S., Dold, G., Trang, D., Simmons, N., Perez, A., et al. (2004). Automated apparatus for quantitation of social approach behaviors in mice. *Genes, Brain, and Behavior, 3*, 303–314.
Nakatani, J., Tamada, K., Hatanaka, F., Ise, S., Ohta, H., Inoue, K., et al. (2009). Abnormal behavior in a chromosome-engineered mouse model for human 15q11-13 duplication seen in autism. *Cell, 137*, 1235–1246.
Neale, B. M., Kou, Y., Liu, L., Ma'ayan, A., Samocha, K. E., Sabo, A., et al. (2012). Patterns and rates of exonic de novo mutations in autism spectrum disorders. *Nature, 485*(7397), 242–245.
Neunuebel, J. P., Taylor, A. L., Arthur, B. J., & Egnor, S. R. (2015). Female mice ultrasonically interact with males during courtship displays. *Elife, 4*, 1–24.
Newman, J. (2007). Neural circuits underlying crying and cry responding in mammals. *Behavioural Brain Research, 182*, 155–165.
Nieman, D. H., Velthorst, E., Becker, H. E., de Haan, L., Dingemans, P. M., Linszen, D. H., et al. (2013). The Strauss and Carpenter Prognostic Scale in subjects clinically at high risk of psychosis. *Acta Psychiatrica Scandinavica, 127*, 53–61.
Noirot, E., National, C., & Recherche, D. (1972). Ultrasounds and maternal behavior in small rodents. *Developmental Psychobiology, 5*, 371–387.
O'Roak, B. J., Vives, L., Girirajan, S., Karakoc, E., Krumm, N., Coe, B. P., et al. (2012). Sporadic autism exomes reveal a highly interconnected protein network of de novo mutations. *Nature, 485*(7397), 246–250.

Oberman, L. M., Winkielman, P., & Ramachandran, V. S. (2009). Slow echo: Facial EMG evidence for the delay of spontaneous, but not voluntary, emotional mimicry in children with autism spectrum disorders. *Developmental Science, 12*, 510–520.

Oerlemans, A. M., Droste, K., van Steijn, D. J., de Sonneville, L. M. J., Buitelaar, J. K., & Rommelse, N. N. J. (2013). Co-segregation of social cognition, executive function and local processing style in children with ASD, their siblings and normal controls. *Journal of Autism and Developmental Disorders, 43*, 2764–2778.

Oerlemans, A. M., van der Meer, J. M. J., van Steijn, D. J., de Ruiter, S. W., de Bruijn, Y. G. E., de Sonneville, L. M. J., et al. (2014). Recognition of facial emotion and affective prosody in children with ASD (+ADHD) and their unaffected siblings. *European Child & Adolescent Psychiatry, 23*, 257–271.

Ogier, M., Wang, H., Hong, E., Wang, Q., Greenberg, M. E., & Katz, D. M. (2007). Brain-derived neurotrophic factor expression and respiratory function improve after ampakine treatment in a mouse model of Rett syndrome. *The Journal of Neuroscience, 27*, 10912–10917.

Osterling, J. A., Dawson, G., & Munson, J. A. (2002). Early recognition of 1-year-old infants with autism spectrum disorder versus mental retardation. *Development and Psychopathology, 14*, 239–251.

Panksepp, J. B. J. (2013). Toward a cross-species understanding of empathy. *Trends in Neurosciences, 36*, 489–496.

Panksepp, J. B., Jochman, K. A., Kim, J. U., Koy, J. J., Wilson, E. D., Chen, Q., et al. (2007). Affiliative behavior, ultrasonic communication and social reward are influenced by genetic variation in adolescent mice. *PLoS One, 2*, e351.

Panksepp, J. B., & Lahvis, G. P. (2007). Social reward among juvenile mice. *Genes, Brain, and Behavior, 6*, 661–671.

Parikshak, N. N. N., Luo, R., Zhang, A., Won, H., Lowe, J. K. K., Chandran, V., et al. (2013). Integrative functional genomic analyses implicate specific molecular pathways and circuits in autism. *Cell, 155*, 1008–1021.

Peça, J., Feliciano, C., Ting, J. T., Wang, W., Wells, M. F., Venkatraman, T. N., et al. (2011). Shank3 mutant mice display autistic-like behaviours and striatal dysfunction. *Nature, 472*, 437–442.

Pedersen, C. A., Gibson, C. M., Rau, S. W., Salimi, K., Smedley, K. L., Casey, R. L., et al. (2011). Intranasal oxytocin reduces psychotic symptoms and improves Theory of Mind and social perception in schizophrenia. *Schizophrenia Research, 132*, 50–53.

Pedersen, C. B., Mors, O., Bertelsen, A., Waltoft, B. L., Agerbo, E., McGrath, J. J., et al. (2014). A comprehensive nationwide study of the incidence rate and lifetime risk for treated mental disorders. *JAMA Psychiatry, 71*, 573–581.

Peñagarikano, O., Abrahams, B. S., Herman, E. I., Winden, K. D., Gdalyahu, A., Dong, H., et al. (2011). Absence of CNTNAP2 leads to epilepsy, neuronal migration abnormalities, and core autism-related deficits. *Cell, 147*, 235–246.

Persico, A. M., & Sacco, R. (2014). Endophenotypes in Autism Spectrum Disorders. In V. B. Patel, V. R. Preedy, & C. R. Martin (Eds.), *Comprehensive guide to autism* (pp. 77–95). New York, NY: Springer.

Picker, J. D., Yang, R., Ricceri, L., & Berger-Sweeney, J. (2006). An altered neonatal behavioral phenotype in Mecp2 mutant mice. *Neuroreport, 17*, 541–544.

Pickles, A., Starr, E., Kazak, S., Bolton, P., Papanikolaou, K., Bailey, A., et al. (2000). Variable expression of the autism broader phenotype: Findings from extended pedigrees. *Journal of Child Psychology and Psychiatry, and Allied Disciplines, 41*, 491–502.

Pinkham, A., Hopfinger, J., & Pelphrey, K. (2008). Neural bases for impaired social cognition in schizophrenia and autism spectrum disorders. *Schizophrenia Research, 99*, 164–175.

Pinto, D., Delaby, E., Merico, D., Barbosa, M., Merikangas, A., Klei, L., et al. (2014). Convergence of genes and cellular pathways dysregulated in autism spectrum disorders. *American Journal of Human Genetics, 94*, 677–694.

Pinto, D., Pagnamenta, A. T., Klei, L., Anney, R., Merico, D., Regan, R., et al. (2010). Functional impact of global rare copy number variation in autism spectrum disorders. *Nature, 466*, 368–372.

Piven, J., Palmer, P., Landa, R., Santangelo, S., Jacobi, D., & Childress, D. (1997). Personality and language characteristics in parents from multiple-incidence autism families. *American Journal of Medical Genetics, 74*, 398–411.

Portfors, C. V. (2007). Types and functions of ultrasonic vocalizations in laboratory rats and mice. *Journal of the American Association for Laboratory Animal Science: JAALAS, 46*, 28–34.

Provenzano, G., Zunino, G., Genovesi, S., Sgadó, P., & Bozzi, Y. (2012). Mutant mouse models of autism spectrum disorders. *Disease Markers, 33*, 225–239.

Purcell, S. M., Moran, J. L., Fromer, M., Ruderfer, D., Solovieff, N., Roussos, P., et al. (2014). A polygenic burden of rare disruptive mutations in schizophrenia. *Nature, 506*(7487), 185–190. doi:10.1038/nature12975.

Purcell, S. M., Wray, N. R., Stone, J. L., Visscher, P. M., O'Donovan, M. C., Sullivan, P. F., et al. (2009). Common polygenic variation contributes to risk of schizophrenia and bipolar disorder. *Nature, 460*, 748–752.

Ramirez-Solis, R., Liu, P., & Bradley, A. (1995). Chromosome engineering in mice. *Nature, 378*(6558), 720–724.

Reser, J. E. (2014). Solitary mammals provide an animal model for autism spectrum disorders. *Journal of Comparative Psychology, 128*(1), 99–113. doi:10.1037/a0034519.

Ritsner, M. S., & Gottesman, I. I. (2009). Where do we stand in the quest for neuropsychiatric biomarkers and endophenotypes and what next? In M. S. Ritsner (Ed.), *The handbook of neuropsychiatric biomarkers, endophenotypes, and genes* (pp. 3–21). Dordrecht: Springer.

Sanders, S. J., Murtha, M. T., Gupta, A. R., Murdoch, J. D., Raubeson, M. J., Willsey, A. J., et al. (2012). De novo mutations revealed by whole-exome sequencing are strongly associated with autism. *Nature, 485*, 1–6.

Sandin, S., Lichtenstein, P., Kuja-Halkola, R., Larsson, H., Hultman, C. M., & Reichenberg, A. (2014). The familial risk of autism. *JAMA, 311*, 1770.

Santos, M., Silva-Fernandes, A., Oliveira, P., Sousa, N., & Maciel, P. (2007). Evidence for abnormal early development in a mouse model of Rett syndrome. *Genes, Brain, and Behavior, 6*, 277–286.

Sarna, J., Dyck, R., & Whishaw, I. (2000). The Dalila effect: C57BL6 mice barber whiskers by plucking. *Behavioural Brain Research, 108*, 39–45.

Sato, A., Kasai, S., Kobayashi, T., Takamatsu, Y., Hino, O., Ikeda, K., et al. (2012a). Rapamycin reverses impaired social interaction in mouse models of tuberous sclerosis complex. *Nature Communications, 3*, 1292.

Sato, D., Lionel, A. C., Leblond, C. S., Prasad, A., Pinto, D., Walker, S., et al. (2012b). SHANK1 deletions in males with autism spectrum disorder. *American Journal of Human Genetics, 90*, 879–887.

Scattoni, M. L. M., Crawley, J., & Ricceri, L. (2009). Ultrasonic vocalizations: A tool for behavioural phenotyping of mouse models of neurodevelopmental disorders. *Neuroscience and Biobehavioral Reviews, 33*, 508–515.

Scattoni, M., McFarlane, H., Zhodzishsky, V., Caldwell, H. K., Young, W. S., Ricceri, L., et al. (2008). Reduced ultrasonic vocalizations in vasopressin 1b knockout mice. *Behavioural Brain Research, 187*, 371–378.

Scearce-Levie, K., Roberson, E. D., Gerstein, H., Cholfin, J. A., Mandiyan, V. S., Shah, N. M., et al. (2008). Abnormal social behaviors in mice lacking Fgf17. *Genes, Brain, and Behavior, 7*, 344–354.

Sebat, J., Lakshmi, B., Malhotra, D., Troge, J., Lese-martin, C., Walsh, T., et al. (2007). Strong association of de novo copy number mutations with autism. *Science, 316*, 18–22.

Seeman, T. E. (1996). Social ties and health: The benefits of social integration. *Annals of Epidemiology, 6*, 442–451.

Šestan, N., State, M., & Sestan, N. (2013). The emerging biology of autism spectrum disorders. *Science, 337*, 1301–1303.

Shahbazian, M., Young, J., Yuva-Paylor, L., Spencer, C., Antalffy, B., Noebels, J., et al. (2002). Mice with truncated MeCP2 recapitulate many Rett syndrome features and display hyperacetylation of histone H3. *Neuron, 35*, 243–254.

Shinawi, M., Schaaf, C. P., Bhatt, S. S., Xia, Z., Patel, A., Cheung, S. W., et al. (2009). A small recurrent deletion within 15q13.3 is associated with a range of neurodevelopmental phenotypes. *Nature Genetics, 41*, 1269–1271.

Shriberg, L. D., Paul, R., McSweeny, J. L., Klin, A. M., Cohen, D. J., & Volkmar, F. R. (2001). Speech and prosody characteristics of adolescents and adults with high-functioning autism and Asperger syndrome. *Journal of Speech, Language, and Hearing Research: JSLHR, 44*, 1097–1115.

Shu, W., Cho, J. Y., Jiang, Y., Zhang, M., Weisz, D., Elder, G. A., et al. (2005). Altered ultrasonic vocalization in mice with a disruption in the Foxp2 gene. *Proceedings of the National Academy of Sciences of the United States of America, 102*, 9643–9648.

Silverman, J. L., Tolu, S. S., Barkan, C. L., & Crawley, J. N. (2010a). Repetitive self-grooming behavior in the BTBR mouse model of autism is blocked by the mGluR5 antagonist MPEP. *Neuropsychopharmacology, 35*, 976–989.

Silverman, J., Yang, M., Lord, C., & Crawley, J. N. (2010b). Behavioural phenotyping assays for mouse models of autism. *Nature Reviews Neuroscience, 11*, 490–502.

Simmons, J. M., & Quinn, K. J. (2014). The NIMH Research Domain Criteria (RDoC) Project: Implications for genetics research. *Mammalian Genome, 25*, 23–31.

Singh, S. K., & Eroglu, C. (2013). Neuroligins provide molecular links between syndromic and nonsyndromic autism. *Science Signaling, 6*, re4.

Spencer, C. M., Graham, D. F., Yuva-Paylor, L. A., Nelson, D. L., & Paylor, R. (2008). Social behavior in Fmr1 knockout mice carrying a human FMR1 transgene. *Behavioral Neuroscience, 122*, 710–715.

Spinka, M., Newberry, R., & Bekoff, M. (2001). Mammalian play: Training for the unexpected. *Quarterly Review of Biology, 76*, 141–168.

Stanley, D. A., & Adolphs, R. (2013). Toward a neural basis for social behavior. *Neuron, 80*, 816–826.

Stark, K. L., Xu, B., Bagchi, A., Lai, W.-S., Liu, H., Hsu, R., et al. (2008). Altered brain microRNA biogenesis contributes to phenotypic deficits in a 22q11-deletion mouse model. *Nature Genetics, 40*, 751–760.

Stefansson, H., Ophoff, R. A., Steinberg, S., Andreassen, O. A., Cichon, S., Rujescu, D., et al. (2009). Common variants conferring risk of schizophrenia. *Nature, 460*, 744–747.

Stefansson, H., Rujescu, D., Cichon, S., Pietiläinen, O. P. H., Ingason, A., Steinberg, S., et al. (2008). Large recurrent microdeletions associated with schizophrenia. *Nature, 455*, 232–236.

Strauss, J. S., & Carpenter, W. T. (1974). The Prediction of Outcome in schizophrenia. II. Relationships between predictor and outcome variables. *JAMA, 31*, 37–42.

Sucksmith, E., Allison, C., Baron-Cohen, S., Chakrabarti, B., & Hoekstra, R. a. (2013). Empathy and emotion recognition in people with autism, first-degree relatives, and controls. *Neuropsychologia, 51*, 98–105.

Sullivan, P. F., Daly, M. J., & O'Donovan, M. (2012a). Genetic architectures of psychiatric disorders: The emerging picture and its implications. *Nature Reviews. Genetics, 13*, 537–551.

Sullivan, P. F., Magnusson, C., Reichenberg, A., Boman, M., Dalman, C., Davidson, M., et al. (2012b). Family history of schizophrenia and bipolar disorder as risk factors for autism. *Archives of General Psychiatry, 69*, 1099–1103.

Szatmari, P., Paterson, A. D., Zwaigenbaum, L., Roberts, W., Brian, J., Liu, X.-Q., et al. (2007). Mapping autism risk loci using genetic linkage and chromosomal rearrangements. *Nature Genetics, 39*, 319–328.

Tarbox, S., & Pogue-Geile, M. (2011). A multivariate perspective on schizotypy and familial association with schizophrenia: A review. *Clinical Psychology Review, 31*, 1169–1182.

Thurmond, J. B. (1975). Technique for producing and measuring territorial aggression using laboratory mice. *Physiology & Behavior, 14*, 879–881.

Tseng, K., Chambers, R., & Lipska, B. (2009). The neonatal ventral hippocampal lesion as a heuristic neurodevelopmental model of schizophrenia. *Behavioural Brain Research, 204*, 295–305.

Wang, X., McCoy, P. A., Rodriguiz, R. M., Pan, Y., Je, H. S., Roberts, A. C., et al. (2011). Synaptic dysfunction and abnormal behaviors in mice lacking major isoforms of Shank3. *Human Molecular Genetics, 20*, 3093–3108.

Watanabe, T., Abe, O., Kuwabara, H., Yahata, N., Takano, Y., Iwashiro, N., et al. (2014). Mitigation of sociocommunicational deficits of autism through oxytocin-induced recovery of medial prefrontal activity. *JAMA Psychiatry, 71*(2), 166–175. doi:10.1001/jamapsychiatry.2013.3181.

White, T., & Gottesman, I. (2012). Brain connectivity and gyrification as endophenotypes for schizophrenia: Weight of the evidence. *Current Topics in Medicinal Chemistry, 12*, 2393–2403.

Winslow, J. T. (2009). Mood and anxiety related phenotypes in mice. *Neuromethods, 42*, 67–84.

Wöhr, M., & Schwarting, R. K. W. (2010). Rodent ultrasonic communication and its relevance for models of neuropsychiatric disorders. *e-Neuroforum, 1*, 71–80.

Yan, Q. J., Rammal, M., Tranfaglia, M., & Bauchwitz, R. P. (2005). Suppression of two major Fragile X Syndrome mouse model phenotypes by the mGluR5 antagonist MPEP. *Neuropharmacology, 49*, 1053–1066.

Yang, M., Bozdagi, O., Scattoni, M. L., Wöhr, M., Roullet, F. I., Katz, A. M., et al. (2012). Reduced excitatory neurotransmission and mild autism-relevant phenotypes in adolescent Shank3 null mutant mice. *The Journal of Neuroscience, 32*, 6525–6541.

Yang, M., & Crawley, J. (2009). Simple behavioral assessment of mouse olfaction. *Current Protocols in Neuroscience, 8*, 1–14.

Yang, M., Scattoni, M. L., Zhodzishsky, V., Chen, T., Caldwell, H., Young, W. S., et al. (2007). Social approach behaviors are similar on conventional versus reverse lighting cycles, and in replications across cohorts, in BTBR T+tf/J, C57BL/6J, and vasopressin receptor 1B mutant mice. *Frontiers in Behavioral Neuroscience, 1*, 1–9.

Young, P. (1959). The role of affective processes in learning and motivation. *Psychological Review, 66*, 104–125.

Young, D. M., Schenk, A. K., Yang, S.-B., Jan, Y. N., & Jan, L. Y. (2010). Altered ultrasonic vocalizations in a tuberous sclerosis mouse model of autism. *Proceedings of the National Academy of Sciences of the United States of America, 107*, 11074–11079.

Zwaigenbaum, L., Bryson, S., Rogers, T., Roberts, W., Brian, J., & Szatmari, P. (2005). Behavioral manifestations of autism in the first year of life. *International Journal of Developmental Neuroscience, 23*, 143–152.

Chapter 9
Rodent Models of Autism, Epigenetics, and the Inescapable Problem of Animal Constraint

Garet P. Lahvis

Current Issues and Topics

Behavioral Features of Autism and Mouse Models

Autism

To identify the deficits associated with autism, clinicians use several assessment tools. Children with ASD are often identified during routine pediatric visits, flagged by parental concerns and by screening tests, such as the M-CHAT (Modified Checklist for Autism in Toddlers) (Robins et al. 2014). A positive screen triggers semi-structured clinical observations of the child and interviews with the primary caregiver. The Autism Diagnostic Observation Schedule (ADOS) provides social structure for a clinician to evaluate a child (Lord et al. 2001). The Autism Diagnostic Interview-Revised (ADI-R) guides a clinical interview with the primary caregiver to obtain a developmental and social history of a child (Lord et al. 1994).

Autism is called a *spectrum* disorder because individuals with an autism diagnosis often differ substantially from one another. Multiple behavioral phenotypes contribute to the social deficits in autism, so each child with an autism diagnosis has a relatively unique set of social capabilities and challenges. An autistic child may be challenged with use of nonverbal behaviors to convey needs, lack an ability to develop peer relationships, be unable to share enjoyment with others, or lack a capacity for empathy. Autism can include a broad range of social deficits, including

G.P. Lahvis (✉)
Department of Behavioral Neuroscience, Oregon Health & Sciences University,
3181 SW Sam Jackson Park Road, L-470, Portland, OR 97212, USA
e-mail: lahvisg@ohsu.edu; lahvisg@gmail.com

J.C. Gewirtz, Y.-K. Kim (eds.), *Animal Models of Behavior Genetics*,
Advances in Behavior Genetics, DOI 10.1007/978-1-4939-3777-6_9

avoidance of gaze toward eyes (Baron-Cohen et al. 1995) and challenges with imitation (Rogers 1999), joint attention (Dawson et al. 2004; Gernsbacher et al. 2008; Toth et al. 2006), coherence and inference within story telling (Colle et al. 2008; Diehl et al. 2006), and synchronized use of gestures with verbal communication (de Marchena and Eigsti 2010). Individuals with autism often lack typical abilities to perceive emotional information from facial expressions (Ashwin et al. 2007; Clark et al. 2008; Humphreys et al. 2007), body gestures (Hadjikhani et al. 2009; Hubert et al. 2007a, b), or via the melody, prosody, or intonation of speech (Diehl 2008; McCann and Peppe 2003; McCann et al. 2007; Shriberg et al. 2001; Van Santen et al. 2010). Autistic children and adults can also have difficulties interpreting emotional cues when they involve multisensory integration (Magnee et al. 2008).

The ability to perceive and feel social cues, facial expressions and gestures expressed in combination with both the semantic content *and* vocal intonations of speech, can be essential for an experience of empathy, "the generation of an affective state more appropriate to the situation of another compared to one's own" (Hoffman 1975). Capacities for empathy can be diminished among children with autism (Bacon et al. 1998; Baron-Cohen and Wheelwright 2004). In some children, these social deficits suggest a lack in social motivation, an inability to derive pleasure from social interactions (Dawson et al. 2005).

Repetition and circumscribed interests also vary among children on the spectrum. One child might flap his arms and jump in a circular movement in anticipation of response from a caregiver while another child might rock all afternoon behind a window with arms tucked into his chest. Similarly, a child might have interests in particular kinds of objects, such as trains, cars, or volcanoes, another child might focus facts, like president's birthdays, and others still may be captivated by ideas, like how the seasons correspond to the earth's rotation (Lam et al. 2008; Lewis and Bodfish 1998; Sasson et al. 2008).

Mouse Models

Like the varied social phenotypes of autism, a spectrum of mouse social phenotypes can be evaluated, including measures of social approach, social motivation, empathy, imitation, observational learning, and communication. Rapid screens are used to identify mouse strains with deficits in social approach while more involved behavioral studies elucidate a broader spectrum of deficits in social neurodevelopment.

Screens used to assess social phenotypes of mouse models of autism typically measure the extent that the test mouse, or "subject," approaches a stimulus mouse, sometimes called the "object," which freely moves about the subject's cage (Panksepp et al. 2007). The object is either fastened to a tether or held inside a small wire cage (Nadler et al. 2004). Variations of *social approach tests* yield different measures, including social investigation (Panksepp et al. 2007), social recognition (Young 2002; Winslow and Insel 2002), approach toward social novelty (Moy et al. 2007) and reciprocal social interactions (Hamilton et al. 2011). Protocols require minimal equipment, often just a cage and a video camera, and are best served by a reverse light cycle, allowing mice to interact socially during their active nighttime hours. Currently, the most rapid automated measures restrict the stimulus mouse to the inside of a wire cage (Nadler et al. 2004), which removes physical contact from the social interaction. In so doing, the test likely measures the degree a mouse seeks social interaction, not whether it derives a rewarding experience embodied by tactile and reciprocal social interactions. Indeed, all social approach tests are difficult to interpret at a theoretical level, as they are unable to dissociate mental states, such as social anxiety, from a lack of social interest.

Failure to develop peer relationships: Some individuals with autism can find it difficult to develop peer relationships, not because they may hesitate to engage in a social interaction, but rather that they do not derive pleasure from social interactions, they experience diminished social motivation (Chevallier et al. 2012).

We can assess social motivation in rats and mice by employing the *social conditioned place preference (social CPP) test*, a measure of whether a mouse values social housing with its peers over housing in social isolation. Prior to testing, subjects are "conditioned," shuttled back and forth between two housing conditions containing two novel beddings, along with associated PVC couplers, either threaded or smooth. Once each day, mice are transferred from one housing condition to the other and each condition is paired with a particular social context, either the presence or absence of other mice. After conditioning and ending with a day of social isolation, subjects are tested in a social CPP test to determine whether they prefer one of the two conditioned beddings.

In the context of a social CPP, if subjects derive a positive affective experience from social housing, this association would present as a preference for the bedding environment with which the positive social affect is paired. Most mouse strains tested prefer conditions associated with social housing to conditions resembling the environments where they were housed alone. In this social CPP test, a mouse model of autism (BALB/cJ) feels indifferent to housing conditioned by access to cagemates versus by confinement alone (Panksepp and Lahvis 2007). Using social CPP, we can ask whether a vigorous test response reflects seeking of a social reward or avoidance of the environment paired with social isolation (Panksepp and Lahvis 2007). Social CPP experiments can be used to gain further insight to mouse subjective experience. For example, we can ask whether indifference to conditions associated with social access is intrinsic to a given strain, in this case BALB/cJ mice, by testing social CPP after conditioning BALB/cJ mice with other more sociable strains. Conversely, we can ask whether social CPP indifference reflects unrewarding social interactions with BALB/cJ mice, testing how a more gregarious mouse strain responds to conditioning with BALB/cJ mice. The social CPP test, like other forms of CPP, is well suited for neuroscience. Recent studies have revealed that social reward is mediated by coordinated activities of oxytocin and serotonin in the nucleus accumbens (Dolen et al. 2013).

Deficits in imitation and emulation: Autism often involves difficulties with imitation (Rogers 1999), abilities that can be assessed in mice via *observational learning tests*, and measures of how well a mouse can learn from others how to manipulate an object within its environment. Mice can learn from their conspecifics how to swing a door open (Collins 1988). In this experiment, subjects observe a demonstrator open a door, hinged at the top and directed to swing in only one direction (right or left), to obtain a food reward. Observers open the door after fewer trials than naïve controls, more likely to swing open the door in the same direction as the demonstrator (Collins 1988). Emulation is a process whereby the result of an observed action is learned through social observation, which then engenders trial-and-error learning. Experiments with puzzle-boxes uncouple imitation from emulation by requiring learners reproduce the goal of the behavior (Galef 2013). Mice can learn from others how to manipulate a puzzle box (Carlier and Jamon 2006; Valsecchi et al. 2002) but to our knowledge these tests have not been used to assess mouse models of autism, despite their relevance.

Empathy: Empathy refers to a psychological phenomenon in which an individual can experience vicariously the affective state of another (Preston and de Waal 2002). Debates persist about the role of cognitive function in empathy, but at its core, empathy is "the generation of an affective state more appropriate to the situation of another compared to one's own" (Hoffman 1975). In popular usage, empathy often adopts acts of compassion or, at least, more nuanced and decidedly pro-social responses, such as a behavioral display of sorrow for the victim. While compassion, which has been demonstrated in rats (Bartal et al. 2011, 2014), may require empathy, the reciprocal is not true; empathy requires only a change in a subject's feelings in response to those of another individual. Levels of empathic ability range from shared emotions to highly abstract forms, such as affective responses to knowledge of distant suffering (e.g., victims of a tsunami).

One form of empathy, the ability of a mouse strain to experience vicarious fear learning, can be measured via a cue-conditioned fear paradigm in which object mice are conditioned next to a subject mouse and we ask whether the subject learns the fear contingency from the object undergoing conditioning (Chen et al. 2009). Subject and object mice are presented with a neutral stimulus, such as a tone (conditioned stimulus, CS). For the object mice, the CS is forward paired with an aversive stimulus, such as an electrical shock (unconditioned stimulus, UCS). With pairings of the CS-UCS, the object mice emit an audible vocalization, a squeak, when the USC is delivered and gradually develop a fear response (they freeze) in response to CS (the tone only) in anticipation of the UCS. Subject mice learn the USC-CS contingency by observing object mice experience the tone-shock contingency. When subsequently placed inside the shock chamber, subjects freeze in response to hearing the tone-only. Gregarious C57Bl/6J mice learn that the tone predicts a distressful stimulus by hearing the vocalizations (squeaks) of the demonstrator mice; they acquire a fear response to the CS conditioned by playbacks of recorded vocalizations (Chen et al. 2009). A mouse model of autism, the BALB/cJ strain, fails to acquire a vicarious fear response.

The vicarious fear learning experiment has been conducted in multiple forms for both mice and rats (Panksepp and Lahvis 2011). Social experience prior to conditioning affects the magnitude of the vicarious fear response, which is increased by the subject mouse's familiarity with the object mouse (Jeon and Shin 2011), diminished when subject mice are housed with non-fearful mice (Guzman et al. 2009), and impaired by extended periods of social isolation (Yusufishaq and Rosenkranz 2013). Consistent with human studies of empathy, subjects are preexposed to a non-contingent experience with the aversive stimulus (the shock) (Atsak et al. 2011; Chen et al. 2009; Kim et al. 2010; Sanders et al. 2013), allowing them to learn the contingency, though vicarious fear learning has also been shown without experience with the shock (Jeon et al. 2010). Elegant work shows that if the activity of the auditory cortex in mouse subjects is silenced so that they cannot hear their own vocalizations in response to the shock, they fail to learn fear when hearing the object mouse conditioned in the shock chamber (Kim et al. 2010).

Like social CPP, the neurobiology of vicarious fear learning can be closely studied. Neural substrates for vicarious fear learning include major roles for dopamine and serotonin (Kim et al. 2014) within the anterior cingulate cortex (ACC) (Jeon et al. 2010) and appear to be lateralized, involving the ACC of the right but not the left hemisphere (Kim et al. 2012). Functional MRI experiments find ACC activity associated with analogous human experiments (Singer et al. 2004, 2006).

A test that may assess both observational learning and empathy involves subjects that learn to imitate objects that are evading biting flies. With time, the object mice learn to bury themselves beneath the bedding to escape (Kavaliers et al. 2001). After observing object mice, subjects then display more rapid burying behaviors in the presence of flies, suggesting that they learned by observation how to manipulate their environment. Subjects also expressed decreased pain sensitivity (Martin et al. 2015), suggesting they sensed the pain experienced by the objects.

Responses of peers: Mice avoid other mice that carry parasites (Kavaliers et al. 2005a, b; Kavaliers and Colwell 1995), and this sensitivity requires at least one functional oxytocin allele (Kavaliers et al. 2005a), a gene that plays a critical role in social interaction (Donaldson and Young 2008). Similarly, social behaviors of test subjects can be assessed according to how reference mice respond to them. Social behaviors of reference mice can distinguish mouse strains expressing autism-like social behaviors from their controls (Benson et al. 2013; Shah et al. 2013; Shahbazian et al. 2002).

Communication: Rodent communication is conveyed through vocalizations, scent, gesture, physical contact, and perhaps facial expressions. Rodents emit vocalizations audible to humans and at frequencies beyond the upper limit of human hearing, ultrasonic vocalizations (USVs), above 20 kHz. An infant mouse emits wriggling calls (~35 kHz) to solicit maternal care within the nest (D'Amato et al. 2005; Ehret and Bernecker 1986) and distress calls (~90 kHz) to solicit retrieval when displaced from the nest (Branchi et al. 1998). Infant USVs and these maternal responses are sensitive to opiates, dopamine, and serotonin (D'Amato et al. 2005; Dastur et al. 1999; Moles et al. 2004), suggesting that changes in affective state are required to emit and respond to these USVs. Adolescent mice emit USVs that correlate with social approach behaviors (Panksepp et al. 2007). Laboratory rats emit 22 and 50 kHz USVs to signal negative and positive affect (Burgdorf et al. 2005; Carden et al. 1993; Harmon et al. 2008) but mouse USVs are not so clearly dissociable (Lahvis et al. 2010; Scattoni et al. 2009). A recent study shows that females emit a 38 kHz call to coordinate paternal pup retrieval (Liu et al. 2013).

Scent marking can be used to express territorial dominance among males (Hurst 1990) and to attract females (Roberts et al. 2014; Thonhauser et al. 2013). Odors from urine can also signal alarm in mice. When exposed to two cotton balls, one with urine from a calm mouse, the other with urine from a mouse exposed to a single shock, the test mouse avoids the urine of a distressed conspecific (Rottman and Snowdon 1972). Specific volatile molecules released by alarmed mice evoke increased systemic corticosterone levels in subjects (Brechbühl et al. 2013). Mouse models of autism can express diminished levels of scent marking (Wöhr et al. 2011a, b) and their pups can lack the typical preference for maternal scent (Kane et al. 2012).

Physical movements can also serve as a form of communication. For instance, when mice are injected with different concentrations of an irritant, such as acetic acid, into their peritoneum, they exhibit levels of writhing behavior that correspond to the levels of irritant. When placed next to one another, their writhing responses become more similar (Langford et al. 2006). When the paws of mice are injected with different concentrations of formalin, their paw licking behaviors also converge when they are placed next to each other (Langford et al. 2006). Emotional contagion, like vicarious fear learning, is sensitive to familiarity (Martin et al. 2015).

More recent studies show that rodent facial expressions can indicate both positive and negative affective states (Kelley and Berridge 2002) and pain in mice (Langford et al. 2010). The role of facial expressions in the communication of emotion remains unclear.

Genetic Risk Factors for Autism

Human Genetics

Twin studies provide ample evidence of autism heritability. The concordance rate (likelihood that the sibling of an autistic child will be diagnosed with autism) can approach 90 % among monozygotic twins, threefold higher than concordance among dizygotic twins and non-twin siblings (Bailey et al. 1995; Hallmayer et al. 2011; Ozonoff et al. 2011; Rosenberg et al. 2009; Steffenburg et al. 1989). High heritability suggests a prominent role for genetic risk factors, though other potential mechanisms of inheritance include maternal transmission of epigenetic marks, maternal RNA, nutrients, bacteria, and antibodies (Meaney 2010). ASD is also nearly five times more prevalent among boys than girls (Baio 2012), implicating a role for sex-based differences in physiology.

Studies have found associations between ASD and genetic polymorphisms (Abrahams and Geschwind 2008; Berg and Geschwind 2012), with several polymorphisms encoding gene products involved in synapse formation or function (for reviews see Bourgeron 2009; Geschwind and Levitt 2007), findings that fit with the concept that autism can be the behavioral consequence of poor connectivity between different brain regions (Belmonte et al. 2004; Courchesne and Pierce 2005;

Geschwind and Levitt 2007). ASD can also be associated with several copy number variants (CNV) (Sebat et al. 2007), and *de novo* mutations (O'Roak et al. 2011; Sanders et al. 2012).

For any given study, allelic variants explain only a small fraction of diagnosed individuals with autism and the study results are rarely replicated (Abrahams and Geschwind 2008; Berg and Geschwind 2012), even in large scale genome-wide studies (Autism Genome Project Consortium et al. 2007; Barrett et al. 1999; Bourgeron 2012; Kumar and Christian 2009; Liu et al. 2008).

Mouse Genetics

Laboratory mice are the most commonly used mammals in genetics research. Alleles can be targeted for unconditional deletion, mutation, or replacement, or be conditionally deleted under specific pharmacological, anatomical, or development conditions. Molecular strategies for targeting alleles and elucidating gene function in mice are evolving at a rapid pace (Josh Huang and Zeng 2013; Pollock et al. 2014).

Several mouse strains harbor mutations associated with autism. Best characterized are mouse models for fragile-X syndrome. Fragile-X results from an extended CPG repeat that disrupts expression of the fragile-X allele (FMR1), unequivocal for more than 80 % of children with the disorder. Approximately one-quarter of males with fragile-X syndrome are diagnosed with autism (Budimirovic and Kaufmann 2011; Hagerman et al. 1986; Harris et al. 2008). Like humans with fragile-X syndrome, gene-targeted mice lacking the FMR1 allele are prone to seizures, express cognitive deficits, and display atypical social behaviors (Brodkin 2008; Comery et al. 1997; Kooy et al. 1996; McNaughton et al. 2008; Spencer et al. 2008). Abnormal behaviors of fmr1 knockout mice can be responsive to experimental drug treatments (Hagerman et al. 2009; Michalon et al. 2012).

A multitude of mouse models have targeted genetic variants associated with ASD. Typical assessments include measures of social approach and rates of USV emission. These mouse models of autism have been extensively reviewed elsewhere (Abrahams and Geschwind 2008; Andres 2002; Bishop and Lahvis 2011; Crawley 2012; Lahvis and Black 2011; Stamou et al. 2013).

Environmental Risk Factors for Autism

Environmental Risk Factors and Autism

The high heritability of autism, particularly among monozygotic twins, suggests genetic risks, but does not discount the impact of environmental risks (Hallmayer et al. 2011), which may be considerable (Landrigan 2010). Twins are exposed to similar environmental factors through shared breast milk, baby food, and home environment, and they share similar genetic susceptibilities to their deleterious effects (Hallmayer et al. 2011). Environmental risk factors in ASD are often suspected to be chemicals in the human environment. Over 85,000 different chemicals are used in commerce today (Grandjean and Landrigan 2006; Rao et al. 2014) and vast numbers of other chemicals are generated as industrial waste (Chang et al. 2011; Milestone et al. 2013; Sun et al. 2011). Many of these chemicals accumulate in humans (Gascon et al. 2011; Hertz-Picciotto et al. 2008; Inoue et al. 2004; Rauh et al. 2012; Roze et al. 2009). During pregnancy, women can offload part of their lifetime accumulation of these chemical across the placenta and via breast milk to the infant (Hooper et al. 2007; Ingelido et al. 2007; von Ehrenstein et al. 2009; Yu et al. 1991), all within a brief and environmentally sensitive window of development (Hooper et al. 2007; Ingelido et al. 2007; von Ehrenstein et al. 2009).

The prevalence of autism varies across geographic regions, further suggesting a causal role for environmental exposure (Bertrand et al. 2001; Halladay et al. 2009). Geographical hotspots for autism include Brick Township, New Jersey, a highly industrialized region (Bertrand et al. 2001), homes adjacent to highways (Volk et al. 2011), and areas exposed to high levels of pesticides (Roberts et al. 2007). The high incidence of ASD in South Korea (Kim et al. 2011) may suggest pollutant exposure risks. South Korea is highly industrial and the population consumes high levels of seafood that accumulate pollutants (Kannan et al. 2004; So et al. 2004) and natural toxins (Choi et al. 2009).

Autistic children can be sensitive to chemical exposure (Ashwood et al. 2009; Geier et al. 2009a, b; James et al. 2009). ASD is associated with diminished activity of phenolsulfotransferase (PST), which plays a role in phenol detoxification and excretion (Alberti et al. 1999; McFadden 1996). Statistical associations between chemical exposures and autism susceptibility can be difficult with enzymes like PST, because its activity is also sensitive to variations in natural chemicals found in vegetables (Chi-Tai and Gow-Chin 2003; Yeh and Yen 2005), so children likely have different risks that vary with a multitude of dietary factors.

The aryl hydrocarbon receptor (AHR) also exemplifies how brain development can be sensitive to the interplay between genetic and environmental risk factors. AHR is a nuclear transcription factor that moderates expression of genes that mediate PCB, dioxin and polyaromatic hydrocarbon depuration and toxicity (Burbach et al. 1992) and it plays a role in brain development and neonatal vascular pruning (Chun-Hua et al. 2009; Lahvis et al. 2000). Sensitivities to the toxic effects of dioxins and poly-aromatic hydrocarbons vary with polymorphisms of the AHR allele (Brokken et al. 2013; Harper et al. 2002; Hung et al. 2013; Kerley-Hamilton et al. 2012; Poland et al. 1994) and, like PST, AHR activity is also influenced by various foods including potatoes, cruciferous vegetables, and grapefruit juice (Bjeldanes et al. 1991; De Waard et al. 2008). AHR exemplifies how a gene can integrate genetic inheritance with environmental exposure to modify gene expression during development. Ligands of the AHR have been implicated as risk factors for autism, diminishing social conditioned place preference (Cromwell et al. 2007).

Environment Risk Factors and Mouse Sociality

Exposures to synthetic chemicals can impair neurological function, particularly during early development (Grandjean and Landrigan 2006; Roze et al. 2009). Synaptic function can be compromised by early exposures to pesticides (Miranda-Contreras et al. 2005; Slotkin and Seidler 2007) and industrial products, such as perfluorooctane sulfonate (PFOS) (Wang et al. 2010; Zeng et al. 2011). Children genetically susceptible to poor synaptic function might be sensitive to chemicals that impair synapse formation and, as a result, be at high risk for ASD (Halladay et al. 2009). For reviews of animal models used to define chemical risks for ASD, see Halladay et al. (2009) and Pessah et al. (2008).

Experiments with laboratory animals typically involve exposures to individual chemicals (Halladay et al. 2009) and thus they do not fully represent the complex exposure history that occurs during development. Human cord blood and milk contain complex chemical mixtures that exert a variety of confounding, additive, and synergistic effects. While testing individual chemicals provides important information about discrete mechanisms of action, it has limited value predicting effects of numerous exposures, varied in dose and chemistry and punctuated over the long span of human development. Further, not all environmental risk factors are chemical. Other suspected factors include parental age at conception (Durkin et al. 2008), prenatal stress (Kinney et al. 2008), altered circadian cycles and melatonin deficiency (Glickman 2010), and gastrointestinal disorders (Grabrucker 2012). When considered as a whole, possible combinations of genetic and environmental risk factors for autism susceptibility appear endless. An alternative approach for identifying mechanisms that integrate genetic and environmental risks may be necessary to direct us toward effective pharmaceutical and environmental remedies.

Epigenetics

Epigenetics and autism: By segregating genetic and environmental influences, we ignore the biological processes that integrate them. Biologists have long known that genetic and environmental factors influence phenotypic traits through alterations in RNA expression and splicing, protein folding, and posttranslational modifications within single cells. Brain development and behavior represent integrations of these basic cellular processes. Epigenetics mechanisms regulate gene expression, the transcriptional processes influenced by both genomic and extra-genomic factors, integrating variations in promoter and enhancer sequence with factors in its molecular surroundings (Bohacek et al. 2013; Meaney 2010). Transcription activity occurs in regions of DNA where chromatin is open and becomes silent or suppressed where DNA is bound to histones. Methylations influence where histones and DNA interact by affecting binding of methyl CpG binding proteins. Methylation promotes histone-DNA coupling whereas demethylation disaggregates these molecules. Other epigenetic mechanisms, such as acetylation and deacetylation, also control where histones and DNA interact.

These epigenetic processes respond to a variety of extracellular factors, including imprinting, sexual differentiation, chemical exposures, and social upbringing (Bourgeron 2012; Kigar and Auger 2013; Meaney 2010). For example, increased levels of systemic testosterone alter DNA methylation patterns in neurons of the neonatal amygdala and the ensuing changes in gene expression have long-term effects on juvenile play behavior (Meaney and McEwen 1986).

Abnormal epigenetic processes are associated with ASD. Evidence comes from comparisons of gene expression patterns among the few monozygotic twins *discordant* for the ASD diagnosis. Two genes, B-cell lymphoma 2 (BCL-2) and retinoic acid-related orphan receptor alpha (RORA), can be differentially methylated in lymphoblastoid cell lines for identical twins not sharing the diagnosis (Nguyen et al. 2010). Levels of RORA and BCL-2 protein in slices of cerebellum and frontal cortex from ASD subjects were found to be diminshed relative to typical controls (Nguyen et al. 2010), consistent with these altered methylation patterns.

Not all genes are equally responsive to epigenetic regulation, but several genetic polymorphisms statistically associated with ASD are responsive to epigenetic influences, which may explain why only some studies find these genetic variants to be associated with ASD. Genes include a GABAA receptor gene (GABRB3) and ubiquitin protein ligase E3A (UBE3A) (Buxbaum et al. 2002; Curran et al. 2006; Grafodatskaya et al. 2010; Hogart et al. 2007), reelin (RELN) (Fatemi et al. 2001; Persico et al. 2006), the oxytocin receptor (OXTR) (Gregory et al. 2009; Kumsta et al. 2013), forkhead box P1 (foxP1) (Chien et al. 2013), and engrailed-2 (EN-2) (James et al. 2013). Indeed, comparisons of lymphocytes and lymphoblastoid cell lines from children with autism versus typical controls indicate large-scale differences in gene expression (Enstrom et al. 2009; Gregg et al. 2008; Hu et al. 2006, 2009). These results are consistent with findings implicating epigenetic dysregulation in other forms of mental illness (Bourgeron 2012; Grafodatskaya et al. 2010; Kigar and Auger 2013; Miyake et al. 2012; Schanen 2006; Zhang and Meaney 2010).

Studies of the X-linked methyl CpG binding protein 2 (MeCP2) also suggest that abnormal epigenetic processes pose a risk for autism. MeCP2 binds DNA and upon interacting with neighboring proteins, controls expression of thousands of genes (Chahrour et al. 2008) by serving as a transcriptional activator and under some conditions as a repressor (Chao and Zoghbi 2012). MeCP2 can also alter gene expression by posttranslation modification (Bellini et al. 2014; Guy et al. 2011).

Loss-of-function mutations and deletions of MeCP2 are associated with 80 % of individuals with Rett syndrome (Cheadle et al. 2000), a disorder that shares clinical similarities with autism. Susceptibility appears to be sensitive to both over- or under- expression of MECP2 (Chao and Zoghbi 2012; Peters et al. 2014). Genetic variants of MeCP2 have been associated with ASD (Carney et al. 2003; Ramocki et al. 2009). Reduced expression of MeCP2 expression was found in postmortem samples of ASD frontal cortex (Nagarajan et al. 2006). Variations in the levels of MeCP2 function, via duplications (Samaco et al. 2012) or loss of MeCP2 expression, can diminish expression of UBE3A and GABAA (Kurian et al. 2007; Samaco et al. 2005), critical molecules involved in brain function.

ASD is also associated with chromosomal instability in regions susceptible to epigenetic influence, including chromosome 15q11-q13, where maternal duplication can be associated with 2% of ASD (Cook et al. 1997; Hogart et al. 2007). Interestingly, GABAA receptor genes, encoded within this region, are biallelically expressed (not imprinted) in typical human cortex, whereas in ASD and Retts their expression can be monoallelic or highly skewed, suggesting epigenetic dysregulation (Hogart et al. 2007). Possible association between specific pollutant exposures, 15q11-13 duplication and risk for ASD (Mitchell et al. 2012) also implicates epigenetic dysregulation in this region.

Indeed, altered methylation patterns result from exposures to chemical risk factors for autism (Cheslack-Postava et al. 2013; Halladay et al. 2009; Hu et al. 2012; Winneke 2011), including metals (Martinez-Zamudio and Ha 2011), industrial organic chemicals (Bollati et al. 2007; Jiang et al. 2014), endocrine-disrupting chemicals such as bisphenol A (Kundakovic et al. 2014; Zhang et al. 2012b), persistent organic pollutants (Lind et al. 2013), and air particulates (Baccarelli and Bollati 2009; Tarantini et al. 2008; Yauk et al. 2008). Drug exposures associated with ASD can also change epigenetic processes during fetal development; best known are diethylstilbestrol, which targets hormone response elements (Li et al. 2014), and valproic acid that inhibits histone deacetylases (Phiel et al. 2001) to substantially alter gene expression (Fukuchi et al. 2009) mediated by MeCP2 (Kim et al. 2016). For expanded discussions of the role of epigenetics in autism and other mental disorders, see Bourgeron (2012), Dudley et al. (2011), Grafodatskaya et al. (2010), Hall and Kelley (2013), Kigar and Auger (2013), Mbadiwe and Millis (2014), Miyake et al. (2012), and Schanen (2006).

Epigenetics and mouse sociality: Mouse models of autism include strains with particular genetic backgrounds, targeted alleles, drug exposures, or brain lesions, yet the most profound variations in mouse social behaviors result from sex differences. Social behaviors of male and female mice appear indistinguishable at weaning and are highly sensitive to differences in genetic background. These patterns of social behavior diverge during adolescence. Relative to females, males begin to express greater levels of novelty seeking (Palanza et al. 2001) and social approach behaviors (Panksepp et al. 2007). Among rats and ground squirrels, juvenile males display more vigorous rough-and-tumble play (Nunes et al. 1999; Olioff and Stewart 1978; Pellis 2002). With reproductive maturity, disparities in play behavior concede to more stereotypical behaviors of male territorial defense and female nurturing (Lynn and Brown 2009).

Sex-related behaviors respond to differences in reproductive hormones even during fetal (Vom Saal and Bronson 1978) and neonatal (Meaney and Stewart 1981) development. For instance, castration diminishes rough-and-tumble play of adolescent males (Pellis et al. 1994), whereas females engage in more robust levels of play behavior if exposed at birth to testosterone (Meaney and Stewart 1981) or testosterone implants within amygdalae (Meaney and McEwen 1986), brain regions that plays an essential role in rough-and-tumble play (Meaney et al. 1981). Metabolites of testosterone bind DNA response elements to modify gene expression and alter brain development and behavior (Auger et al. 2011; Morris et al. 2004).

Hormones influence sex-related brain development by altering gene expression patterns through binding proteins like MeCP2 (Kurian et al. 2007, 2008). Social behaviors are sensitive to targeting of MeCP2. Mice harboring genetic duplication of MeCP2 express diminished social approach (Samaco et al. 2005, 2012). Mouse strains with Mecp2 deficiency emit fewer neonatal distress calls when removed from the nest (Picker et al. 2006) and express reduced levels of male social play behavior (Kurian et al. 2008; Pearson et al. 2012), increased male aggression (Fyffe et al. 2008), abnormal durations of conflict behavior (Moretti et al. 2005), heightened interactions with unfamiliar mice (Schaevitz et al. 2010), and abnormal nesting behavior (Moretti et al. 2005). Relative to controls, mice with MeCP2 truncation are also less attractive to wild types (Shahbazian et al. 2002). Mice engineered to lack methyl-CpG binding protein 1 (MBD1), also a methyl-CpG binding protein, engage in diminished social approach (Allan et al. 2008). Again, phenotypes vary from study to study, in part due to differences in background strain (Tantra et al. 2014), testing procedures, design of genetic construct, pathogen exposure, and food source (Lahvis and Bradfield 1998).

Genes under MeCP2 control can play prominent roles in sexually differentiated behavior. For instance, injection of small interfering (si) RNA into the amygdala of neonatal rats reduces male expression of vasopressin and diminishes adolescent social play behaviors of males but not females (Forbes-Lorman et al. 2012; Kurian et al. 2008). Oxytocin and vasopressin are neurotransmitters that influence sex-specific social behaviors, including pair bonding in prairie voles (McGraw and Young 2010; Young et al. 2001). For reviews of how epigenetic mechanisms contribute to differences in sex-linked social behavior, see Auger et al. (2011), Kigar and Auger (2013), and Matsuda (2014).

Within the home cage environment, variations in maternal care can alter these sex-dependent developmental trajectories (Parent and Meaney 2008). For instance, variable levels of a dam's licking behavior directly affect DNA methylation of the ERα promoter region of her pups, altering their expression of juvenile play behaviors and adult aggression (Auger and Olesen 2009; Champagne et al. 2006; Matsuda 2014).

Home Cage Environment

Following early work by Donald Hebb and the Forgays (Forgays and Forgays 1952), Mark Rosenzweig's laboratory asked how differences in environmental complexity influence brain anatomy, physiology, and behavior. Experimental rats were housed inside complex environments consisting of a large cage with wooden "toys" and given access to a small wooden maze used for a nesting box. Each day investigators replaced pairs of toys and allowed experimental rats 30-min access to a Hebb-Williams maze that was modified daily into a new configuration. Control rats were housed alone inside "impoverished" environments approximating the dimensions of our current standard rat cages. Housing with environmental complexity resulted in increased cholinesterase activity within cortical versus subcortical regions of the rat brain (Krech et al. 1960) reflecting in part increased cortical weight (Bennett et al. 1969; Rosenzweig et al. 1962). If environmental complexity were provided to adult rats *after* being raised in impoverished environments, the treatment effectively reversed the developmental effects of impoverishment (Rosenzweig et al. 1962).

Environmental complexity in these early experiments provided rats with both *spatial* heterogeneity (toys and mazes) and *temporal* variety, including daily changes in toy availability and maze configuration. Now assessments of the environmental influences on brain development are typically limited to variations in spatial complexity. Increases in environmental "enrichment" mean that rodents are housed in larger home cages containing objects such as toys, tunnels, nesting material, and running wheels. In some experiments, food locations are changed to offer some level of temporal variation (Van Praag et al. 2000). In addition to "enrichment," environmental complexity can also be introduced by raising offspring in communal nests (Branchi et al. 2006; Gracceva et al. 2009) or via brief exposures to novel physical environments (Tang et al. 2006).

Surprisingly, even modest increases in spatial heterogeneity promote regional increases in the densities of neurons, oligodendrocytes, astrocytes (Szeligo and Leblond 1977), capillaries (Black et al. 1987), and synapses (Globus et al. 1973; Greenough et al. 1985#7308). Histological responses to environmental "enrichment" varied with brain region (Greenough et al. 1973) and according to neuronal cell population (Juraska et al. 1980; Kempermann et al. 1997). Adult exposures to environmental "enrichment" affected how neurotransmitters respond to acute stress, altering acetylcholine levels, but not dopamine levels, in the PFC (Del Arco et al. 2007). Brain morphology also changed. With adult mice exposed to 3 weeks of maze reconfiguration, in vivo MRI scans demonstrated rapid and astounding neural plasticity with growth in regions specifically associated with spatial memory, navigation, and sensorimotor experience (Scholz et al. 2015).

Since changes in brain morphology and composition were regional, they suggested responses to specific neurological stimuli, discrete aspects of mental experience, rather than global responses to changes in systemic physiology or nutrition (Greenough et al. 1973). Indeed modest "enrichment" relative to a standard shoebox cage alters the relative proportionality of brain regions and cell types in those regions, conferring not just quantitative changes in brain function, *but qualitative redesign.*

How might qualitative redesign influence rodent experiments? Housing of gene-targeted mice in "enriched" environments can entirely reverse the biological responses of a targeted allele housed in impoverished conditions. For instance, when raised in standard cages, phospholipase C-β1 (CPLC-β1) knockout mice express deficits in sensorimotor gating and hyperactivity, behavioral markers for schizophrenia. When these mice are weaned into "enriched" cages, these phenotypes are completely reversed (McOmish et al. 2008). Environmental "enrichment" rescues the memory deficit of mice lacking polysialytransferase ST8SiaIV (Zerwas et al. 2016), mitigates the neuronal deficits of a transgenic model of Huntington's disease (Lazic et al. 2006), reverses both β-amyloid deposition and memory deficits in transgenic mice expressing high levels of amyloid precursor protein (Maesako et al. 2012), reverses coordination deficits in a mouse model of Rett syndrome (Kondo et al. 2008), and ameliorates the deficits in a model of Alzheimer's disease (Jankowsky et al. 2005). In this context, it is not surprising that environmental "enrichment" also rescues phenotypes relevant to autism, as recently shown in the BTBR mouse model of autism (Reynolds et al. 2013). Enrichment is ironically suggested as a "therapy" to reverse autism-relevant behavioral phenotypes in mice, because social enrichment is a useful treatment for autism.

Environmental "enrichment" confers resiliency and recovery to rodents with brain lesions and seizures (Koh et al. 2007; Passineau et al. 2001; Will et al. 2004) and reverses the deleterious neurological effects of chemical and drug exposures. For instance, in rodents exposed to 1-methyl-4-phenyl-1,2,3,6-tetrahydropyridine (MPTP) (a model of Parkinsonism), environmental "enrichment" protects dopaminergic neurons, increases expression of neurotrophic factors (Faherty et al. 2005), promotes neuronal and behavioral recovery from MPTP exposure (Goldberg et al. 2011) and reverses the effects of MPTP on motor impairment (Goldberg et al. 2012). "Enrichment" also reverses lead-induced deficits in spatial memory and N-methyl-d-aspartate receptor subunit 1 (NR1) expression (Guilarte et al. 2003), as well as sevoflurane-induced memory impairments (Shih et al. 2012). Rearing rats in an "enriched" environments renders them less susceptible to drugs of abuse (Stairs and Bardo 2009), ameliorating self-administration of amphetamines (Bardo et al. 2001), reversing vulnerability to cocaine reward and addiction (Nader et al. 2014; Solinas et al. 2008) (Zakharova et al. 2009), and providing resiliency to the rewarding effects of heroin (El Rawas et al. 2009; Solinas et al. 2010).

Environmental "enrichment" also reverses nearly every behavioral phenotype expressed by rats exposed *in utero* to valproic acid, the most commonly used environmental model of autism in rodents. Phenotypic reversal includes pain sensitivity, prepulse inhibition, repetitive and exploratory behaviors, expression of anxiety, as well as several measures of social behavior: duration of social interaction, number of successful pins in juvenile rat rough-and-tumble play, and both latency and duration of social exploration during adulthood (Schneider et al. 2006). Indeed, the BTBR mouse strain shows marked improvements in sociability simply by being housed with more sociable mouse strains (Yang et al. 2011).

Communal nesting also has a marked influence on adolescent and adult social behavior. For instance, when pups are reared in a communal nest of three mothers, rather than the standard single mother, they receive greater levels of maternal care (Branchi et al. 2006; Curley et al. 2009) and as adults are more responsive to social context (Branchi and Alleva 2006) and more socially interactive (D'Andrea et al. 2007). Females raised in a communal nest are less responsive to social novelty (Gracceva et al. 2009) and males more readily adopt their positions within an adult social hierarchy (Branchi et al. 2006). Such social behavioral changes are associated with altered expression of nerve growth factor (NGF) and brain-derived neurotrophic factor (BDNF) in the hippocampus and hypothalamus (Branchi et al. 2006).

Spatial environmental "enrichment" can change adolescent social behaviors such as play, engendering recovery from previous environmental influences that depress play behavior, such as prenatal stress (Morley-Fletcher et al. 2003). Exposure to spatial enrichment can reverse the affects of prenatal and maternal care on rodent spatial learning and object recognition (Bredy et al. 2003, 2004), responses to novelty (Francis et al. 2002), as well as detrimental effects of postnatal handling on a two-way active avoidance task (Escorihuela et al. 1994) and depressive-like symptoms in rats that are exposed to a learned helplessness paradigm (Richter et al. 2013), interestingly without reversing depression of corticotrophin-releasing factor (Francis et al. 2002).

Mice housed in "enriched" environments express lower levels of anxiety-like behaviors associated with diminished expression of CRF receptor 1 in the basolateral amygdala (Sztainberg et al. 2010) and increases expression of α-amino-3-hydroxy-5-methyl-4-isoxazolepropionic acid (AMPA)-type receptors GluR2 and GluR4 (Naka et al. 2005) and BDNF (Chourbaji et al. 2012) in the hippocampus. Decreased expression of NGF and increased BDNF expression in the hypothalamus accompanies an affiliative sociality that results in dominance (Pietropaolo et al. 2004).

Many genes critical for brain development and function are differentially expressed soon after exposure to "enrichment" and for weeks thereafter (Rampon et al. 2000; Thiriet et al. 2008; Mychasiuk et al. 2012). Such alterations in gene expression are likely mediated by a variety of processes. For instance, expression of BDNF in the nucleus accumbens responds to cyclic adenosine monophosphate response element binding (CREB) activity (Green et al. 2010; Nader et al. 2014).

Environmental "enrichment" has mixed results in reversing the behavioral impairments in mice lacking a functional MeCP2 allele (Kondo et al. 2008; Lonetti et al. 2010; Nag et al. 2009). As noted earlier, MeCP2 plays a pivotal role in moderating the effects of environmental "enrichment" on gene expression, so its targeted deletion might be expected to impair "enrichment's" ability to reverse the deleterious effects. For reviews that further address the influences of environmental "enrichment," particularly maternal care, on epigenetic processes, social behavior, and models of mental illness, see Branchi (2009), Green et al. (2010), Meaney and Szyf (2005), Nithianantharajah and Hannan (2006), Van Praag et al. (2000), and Williams et al. (2001).

Conclusions and Future Directions

ASD is primarily a deficit of social functioning, yet each child with an ASD diagnosis displays a unique composition of social disabilities. Such variation reflects the broad array of social abilities necessary for healthy interactions within a social context. Healthy social functioning requires social motivation, self-regulation of social emotions, perspective taking, shared enjoyment, joint attention, emotional empathy, imitation, gestural perception, and receptive and expressive prosody. Individuals with autism can express any combination of these impairments and many of these social abilities can be robustly assessed in mouse models. Though substantial progress has been achieved in developing mouse models of ASD and the behavioral tests necessary to test them, the cage environments used to house laboratory rodents impede our efforts to discover the genetic and environmental risk factors contributing to ASD and its underlying neurobiological mechanisms.

Common sense would argue that a rodent's response to a social stimulus is influenced by prior encounters with social refuge, the agency to make decisions that afford consequences, and variations in affective experience. Though small cages are convenient for husbandry, mice bred under these highly restrictive conditions lack access to a vast complexity of olfactory, tactile, auditory, and visual stimuli typical of uncaged mammalian experience (Latham and Mason 2004).

Rearing of mice in shoebox cages is not analogous to raising children in impoverished homes and community settings. Standard laboratory cages deny mice access to all social refuge, which would be commonplace in natural settings and in human experience, even in poor communities. The unfortunate use of the word "enrichment" has led some to conclude that it serves as a form of "therapy." "Enriched" caging offers marginal increases in spatial complexity and no improvements to temporal variation. As mentioned by others, the term environmental "enrichment" should be considered only relative to the standard laboratory housing conditions (Van Praag et al. 2000), *not to what a rodent would naturally experience*. Thus, "enriched" caging is not therapy but a housing condition that ameliorates the deleterious artifacts emerging from an extreme poverty of environmental stimulation. Widespread concern persists that animal studies inflate biological effects that cannot be translated for human trials (Hackam and Redelmeier 2006; Pound et al. 2004; Tsilidis et al. 2013). As shown in this review, inflated effects of genetic and environmental manipulations result, in part, from rearing laboratory animals under their standard housing conditions.

Social behaviors are propelled by motivations to nurture young, to play, to acquire mating opportunities and defend territories. Specific circuits underlie these motivated behaviors. Under natural conditions, comforting and aversive experiences continually fluctuate with variations in food and shelter availability and quality, variations in temperature and humidity, continual changes in the composition of the social community and its ever-changing hierarchy, and ongoing oscillations between bounty and the catastrophe of floods, drought, and temperature extremes. Such temporal variety, dramatic relative to the cage environment, activates both cognitive and affective circuits, strengthening neurobiological substrates that support relationships between affective experience and the cognitive decisions that maximize comfort and minimize adversity. Just as increased spatial enrichment develops more articulated circuitry in brain regions that play a role in spatial navigation (Duffy et al. 2001; Williams et al. 2001), temporal variations in rewards and punishments and opportunities for rodents to find solutions that optimize their comfort will help sculpt circuitry to navigate social rewards and social distress. Were these circuits optimized by natural or semi-natural experiences, they might enhance abilities of a rodent to perceive and respond to yet more complex social situations, such as vicarious fear learning, situations that require substantial integration between cognitive and affective perception (Jurado-Parras et al. 2012) and sensitive to social conditions during development (Yusufishaq and Rosenkranz 2013; Panksepp and Lahvis 2016). Without temporal variations in developmental exposure to reward and punishment, it remains unclear the extent to which a test subject's response to a specific reward is relevant.

In this regard, newborns who fail to receive adequate light stimulation resulting, for example, from cataracts later removed, can experience life-long blindness or impairments in visual abilities (Ostrovsky et al. 2006). Normal visual development *requires* normal visual stimuli. Studies of rodent and primate social and emotional regulation should not assume that their cognitive and affective circuitry is representative of typical development when underlying circuits go unchallenged inside a cage throughout development and adult life.

A common argument for standard laboratory animal caging is that even under conditions of environmental impoverishment, inferences regarding underlying neurological processes are valid. Such an argument likely holds for the relationships between cells or molecules in the brain but not for systems-level indices of health and illness, such as variations in social behavior. As described above and long noted in the literature (Juraska et al. 1983), a brain that develops inside a shoebox cage is qualitatively different from that of an "enriched" cage, evidenced by regional changes in brain excitability, by proliferation of selected populations of neurons but not others, by changing ratios of nonneuronal cell populations, and by variations in the expression of some genes and neurotransmitters but not others. These findings suggest that the brain of a mouse in a shoebox cage does not resemble the brain it is intended to model.

With few challenges inside a cage, rodent behaviors are susceptible to nuanced differences between laboratories, such as audible and ultrasonic noise or subtle differences in colony odors. This may explain why strain comparisons of mouse behaviors are difficult to reproduce across laboratories when standard shoebox-sized cages are used (Crabbe et al. 1999; Richter et al. 2010) and more reliable when housing conditions are varied (Richter et al. 2009, 2010; Würbel 2002). This may also explain why many animal research studies fail to yield meaningful results when translated to human trials (Hackam and Redelmeier 2006; Pound et al. 2004; Tsilidis et al. 2013).

What are the next steps? Traditional approaches to environmental "enrichment" offer multiple objects, tunnels, a running wheel, and large enough to support communal nesting. From the perspective of an individual rodent, this kind of "enrichment" might allow for moments of social refuge to experience objects, tunnels and running wheels when wandering from the other members of the cage. Such refuge would allow for some level of temporal variation in affective experience, at least with respect to social reward. These ideas are perhaps suggested by studies showing that "enrichment" modifies affective circuitry and consistently diminishes rodent susceptibility to drugs of abuse (Bardo et al. 2001; Gipson et al. 2011; Stairs and Bardo 2009). Opportunities for a rodent to act on its environment and have agency over its own experience, if only ambling among objects or within the more enlightened visible burrow systems (Blanchard et al. 2001), may augment development of reward circuitry but these designed increases in spatial heterogeneity still fail to provide temporal variety and opportunities for agency relevant to human experience and the natural experience of a rodent.

Early work of Rosenweig (Krech et al. 1960) included home cage environments that offered daily access to a reconfiguration of a Hebb maze and to daily changes of objects within a cage, thereby providing both spatial *and* temporal variation. To further augment the temporal variation in these environments, rodents might be provided with alternate access to highly and less palatable foods, thereby familiarizing them with varied affective experiences. Aversive conditions would be valuable. For instance, addition of predator scent has a profound effect on social behavior of juvenile rodents (Kendig et al. 2011). Housing that includes spatial heterogeneity and temporal variation in the access to rewards would at least bear resemblance to the temporal and spatial variety naturally experienced by a rodent or a human.

Another possibility for housing includes semi-natural environments akin to the complex environments designed by Peter Crowcroft in the 1950s: large unheated enclosures for mice containing hundreds of objects and housing structures, and capable of containing complex social hierarchies that involve multiple territories, litters, and social arrangements (Crowcroft 1966; Crowcroft and Rowe 1963). These expanded environments, shared sometimes by hundreds of mice, provided rodents with opportunities for decision-making in the context of mixed food sources, social refuge, and established dominance hierarchies. The visible burrow system (Pobbe et al. 2010) might also be more fully utilized for housing, adapted to include temporal variations in environmental stimuli. Under these expanded husbandry conditions, individual mice could be remotely monitored for many of the behaviors we now measure in standard behavioral tests run in the laboratory. Under any semi-naturalistic conditions, mice could be collected to determine how specific genetic or environmental treatments influence brain anatomy, physiology, and underlying patterns of gene expression. Further, with modern options for biotelemetry, investigators can now remotely record EEG as well as action and field potentials (Vyssotski et al. 2006), hemodynamics (Bussey et al. 2014), and neuronal activity (Zhang et al. 2012). Radiotelemetry could be employed with an array of current molecular tools to gain insight to brain function in environments more closely analogous to human conditions.

Sentinel organisms are animals that live freely in the wild and they might also be employed to elucidate causes of autism. Sentinels have been used successfully for environmental assessments of chemical exposures and effects (Basu et al. 2007; Fox 2001; Ramalhinho et al. 2012). By combining high-throughput sequencing analysis with rodents at study sites targeted for their potential relevance to human health (adjacency to highways, crops sprayed with pesticides, complex industrial effluents), we might fair a better chance of identifying gene-by-environment interactions that contribute to ASD.

In highlighting social neuroscience, we urge for reconsideration of how neuroscience and studies of other systems affected by caging, such as immunology (Beura et al. 2016), conceives of the environmental experiences of laboratory animals. In most current studies, rodent and primate models of mental disability compare experimental and control groups under unnatural levels of environmental deprivation. Even so-called "enriched" cages, with its various objects added, provide animal subjects with a remarkable paucity of spatial and temporal variation relative to what they would encounter in natural or semi-natural environments. Environmental "enrichment" is not *enrichment*, rather a nuanced step from an abject poverty of experience. The rodent brain, like the primate brain, has an evolved capacity for active decision-making through an extraordinarily complex array of natural contingencies, rewards and punishments. By contrast, with few options for choice, relative to wild conspecifics of the humans they presumably model, laboratory test subjects mature with incomplete substrates for brain development and function. Provisioning rodents with expanded spatial, temporal, and affective enrichment, combined with new technologies in animal monitoring, will help us predict the outcomes of clinical trials, yielding novel opportunities for progress in behavioral neuroscience.

References

Abrahams, B. S., & Geschwind, D. H. (2008). Advances in autism genetics: On the threshold of a new neurobiology. *Nature Reviews Genetics, 9*(5), 341–355.

Alberti, A., Pirrone, P., Elia, M., Waring, R. H., & Romano, C. (1999). Sulphation deficit in "low-functioning" autistic children: A pilot study. *Biological Psychiatry, 46*(3), 420–424.

Allan, A. M., Liang, X., Luo, Y., Pak, C., Li, X., Szulwach, K. E., et al. (2008). The loss of methyl-CpG binding protein 1 leads to autism-like behavioral deficits. *Human Molecular Genetics, 17*(13), 2047–2057.

Andres, C. (2002). Molecular genetics and animal models in autistic disorder. *Brain Research Bulletin, 57*(1), 109–119.

Ashwin, C., Baron-Cohen, S., Wheelwright, S., O'Riordan, M., & Bullmore, E. T. (2007). Differential activation of the amygdala and the 'social brain' during fearful face-processing in Asperger Syndrome. *Neuropsychologia, 45*(1), 2–14.

Ashwood, P., Schauer, J., Isaac, N., Pessah, I. N., & Van de Water, J. (2009). Preliminary evidence of the in vitro effects of BDE-47 on innate immune responses in children with autism spectrum disorders. *Journal of Neuroimmunology, 208*, 130–135.

Atsak, P., Orre, M., Bakker, P., Cerliani, L., Roozendaal, B., Gazzola, V., et al. (2011). Experience modulates vicarious freezing in rats: A model for empathy. *Stress and Cognition, 6*, 17.

Auger, A. P., Jessen, H. M., & Edelmann, M. N. (2011). Epigenetic organization of brain sex differences and juvenile social play behavior. *Hormones and Behavior, 59*(3), 358–363.

Auger, A. P., & Olesen, K. M. (2009). Brain sex differences and the organisation of juvenile social play behaviour. *Journal of Neuroendocrinology, 21*(6), 519–525.

Autism Genome Project Consortium, Szatmari, P., Paterson, A. D., Zwaigenbaum, L., Roberts, W., Brian, J., et al. (2007). Mapping autism risk loci using genetic linkage and chromosomal rearrangements. .[erratum appears in Nat Genet. 2007 Oct;39(10):1285 Note: Meyer, Kacie J [added]; Koop, Frederike [corrected to Koop, Frederieke]; Langemeijer, Marjolijn [corrected to Langemeijer, Marjolein]; Hijimans, Channa [corrected to Hijmans, Channa]]. *Nature Genetics, 39*(3), 319–328.

Baccarelli, A., & Bollati, V. (2009). Epigenetics and environmental chemicals. *Current Opinion in Pediatrics, 21*(2), 243.

Bacon, A., Fein, D., Morris, R., Waterhouse, L., & Allen, D. (1998). The responses of autistic children to the distress of others. *Journal of Autism and Developmental Disorders, 28*(2), 129–142.

Bailey, A., Le Couteur, A., Gottesman, I., Bolton, P., Simonoff, E., Yuzda, E., et al. (1995). Autism as a strongly genetic disorder: Evidence from a British twin study. *Psychological Medicine, 25*(01), 63–77.

Baio, J. (2012). Prevalence of Autism Spectrum Disorders—Autism and Developmental Disabilities Monitoring Network, 14 Sites, United States, 2008. *MMWR Surveillance Summaries, 61*(3), 1–19.

Bardo, M., Klebaur, J., Valone, J., & Deaton, C. (2001). Environmental enrichment decreases intravenous self-administration of amphetamine in female and male rats. *Psychopharmacology, 155*(3), 278–284.

Baron-Cohen, S., Campbell, R., Karmiloff-Smith, A., Grant, J., & Walker, J. (1995). Are children with autism blind to the mentalistic significance of the eyes? *British Journal of Developmental Psychology, 13*(4), 379–398. doi:10.1111/j.2044-835X.1995.tb00687.x.

Baron-Cohen, S., & Wheelwright, S. (2004). The empathy quotient: An investigation of adults with Asperger syndrome or high functioning autism, and normal sex differences. *Journal of Autism and Developmental Disorders, 34*(2), 163–175. doi:10.1023/b:jadd.0000022607.19833.00.

Barrett, S., Beck, J. C., Bernier, R., Bisson, E., Braun, T. A., Casavant, T. L., et al. (1999). An autosomal genomic screen for autism. Collaborative linkage study of autism. *American Journal of Medical Genetics, 88*(6), 609–615.

Bartal, I. B.-A., Decety, J., & Mason, P. (2011). Empathy and pro-social behavior in rats. *Science, 334*(6061), 1427–1430.

Bartal, I. B.-A., Rodgers, D. A., Sarria, M. S. B., Decety, J., & Mason, P. (2014). Pro-social behavior in rats is modulated by social experience. *Elife, 3*, e01385.

Basu, N., Scheuhammer, A. M., Bursian, S. J., Elliott, J., Rouvinen-Watt, K., & Chan, H. M. (2007). Mink as a sentinel species in environmental health. *Environmental Research, 103*(1), 130–144.

Bellini, E., Pavesi, G., Barbiero, I., Bergo, A., Chandola, C., Nawaz, M. S., et al. (2014). MeCP2 post-translational modifications: A mechanism to control its involvement in synaptic plasticity and homeostasis? *Frontiers in Cellular Neuroscience, 8*, 236.

Belmonte, M. K., Allen, G., Beckel-Mitchener, A., Boulanger, L. M., Carper, R. A., & Webb, S. J. (2004). Autism and abnormal development of brain connectivity. *Journal of Neuroscience, 24*(42), 9228–9231.

Bennett, E. L., Rosenzweig, M. R., & Diamond, M. C. (1969). Rat brain: Effects of environmental enrichment on wet and dry weights. *Science, 163*(3869), 825–826.

Benson, A. D., Burket, J. A., & Deutsch, S. I. (2013). Balb/c mice treated with d-cycloserine arouse increased social interest in conspecifics. *Brain Research Bulletin, 99*, 95–99.

Berg, J. M., & Geschwind, D. H. (2012). Autism genetics: Searching for specificity and convergence. *Genome Biology, 13*(7), 247. doi:10.1186/gb-2012-1113-1187-1247.

Bertrand, J., Mars, A., Boyle, C., Bove, F., Yeargin-Allsopp, M., & Decoufle, P. (2001). Prevalence of autism in a United States population: The Brick Township, New Jersey, investigation. *Pediatrics, 108*(5), 1155–1161.

Beura, L. K., Hamilton, S. E., Bi, K., Schenkel, J. M., Odumade, O. A., Casey, K. A., et al. (2016). Normalizing the environment recapitulates adult human immune traits in laboratory mice. *Nature, 532*, 512–516.

Bishop, S. L., & Lahvis, G. P. (2011). The autism diagnosis in translation: Shared affect in children and mouse models of ASD. *Autism Research, 4*(5), 317–335.

Bjeldanes, L. F., Kim, J. Y., Grose, K. R., Bartholomew, J. C., & Bradfield, C. A. (1991). Aromatic hydrocarbon responsiveness-receptor agonists generated from indole-3-carbinol in vitro and in vivo: Comparisons with 2,3,7,8-tetrachlorodibenzo-p-dioxin. *Proceedings of the National Academy of Sciences, 88*(21), 9543–9547.

Black, J. E., Sirevaag, A. M., & Greenough, W. T. (1987). Complex experience promotes capillary formation in young rat visual cortex. *Neuroscience Letters, 83*(3), 351–355.

Blanchard, R. J., Dulloog, L., Markham, C., Nishimura, O., Compton, J. N., Jun, A., et al. (2001). Sexual and aggressive interactions in a visible burrow system with provisioned burrows. *Physiology & Behavior, 72*(1), 245–254.

Bohacek, J., Gapp, K., Saab, B. J., & Mansuy, I. M. (2013). Transgenerational epigenetic effects on brain functions. *Biological Psychiatry, 73*(4), 313–320.

Bollati, V., Baccarelli, A., Hou, L., Bonzini, M., Fustinoni, S., Cavallo, D., et al. (2007). Changes in DNA methylation patterns in subjects exposed to low-dose benzene. *Cancer Research, 67*(3), 876–880.

Bourgeron, T. (2009). A synaptic trek to autism. *Current Opinion in Neurobiology, 19*(2), 231–234. doi:10.1016/j.conb.2009.06.003.

Bourgeron, T. (2012). Genetics and epigenetics of autism spectrum disorders. In C. Paolo Sassone & C. Yves (Eds.), *Epigenetics, brain and behavior* (pp. 105–132). Berlin: Springer.

Branchi, I. (2009). The mouse communal nest: Investigating the epigenetic influences of the early social environment on brain and behavior development. *Neuroscience & Biobehavioral Reviews, 33*(4), 551–559. doi:10.1016/j.neubiorev.2008.03.011.

Branchi, I., & Alleva, E. (2006). Communal nesting, an early social enrichment, increases the adult anxiety-like response and shapes the role of social context in modulating the emotional behavior. *Behavioural Brain Research, 172*(2), 299–306.

Branchi, I., D'Andrea, I., Fiore, M., Di Fausto, V., Aloe, L., & Alleva, E. (2006). Early social enrichment shapes social behavior and nerve growth factor and brain-derived neurotrophic factor levels in the adult mouse brain. *Biological Psychiatry, 60*(7), 690–696. doi:10.1016/j.biopsych.2006.01.005.

Branchi, I., Santucci, D., Vitale, A., & Alleva, E. (1998). Ultrasonic vocalizations by infant laboratory mice: A preliminary spectrographic characterization under different conditions. *Developmental Psychobiology, 33*(3), 249–256.

Brechbühl, J., Moine, F., Klaey, M., Nenniger-Tosato, M., Hurni, N., Sporkert, F., et al. (2013). Mouse alarm pheromone shares structural similarity with predator scents. *Proceedings of the National Academy of Sciences, 110*(12), 4762–4767.

Bredy, T., Humpartzoomian, R., Cain, D., & Meaney, M. (2003). Partial reversal of the effect of maternal care on cognitive function through environmental enrichment. *Neuroscience, 118*(2), 571–576.

Bredy, T. W., Zhang, T. Y., Grant, R. J., Diorio, J., & Meaney, M. J. (2004). Peripubertal environmental enrichment reverses the effects of maternal care on hippocampal development and glutamate receptor subunit expression. *European Journal of Neuroscience, 20*(5), 1355–1362.

Brodkin, E. S. (2008). Social behavior phenotypes in fragile X syndrome, autism, and the Fmr1 knockout mouse: Theoretical comment on McNaughton et al. (2008). *Behavioral Neuroscience, 122*(2), 483–489.

Brokken, L. J. S., Lundberg-Giwercman, Y., Rajpert-De Meyts, E., Eberhard, J., Stahl, O., Cohn-Cedermark, G., et al. (2013). Association between polymorphisms in the aryl hydrocarbon receptor repressor gene and disseminated testicular germ cell cancer. *Frontiers in Endocrinology, 4*, 4.

Budimirovic, D. B., & Kaufmann, W. E. (2011). What can we learn about autism from studying fragile X syndrome? *Developmental Neuroscience, 33*(5), 379.

Burbach, K. M., Poland, A., & Bradfield, C. A. (1992). Cloning of the Ah-receptor cDNA reveals a distinctive ligand-activated transcription factor. *Proceedings of the National Academy of Sciences, 89*(17), 8185–8189.

Burgdorf, J., Panksepp, J., Brudzynski, S. M., Kroes, R., & Moskal, J. R. (2005). Breeding for 50-kHz positive affective vocalization in rats. *Behavior Genetics, 35*(1), 67–72.

Bussey, C., de Leeuw, A., Cook, R., Ashley, Z., Schofield, J., & Lamberts, R. (2014). Dual implantation of a radio-telemeter and vascular access port allows repeated hemodynamic and pharmacological measures in conscious lean and obese rats. *Laboratory Animals, 48*(3), 250–260.

Buxbaum, J., Silverman, J., Smith, C., Greenberg, D., Kilifarski, M., Reichert, J., et al. (2002). Association between a GABRB3 polymorphism and autism. *Molecular Psychiatry, 7*(3), 311–316.

Carden, S. E., Bortot, A. T., & Hofer, M. A. (1993). Ultrasonic vocalizations are elicited from rat pups in the home cage by pentylenetetrazol and U50,488, but not naltrexone. *Behavioral Neuroscience, 107*(5), 851–859.

Carlier, P., & Jamon, M. (2006). Observational learning in C57BL/6j mice. *Behavioural Brain Research, 174*(1), 125–131.
Carney, R. M., Wolpert, C. M., Ravan, S. A., Shahbazian, M., Ashley-Koch, A., Cuccaro, M. L., et al. (2003). Identification of MeCP2 mutations in a series of females with autistic disorder. *Pediatric Neurology, 28*(3), 205–211.
Chahrour, M., Jung, S. Y., Shaw, C., Zhou, X., Wong, S. T., Qin, J., et al. (2008). MeCP2, a key contributor to neurological disease, activates and represses transcription. *Science, 320*(5880), 1224–1229.
Champagne, F. A., Weaver, I. C., Diorio, J., Dymov, S., Szyf, M., & Meaney, M. J. (2006). Maternal care associated with methylation of the estrogen receptor-α1b promoter and estrogen receptor-α expression in the medial preoptic area of female offspring. *Endocrinology, 147*(6), 2909–2915.
Chang, H., Wan, Y., Wu, S., Fan, Z., & Hu, J. (2011). Occurrence of androgens and progestogens in wastewater treatment plants and receiving river waters. *Comparison to Estrogens, 45*(2), 732–740.
Chao, H.-T., & Zoghbi, H. Y. (2012). MeCP2: Only 100 % will do. *Nature Neuroscience, 15*(2), 176–177.
Cheadle, J. P., Gill, H., Fleming, N., Maynard, J., Kerr, A., Leonard, H., et al. (2000). Long-read sequence analysis of the MECP2 gene in Rett syndrome patients: Correlation of disease severity with mutation type and location. *Human Molecular Genetics, 9*(7), 1119–1129.
Chen, Q., Panksepp, J. B., & Lahvis, G. P. (2009). Empathy is moderated by genetic background in mice. *PLoS One [Electronic Resource], 4*(2), e4387.
Cheslack-Postava, K., Rantakokko, P. V., Hinkka-Yli-Salomäki, S., Surcel, H.-M., McKeague, I. W., Kiviranta, H. A., et al. (2013). Maternal serum persistent organic pollutants in the Finnish Prenatal Study of Autism: A pilot study. *Neurotoxicology and Teratology, 38*, 1–5.
Chevallier, C., Kohls, G., Troiani, V., Brodkin, E. S., & Schultz, R. T. (2012). The social motivation theory of autism. *Trends in Cognitive Sciences, 16*(4), 231–239.
Chien, W.-H., Gau, S. S.-F., Chen, C.-H., Tsai, W.-C., Wu, Y.-Y., Chen, P.-H., et al. (2013). Increased gene expression of FOXP1 in patients with autism spectrum disorders. *Molecular Autism, 4*(1), 23.
Chi-Tai, Y., & Gow-Chin, Y. (2003). Effects of phenolic acids on human phenolsulfotransferases in relation to their antioxidant activity. *Journal of Agricultural and Food Chemistry, 26*(51), 1474–1479.
Choi, K. D., Lee, J. S., Lee, J. O., Oh, K. S., & Shin, I. S. (2009). Investigation of domoic acid in shellfish collected from Korean fish retail outlets. *Food Science and Biotechnology, 18*(4), 842–848.
Chourbaji, S., Hörtnagl, H., Molteni, R., Riva, M., Gass, P., & Hellweg, R. (2012). The impact of environmental enrichment on sex-specific neurochemical circuitries–effects on brain-derived neurotrophic factor and the serotonergic system. *Neuroscience, 220*, 267–276.
Chun-Hua, L., Chien-Chang, C., Chih-Ming, C., Chen-Yu, W., Chia-Chi, H., Julia, Y. C., et al. (2009). Knockdown of the aryl hydrocarbon receptor attenuates excitotoxicity and enhances NMDA-induced BDNF expression in cortical neurons. *Journal of Neurochemistry, 111*(3), 777–789.
Clark, T. F., Winkielman, P., & McIntosh, D. N. (2008). Autism and the extraction of emotion from briefly presented facial expressions: Stumbling at the first step of empathy. *Emotion, 8*(6), 803–809.
Colle, L., Baron-Cohen, S., Wheelwright, S., & Lely, H. J. (2008). Narrative discourse in adults with high-functioning autism or Asperger syndrome. *Journal of Autism and Developmental Disorders, 38*(1), 28–40. doi:10.1007/s10803-007-0357-5.
Collins, R. L. (1988). Observational learning of a left-right behavioral asymmetry in mice (Mus musculus). *Journal of Comparative Psychology, 102*(3), 222–224.
Comery, T. A., Harris, J. B., Willems, P. J., Oostra, B. A., Irwin, S. A., Weiler, I. J., et al. (1997). Abnormal dendritic spines in fragile X knockout mice: Maturation and pruning deficits. *Proceedings of the National Academy of Sciences of the United States of America, 94*(10), 5401–5404.

Cook, E. H., Jr., Lindgren, V., Leventhal, B. L., Courchesne, R., Lincoln, A., Shulman, C., et al. (1997). Autism or atypical autism in maternally but not paternally derived proximal 15q duplication. *American Journal of Human Genetics, 60*(4), 928.

Courchesne, E., & Pierce, K. (2005). Why the frontal cortex in autism might be talking only to itself: Local over-connectivity but long-distance disconnection. *Current Opinion in Neurobiology, 15*(2), 225–230. doi:10.1016/j.conb.2005.03.001.

Crabbe, J. C., Wahlsten, D., & Dudek, B. C. (1999). Genetics of mouse behavior: Interactions with laboratory environment. *Science, 284*(5420), 1670–1672. doi:10.1126/science.284.5420.1670.

Crawley, J. N. (2012). Translational animal models of autism and neurodevelopmental disorders. *Dialogues in Clinical Neuroscience, 14*(3), 293.

Cromwell, H. C., Johnson, A., McKnight, L., Horinek, M., Asbrock, C., Burt, S., et al. (2007). Effects of polychlorinated biphenyls on maternal odor conditioning in rat pups. *Physiology & Behavior, 91*(5), 658–666.

Crowcroft, P. (1966). *Mice all over*. London: Foulis.

Crowcroft, P., & Rowe, F. P. (1963). Social organization and territorial behaviour in the wild house mouse (*Mus musculus L.*). *Proceedings of the Zoological Society of London, 140*, 517–531.

Curley, J. P., Davidson, S., Bateson, P., & Champagne, F. A. (2009). Social enrichment during postnatal development induces transgenerational effects on emotional and reproductive behavior in mice. *Frontiers in Behavioral Neuroscience, 3*, 25. doi:10.3389/neuro.08.025.2009.

Curran, S., Powell, J., Neale, B., Dworzynski, K., Li, T., Murphy, D., et al. (2006). An association analysis of candidate genes on chromosome 15 q11-13 and autism spectrum disorder. *Molecular Psychiatry, 11*(8), 709–713.

D'Amato, F. R., Scalera, E., Sarli, C., & Moles, A. (2005). Pups call, mothers rush: Does maternal responsiveness affect the amount of ultrasonic vocalizations in mouse pups? *Behavior Genetics, 35*(1), 103–112.

D'Andrea, I., Alleva, E., & Branchi, I. (2007). Communal nesting, an early social enrichment, affects social competences but not learning and memory abilities at adulthood. *Behavioural Brain Research, 183*(1), 60–66.

Dastur, F. N., McGregor, I. S., & Brown, R. E. (1999). Dopaminergic modulation of rat pup ultrasonic vocalizations. *European Journal of Pharmacology, 382*(2), 53–67.

Dawson, G., Toth, K., Abbott, R., Osterling, J., Munson, J., Estes, A., et al. (2004). Early social attention impairments in autism: Social orienting, joint attention, and attention to distress. *Developmental Psychology, 40*(2), 271–283.

Dawson, G., Webb, S. J., & McPartland, J. (2005). Understanding the nature of face processing impairment in autism: Insights from behavioral and electrophysiological studies. *Developmental Neuropsychology, 27*(3), 403–424.

de Marchena, A., & Eigsti, I.-M. (2010). Conversational gestures in autism spectrum disorders: Asynchrony but not decreased frequency. *Autism Research, 3*(6), 311–322. doi:10.1002/aur.159.

De Waard, W. J., Aarts, J. M. M. J. G., Peijnenburg, A. C. M., De Kok, T. M. C. M., Van Schooten, F.-J., & Hoogenboom, L. A. P. (2008). Ah receptor agonist activity in frequently consumed food items. *Food Additives & Contaminants: Part A: Chemistry, Analysis, Control, Exposure & Risk Assessment, 25*(6), 779–787.

Del Arco, A., Segovia, G., Garrido, P., de Blas, M., & Mora, F. (2007). Stress, prefrontal cortex and environmental enrichment: Studies on dopamine and acetylcholine release and working memory performance in rats. *Behavioural Brain Research, 176*(2), 267–273.

Diehl, J. (2008). *Prosody comprehension in high-functioning autism*. Dissertation Abstracts International: Section B: The Sciences and Engineering.

Diehl, J., Bennetto, L., & Young, E. C. (2006). Story recall and narrative coherence of high-functioning children with autism spectrum disorders. *Journal of Abnormal Child Psychology, 34*(1), 83–98. doi:10.1007/s10802-005-9003-x.

Dolen, G., Darvishzadeh, A., Huang, K. W., & Malenka, R. C. (2013). Social reward requires coordinated activity of nucleus accumbens oxytocin and serotonin. *Nature, 501*(7466), 179–184.

Donaldson, Z. R., & Young, L. J. (2008). Oxytocin, vasopressin, and the neurogenetics of sociality. *Science, 322*(5903), 900–904.
Dudley, K. J., Li, X., Kobor, M. S., Kippin, T. E., & Bredy, T. W. (2011). Epigenetic mechanisms mediating vulnerability and resilience to psychiatric disorders. *Neuroscience & Biobehavioral Reviews, 35*(7), 1544–1551. doi:10.1016/j.neubiorev.2010.12.016.
Duffy, S. N., Craddock, K. J., Abel, T., & Nguyen, P. V. (2001). Environmental enrichment modifies the PKA-dependence of hippocampal LTP and improves hippocampus-dependent memory. *Learning and Memory, 8*(1), 26–34.
Durkin, M. S., Maenner, M. J., Newschaffer, C. J., Lee, L.-C., Cunniff, C. M., Daniels, J. L., et al. (2008). Advanced parental age and the risk of autism spectrum disorder. *American Journal of Epidemiology, 168*(11), 1268–1276.
Ehret, G., & Bernecker, C. (1986). Low-frequency sound communication by mouse pups (Mus musculus): Wriggling calls release maternal behavior. *Animal Behaviour, 34*(3), 821–830.
El Rawas, R., Thiriet, N., Lardeux, V., Jaber, M., & Solinas, M. (2009). Environmental enrichment decreases the rewarding but not the activating effects of heroin. *Psychopharmacology, 203*(3), 561–570.
Enstrom, A. M., Lit, L., Onore, C. E., Gregg, J. P., Hansen, R. L., Pessah, I. N., et al. (2009). Altered gene expression and function of peripheral blood natural killer cells in children with autism. *Brain, Behavior, and Immunity, 23*(1), 124–133.
Escorihuela, R. M., Tobeña, A., & Fernández-Teruel, A. (1994). Environmental enrichment reverses the detrimental action of early inconsistent stimulation and increases the beneficial effects of postnatal handling on shuttlebox learning in adult rats. *Behavioural Brain Research, 61*(2), 169–173.
Faherty, C. J., Shepherd, K. R., Herasimtschuk, A., & Smeyne, R. J. (2005). Environmental enrichment in adulthood eliminates neuronal death in experimental Parkinsonism. *Molecular Brain Research, 134*(1), 170–179.
Fatemi, S. H., Stary, J. M., Halt, A. R., & Realmuto, G. R. (2001). Dysregulation of Reelin and Bcl-2 proteins in autistic cerebellum. *Journal of Autism and Developmental Disorders, 31*(6), 529–535.
Forbes-Lorman, R. M., Rautio, J. J., Kurian, J. R., Auger, A. P., & Auger, C. J. (2012). Neonatal MeCP2 is important for the organization of sex differences in vasopressin expression. *Epigenetics, 7*(3), 230–238.
Forgays, D. G., & Forgays, J. W. (1952). The nature of the effect of free-environmental experience in the rat. *Journal of Comparative and Physiological Psychology, 45*(4), 322.
Fox, G. A. (2001). Wildlife as sentinels of human health effects in the Great Lakes—St. Lawrence basin. *Environmental Health Perspectives, 109*(Suppl. 6), 853.
Francis, D. D., Diorio, J., Plotsky, P. M., & Meaney, M. J. (2002). Environmental enrichment reverses the effects of maternal separation on stress reactivity. *The Journal of Neuroscience, 22*(18), 7840–7843.
Fukuchi, M., Nii, T., Ishimaru, N., Minamino, A., Hara, D., Takasaki, I., et al. (2009). Valproic acid induces up-or down-regulation of gene expression responsible for the neuronal excitation and inhibition in rat cortical neurons through its epigenetic actions. *Neuroscience Research, 65*(1), 35–43.
Fyffe, S. L., Neul, J. L., Samaco, R. C., Chao, H.-T., Ben-Shachar, S., Moretti, P., et al. (2008). Deletion of Mecp2 in Sim1-expressing neurons reveals a critical role for MeCP2 in feeding behavior, aggression, and the response to stress. *Neuron, 59*(6), 947–958.
Gascon, M., Vrijheid, M., Martinez, D., Forns, J., Grimalt, J. O., Torrent, M., et al. (2011). Effects of pre and postnatal exposure to low levels of polybromodiphenyl ethers on neurodevelopment and thyroid hormone levels at 4 years of age. *Environment International, 37*(3), 605–611. doi:10.1016/j.envint.2010.12.005.
Galef, B. G. (2013). Imitation and local enhancement: Detrimental effects of consensus definitions on analyses of social learning in animals. *Behavioural Processes, 100*, 123–130.
Geier, D. A., Kern, J. K., Garver, C. R., Adams, J. B., Audhya, T., Nataf, R., et al. (2009a). Biomarkers of environmental toxicity and susceptibility in autism. *Journal of the Neurological Sciences, 280*(1–2), 101–108.

Geier, D. A., Kern, J. K., Garver, C. R., Adams, J. B., Audhya, T., & Geier, M. R. (2009b). A prospective study of transsulfuration biomarkers in autistic disorders. [Erratum appears in Neurochem Res 2009 Feb;34(2):394]. *Neurochemical Research, 34*(2), 386–393.

Gernsbacher, M. A., Stevenson, J. L., Khandakar, S., & Goldsmith, H. H. (2008). Why does joint attention look atypical in autism? *Child Development Perspectives, 2*(1), 38–45. doi:10.1111/j.1750-8606.2008.00039.x.

Geschwind, D. H., & Levitt, P. (2007). Autism spectrum disorders: Developmental disconnection syndromes. *Current Opinion in Neurobiology, 17*(1), 103–111.

Gipson, C., Beckmann, J., El-Maraghi, S., Marusich, J., & Bardo, M. (2011). Effect of environmental enrichment on escalation of cocaine self-administration in rats. *Psychopharmacology, 214*(2), 557–566. doi:10.1007/s00213-010-2060-z.

Glickman, G. (2010). Circadian rhythms and sleep in children with autism. *Neuroscience & Biobehavioral Reviews, 34*(5), 755–768.

Globus, A., Rosenzweig, M. R., Bennett, E. L., & Diamond, M. C. (1973). Effects of differential experience on dendritic spine counts in rat cerebral cortex. *Journal of Comparative and Physiological Psychology, 82*(2), 175.

Goldberg, N. R., Fields, V., Pflibsen, L., Salvatore, M. F., & Meshul, C. K. (2012). Social enrichment attenuates nigrostriatal lesioning and reverses motor impairment in a progressive 1-methyl-2-phenyl-1, 2, 3, 6-tetrahydropyridine (MPTP) mouse model of Parkinson's disease. *Neurobiology of Disease, 45*(3), 1051–1067.

Goldberg, N., Haack, A., & Meshul, C. (2011). Enriched environment promotes similar neuronal and behavioral recovery in a young and aged mouse model of Parkinson's disease. *Neuroscience, 172*, 443–452.

Grabrucker, A. M. (2012). Environmental factors in autism. *Frontiers in Psychiatry, 3*, 118.

Gracceva, G., Venerosi, A., Santucci, D., Calamandrei, G., & Ricceri, L. (2009). Early social enrichment affects responsiveness to different social cues in female mice. *Behavioural Brain Research, 196*(2), 304–309.

Grafodatskaya, D., Chung, B., Szatmari, P., & Weksberg, R. (2010). Autism spectrum disorders and epigenetics. *Journal of the American Academy of Child and Adolescent Psychiatry, 49*(8), 794–809.

Grandjean, P., & Landrigan, P. J. (2006). Developmental neurotoxicity of industrial chemicals. *The Lancet, 368*(9553), 2167–2178.

Green, T. A., Alibhai, I. N., Roybal, C. N., Winstanley, C. A., Theobald, D. E., Birnbaum, S. G., et al. (2010). Environmental enrichment produces a behavioral phenotype mediated by low cyclic adenosine monophosphate response element binding (CREB) activity in the nucleus accumbens. *Biological Psychiatry, 67*(1), 28–35.

Greenough, W. T., Hwang, H., & Gorman, C. (1985). Evidence for active synapse formation or altered postsynaptic metabolism in visual cortex of rats reared in complex environments. *Proceedings of the National Academy of Sciences, 82*(13), 4549–4552.

Greenough, W. T., Volkmar, F. R., & Juraska, J. M. (1973). Effects of rearing complexity on dendritic branching in frontolateral and temporal cortex of the rat. *Experimental Neurology, 41*(2), 371–378.

Gregg, J. P., Lit, L., Baron, C. A., Hertz-Picciotto, I., Walker, W., Davis, R. A., et al. (2008). Gene expression changes in children with autism. *Genomics, 91*(1), 22–29.

Gregory, S. G., Connelly, J. J., Towers, A. J., Johnson, J., Biscocho, D., Markunas, C. A., et al. (2009). Genomic and epigenetic evidence for oxytocin receptor deficiency in autism. *BMC Medicine, 7*(1), 62.

Guilarte, T. R., Toscano, C. D., McGlothan, J. L., & Weaver, S. A. (2003). Environmental enrichment reverses cognitive and molecular deficits induced by developmental lead exposure. *Annals of Neurology, 53*(1), 50–56.

Guy, J., Cheval, H., Selfridge, J., & Bird, A. (2011). The role of MeCP2 in the brain. *Annual Review of Cell and Developmental Biology, 27*, 631–652.

Guzman, Y. F., Tronson, N. C., Guedea, A., Huh, K. H., Gao, C., & Radulovic, J. (2009). Social modeling of conditioned fear in mice by non-fearful conspecifics. *Behavioural Brain Research, 201*(1), 173–178.

Hackam, D. G., & Redelmeier, D. A. (2006). Translation of research evidence from animals to humans. *JAMA, 296*(14), 1727–1732. doi:10.1001/jama.296.14.1731.

Hadjikhani, N., Joseph, R. M., Manoach, D. S., Naik, P., Snyder, J., Dominick, K., et al. (2009). Body expressions of emotion do not trigger fear contagion in autism spectrum disorder. *Social Cognitive and Affective Neuroscience, 4*(1), 70–78.

Hagerman, R. J., Berry-Kravis, E., Kaufmann, W. E., Ono, M. Y., Tartaglia, N., Lachiewicz, A., et al. (2009). Advances in the treatment of fragile X syndrome. *Pediatrics, 123*(1), 378–390.

Hagerman, R. J., Jackson, A. W., 3rd, Levitas, A., Rimland, B., & Braden, M. (1986). An analysis of autism in fifty males with the fragile X syndrome. *American Journal of Medical Genetics, 23*(1–2), 359–374.

Hall, L., & Kelley, E. (2013). The contribution of epigenetics to understanding genetic factors in autism. *Autism, 18*(8), 872–881.

Halladay, A. K., Amaral, D., Aschner, M., Bolivar, V. J., Bowman, A., DiCicco-Bloom, E., et al. (2009). Animal models of autism spectrum disorders: Information for neurotoxicologists. *Neurotoxicology, 30*(5), 811–821.

Hallmayer, J., Cleveland, S., Torres, A., Phillips, J., Cohen, B., Torigoe, T., et al. (2011). Genetic heritability and shared environmental factors among twin pairs with autism. *Archives of General Psychiatry, 68*(11), 1095–1102.

Hamilton, S. M., Spencer, C. M., Harrison, W. R., Yuva-Paylor, L. A., Graham, D. F., Daza, R. A. M., et al. (2011). Multiple autism-like behaviors in a novel transgenic mouse model. *Behavioural Brain Research, 218*(1), 29–41. doi:10.1016/j.bbr.2010.11.026.

Harmon, K. M., Cromwell, H. C., Burgdorf, J., Moskal, J. R., Brudzynski, S. M., Kroes, R. A., et al. (2008). Rats selectively bred for low levels of 50 kHz ultrasonic vocalizations exhibit alterations in early social motivation. *Developmental Psychobiology, 50*(4), 322–331.

Harper, P. A., Wong, J. M. Y., Lam, M. S. M., & Okey, A. B. (2002). Polymorphisms in the human AH receptor. *Chemico-Biological Interactions, 141*(1–2), 161–187. doi:10.1016/S0009-2797(02)00071-6.

Harris, S. W., Hessl, D., Goodlin-Jones, B., Ferranti, J., Bacalman, S., Barbato, I., et al. (2008). Autism profiles of males with fragile X syndrome. *American Journal on Mental Retardation, 113*(6), 427–438.

Hertz-Picciotto, I., Park, H. Y., Dostal, M., Kocan, A., Trnovec, T., & Sram, R. (2008). Prenatal exposures to persistent and non-persistent organic compounds and effects on immune system development. *Basic & Clinical Pharmacology & Toxicology, 102*(2), 146–154.

Hoffman, M. L. (1975). Developmental synthesis of affect and cognition and its interplay for altruistic motivation. *Developmental Psychology, 11*, 607–622.

Hogart, A., Nagarajan, R. P., Patzel, K. A., Yasui, D. H., & LaSalle, J. M. (2007). 15q11-13 GABAA receptor genes are normally biallelically expressed in brain yet are subject to epigenetic dysregulation in autism-spectrum disorders. *Human Molecular Genetics, 16*(6), 691–703.

Hooper, K., She, J., Sharp, M., Chow, J., Jewell, N., Gephart, R., et al. (2007). Depuration of polybrominated diphenyl ethers (PBDEs) and polychlorinated biphenyls (PCBs) in breast milk from California first-time mothers (Primiparae). *Environmental Health Perspectives, 115*(9), 1271–1275.

Hu, V. W., Frank, B. C., Heine, S., Lee, N. H., & Quackenbush, J. (2006). Gene expression profiling of lymphoblastoid cell lines from monozygotic twins discordant in severity of autism reveals differential regulation of neurologically relevant genes. *BMC Genomics, 7*(1), 118.

Hu, Q., Franklin, J. N., Bryan, I., Morris, E., Wood, A., & DeWitt, J. C. (2012). Does developmental exposure to perflurooctanoic acid (PFOA) induce immunopathologies commonly observed in neurodevelopmental disorders? *Neurotoxicology, 33*(6), 1491–1498. doi:10.1016/j.neuro.2012.10.016.

Hu, V. W., Sarachana, T., Kim, K. S., Nguyen, A., Kulkarni, S., Steinberg, M. E., et al. (2009). Gene expression profiling differentiates autism case–controls and phenotypic variants of autism spectrum disorders: Evidence for circadian rhythm dysfunction in severe autism. *Autism Research, 2*(2), 78–97.

Hubert, B., Wicker, B., Moore, D., Monfardini, E., Duverger, H., Da Fonseca, D., et al. (2007a). Brief report: Recognition of emotional and non-emotional biological motion in individuals with autistic spectrum disorders. *Journal of Autism and Developmental Disorders, 37*(7), 1386–1392.

Hubert, B., Wicker, B., Moore, D., Monfardini, E., Duverger, H., Da Fonseca, D., et al. (2007b). "Brief report: Recognition of emotional and non-emotional biological motion in individuals with autistic spectrum disorders": Erratum. *Journal of Autism and Developmental Disorders, 37*(7), 1393.

Humphreys, K., Minshew, N., Leonard, G. L., & Behrmanna, M. (2007). A fine-grained analysis of facial expression processing in high-functioning adults with autism. *Neuropsychologia, 45*(4), 685–695.

Hung, W.-T., Lambert, G. H., Huang, P.-W., Patterson, D. G., Jr., & Guo, Y. L. (2013). Genetic susceptibility to dioxin-like chemicals' induction of cytochrome P4501A2 in the human adult linked to specific AhRR polymorphism. *Chemosphere, 90*(9), 2358–2364. doi:10.1016/j.chemosphere.2012.10.026.

Hurst, J. L. (1990). Urine marking in populations of wild house mice Mus domesticus Rutty. I. Communication between males. *Animal Behaviour, 40*(2), 209–222.

Ingelido, A. M., Ballard, T., Dellatte, E., di Domenico, A., Ferri, F., Fulgenzi, A. R., et al. (2007). Polychlorinated biphenyls (PCBs) and polybrominated diphenyl ethers (PBDEs) in milk from Italian women living in Rome and Venice. *Chemosphere, 67*(9), S301–S306. Epub 2007 Jan 2025.

Inoue, K., Okada, F., Ito, R., Kato, S., Sasaki, S., Nakajima, S., et al. (2004). Perfluorooctane sulfonate (PFOS) and related perfluorinated compounds in human maternal and cord blood samples: Assessment of PFOS exposure in a susceptible population during pregnancy. *Environmental Health Perspectives, 112*(11), 1204–1207.

James, S. J., Melnyk, S., Fuchs, G., Reid, T., Jernigan, S., Pavliv, O., et al. (2009). Efficacy of methylcobalamin and folinic acid treatment on glutathione redox status in children with autism. *American Journal of Clinical Nutrition, 89*(1), 425–430.

James, S., Shpyleva, S., Melnyk, S., Pavliv, O., & Pogribny, I. (2013). Complex epigenetic regulation of Engrailed-2 (EN-2) homeobox gene in the autism cerebellum. *Translational Psychiatry, 3*(2), e232.

Jankowsky, J. L., Melnikova, T., Fadale, D. J., Xu, G. M., Slunt, H. H., Gonzales, V., et al. (2005). Environmental enrichment mitigates cognitive deficits in a mouse model of Alzheimer's disease. *The Journal of Neuroscience, 25*(21), 5217–5224.

Jeon, D., Kim, S., Chetana, M., Jo, D., Ruley, H. E., Lin, S.-Y., et al. (2010). Observational fear learning involves affective pain system and Cav1. 2 Ca2+ channels in ACC. *Nature Neuroscience, 13*(4), 482–488.

Jeon, D., & Shin, H. S. (2011). A mouse model for observational fear learning and the empathetic response. *Current Protocols in Neuroscience, 8*, Unit 8.27.

Jiang, Y., Chen, J., Tong, J., & Chen, T. (2014). Trichloroethylene-induced gene expression and DNA methylation changes in B6C3F1 mouse liver. *PLoS One, 9*(12), e116–179.

Josh Huang, Z., & Zeng, H. (2013). Genetic approaches to neural circuits in the mouse. *Annual Review of Neuroscience, 36*, 183–215.

Jurado-Parras, M. T., Gruart, A., & Delgado-García, J. M. (2012). Observational learning in mice can be prevented by medial prefrontal cortex stimulation and enhanced by nucleus accumbens stimulation. *Learning & Memory, 19*(3), 99–106.

Juraska, J. M., Greenough, W. T., Elliott, C., Mack, K. J., & Berkowitz, R. (1980). Plasticity in adult rat visual cortex: An examination of several cell populations after differential rearing. *Behavioral and Neural Biology, 29*(2), 157–167.

Juraska, J. M., Greenough, W. T., & Conlee, J. W. (1983). Differential rearing affects responsiveness of rats to depressant and convulsant drugs. *Physiology & Behavior, 31*(5), 711–715.

Kane, M. J., Angoa-Peréz, M., Briggs, D. I., Sykes, C. E., Francescutti, D. M., Rosenberg, D. R., et al. (2012). Mice genetically depleted of brain serotonin display social impairments, com-

munication deficits and repetitive behaviors: Possible relevance to autism. *PLoS One, 7*(11), e48975.
Kannan, K., Corsolini, S., Falandysz, J., Fillmann, G., Kumar, K. S., Loganathan, B. G., et al. (2004). Perfluorooctanesulfonate and related fluorochemicals in human blood from several countries. *Environmental Science & Technology, 38*(17), 4489–4495.
Kavaliers, M., Choleris, E., Agmo, A., Muglia, L. J., Ogawa, S., & Pfaff, D. W. (2005a). Involvement of the oxytocin gene in the recognition and avoidance of parasitized males by female mice. *Animal Behaviour, 70*(3), 693–702.
Kavaliers, M., Choleris, E., & Colwell, D. D. (2001). Learning from others to cope with biting flies: Social learning of fear-induced conditioned analgesia and active avoidance. *Behavioral Neuroscience, 115*(3), 661–674.
Kavaliers, M., Choleris, E., & Pfaff, D. W. (2005b). Recognition and avoidance of the odors of parasitized conspecifics and predators: Differential genomic correlates. *Neuroscience & Biobehavioral Reviews, 29*(8), 1347–1359.
Kavaliers, M., & Colwell, D. D. (1995). Odours of parasitized males induce aversive responses in female mice. *Animal Behaviour, 50*(5), 1161–1169.
Kelley, A. E., & Berridge, K. C. (2002). The neuroscience of natural rewards: Relevance to addictive drugs. *The Journal of Neuroscience, 22*(9), 3306–3311.
Kempermann, G., Kuhn, H. G., & Gage, F. H. (1997). More hippocampal neurons in adult mice living in an enriched environment. *Nature, 386*(6624), 493–495.
Kendig, M. D., Bowen, M. T., Kemp, A. H., & McGregor, I. S. (2011). Predatory threat induces huddling in adolescent rats and residual changes in early adulthood suggestive of increased resilience. *Behavioural Brain Research, 225*(2), 405–414.
Kerley-Hamilton, J. S., Trask, H. W., Ridley, C. J. A., DuFour, E., Lesseur, C., Ringelberg, C. S., et al. (2012). Inherent and benzo[a]pyrene-induced differential aryl hydrocarbon receptor signaling greatly affects life span, atherosclerosis, cardiac gene expression, and body and heart growth in mice. *Toxicological Sciences, 126*(2), 391–404.
Kigar, S., & Auger, A. (2013). Epigenetic mechanisms may underlie the aetiology of sex differences in mental health risk and resilience. *Journal of Neuroendocrinology, 25*(11), 1141–1150.
Kim, K. C., Choi, C. S., Kim, J.-W., Han, S.-H., Cheong, J. H., Ryu, J. H., et al. (2016). MeCP2 modulates sex differences in the postsynaptic development of the valproate animal model of autism. *Molecular Neurobiology, 53*(1), 40–56.
Kim, E. J., Kim, E. S., Covey, E., & Kim, J. J. (2010). Social transmission of fear in rats: The role of 22-kHz ultrasonic distress vocalization. *PLoS One, 5*(12), e15077.
Kim, B. S., Lee, J., Bang, M., Am Seo, B., Khalid, A., Jung, M. W., et al. (2014). Differential regulation of observational fear and neural oscillations by serotonin and dopamine in the mouse anterior cingulate cortex. *Psychopharmacology, 231*(22), 4371–4381.
Kim, Y. S., Leventhal, B. L., Koh, Y. J., Fombonne, E., Laska, E., Lim, E. C., et al. (2011). Prevalence of autism spectrum disorders in a total population sample. *American Journal of Psychiatry, 168*(9), 904–912.
Kim, S., Mátyás, F., Lee, S., Acsády, L., & Shin, H.-S. (2012). Lateralization of observational fear learning at the cortical but not thalamic level in mice. *Proceedings of the National Academy of Sciences, 109*(38), 15497–15501.
Kinney, D. K., Munir, K. M., Crowley, D. J., & Miller, A. M. (2008). Prenatal stress and risk for autism. *Neuroscience & Biobehavioral Reviews, 32*(8), 1519–1532.
Koh, S., Magid, R., Chung, H., Stine, C. D., & Wilson, D. N. (2007). Depressive behavior and selective down-regulation of serotonin receptor expression after early-life seizures: Reversal by environmental enrichment. *Epilepsy & Behavior: E&B, 10*(1), 26.
Kondo, M., Gray, L. J., Pelka, G. J., Christodoulou, J., Tam, P. P., & Hannan, A. J. (2008). Environmental enrichment ameliorates a motor coordination deficit in a mouse model of Rett syndrome–Mecp2 gene dosage effects and BDNF expression. *European Journal of Neuroscience, 27*(12), 3342–3350.

Kooy, R. F., D'Hooge, R., Reyniers, E., Bakker, C. E., Nagels, G., De Boulle, K., et al. (1996). Transgenic mouse model for the fragile X syndrome. *American Journal of Medical Genetics, 64*(2), 241–245.

Krech, D., Rosenzweig, M. R., & Bennett, E. L. (1960). Effects of environmental complexity and training on brain chemistry. *Journal of Comparative and Physiological Psychology, 53*(6), 509.

Kumar, R. A., & Christian, S. L. (2009). Genetics of autism spectrum disorders. *Current Neurology and Neuroscience Reports, 9*(3), 188–197.

Kumsta, R., Hummel, E., Chen, F. S., & Heinrichs, M. (2013). Epigenetic regulation of the oxytocin receptor gene: Implications for behavioral neuroscience. *Frontiers in Neuroscience, 7*.

Kundakovic, M., Gudsnuk, K., Herbstman, J. B., Tang, D., Perera, F. P., & Champagne, F. A. (2014). DNA methylation of BDNF as a biomarker of early-life adversity. *Proceedings of the National Academy of Sciences, 112*(22), 6807–6813.

Kurian, J. R., Bychowski, M. E., Forbes-Lorman, R. M., Auger, C. J., & Auger, A. P. (2008). Mecp2 organizes juvenile social behavior in a sex-specific manner. *The Journal of Neuroscience, 28*(28), 7137–7142.

Kurian, J. R., Forbes-Lorman, R. M., & Auger, A. P. (2007). Sex difference in mecp2 expression during a critical period of rat brain development. *Epigenetics, 2*(3), 173–178.

Lahvis, G. P., Alleva, E., & Scattoni, M. L. (2010). Translating mouse vocalizations: Prosody and frequency modulation. *Genes, Brain and Behavior, 10*(1), 4–16.

Lahvis, G. P., & Black, L. M. (2011). Social interactions in the clinic and the cage: Toward a more valid mouse model of autism. In R. Jacob (Ed.), *Animal models of behavioral analysis* (Vol. 50, pp. 153–192). Totowa: Humana Press.

Lahvis, G. P., & Bradfield, C. A. (1998). Ahr null alleles: Distinctive or different? *Biochemical Pharmacology, 56*(7), 781–787.

Lahvis, G. P., Lindell, S. L., Thomas, R. S., McCuskey, R. S., Murphy, C., Glover, E., et al. (2000). Portosystemic shunting and persistent fetal vascular structures in aryl hydrocarbon receptor-deficient mice. *Proceedings of the National Academy of Sciences of the United States of America, 97*(19), 10442–10447.

Lam, K. S., Bodfish, J. W., & Piven, J. (2008). Evidence for three subtypes of repetitive behavior in autism that differ in familiality and association with other symptoms. *Journal of Child Psychology & Psychiatry & Allied Disciplines, 49*(11), 1193–1200.

Landrigan, P. J. (2010). What causes autism? Exploring the environmental contribution. *Current Opinion in Pediatrics, 22*(2), 219–225.

Langford, D. J., Bailey, A. L., Chanda, M. L., Clarke, S. E., Drummond, T. E., Echols, S., et al. (2010). Coding of facial expressions of pain in the laboratory mouse. *Nature Methods, 7*(6), 447–449.

Langford, D. J., Crager, S. E., Shehzad, Z., Smith, S. B., Sotocinal, S. G., Levenstadt, J. S., et al. (2006). Social modulation of pain as evidence for empathy in mice. *Science, 312*(5782), 1967–1970.

Latham, N., & Mason, G. (2004). From house mouse to mouse house: The behavioural biology of free-living Mus musculus and its implications in the laboratory. *Applied Animal Behaviour Science, 86*(3), 261–289.

Lazic, S., Grote, H., Blakemore, C., Hannan, A., van Dellen, A., Phillips, W., et al. (2006). Neurogenesis in the R6/1 transgenic mouse model of Huntington's disease: Effects of environmental enrichment. *The European Journal of Neuroscience, 23*(7), 1829.

Lewis, M. H., & Bodfish, J. W. (1998). Repetitive behavior disorders in autism. *Mental Retardation and Developmental Disabilities Research Reviews, 4*(2), 80–89. doi:10.1002/(sici)1098-2779(1998)4:2<80::aid-mrdd4>3.0.co;2-0.

Li, Y., Hamilton, K. J., Lai, A. Y., Burns, K. A., Li, L., Wade, P. A., et al. (2014). Diethylstilbestrol (DES)-stimulated hormonal toxicity is mediated by ERα alteration of target gene methylation patterns and epigenetic modifiers (DNMT3A, MBD2, and HDAC2) in the mouse seminal vesicle. *Environmental Health Perspectives, 122*(3), 262.

Lind, L., Penell, J., Luttropp, K., Nordfors, L., Syvänen, A.-C., Axelsson, T., et al. (2013). Global DNA hypermethylation is associated with high serum levels of persistent organic pollutants in an elderly population. *Environment International, 59*, 456–461.

Liu, H.-X., Lopatina, O., Higashida, C., Fujimoto, H., Akther, S., Inzhutova, A., et al. (2013). Displays of paternal mouse pup retrieval following communicative interaction with maternal mates. *Nature Communications, 4*, 1346.

Liu, X.-Q., Paterson, A. D., Szatmari, P., & Autism Genome Project Consortium. (2008). Genome-wide linkage analyses of quantitative and categorical autism subphenotypes. *Biological Psychiatry, 64*(7), 561–570.

Lonetti, G., Angelucci, A., Morando, L., Boggio, E. M., Giustetto, M., & Pizzorusso, T. (2010). Early environmental enrichment moderates the behavioral and synaptic phenotype of MeCP2 null mice. *Biological Psychiatry, 67*(7), 657–665.

Lord, C., Leventhal, B. L., & Cook, E. H., Jr. (2001). Quantifying the phenotype in autism spectrum disorders. *American Journal of Medical Genetics, 105*(1), 36–38.

Lord, C., Rutter, M., & Le Couteur, A. (1994). Autism Diagnostic Interview-Revised: A revised version of a diagnostic interview for caregivers of individuals with possible pervasive developmental disorders. *Journal of Autism and Developmental Disorders, 24*(5), 659–685.

Lynn, D. A., & Brown, G. R. (2009). The ontogeny of exploratory behavior in male and female adolescent rats (Rattus norvegicus). *Developmental Psychobiology, 51*(6), 513–520. doi:10.1002/dev.20386.

Maesako, M., Uemura, K., Kubota, M., Kuzuya, A., Sasaki, K., Asada, M., et al. (2012). Environmental enrichment ameliorated high-fat diet-induced Aβ deposition and memory deficit in APP transgenic mice. *Neurobiology of Aging, 33*(5), 1011. e1011–1011. e1023.

Magnee, M. J., de Gelder, B., van Engeland, H., & Kemner, C. (2008). Atypical processing of fearful face-voice pairs in pervasive developmental disorder: An ERP study. *Clinical Neurophysiology, 119*(9), 2004–2010.

Martin, L. J., Hathaway, G., Isbester, K., Mirali, S., Acland, E. L., Niederstrasser, N., et al. (2015). Reducing social stress elicits emotional contagion of pain in mouse and human strangers. *Current Biology, 25*(3), 326–332. doi:10.1016/j.cub.2014.11.028.

Martinez-Zamudio, R., & Ha, H. C. (2011). Environmental epigenetics in metal exposure. *Epigenetics, 6*(7), 820–827.

Matsuda, K. I. (2014). Epigenetic changes in the estrogen receptor α gene promoter: Implications in sociosexual behaviors. *Frontiers in Neuroscience, 8*, 344.

Mbadiwe, T., & Millis, R. M. (2014). In C Payne (Ed.), *Epigenetic mechanisms in autism spectrum disorders, epigenetics and epigenomics*. InTech, doi:10.5772/57195. Retrieved from http://www.intechopen.com/books/epigenetics-and-epigenomics/epigenetic-mechanisms-in-autism-spectrum-disorders.

McCann, J., & Peppe, S. (2003). Prosody in autism spectrum disorders: A critical review. *International Journal of Language & Communication Disorders, 38*(4), 325–350.

McCann, J., Peppe, S., Gibbon, F. E., O'Hare, A., & Rutherford, M. (2007). Prosody and its relationship to language in school-aged children with high-functioning autism. *International Journal of Language & Communication Disorders, 42*(6), 682–702.

McFadden, S. A. (1996). Phenotypic variation in xenobiotic metabolism and adverse environmental response: Focus on sulfur-dependent detoxification pathways. *Toxicology, 111*(1–3), 43–65.

McGraw, L. A., & Young, L. J. (2010). The prairie vole: An emerging model organism for understanding the social brain. *Trends in Neurosciences, 33*(2), 103–109. doi:10.1016/j.tins.2009.11.006.

McNaughton, C. H., Moon, J., Strawderman, M. S., Maclean, K. N., Evans, J., & Strupp, B. J. (2008). Evidence for social anxiety and impaired social cognition in a mouse model of fragile X syndrome. *Behavioral Neuroscience, 122*(2), 293–300.

McOmish, C., Burrows, E., Howard, M., Scarr, E., Kim, D., Shin, H., et al. (2008). Phospholipase C-β1 knockout mice exhibit endophenotypes modeling schizophrenia which are rescued by environmental enrichment and clozapine administration. *Molecular Psychiatry, 13*(7), 661–672.

Meaney, M. J. (2010). Epigenetics and the biological definition of gene x environment interactions. *Child Development, 81*(1), 41–79.

Meaney, M. J., Dodge, A. M., & Beatty, W. W. (1981). Sex-dependent effects of amygdaloid lesions on the social play of prepubertal rats. *Physiology & Behavior, 26*(3), 467–472.

Meaney, M. J., & McEwen, B. S. (1986). Testosterone implants into the amygdala during the neonatal period masculinize the social play of juvenile female rats. *Brain Research, 398*(2), 324–328.
Meaney, M. J., & Stewart, J. (1981). Neonatal androgens influence the social play of prepubescent rats. *Hormones and Behavior, 15*(2), 197–213.
Meaney, M. J., & Szyf, M. (2005). Maternal care as a model for experience-dependent chromatin plasticity? *Trends in Neurosciences, 28*(9), 456–463.
Michalon, A., Sidorov, M., Ballard, T. M., Ozmen, L., Spooren, W., Wettstein, J. G., et al. (2012). Chronic pharmacological mGlu5 inhibition corrects fragile X in adult mice. *Neuron, 74*(1), 49–56.
Milestone, C. B., Maclatchy, D. L., & Hewitt, M. L. (2013). Process for refining chemicals from pulp and paper mill wastewaters. US Patent 20,130,072,724.
Miranda-Contreras, L., Davila-Ovalles, R., Benitez-Diaz, P., Pena-Contreras, Z., & Palacios-Pru, E. (2005). Effects of prenatal paraquat and mancozeb exposure on amino acid synaptic transmission in developing mouse cerebellar cortex. *Developmental Brain Research, 160*(1), 19–27. doi:10.1016/j.devbrainres.2005.08.001.
Mitchell, M. M., Woods, R., Chi, L. H., Schmidt, R. J., Pessah, I. N., Kostyniak, P. J., et al. (2012). Levels of select PCB and PBDE congeners in human postmortem brain reveal possible environmental involvement in 15q11-q13 duplication autism spectrum disorder. *Environmental and Molecular Mutagenesis, 53*(8), 589–598.
Miyake, K., Hirasawa, T., Koide, T., & Kubota, T. (2012). Epigenetics in autism and other neurodevelopmental diseases. In S. I. Ahmed (Ed.), *Neurodegenerative diseases* (pp. 91–98). New York: Springer.
Moles, A., Kieffer, B. L., & D'Amato, F. R. (2004). Deficit in attachment behavior in mice lacking the mu-opioid receptor gene. *Science, 304*(5679), 1983–1986.
Moretti, P., Bouwknecht, J. A., Teague, R., Paylor, R., & Zoghbi, H. Y. (2005). Abnormalities of social interactions and home-cage behavior in a mouse model of Rett syndrome. *Human Molecular Genetics, 14*(2), 205–220.
Morley-Fletcher, S., Rea, M., Maccari, S., & Laviola, G. (2003). Environmental enrichment during adolescence reverses the effects of prenatal stress on play behaviour and HPA axis reactivity in rats. *European Journal of Neuroscience, 18*(12), 3367–3374.
Morris, J. A., Jordan, C. L., & Breedlove, S. M. (2004). Sexual differentiation of the vertebrate nervous system. *Nature Neuroscience, 7*(10), 1034–1039.
Moy, S. S., Nadler, J. J., Young, N. B., Perez, A., Holloway, L. P., Barbaro, R. P., et al. (2007). Mouse behavioral tasks relevant to autism: Phenotypes of 10 inbred strains. *Behavioural Brain Research, 176*(1), 4–20. doi:10.1016/j.bbr.2006.07.030.
Mychasiuk, R., Zahir, S., Schmold, N., Ilnytskyy, S., Kovalchuk, O., & Gibb, R. (2012). Parental enrichment and offspring development: Modifications to brain, behavior and the epigenome. *Behavioural Brain Research, 228*(2), 294–298. doi:10.1016/j.bbr.2011.11.036.
Nader, J., Claudia, C., El Rawas, R., Favot, L., Jaber, M., Thiriet, N., et al. (2014). Loss of environmental enrichment increases vulnerability to cocaine addiction. *Neuropsychopharmacology, 39*(3), 780.
Nadler, J., Moy, S., Dold, G., Trang, D., Simmons, N., Perez, A., et al. (2004). Automated apparatus for quantitation of social approach behaviors in mice. *Genes, Brain and Behavior, 3*(5), 303–314.
Nag, N., Moriuchi, J. M., Peitzman, C. G., Ward, B. C., Kolodny, N. H., & Berger-Sweeney, J. E. (2009). Environmental enrichment alters locomotor behaviour and ventricular volume in Mecp2 1lox mice. *Behavioural Brain Research, 196*(1), 44–48.
Nagarajan, R., Hogart, A., Gwye, Y., Martin, M. R., & LaSalle, J. M. (2006). Reduced MeCP2 expression is frequent in autism frontal cortex and correlates with aberrant MECP2 promoter methylation. *Epigenetics, 1*(4), 172–182.
Naka, F., Narita, N., Okado, N., & Narita, M. (2005). Modification of AMPA receptor properties following environmental enrichment. *Brain and Development, 27*(4), 275–278.
Nguyen, A., Rauch, T. A., Pfeifer, G. P., & Hu, V. W. (2010). Global methylation profiling of lymphoblastoid cell lines reveals epigenetic contributions to autism spectrum disorders and a novel autism candidate gene, RORA, whose protein product is reduced in autistic brain. *The FASEB Journal, 24*(8), 3036–3051.

Nithianantharajah, J., & Hannan, A. J. (2006). Enriched environments, experience-dependent plasticity and disorders of the nervous system. *Nature Reviews Neuroscience, 7*(9), 697–709.

Nunes, S., Muecke, E.-M., Anthony, J. A., & Batterbee, A. S. (1999). Endocrine and energetic mediation of play behavior in free-living belding's ground squirrels. *Hormones and Behavior, 36*(2), 153–165. doi:10.1006/hbeh.1999.1538.

O'Roak, B. J., Deriziotis, P., Lee, C., Vives, L., Schwartz, J. J., Girirajan, S., et al. (2011). Exome sequencing in sporadic autism spectrum disorders identifies severe de novo mutations. *Nature Genetics, 43*(6), 585–589. doi:10.1038/ng.835.

Olioff, M., & Stewart, J. (1978). Sex differences in the play behavior of prepubescent rats. *Physiology & Behavior, 20*(2), 113–115. doi:10.1016/0031-9384(78)90060-4.

Ostrovsky, Y., Andalman, A., & Sinha, P. (2006). Vision following extended congenital blindness. *Psychological Science, 17*(12), 1009–1014. doi:10.1111/j.1467-9280.2006.01827.x.

Ozonoff, S., Young, G. S., Carter, A., Messinger, D., Yirmiya, N., Zwaigenbaum, L., et al. (2011). Recurrence risk for autism spectrum disorders: A baby siblings research consortium study. *Pediatrics, 128*(3), e488–e495.

Palanza, P., Morley-Fletcher, S., & Laviola, G. (2001). Novelty seeking in periadolescent mice: Sex differences and influence of intrauterine position. *Physiology & Behavior, 72*(1–2), 255–262. doi:10.1016/S0031-9384(00)00406-6.

Panksepp, J. B., Jochman, K., Kim, J. U., Koy, J. J., Wilson, E. D., Chen, Q., et al. (2007). Affiliative behavior, ultrasonic communication and social reward are influenced by genetic variation in adolescent mice. *PLoS One [Electronic Resource], 2*, e351. doi:10.1371/journal.pone.0000351.

Panksepp, J. B., & Lahvis, G. P. (2007). Social reward among juvenile mice. *Genes, Brain and Behavior, 6*(7), 661–671. doi:10.1111/j.1601-183X.2006.00295.x.

Panksepp, J. B., & Lahvis, G. P. (2011). Rodent empathy and affective neuroscience. *Neuroscience & Biobehavioral Reviews, 35*(9), 1864–1875. doi:10.1016/j.neubiorev.2011.05.013.

Panksepp, J. B., & Lahvis, G. P. (2016). Differential influence of social versus isolate housing on vicarious fear learning in adolescent mice. *Behavioral Neuroscience, 130*(2), 206.

Parent, C. I., & Meaney, M. J. (2008). The influence of natural variations in maternal care on play fighting in the rat. *Developmental Psychobiology, 50*(8), 767–776.

Passineau, M. J., Green, E. J., & Dietrich, W. D. (2001). Therapeutic effects of environmental enrichment on cognitive function and tissue integrity following severe traumatic brain injury in rats. *Experimental Neurology, 168*(2), 373–384.

Pearson, B., Defensor, E., Pobbe, R., Yamamoto, L., Bolivar, V., Blanchard, D., et al. (2012). Mecp2 truncation in male mice promotes affiliative social behavior. *Behavior Genetics, 42*(2), 299–312.

Pellis, S. (2002). Sex differences in play fighting revisited: Traditional and nontraditional mechanisms of sexual differentiation in rats. *Archives of Sexual Behavior, 31*(1), 17–26. doi:10.1023/a:1014070916047.

Pellis, S. M., Pellis, V. C., & McKenna, M. M. (1994). Feminine dimension in the play fighting of rats (Rattus norvegicus) and its defeminization neonatally by androgens. *Journal of Comparative Psychology, 108*(1), 68.

Persico, A., Levitt, P., & Pimenta, A. (2006). Polymorphic GGC repeat differentially regulates human reelin gene expression levels. *Journal of Neural Transmission, 113*(10), 1373–1382.

Pessah, I. N., Seegal, R. F., Lein, P. J., LaSalle, J., Yee, B. K., Van De Water, J., et al. (2008). Immunologic and neurodevelopmental susceptibilities of autism. *Neurotoxicology, 29*(3), 532–545.

Peters, S. U., Gordon, R. L., & Key, A. P. (2014). Induced gamma oscillations differentiate familiar and novel voices in children with MECP2 duplication and Rett syndromes. *Journal of Child Neurology, 30*(2), 145–152.

Phiel, C. J., Zhang, F., Huang, E. Y., Guenther, M. G., Lazar, M. A., & Klein, P. S. (2001). Histone deacetylase is a direct target of valproic acid, a potent anticonvulsant, mood stabilizer, and teratogen. *Journal of Biological Chemistry, 276*(39), 36734–36741.

Picker, J. D., Yang, R., Ricceri, L., & Berger-Sweeney, J. (2006). An altered neonatal behavioral phenotype in Mecp2 mutant mice. *Neuroreport, 17*(5), 541–544.

Pietropaolo, S., Branchi, I., Cirulli, F., Chiarotti, F., Aloe, L., & Alleva, E. (2004). Long-term effects of the periadolescent environment on exploratory activity and aggressive behaviour in mice: Social versus physical enrichment. *Physiology & Behavior, 81*(3), 443–453. doi:10.1016/j.physbeh.2004.02.022.

Pobbe, R. L., Pearson, B. L., Defensor, E. B., Bolivar, V. J., Blanchard, D. C., & Blanchard, R. J. (2010). Expression of social behaviors of C57BL/6J versus BTBR inbred mouse strains in the visible burrow system. *Behavioural Brain Research, 214*(2), 443–449.

Poland, A., Palen, D., & Glover, E. (1994). Analysis of the four alleles of the murine aryl hydrocarbon receptor. *Molecular Pharmacology, 46*(5), 915–921.

Pollock, J. D., Wu, D.-Y., & Satterlee, J. S. (2014). Molecular neuroanatomy: A generation of progress. *Trends in Neurosciences, 37*(2), 106–123.

Pound, P., Ebrahim, S., Sandercock, P., Bracken, M. B., & Roberts, I. (2004). Where is the evidence that animal research benefits humans? *BMJ, 328*(7438), 514–517.

Preston, S. D., & de Waal, F. B. (2002). Empathy: Its ultimate and proximate bases. *Behavioral and Brain Sciences, 25*(1), 1–20.

Ramalhinho, M. G., Mathias, M. L., & Muccillo-Baisch, A. L. (2012). Physiological damage in Algerian mouse Mus spretus (Rodentia: muridae) exposed to crude oil. *Journal of BioScience and Biotechnology, 1*(2), 125–133.

Ramocki, M. B., Peters, S. U., Tavyev, Y. J., Zhang, F., Carvalho, C., Schaaf, C. P., et al. (2009). Autism and other neuropsychiatric symptoms are prevalent in individuals with MeCP2 duplication syndrome. *Annals of Neurology, 66*(6), 771–782.

Rampon, C., Jiang, C. H., Dong, H., Tang, Y.-P., Lockhart, D. J., Schultz, P. G., et al. (2000). Effects of environmental enrichment on gene expression in the brain. *Proceedings of the National Academy of Sciences, 97*(23), 12880–12884.

Rao, D. B., Jortner, B. S., & Sills, R. C. (2014). Animal models of peripheral neuropathy due to environmental toxicants. *Ilar Journal, 54*(3), 315–323.

Rauh, V. A., Perera, F. P., Horton, M. K., Whyatt, R. M., Bansal, R., Hao, X., et al. (2012). Brain anomalies in children exposed prenatally to a common organophosphate pesticide. *Proceedings of the National Academy of Sciences, 109*(20), 7871–7876. doi:10.1073/pnas.1203396109.

Reynolds, S., Urruela, M., & Devine, D. P. (2013). Effects of environmental enrichment on repetitive behaviors in the BTBR T+tf/J mouse model of autism. *Autism Research: Official Journal of the International Society for Autism Research, 6*(5), 337–343. doi:10.1002/aur.1298.

Richter, S. H., Garner, J. P., Auer, C., Kunert, J., & Wurbel, H. (2010). Systematic variation improves reproducibility of animal experiments. *Nature Methods, 7*(3), 167–168.

Richter, S. H., Garner, J. P., & Wurbel, H. (2009). Environmental standardization: Cure or cause of poor reproducibility in animal experiments? *Nature Methods, 6*(4), 257–261.

Richter, S. H., Zeuch, B., Riva, M. A., Gass, P., & Vollmayr, B. (2013). Environmental enrichment ameliorates depressive-like symptoms in young rats bred for learned helplessness. *Behavioural Brain Research, 252*, 287–292.

Roberts, S. A., Davidson, A. J., Beynon, R. J., & Hurst, J. L. (2014). Female attraction to male scent and associative learning: The house mouse as a mammalian model. *Animal Behaviour, 97*, 313–321.

Roberts, E. M., English, P. B., Grether, J. K., Windham, G. C., Somberg, L., & Wolff, C. (2007). Maternal residence near agricultural pesticide applications and autism spectrum disorders among children in the California Central Valley. *Environmental Health Perspectives, 115*(10), 1482–1489.

Robins, D. L., Casagrande, K. s., Barton, M., Chen, C.-M. A., Dumont-Mathieu, T., & Fein, D. (2014). Validation of the modified checklist for autism in toddlers, revised with follow-up (M-CHAT-R/F). *Pediatrics, 133*(1), 37–45.

Rogers, S. J. (1999). An examination of the imitation deficit in autism. In J. N. G. Butterworth (Ed.), *Imitation in infancy* (pp. 254–283). New York, NY: Cambridge University Press.

Rosenberg, R., Law, J., Yenokyan, G., McGready, J., Kaufmann, W. E., & Law, P. A. (2009). Characteristics and concordance of autism spectrum disorders among 277 twin pairs. *Archives of Pediatrics & Adolescent Medicine, 163*(10), 907–914. doi:10.1001/archpediatrics.2009.98.

Rosenzweig, M. R., Krech, D., Bennett, E. L., & Diamond, M. C. (1962a). Effects of environmental complexity and training on brain chemistry and anatomy: A replication and extension. *Journal of Comparative and Physiological Psychology, 55*(4), 429.

Rosenzweig, M. R., Krech, D., Bennett, E. L., & Zolman, J. F. (1962b). Variation in environmental complexity and brain measures. *Journal of Comparative and Physiological Psychology, 55*(6), 1092.

Rottman, S. J., & Snowdon, C. T. (1972). Demonstration and analysis of an alarm pheromone in mice. *Journal of Comparative and Physiological Psychology, 81*(3), 483–490.

Roze, E., Meijer, L., Bakker, A., Van Braeckel, K. N., Sauer, P. J., & Bos, A. F. (2009). Prenatal exposure to organohalogens, including brominated flame retardants, influences motor, cognitive, and behavioral performance at school age. *Environmental Health Perspectives, 117*(12), 1953–1958. Epub 2009 Aug 1931.

Samaco, R. C., Hogart, A., & LaSalle, J. M. (2005). Epigenetic overlap in autism-spectrum neurodevelopmental disorders: MECP2 deficiency causes reduced expression of UBE3A and GABRB3. *Human Molecular Genetics, 14*(4), 483–492.

Samaco, R. C., Mandel-Brehm, C., McGraw, C. M., Shaw, C. A., McGill, B. E., & Zoghbi, H. Y. (2012). Crh and Oprm1 mediate anxiety-related behavior and social approach in a mouse model of MECP2 duplication syndrome. *Nature Genetics, 44*(2), 206–211.

Sanders, J., Mayford, M., & Jeste, D. (2013). Empathic fear responses in mice are triggered by recognition of a shared experience. *PLoS One, 8*(9), e74609.

Sanders, S. J., Murtha, M. T., Gupta, A. R., Murdoch, J. D., Raubeson, M. J., Willsey, A. J., et al. (2012). De novo mutations revealed by whole-exome sequencing are strongly associated with autism. *Nature, 485*(7397), 237–241. doi:10.1038/nature10945.

Sasson, N. J., Turner-Brown, L. M., Holtzclaw, T. N., Lam, K. S. L., & Bodfish, J. W. (2008). Children with autism demonstrate circumscribed attention during passive viewing of complex social and nonsocial picture arrays. *Autism Research, 1*(1), 31–42. doi:10.1002/aur.4.

Scattoni, M. L., Crawley, J., & Ricceri, L. (2009). Ultrasonic vocalizations: A tool for behavioural phenotyping of mouse models of neurodevelopmental disorders. *Neuroscience & Biobehavioral Reviews, 33*(4), 508–515.

Schaevitz, L. R., Moriuchi, J. M., Nag, N., Mellot, T. J., & Berger-Sweeney, J. (2010). Cognitive and social functions and growth factors in a mouse model of rett syndrome. *Physiology & Behavior, 100*(3), 255–263.

Schanen, N. C. (2006). Epigenetics of autism spectrum disorders. *Human Molecular Genetics, 15*(Suppl. 2), R138–R150.

Schneider, T., Turczak, J., & Przewłocki, R. (2006). Environmental enrichment reverses behavioral alterations in rats prenatally exposed to valproic acid: Issues for a therapeutic approach in autism. *Neuropsychopharmacology, 31*(1), 36–46.

Scholz, J., Allemang-Grand, R., Dazai, J., & Lerch, J. (2015). Environmental enrichment is associated with rapid volumetric brain changes in adult mice. *Neuroimage, 109*, 190–198.

Sebat, J., Lakshmi, B., Malhotra, D., Troge, J., Lese-Martin, C., Walsh, T., et al. (2007). Strong association of de novo copy number mutations with autism. *Science, 316*(5823), 445–449.

Shah, C. R., Forsberg, C. G., Kang, J. Q., & Veenstra-VanderWeele, J. (2013). Letting a typical mouse judge whether mouse social interactions are atypical. *Autism Research, 6*(3), 212–220.

Shahbazian, M. D., Young, J. I., Yuva-Paylor, L. A., Spencer, C. M., Antalffy, B. A., Noebels, J. L., et al. (2002). Mice with truncated MeCP2 recapitulate many rett syndrome features and display hyperacetylation of histone H3. *Neuron, 35*, 243–254.

Shih, J., May, L. D., Gonzalez, H. E., Lee, E. W., Alvi, R. S., Sall, J. W., et al. (2012). Delayed environmental enrichment reverses sevoflurane-induced memory impairment in rats. *Anesthesiology, 116*(3), 586.

Shriberg, L. D., Paul, R., McSweeny, J. L., Klin, A., & Cohen, D. J. (2001). Speech and prosody characteristics of adolescents and adults with high-functioning autism and Asperger syndrome. *Journal of Speech Language and Hearing Research, 44*(5), 1097–1115.

Singer, T., Seymour, B., O'Doherty, J., Kaube, H., Dolan, R. J., & Frith, C. D. (2004). Empathy for pain involves the affective but not sensory components of pain. *Science, 303*(5661), 1157–1162.

Singer, T., Seymour, B., O'Doherty, J. P., Stephan, K. E., Dolan, R. J., & Frith, C. D. (2006). Empathic neural responses are modulated by the perceived fairness of others. *Nature, 439*(7075), 466–469.
Slotkin, T. A., & Seidler, F. J. (2007). Prenatal chlorpyrifos exposure elicits presynaptic serotonergic and dopaminergic hyperactivity at adolescence: Critical periods for regional and sex-selective effects. *Reproductive Toxicology, 23*(3), 421–427. doi:10.1016/j.reprotox.2006.07.010.
So, M. K., Taniyasu, S., Yamashita, N., Giesy, J. P., Zheng, J., Fang, Z., et al. (2004). Perfluorinated compounds in coastal waters of Hong Kong, South China, and Korea. *Environmental Science & Technology, 38*(15), 4056–4063.
Solinas, M., Chauvet, C., Thiriet, N., El Rawas, R., & Jaber, M. (2008). Reversal of cocaine addiction by environmental enrichment. *Proceedings of the National Academy of Sciences, 105*(44), 17145–17150.
Solinas, M., Thiriet, N., Chauvet, C., & Jaber, M. (2010). Prevention and treatment of drug addiction by environmental enrichment. *Progress in Neurobiology, 92*(4), 572–592.
Spencer, C. M., Graham, D. F., Yuva-Paylor, L. A., Nelson, D. L., & Paylor, R. (2008). Social behavior in Fmr1 knockout mice carrying a human FMR1 transgene. *Behavioral Neuroscience, 122*(3), 710–715.
Stairs, D. J., & Bardo, M. T. (2009). Neurobehavioral effects of environmental enrichment and drug abuse vulnerability. *Pharmacology Biochemistry and Behavior, 92*(3), 377–382.
Stamou, M., Streifel, K. M., Goines, P. E., & Lein, P. J. (2013). Neuronal connectivity as a convergent target of gene × environment interactions that confer risk for Autism Spectrum Disorders. *Neurotoxicology and Teratology, 36*, 3–16.
Steffenburg, S., Gillberg, C., Hellgren, L., Andersson, L., Gillberg, I. C., Jakobsson, G., et al. (1989). A twin study of autism in Denmark, Finland, Iceland, Norway and Sweden. *Journal of Child Psychology and Psychiatry, 30*(3), 405–416. doi:10.1111/j.1469-7610.1989.tb00254.x.
Sun, H., Li, F., Zhang, T., Zhang, X., He, N., Song, Q., et al. (2011). Perfluorinated compounds in surface waters and WWTPs in Shenyang, China: Mass flows and source analysis. *Water Research, 45*(15), 4483–4490.
Szeligo, F., & Leblond, C. (1977). Response of the three main types of glial cells of cortex and corpus callosum in rats handled during suckling or exposed to enriched, control and impoverished environments following weaning. *Journal of Comparative Neurology, 172*(2), 247–263.
Sztainberg, Y., Kuperman, Y., Tsoory, M., Lebow, M., & Chen, A. (2010). The anxiolytic effect of environmental enrichment is mediated via amygdalar CRF receptor type 1. *Molecular Psychiatry, 15*(9), 905–917.
Tang, A. C., Akers, K. G., Reeb, B. C., Romeo, R. D., & McEwen, B. S. (2006). Programming social, cognitive, and neuroendocrine development by early exposure to novelty. *Proceedings of the National Academy of Sciences, 103*(42), 15716–15721.
Tantra, M., Hammer, C., Kästner, A., Dahm, L., Begemann, M., Bodda, C., et al. (2014). Mild expression differences of MECP2 influencing aggressive social behavior. *EMBO Molecular Medicine, 6*(5), 662–684. doi:10.1002/emmm.201303744.
Tarantini, L., Bonzini, M., Apostoli, P., Pegoraro, V., Bollati, V., Marinelli, B., et al. (2008). Effects of particulate matter on genomic DNA methylation content and iNOS promoter methylation. *Environmental Health Perspectives, 117*(2), 217–222.
Thiriet, N., Amar, L., Toussay, X., Lardeux, V., Ladenheim, B., Becker, K. G., et al. (2008). Environmental enrichment during adolescence regulates gene expression in the striatum of mice. *Brain Research, 1222*, 31–41. doi:10.1016/j.brainres.2008.05.030.
Thonhauser, K. E., Raveh, S., Hettyey, A., Beissmann, H., & Penn, D. J. (2013). Scent marking increases male reproductive success in wild house mice. *Animal Behaviour, 86*(5), 1013–1021.
Toth, K., Munson, J., Meltzoff, A., & Dawson, G. (2006). Early predictors of communication development in young children with autism spectrum disorder: Joint attention, imitation, and toy play. *Journal of Autism and Developmental Disorders, 36*(8), 993–1005. doi:10.1007/s10803-006-0137-7.
Tsilidis, K. K., Panagiotou, O. A., Sena, E. S., Aretouli, E., Evangelou, E., Howells, D. W., et al. (2013). Evaluation of excess significance bias in animal studies of neurological diseases. *PLoS Biology, 11*(7), e1001609.

Valsecchi, P., Bosellini, I., Sabatini, F., Mainardi, M., & Fiorito, G. (2002). Behavioral analysis of social effects on the problem-solving ability in the house mouse. *Ethology, 108*(12), 1115–1134.

Van Praag, H., Kempermann, G., & Gage, F. H. (2000). Neural consequences of environmental enrichment. *Nature Reviews Neuroscience, 1*(3), 191–198.

Van Santen, J. P. H., Prud'hommeaux, E. T., Black, L. M., & Mitchell, M. (2010). Computational prosodic markers for autism. *Autism, 14*(3), 215–236.

Volk, H. E., Hertz-Picciotto, I., Delwiche, L., Lurmann, F., & McConnell, R. (2011). Residential proximity to freeways and autism in the CHARGE study. *Environmental Health Perspectives, 119*(6), 873.

Vom Saal, F. S., & Bronson, F. (1978). In utero proximity of female mouse fetuses to males: Effect on reproductive performance during later life. *Biology of Reproduction, 19*(4), 842–853.

von Ehrenstein, O. S., Fenton, S. E., Kato, K., Kuklenyik, Z., Calafat, A. M., & Hines, E. P. (2009). Polyfluoroalkyl chemicals in the serum and milk of breastfeeding women. *Reproductive Toxicology, 27*(3–4), 239–245.

Vyssotski, A. L., Serkov, A. N., Itskov, P. M., Dell'Omo, G., Latanov, A. V., Wolfer, D. P., et al. (2006). Miniature neurologgers for flying pigeons: Multichannel EEG and action and field potentials in combination with GPS recording. *Journal of Neurophysiology, 95*(2), 1263–1273.

Wang, F., Liu, W., Jin, Y., Dai, J., Yu, W., Liu, X., et al. (2010). Transcriptional effects of prenatal and neonatal exposure to PFOS in developing rat brain. *Environmental Science & Technology, 44*(5), 1847–1853. doi:10.1021/es902799f.

Will, B., Galani, R., Kelche, C., & Rosenzweig, M. R. (2004). Recovery from brain injury in animals: Relative efficacy of environmental enrichment, physical exercise or formal training (1990–2002). *Progress in Neurobiology, 72*(3), 167–182.

Williams, B. M., Luo, Y., Ward, C., Redd, K., Gibson, R., Kuczaj, S. A., et al. (2001). Environmental enrichment: Effects on spatial memory and hippocampal CREB immunoreactivity. *Physiology & Behavior, 73*(4), 649–658.

Winneke, G. (2011). Developmental aspects of environmental neurotoxicology: Lessons from lead and polychlorinated biphenyls. *Journal of the Neurological Sciences, 308*(1–2), 9–15. doi:10.1016/j.jns.2011.05.020.

Winslow, J. T., & Insel, T. R. (2002). The social deficits of the oxytocin knockout mouse. *Neuropeptides, 36*(2–3), 221–229.

Wöhr, M., Roullet, F. I., Hung, A. Y., Sheng, M., & Crawley, J. N. (2011a). Communication impairments in mice lacking Shank1: Reduced levels of ultrasonic vocalizations and scent marking behavior. *PLoS One, 6*(6), e20631.

Wöhr, M., Roullet, F. I., & Crawley, J. N. (2011b). Reduced scent marking and ultrasonic vocalizations in the BTBR T+ tf/J mouse model of autism. *Genes, Brain and Behavior, 10*(1), 35–43.

Würbel, H. (2002). Behavioral phenotyping enhanced—beyond (environmental) standardization. *Genes, Brain and Behavior, 1*(1), 3–8. doi:10.1046/j.1601-1848.2001.00006.x.

Yang, M., Perry, K., Weber, M. D., Katz, A. M., & Crawley, J. N. (2011). Social peers rescue autism-relevant sociability deficits in adolescent mice. *Autism Research, 4*(1), 17–27.

Yauk, C., Polyzos, A., Rowan-Carroll, A., Somers, C. M., Godschalk, R. W., Van Schooten, F. J., et al. (2008). Germ-line mutations, DNA damage, and global hypermethylation in mice exposed to particulate air pollution in an urban/industrial location. *Proceedings of the National Academy of Sciences, 105*(2), 605–610.

Yeh, C. T., & Yen, G. C. (2005). Effect of vegetables on human phenolsulfotransferases in relation to their antioxidant activity and total phenolics. *Free Radical Research, 39*(8), 893–904.

Young, L. J. (2002). The neurobiology of social recognition, approach, and avoidance. *Biological Psychiatry, 51*(1), 18–26.

Young, L. J., Lim, M. M., Gingrich, B., & Insel, T. R. (2001). Cellular mechanisms of social attachment. *Hormones and Behavior, 40*(2), 133–138. doi:10.1006/hbeh.2001.1691.

Yu, M. L., Hsu, C. C., Gladen, B. C., & Rogan, W. J. (1991). In utero PCB/PCDF exposure: Relation of developmental delay to dysmorphology and dose. *Neurotoxicology and Teratology, 13*(2), 195–202.

Yusufishaq, S., & Rosenkranz, J. A. (2013). Post-weaning social isolation impairs observational fear conditioning. *Behavioural Brain Research, 242*, 142–149.

Zakharova, E., Miller, J., Unterwald, E., Wade, D., & Izenwasser, S. (2009). Social and physical environment alter cocaine conditioned place preference and dopaminergic markers in adolescent male rats. *Neuroscience, 163*(3), 890–897. doi:10.1016/j.neuroscience.2009.06.068.

Zeng, H.-C., Zhang, L., Li, Y.-Y., Wang, Y.-J., Xia, W., Lin, Y., et al. (2011). Inflammation-like glial response in rat brain induced by prenatal PFOS exposure. *Neurotoxicology, 32(*(1), 130–139. doi:10.1016/j.neuro.2010.10.001.

Zerwas, M., Trouche, S., Richetin, K., Escudé, T., Halley, H., Gerardy-Schahn, R., et al. (2016). Environmental enrichment rescues memory in mice deficient for the polysialytransferase ST8SiaIV. *Brain Structure and Function, 221*(3), 1591–1605.

Zhang, D., Dong, Y., Li, M., & Wang, H. (2012a). A radio-telemetry system for navigation and recording neuronal activity in free-roaming rats. *Journal of Bionic Engineering, 9*(4), 402–410.

Zhang, T.-Y., & Meaney, M. J. (2010). Epigenetics and the environmental regulation of the genome and its function. *Annual Review of Psychology, 61*, 439–466.

Zhang, X.-F., Zhang, L.-J., Feng, Y.-N., Chen, B., Feng, Y.-M., Liang, G.-J., et al. (2012b). Bisphenol A exposure modifies DNA methylation of imprint genes in mouse fetal germ cells. *Molecular Biology Reports, 39*(9), 8621–8628.

Chapter 10
Animal Models of Addiction: Genetic Influences

Nathan A. Holtz and Marilyn E. Carroll

Introduction

Despite similar opportunities to abuse illicit drugs, some individuals develop substance use disorders while others do not (Bock and Whelan 1992; Murphy et al. 1989; Pomerleau 1995; Shiffman 1989). This phenomenon of differential liability is an issue that is central to our understanding of substance abuse and its prevention. It is clear that many factors contribute to these divergent vulnerability profiles (e.g., cultural and socioeconomic conditions, psychiatric liability, genetic predisposition), and of these factors, genetic variance has substantial influence (Agrawal et al. 2012; Swendsen and Le Moal 2011). For instance, adoption and twin studies suggest that 33–79 % of the variance in the occurrence of substance use disorders between individuals can be accounted for by genetic factors (for review, see Agrawal and Lynskey 2008).

Animal research has supported this notion using a variety of species and experimental techniques. One widely used method employs complex breeding strategies and genetic manipulations in mice to isolate the role of particular genes in functional aspects of addiction vulnerability. The scope of this literature demands an independent focus on such methods, and for that we refer readers elsewhere (Crabbe 2002, 2008; Hall et al. 2012; Le Foll et al. 2009). The present chapter will focus on the use of relatively simple yet powerful techniques for examining the role of genetics in addiction liability and related behavioral traits in the rat. The traits or

N.A. Holtz, Ph.D.
Department of Psychiatry and Behavioral Sciences, University of Washington,
1959 NE Pacific, Box 356560, Seattle, WA 98915, USA
e-mail: holtzn@uw.edu

M.E. Carroll (✉)
Department of Psychiatry, University of Minnesota,
2450 Riverside Ave. S., Minneapolis, MN 55454, USA
e-mail: mcarroll@umn.edu

J.C. Gewirtz, Y.-K. Kim (eds.), *Animal Models of Behavior Genetics*,
Advances in Behavior Genetics, DOI 10.1007/978-1-4939-3777-6_10

individual differences that will be discussed in this review with respect to drug abuse vulnerability include novelty reactivity, sweet preference, impulsivity, physical activity, emotional reactivity and reward-stimulus attribution. Sex/gender is another major individual difference that is related to all aspects of drug abuse; however, this factor and its accompanying hormonal influences have been previously discussed in recent reviews (Anker and Carroll 2011; Carroll and Anker 2010). It is important to note, however, that the vulnerability factors that will be discussed interact with sex differences, and these factors are additive.

The *Topics* section will first provide some background on the use of rats in examining the role of genetics in substance use disorders, and the different methods that have been used to this end (i.e., selective breeding, selection, between-strain comparisons). It will give some examples of bidirectional criteria for the selection or selective breeding of these animals based on traits related to substance use liability, and other psychiatric disorders, in humans. To illustrate the degree to which different drug-vulnerable and -resilient phenotype dichotomies share other bidirectional behavioral profiles, these models will also be compared and contrasted with regard to responding for drug or non-drug reward, as well as responding to aversive events. A question addressed by this section asks whether these animal models provide evidence for one or multiple addiction-prone phenotypes. Finally, the *Future Directions* section will describe important areas in which little research been done, namely that which may establish the use of specific individual differences as endophenotypes to help predict responses to particular therapeutic interventions. An endophenotype is a measurable intermediary between a disease (i.e., substance dependence) and its heritable biological mechanisms (Gottesman and Gould 2003). Ultimately, the goal of this research is to guide more effective, personalized treatment strategies.

Topics: Background into the Behavioral Genetics of Addiction

Observations of differential drug use vulnerability in non-human mammals precede recent developments in the systematic application of animal models of drug self-administration (Higgins 2003). For example, Darwin noted in *The Descent of Man, and Selection in Relation to Sex* (Darwin 1874) that while "many kinds of monkeys have a strong taste for tea, coffee, and spirituous liquor… An American monkey, an Ateles, after getting drunk on brandy, would never touch it again, and thus was wiser than many men." Current research has expanded on this prescient observation using a variety of methodologies. One approach involves assessing animals from genetically heterogeneous outbred stocks for differences in drug-related measures, such as drug-induced locomotor behavior or drug self-administration. One assumption of this method is that rats are produced from more-or-less random mating pairs, and can therefore allow for inferences to human populations. This assumption has been supported with a study by Deroche-Gamonet et al. (2004), in which a large number of outbred rats were assessed with various criteria for substance use disorders that are used in humans; such as, increased motivation to take drug (progressive ratio responding), relapse after quit attempts (reinstatement), and persistence despite negative consequences (resilience of shock-punished cocaine self-administration). Results indicated a large amount of variability between the animals in meeting different criteria. Importantly, as in the case with humans, only a small subset of rats (i.e., ~15 %) that were exposed to the drug met all of these criteria for dependence (American Psychiatric Association, and Task Force on DSM-IV 1994). Presumably, these differences in vulnerability were largely mediated by genetic differences between individual rats, as experimental procedures were consistent between all animals.

One prediction arising from these data is that heterogeneous animals that display drug abuse vulnerability also exhibit behaviors that model genetically mediated traits found to accompany drug abuse vulnerability in humans. Rats screened for behavioral traits from outbred, heterogeneous populations (not selectively bred) are referred to in this review as "selected" rats (e.g., "rats selected for this trait"). One advantage of this method is that it approximates the heterogeneity of human populations. A disadvantage, however, is that screening procedures may have confounding effects on subsequent assessments (Marusich et al. 2011a). Another prediction is that the features of high and low drug use susceptibility and related behaviors (or physiological markers) can be used as selective breeding criteria to establish high and low drug-vulnerable phenotypes. This review will refer to rats derived from selective breeding as "selectively bred" lines (e.g., "rats selectively bred for this trait"). With these selective breeding procedures, experimenters typically avoid inbreeding in long-running high- and low-vulnerable lines by prohibiting certain mating pairs within phenotypes (sibling, half-sibling, cousin, etc.). A related prediction is that different rat strains (i.e., Lewis, Fischer 344, etc.) will exhibit variance in drug abuse vulnerability and related behaviors as a consequence of intentional inbreeding or the genetic isolation that arises from commercial breeding practices. An advantage to selectively bred lines and inbred strains is that behaviors of interest are more stable because, in principle, they more heavily mediated by genotype. One disadvantage is that, as discussed later in the present chapter, these high and low phenotypes may only represent a small proportion of the extreme ends of a heterogeneous population. The following subsections will illustrate how these predictions have been tested and generally supported using these three basic approaches: behavioral selection from outbred stocks, selective breeding based on extremely high or low behavioral measures, and between-strain comparisons.

Topics: Behavioral Phenotypes

Drug Consumption

Initial studies conducted in rats to examine the influence of genes on substance use vulnerability used extremely high or low intake of or sensitivity to alcohol as the criteria for selective breeding. This strategy resulted in multiple alcohol-preferring and non-preferring phenotypes (AA/ANA, alcohol-accepting and non-accepting;

HAC/LAC, high and low alcohol consumers; P/NP, alcohol-preferring and non-preferring) (Bell et al. 2006; Deitrich 1993; Foroud et al. 2003; Kiianmaa et al. 1991; Li et al. 1993, 1994; McBride and Li 1998), many of which continue to remain active and stable after many generations (Roman et al. 2012). These lines have provided strong convergent evidence for the heritability of substance abuse vulnerability, and have been extensively investigated to better understand its neurobiological underpinnings and associated behavioral characteristics (Froehlich 2010; Roman et al. 2012). Other selective breeding programs have shown that genetic susceptibility is not limited to alcohol, with lines differentiated by their intake or response to other drugs of abuse, such as such as cocaine (He et al. 2008), diazepam (Gallaher et al. 1987), methamphetamine (Atkins et al. 2001; Wheeler et al. 2009), nicotine (Smolen et al. 1994), and opiates (Belknap et al. 1983). Interestingly, an early study involving the selective breeding of high and low morphine-consuming rats showed that the high-consuming animals also had a greater preference for ethanol (Nichols and Hsiao 1967). Similarly, increased vulnerability to drugs other than the drug rats have been selected for has also been reported (de Fiebre et al. 2002; Gordon et al. 1993; Le et al. 2006) (but see Acewicz et al. 2012; Carlson and Perez 1997), suggesting that some common mechanisms may be involved in the genetic mediation of liability to all abused substances.

Novel Reactivity, Novelty Preference, and Sensation/ Novelty-Seeking

Humans with substance use disorders rate higher on measures of the sensation/novelty-seeking trait compared to the non-abusing population (Ersche et al. 2012). An important question follows from this observation: does drug abuse itself increase sensation-seeking behavior, or does the sensation-seeking phenotype also tend to co-occur with the addiction-prone phenotype irrespective of drug exposure? Piazza and coworkers (1989) investigated this question by first assessing locomotor activity in a population of outbred rats after being placed in a novel environment. Exploration or activity in a novel environment has been proposed to be an animal model of human sensation-seeking (Blanchard et al. 2009; Dellu et al. 1996; Howard et al. 1997). The animals were then assessed for amphetamine self-administration, and locomotor activity in the novel environment was positively associated with amphetamine consumption, suggesting that this behavioral predisposition was present in drug-prone rats prior to drug exposure.

Outbred rats screened for high and low novel environment activity have been termed High- (HR) and Low Responders (LR), respectively. In addition to amphetamine self-administration, HR rats display elevated amphetamine—(Deminiere et al. 1989), cocaine—(Hooks et al. 1991), ethanol—(Gingras and Cools 1996), and morphine-induced locomotor activity (Deroche et al. 1993). HR rats do not self-administer more cocaine than LR rats (Belin et al. 2008), although rats selectively bred for extremely high or low expression of this trait (bHR vs. bLR), do differ (bHR>bLR) in acquisition rates (Davis et al. 2008) and in their motivation to self-administer cocaine (Cummings et al. 2011). In general, the literature shows that HR and bHR rats are more drug-prone than their LR and bLR counterparts (Kabbaj 2006), although this relationship might vary with the strain of the progenitor population and experimental conditions (Cools and Gingras 1998; Izidio and Ramos 2007).

Recently, addiction vulnerability has been explored with a trait related to novel environment reactivity (Belin et al. 2011). This trait is termed high novelty preference, as it is assessed with a free choice paradigm in which the amount of time an animal chooses to spend in a novel (vs. familiar) environment is measured. While the novelty preference and novel environment reactivity paradigms share some similarities, it is important to underscore that they are measuring behavior under two different circumstances (choice in environment vs. no choice in environment), and are thus likely measuring two distinct but overlapping traits. Accordingly, rats selected for the high novelty-preferring phenotype (HNP) show greater vulnerability to cocaine self-administration and cocaine-seeking across a number of measures compared to the low novelty-preferring phenotype (LNP) (Belin et al. 2011), while HR and LR rats show no difference in cocaine self-administration (Belin et al. 2008).

Sweet Intake

Avidity for sweet dietary substances is another behavioral trait that is positively associated with substance use disorders in human populations. For instance, those who abuse alcohol (Kampov-Polevoy et al. 1997, 2001; Wronski et al. 2007), cocaine (Janowsky et al. 2003), nicotine (Pepino and Mennella 2007; Pomerleau et al. 1991), and opioids (Weiss 1982) experience greater hedonic effects of sweetened dietary substances than those who do not abuse these drugs. Like addiction vulnerability, variability in response to sweets is a stable, genetically mediated trait (Desor and Beauchamp 1987; Keskitalo et al. 2007a, b, 2008; Reed et al. 1997; Uhl et al. 2009), and research with rats suggests that these two traits may be moderated by common mechanisms. For instance, rats screened for high sucrose-feeding (HSF) consumed more amphetamine and acquired cocaine self-administration at faster rates than their low-sucrose feeding (LSF) counterparts (DeSousa et al. 2000; Gosnell 2000). Additionally, outbred rats screened for high intake of a solution containing the artificial sweetener saccharin (sweet-likers; SL) consumed more ethanol (Gahtan et al. 1996) and morphine (Gosnell et al. 1995), but not cocaine (Gahtan et al. 1996; Gosnell et al. 1998), than rats screened for low saccharin intake (sweet-dislikers: SDL). Furthermore, rats selectively bred for high saccharin intake (HiS) are more drug-prone than rats bred for low saccharin intake (LoS) (Carroll et al. 2008). HiS rats consume more ethanol (Dess et al. 1998) and cocaine (Holtz and Carroll 2013), acquire (Perry et al. 2007a) and escalate (Perry et al. 2006) cocaine self-administration at faster rates, and show greater cocaine-induced locomotor activity (Carroll et al. 2007) and drug-primed reinstatement of cocaine-seeking (Perry et al. 2006). Together, these studies suggest that avidity for sweets also serves as an endophenotype for substance abuse vulnerability.

Impulsivity

Impulsivity is a heritable, multifaceted feature common to a number of psychiatric disorders, including substance use disorders (Perry and Carroll 2008). Chapter 3 of this book is dedicated to the use of animals to model behavioral genetics of impulsivity; therefore, we present only a brief discussion of the topic here. One aspect of this trait is impulsive choice, which is assessed in humans and animals with the delay discounting procedure. During this task, the subject may choose between a small, immediate reward, and a larger reward that is delivered after some delay. Human addicts have a higher preference for the small, immediate reward, indicating that they show greater choice impulsivity (MacKillop 2013). Rats screened for high choice impulsivity (HiI) are also more drug-prone than rats screened for low choice impulsivity (LoI) (Anker et al. 2009a; Perry et al. 2008).

Another component of impulsivity that is associated with drug use in humans and animals is impulsive action, or the impaired ability to restrain goal-seeking behavior or terminate an ongoing, potentially deleterious action. A common procedure for assessing this in the rat is the 5-choice serial reaction time task (5-CSRTT) (Robinson et al. 2009). In this task, the rat must withhold a response during a stimulus presentation in order to receive a food reward. Rats that are unable to withhold inappropriate responding in this paradigm are characterized as high action-impulsive (HI), and they are more drug-prone than low action-impulsive rats (LI) (Jupp et al. 2013). Another example of action-impulsivity discussed in the present chapter is the Go/No-go task, in which reward (Go) or nonreward (No-go) are predicted by discrete stimuli. Failure to withhold responding to the No-go stimulus serves as a metric of impulsive action, similar to the Stop-Signal Reaction Time Task (SSRT) used in clinical research (Perry and Carroll 2008). Although performance in these tasks may be mediated by different mechanisms (Eagle et al. 2008), humans with substance use disorders also exhibit impaired response inhibition as measured by the Go/No-go task (Pike et al. 2011).

Physical Activity and Wheel-Running

Avidity for physical activity or exercise is a genetically mediated characteristic (Bauman et al. 2012) that may also serve as a phenotypic marker for vulnerability to substance use disorders (Olsen 2011). Despite a dearth of clinical evidence, a couple of rat studies have supported the proposition. Ferreira et al. (2006) found that wheel-running in a heterogeneous rat population was positively related to amphetamine-induced locomotor activity, and Larson and Carroll (2005) found that rats screened for high wheel-running (HiR) self-administered more cocaine and showed greater drug-primed reinstatement of cocaine-seeking compared with rats screened for low wheel-running (LoR). Together these studies provide foundational evidence that avidity for physical activity may predict addiction susceptibility.

Sign-Tracking and Incentive Salience

Incentive salience was originally proposed less as a trait than a theoretical construct seeking to explain the role of neurobiological substrates, namely the mesolimbic dopamine (DA) system, in the attribution of reward-related stimuli with desirable or motivational properties (opposed to hedonic or reward-prediction error theories of DA) (Berridge 2012). As nearly all drugs of abuse involve activation of the DA system, addiction in this framework is conceptualized as a dysfunction of incentive salience attribution, such that drug-related stimuli become compulsively desired after prolonged drug use (Tomie et al. 2008). Preclinical research with rats has been able to assess individual differences in the propensity to attribute incentive salience by exploiting the behavioral phenomenon of sign-tracking (also called autoshaping) (Breland and Breland 1961; Tomie et al. 1989). One way of measuring sign-tracking in rats is by extending a retractable lever for a brief period. The lever is then retracted, followed immediately by the delivery of a food pellet in an adjacent receptacle. An important component to this procedure is that the animal is not required to press the lever to receive a food pellet; the lever only serves as a conditioned stimulus that indicates an impending reward (i.e., food delivery is noncontingent). In heterogeneous rat populations, a certain subset of animals, called sign-trackers (ST), will make numerous contacts (e.g., biting, pressing) with the lever during its presentation. This is thought to be a behavioral indication of incentive salience, as the reward-predictive stimulus, or the lever in this instance, develops desirable properties. Other rats, called goal-trackers (GT), tend to contact the reward site (e.g., the food receptacle), following the lever presentation. As predicted by the incentive salience theory, rats screened as ST display greater sensitization to cocaine-induced locomotor activity than GT rats (Flagel et al. 2008). Rats screened as ST also self-administer more cocaine and exhibit more drug-primed reinstatement of cocaine-seeking and cocaine-seeking despite punishment compared to GT (Saunders and Robinson 2011; Saunders et al. 2013). Despite the absence of a reliable assay for exaggerated incentive salience attribution in drug-abusing humans, we will later discuss how other traits, such as novelty-seeking, appear to be strongly associated with this putative phenotypic dimension.

Active Avoidance, Forced Swim Test Mobility, and Emotional Reactivity

Emotional reactivity following aversive events is another dispositional construct that has been associated with several psychiatric disorders, including substance use vulnerability (Aldao et al. 2010; Carmody et al. 2007; Sinha 2011). One way emotional reactivity has been modeled in rats is with the active avoidance paradigm (Cummings et al. 2011). With this procedure, the animal is placed in a two-way shuttle box and exposed to an electric shock presented in an alternating fashion in one of the two chambers comprising the box. The animal is able to avoid the shock by "shuttling" from one side of the box to the other. While some animals readily acquire active avoidance behavior, other animals adopt passive coping responses to the shocks (e.g., freezing). Selective breeding of these active and passive phenotypes has resulted in the Roman high- (RHA) and low- (RLA) avoidance rat lines, which exhibit drug-prone and -resilient profiles, respectively.

RHA rats acquire cocaine self-administration at faster rates, show greater extinction, reinstatement, and reacquisition of cocaine-seeking (Fattore et al. 2009), and RHA rats consume more ethanol (Guitart-Masip et al. 2006; Manzo et al. 2012) compared to RLA rats. RHA rats also show greater cocaine-, amphetamine-, and morphine-induced locomotor activity (Giorgi et al. 1997; Lecca et al. 2004), and higher measures of behavioral sensitization following cocaine (Giorgi et al. 2005; Haney et al. 1994) and morphine (Piras et al. 2003) administration compared to RLA rats.

Another test of emotional reactivity employs the aversive forced swim test. Rats tend to adopt either a passive, immobile coping strategy (i.e., floating), or an active escape strategy (i.e., swimming, attempts to climb out of swimming apparatus). The passive coping strategy is considered to be an animal model of depression, as antidepressants decrease this behavior and increase active coping (Weiss et al. 1998). In humans, depression is a heritable disorder that exhibits high comorbidity with substance abuse (Pettinati et al. 2013). However, rat models addressing this phenomenon have been mixed. In contrast to rats bred for avoidance behavior in the shuttle box test, rats selectively bred for active coping in the forced swim test are not consistently more drug-prone than their passive counterparts. For instance, rats bred for susceptibility (SUS) to the swim-immobility enhancing effects of a pre-swim test stressor consume more ethanol than those bred for stressor resistance (Weiss et al. 2008; West and Weiss 2006). Additionally, rats bred for greater immobility (swim-low, SwLo) in the absence of a pretest stressor orally ingest greater amounts of stimulants compared to those bred for low immobility (swim-high, SwHi) in the forced swim test (Weiss et al. 2008), although the relationship is opposite when the delivery of these drugs is response-contingent (Lin et al. 2012). Furthermore, there is no difference in ethanol consumption between SwLo and SwHi animals (Weiss et al. 2008). While originally bred with another bidirectional criterion (hypothermic response to the 5-HT agonist 8-OH-DPAT) (Overstreet and Djuric 2001), the Flinders sensitive (FSL) and resistant (FRL) lines display robust differences in passive swim test performance (FSL > FRL), and have therefore been extensively studied as an animal model of genetically mediated differences in depression (Overstreet et al. 2005; Overstreet and Wegener 2013). Cocaine self-administration and cocaine-induced locomotor activity were lower in the FSL vs. FRL rats (Fagergren et al. 2005). Together, these data suggest that performance in the forced swim test may have some bearing on drug abuse vulnerability, although this relationship may be especially sensitive to differences in experimental procedure.

Strain

The Lewis (LEW) and Fischer (F344) rats are long-running, inbred strains that display a wide variation over a host of physiological characteristics. As such, they are extensively used to model opposing ends of various pathological spectra, including drug use vulnerability. For instance, LEW rats consume more ethanol (Suzuki et al. 1988), cocaine (Kruzich and Xi 2006b), methamphetamine (Kruzich and Xi 2006a), heroin (Picetti et al. 2012), and morphine (Garcia-Lecumberri et al. 2011) compared to F344 rats. It is noteworthy that data regarding cocaine-seeking and cocaine consumption is mixed with these strains. For example, some studies rate LEW rats higher on these measures (Kruzich and Xi 2006b; Miguens et al. 2011; Picetti et al. 2010), while others rate F344 rats higher (Christensen et al. 2009; Freeman et al. 2009; Kosten et al. 2007).

Another inbred strain are the Fawn-hooded (FH) rats, which, given their passive performance in the forced swim test, have served as an animal model of depression and comorbid alcoholism. Accordingly, the FH rats exhibit increased ethanol consumption compared to their normative comparison group, the inbred ACI rat strain (Rezvani et al. 2002). Importantly, the ACI strain shows more mobility during the forced swim test compared to the FH strain. FH rats also consume more ethanol compared to the inbred Wistar-Kyoto (WKY) rats (Chen et al. 1998; Lodge and Lawrence 2003; Rezvani et al. 2002). However, the locomotor enhancing effects of cocaine and amphetamine were decreased in FH rats compared to outbred Wistar rats (Hall et al. 2001), suggesting that FH rats are relatively less vulnerable to psychostimulants. One important consideration is that a unifying interpretation of these data is impeded by the use of different control groups (e.g., ACI, Wistar, Wistar-Kyoto, etc.) between these experiments. Future studies may benefit by using only one control strain.

The spontaneously hypertensive (SHR) inbred rat strain originated from a breeding procedure in which elevated blood pressure was the selection criterion. Due to their relative increases in impulsivity and other behaviors, the SHR rats have been considered as an animal model of attention deficit/hyperactivity disorder (ADHD) compared to their normotensive WKY controls (Meneses et al. 2011) (but see Fox et al. 2008). SHR rats consume more ethanol (Khanna et al. 1990) than WKY rats, and are generally more sensitive to the rewarding effects of psychostimulants, cannabinoids, and opioids compared to various other rats strains (Vendruscolo et al. 2009). There is disagreement regarding the proper control strain for the SHR rat (Alsop 2007; Sagvolden et al. 2009; van den Bergh et al. 2006), and the relationship between the behavioral features of this phenotype and drug abuse vulnerability should be interpreted with caution.

Topics: Comparing and Contrasting Phenotypes

This review has given examples of rat models that have been used to investigate the heritable nature of both liability to substance use disorders and behavioral features that tend to co-occur in substance using populations. Overall, these data suggest that many of the behavioral profiles exhibited by human substance users, such as sensation-seeking and impulsivity, reflect behavioral phenotypes that are present prior to drug exposure. These high- and low-vulnerable models have been used to identify endophenotypes that increase our understanding of the neurobiological underpinnings of addiction susceptibility. This knowledge may guide diagnoses and personalized treatment strategies, and it will be important to understand how the particular traits relate to a common substance use liability factor, or if they represent distinct components of unique neurobiological profiles that share the common behavioral outcome of substance use. The following discussion will address this issue by comparing and contrasting the drug-prone and -resistant phenotypes described thus far with regard to behavioral indices of substance use liability other than that for which they were screened or selectively bred. For example, one question asks if the more drug-vulnerable HR rats show greater avidity for sweets compared to LR rats. If so, this would suggest that sensation-seeking, sweet preference, and substance use liability may share some common genetically mediated mechanisms. However, equal (HR = LR) or opposite (HR < LR) avidity for sweets would support the conclusion that these traits mark etiologically distinct substance use disorders that may be amenable to particular therapeutic interventions.

Novel Environment Reactivity

Table 10.1 illustrates how drug-prone and resistant phenotypes perform when introduced to a novel environment. In accordance with the HR/LR model, the drug-vulnerable HI (Molander et al. 2011), HiS (Dess and Minor 1996), RHA (Pisula 2003), and SHR (Weiss et al. 1998) rats all exhibit greater novel reactivity compared to their drug-resistant counterpart (LI, LoS, RLA, and WKY, respectively). In contrast, the ethanol-vulnerable AA and P rats did not exhibit differ in this measure compared to the ethanol-protected ANA and NP rats (Badishtov et al. 1995). However, with the generally drug-prone phenotypic models of depression, the FSL and SwLo, the less vulnerable rats (FRL, SwHi) exhibited greater novel environment activity (Fagergren et al. 2005; van den Bergh et al. 2006). Similarly, two studies showed that the generally drug-prone LEW strain exhibits less reactivity than the relatively drug-resistant F344 strain (Chaouloff et al. 1995; Miserendino et al. 2003), although other studies have shown the opposite (LEW > F344) (Ambrosio et al. 1995; Camp et al. 1994) or no relationship (Kosten et al. 2007). Kosten and Ambrosio (2002) proposed that these conflicting data are likely the result of procedural differences between experiments. Interestingly, LEW rats exhibit greater preference for a novel environment than F344 rats (Miserendino et al. 2003).

Table 10.1 Summary of studies comparing novel environment reactivity in drug-prone and -resistant phenotypes

Drug-vulnerable behavior	Phenotype comparison (drug-prone vs. drug-resistant)	Reference
Novel environment reactivity	AA=ANA	Badishtov et al. (1995)
	FSL<FRL	Fagergren et al. (2005)
	HI>LI	Molander et al. (2011)
	HI=LI	Belin et al. (2008)
	HiS>LoS	Dess and Minor (1996)
	LEW>F344	Ambrosio et al. (1995)
	LEW<F344	Camp et al. (1994)
	LEW=F344	Chaouloff et al. (1995)
		Miserendino et al. (2003)
		Kosten et al. (2007)
	P=NP	Badishtov et al. (1995)
	RHA>RLA	Pisula (2003)
	SwLo<SwHi	van den Bergh et al. (2006)
	SHR>WKY	Weiss et al. (1998)

AA and *ANA* alcohol-accepting and alcohol non-accepting, *LEW* and *F344* Lewis and Fischer 344, *FSL* and *FRL* Flinders sensitive line and resistant line, *HI* and *LI* high and low impulsivity (5-CSRTT), *HiS* and *LoS* high and low saccharin intake, *P* and *NP* alcohol-preferring and non-preferring, *RHA* and *RLA* Roman high and low avoidance, *SHR* and *WKY*, spontaneously hypertensive and Wistar-Kyoto, *SwHi* and *SwLo* hi and low forced swim activity

Sweet Intake

Table 10.2 shows that rats selectively bred for high alcohol consumption (AA, P) show a greater avidity for sweetened dietary substances compared to rats bred for low alcohol consumption (ANA, NP) (Overstreet et al. 1993; Riley et al. 1977; Sinclair et al. 1992; Stewart et al. 1994). Similarly, FSL, HI, HR, RHA, SHR, and SUS rats consume more sweet substances compared to their drug-resilient counterparts (Diergaarde et al. 2009; Dommett and Rostron 2013; Fernandez-Teruel et al. 2002; Klebaur et al. 2001; Marusich et al. 2011b; Overstreet et al. 1993; Pucilowski et al. 1993; West and Weiss 2006). In contrast, HiI rats and LEW rats do not differ from LoI and F344 rats, respectively, when tested for saccharin consumption under the same procedures used to assess saccharin preference and selectively bred HiS and LoS rats (Holtz et al., in preparation; Perry and Carroll, unpublished data). However, Tordoff et al. (2008) showed that LEW rats consume more saccharin solution compared to F344 rats, but only at the smallest of the many concentrations that were tested.

Table 10.2 Summary of studies comparing sweet preference/consumption in drug-prone and -resistant phenotypes

Drug-vulnerable behavior	Phenotype comparison (drug-prone vs. drug-resistant)	Reference
Sweet preference/consumption	AA > ANA	Sinclair et al. (1992)
	FSL > FRL	Overstreet et al. (1993)
		Pucilowski et al. (1993)
	HI > LI	Diergaarde et al. (2009)
	HiI = LoI	Perry et al. (unpublished data)
	HR > LR	Klebaur et al. (2001)
	LEW > F344	Tordoff et al. (2008)
	LEW = F344	Holtz et al. (in preparation)
	P > NP	Sinclair et al. (1992)
		Stewart et al. (1994) Overstreet et al. (1993)
		Riley et al. (1977)
	RHA > RLA	Fernandez-Teruel et al. (2002)
	SHR > WKY	Dommett and Rostron (2013)
		Marusich et al. (2011b)
	SUS > RES	West and Weiss (2006)

AA and *ANA* alcohol-accepting and alcohol non-accepting, *FSL* and *FRL* Flinders sensitive and resistant line, *HI* and *LI* high and low impulsive (5-CSRTT), *HiI* and *LoI* high and low impulsive (delay discounting), *HR* and *LR* high and low responders, *LEW* and *F344* Lewis and Fischer 344, *P* and *NP* alcohol-preferring and non-preferring, *RHA* and *RLA* Roman high and low avoidance, *SHR* and *WKY* spontaneously hypertensive and Wistar-Kyoto, *SUS* and *RES* susceptible and resistant to stress-induced forced swim immobility

Impulsivity

Broos et al. (2012) found no differences between rats screened for high or low choice impulsivity (HiI vs. LoI) when assessed for action-impulsivity using the 5-CSRTT (Table 10.3). Belin et al. (2008) similarly found no differences between HR and LR rats with this measure. In contrast, Flagel et al. (2010) found that bHR rats are less impulsive within a choice paradigm compared to bLR rats. Interestingly, ST exhibits less impulsive choice than GT, which is in opposition to their action-impulsivity profile (ST > GT) (Lovic et al. 2011). In contrast, ST and RHA rats exhibit more impulsive action compared to their drug-resilient counterparts (Lovic et al. 2011; Moreno et al. 2010). With the Go/No-go procedure, HiI rats were less action-impulsive than LoI rats when i.v. cocaine was the reward (Anker et al. 2009b). On the other hand, HiS rats made more No-go responses than LoS rats when food was the reward (Anker et al. 2008). HAC, HiS, LEW, and RHA rats all

Table 10.3 Summary of studies comparing impulsivity in drug-prone and -resistant phenotypes

Drug-vulnerable behavior	Phenotype comparison (drug-prone vs. drug-resistant)	Reference
Impulsive action: 5CSRTT or 2CSRTT*	HiI=LoI	Broos et al. (2012)
	HR=LR	Belin et al. (2008)
	RHA>RLA	Moreno et al. (2010)
	ST>GT*	Lovic et al. (2011)
Impulsive action: Go/No-Go	HiI<LoI	Anker et al. (2009b)
	HiS>LoS	Anker et al. (2008)
Impulsive choice: delay discounting	bHR<bLR	Flagel et al. (2010)
	HAC>LAC	Wilhelm and Mitchell (2008)
	HI>LI	Robinson et al. (2009)
	HI=LI	Broos et al. (2012)
	HiS>LoS	Perry et al. (2007b)
	LEW>F344	Madden et al. (2008)
		Garcia-Lecumberri et al. (2011)
		Anderson and Diller (2010)
		Huskinson et al. (2010)
	RHA>RLA	Moreno et al. (2010)
	SHR=WKY	Garcia and Kirkpatrick (2012)
	ST<GT	Lovic et al. (2011)

bHR and *bLR* bred high and low responders, *HAC* and *LAC* high and low alcohol consumers, *HI* and *LI* high and low impulsive (5-CSRTT), *HiI* and *LoI* high and low impulsive (delay discounting), *HiS* and *LoS* high and low saccharin intake, *HR* and *LR* high and low responders, *LEW* and *F344* Lewis and Fischer 344, *RHA* and *RLA* Roman high and low avoidance, *SHR* and *WKY* spontaneously hypertensive and Wistar-Kyoto, *ST* and *GT* sign- and goal-trackers

rate higher in choice impulsivity than their drug-resilient phenotypes (Anderson and Diller 2010; Garcia-Lecumberri et al. 2011; Huskinson et al. 2010; Madden et al. 2008; Moreno et al. 2010; Perry et al. 2007b; Wilhelm and Mitchell 2008). One study showed that rats screened for high action-impulsivity (HI) showed greater impulsive choice behavior compared to LI rats (Robinson et al. 2009), yet another study showed no difference (Broos et al. 2012). SHR rats also do not differ on this measure than their WKY control group (Garcia and Kirkpatrick 2012).

Physical Activity and Wheel-Running

Under baseline conditions (e.g., food satiated, drug-free, etc.), the drug-prone LEW and HAC rats exhibit more wheel-running compared to F344 and LAC rats, respectively (Table 10.4) (Riley et al. 1977; Werme et al. 2000), while the HiS and HR

Table 10.4 Summary of studies comparing wheel-running in drug-prone and -resistant phenotypes

Drug-vulnerable behavior	Phenotype comparison (drug-prone vs. drug-resistant)	Reference
Wheel-Running	FSL < FRL	Bjornebekk et al. (2005)
	LEW > F344	Werme et al. (2000)
	HAC > LAC	Riley et al. (1977)
	HiS = LoS	Dess et al. (2000)
	HR = LR	Ferreira et al. (2006)
	SHR < WKY	Kingwell et al. (1998)

FSL and *FRL* Flinders sensitive and resistant line, *LEW* and *F344* Lewis and Fischer 344, *HAC* and *LAC* high- and low alcohol consuming, *HiS* and *LoS* high and low saccharin intake, *HR* and *LR* high and low responders, *SHR* and *WKY* spontaneously hypertensive and Wistar-Kyoto

lines show similar wheel-running avidity compared to their low-vulnerable counterparts (Dess et al. 2000; Ferreira et al. 2006). In contrast, the vulnerable FSL and SHR rats showed correspondingly less wheel-running compared to FRL and WKY rats (Bjornebekk et al. 2005; Kingwell et al. 1998).

Sign-Tracking and Incentive Salience

The more drug-prone HiI and LEW rats exhibit more sign-tracking compared to LoI, and F344 rats, respectively (Table 10.5) (Kearns et al. 2006; Tomie et al. 1998). Rats bred for high response in a novel environment (bHR) exhibit greater sign-tracking than bLR rats (Flagel et al. 2010, 2011); however, outbred rats screened for high novelty response (HR) do not differ from their LR counterparts (Robinson et al. 2009). Conversely, we have recently found that bred high (HiS) and low (LoS) saccharin-consuming rats do not differ in this measure, whereas outbred rats screened for saccharin intake do (SL > SDL) (Holtz et al., in preparation). These results suggest that selectively bred phenotypes may not necessarily represent the extreme ends of a relatively heterogeneous population distribution with regard to a particular behavior.

Forced Swim Test Immobility

Table 10.6 shows that LEW rats exhibit greater immobility in the forced swim test compared to F344 (Armario et al. 1995), whereas the more drug-prone P, RHA, and SHR rats exhibit less of this behavior compared to their drug-resilient counterparts (NP, RLA, WKY, respectively) (Armario et al. 1995; Diaz-Moran et al. 2012; Godfrey et al. 1997; Piras et al. 2010). HAD rats show no difference in this measure compared to LAD rats (Godfrey et al. 1997). HR and LR rats are also similar (Antoniou et al. 2008; Jama et al. 2008; White et al. 2007), supporting the notion that this test measures a behavior that is distinct from high and low motor activity.

Table 10.5 Summary of studies comparing incentive salience (sign-tracking) in drug-prone and -resistant phenotypes

Drug-vulnerable behavior	Phenotype comparison (drug-prone vs. drug-resistant)	Reference
Incentive salience: sign-tracking	bHR > bLR	Flagel et al. (2010, 2011)
	HiI > LoI	Tomie et al. (1998)
	HiS = LoS	Holtz et al. (in preparation)
	HR = LR	Robinson et al. (2009)
	LEW > F344	Kearns et al. (2006)
	SL > SDL	Holtz et al. (in preparation)

bHR and *bLR* bred high and low responders, *HiI* and *LoI* high and low impulsive (delay discounting), *HiS* and *LoS* high and low saccharin intake, *HR* and *LR* high and low responders, *LEW* and *F344* Lewis and Fischer 344, *SL* and *SDL* sweet-likers and sweet-dislikers

Table 10.6 Summary of studies comparing immobility during the forced swim test in drug-prone and -resistant phenotypes

Drug-vulnerable behavior	Phenotype comparison (drug-prone vs. drug-resistant)	Reference
Forced swim immobility	HAD = LAD	Godfrey et al. (1997)
	HR = LR	Antoniou et al. (2008)
		Jama et al. (2008)
		White et al. (2007)
	LEW > F344	Armario et al. (1995)
	P < NP	Godfrey et al. (1997)
	RHA < RLA	Diaz-Moran et al. (2012)
		Piras et al. (2010)
	SHR < WKY	Armario et al. (1995)

HAD and *LAD* high and low alcohol drinking, *HR* and *LR* high and low responders, *LEW* and *F344* Lewis and Fischer 344, *P* and *NP* alcohol-preferring and non-preferring, *RHA* and *RLA* Roman high and low avoidance, *SHR* and *WKY* spontaneously hypertensive and Wistar-Kyoto

Summary

The literature described in this section illustrates the role of genetics in mediating stable behavioral traits that predict drug vulnerability or resistance prior to drug exposure. These studies also suggest multiple independent drug-vulnerable behavioral phenotypic factors, as opposed to one unified drug-vulnerable phenotype. Some vulnerability markers appear to co-occur, such as avidity for sweets (e.g., HR > LR in sweet preference) and high novel environment reactivity (e.g., HiS > LoS in novel environment reactivity), suggesting that these traits might be mediated by genetic mechanisms that are in common or that otherwise tend to co-occur. This explanation may also account for instances in which high- and low-vulnerable phenotypes do not differ, such as how HR = LR and HI = LI with regard to impulsive action and novel environment reactivity, respectively. These data suggest that novel environment reactivity and action-impulsivity are genetically independent phenomena, both engendering unique drug-vulnerable profiles. Belin et al. (2008), for example, found that HR rats acquire cocaine self-administration faster than LR rats, but did not show differences in escalation of cocaine intake. On the other hand, HI rats exhibited greater escalation of cocaine intake compared to LI rats without differing in rates of acquisition. Together, these studies demonstrate that vulnerability to substance use disorders is associated with a "constellation" of behavioral traits and emphasize the need for animal models that concurrently capture multiple drug-prone or -resistant phenotypes.

Although not incongruent within the framework of overlapping yet distinctive vulnerability factors, it is difficult to reconcile how one set of phenotypes display a positive linear relationship with a given behavior and drug vulnerability, while another shows a negative relationship. For example, the drug-vulnerable LEW rats are more immobile in the forced swim test than the less drug-resilient F344 rats. In contrast, the drug-vulnerable SHR rats show less forced swim immobility compared to the drug-resilient WKY rats. One interpretation of this converse relationship is that forced swim test performance has little bearing on genetic vulnerability to drugs. This appears unlikely given the commonality of systems involved in both of these behaviors (Mikrouli et al. 2011; Roth-Deri et al. 2009).

An alternative explanation employs a non-monotonic model, such as that adapted from Yerkes and Dodson's (1908) model of stimulus strength and habit formation by Kosten and Ambrosio (2002) to account for the opposing linear relationships of HPA-axis activity and drug use vulnerability observed in LEW vs. F344 and HR vs. LR rats. Figure 10.1 shows an adaptation of this model that illustrates an inverted-U-shaped curve, which is based on the additional observations that (1) WKY rats display greater immobility than LEW and SHR rats (Armario et al. 1995; Marti and Armario 1996), (2) LEW exhibit more immobility than SHR rats (Hinojosa et al. 2006), and (3) SHR rats are more immobile than F344 rats (Armario et al. 1995). Note that when considering the four strains together, it is the intermediately immobile that are predicted to be the most drug-prone. This type of model can help us understand how the neurobiological substrates that confer extremely low or high phenotypic traits also mediate drug-resilience, and can emphasize the importance of simultaneously characterizing more than one dyad across a given behavioral dimension.

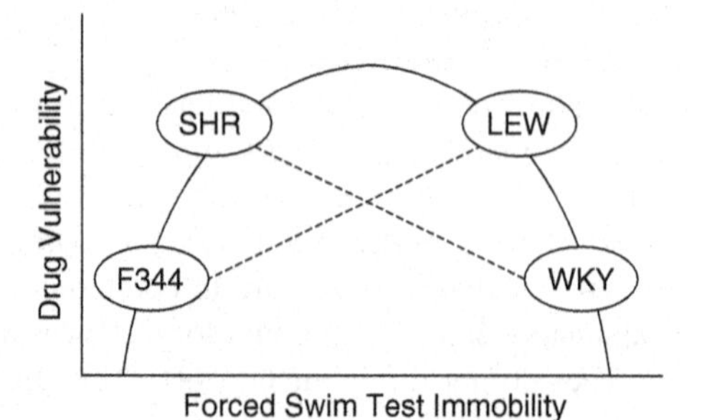

Fig. 10.1 A hypothetical, non-monotonic, inverted-U-shaped function describing intensity of drug vulnerability on the Y-axis against forced swim test immobility on the X-axis in F344, SHR, LEW, and WKY rats

Indeed, it has been found that, when comparing many different pairs of rats selectively bred for high and low ethanol consumption on tests of anxiety, there was a large range of variability when comparing the high-vulnerable phenotypes together (Roman et al. 2012). Since animal models may ultimately serve in establishing behavioral endophenotypes for guiding treatment, accounting for this kind of variance may enhance their comparative impact.

Future Directions

An understanding of how to predict addiction severity and customize treatments based on individual vulnerability should lead to better management of drug abuse treatments. For example, as treatment completion is positively associated with subsequent abstinence (Tzilos et al. 2009), clinical obstacles include patient dropout and lack of compliance (Baekeland and Lundwall 1975; Laudet et al. 2009). Furthermore, the more rapidly these treatment efforts produce positive outcomes, the more likely patients are to continue treatment (Baekeland and Lundwall 1975; Stark 1992). Given the heterogeneity in pharmacological response, quick, noninvasive measures of stable behavioral markers that predict sensitivity to a treatment agent could result in more rapid, positive outcomes and long-term adherence to the program.

While animal models of drug-prone and -resilient phenotypes could help inform which behavioral markers predict differential pharmacological treatment sensitivity, little preclinical work has been done in this regard. Initial results from our laboratory indicate that baclofen (Holtz and Carroll 2011) or progesterone (Anker et al. 2012) treatment is more effective at reducing cocaine self-administration in LoS rats compared to HiS rats. The efficacy of predicting treatment outcomes with assessments of sweet preference has also been shown clinically. For instance, Garbutt et al. (2009) and Laaksonen et al. (2011) demonstrated that alcoholics exhibiting high alcohol craving and a high sweet preference had more days of abstinence with naltrexone treatment than those with a lower sweet preference. Similar studies with other animal models of high and low drug abuse vulnerability may predict outcome. For example, clinical studies have shown that sweet liking predicts alcoholism, only if it co-occurs with sensation-seeking (Kampov-Polevoy et al. 1998; Lange et al. 2010). Questions remain regarding whether the predictive value of this trait in the attenuating effects of naltrexone on alcohol consumption is less powerful than sensation-seeking trait, how these factors interact, and how they are moderated by other phenotypic variables, such as impulsivity. Impulsivity is another major factor that is emerging as a predictor of behavioral and pharmacological treatments for nicotine and cocaine dependence in humans (MacKillop and Kahler 2009). Assessing pharmacological treatment sensitivity in models other than the HiS/LoS rats, such as the HR/LR and HiI/LoI models, amongst others, could help justify and guide clinical investigations to address these issues.

It is important that the hypotheses of studies like Holtz and Carroll (2011) and Anker et al. (2012) are also examined in the clinical laboratory. These rat studies referred to reducing the escalation of drug intake; however, outcomes may have differed if other aspects of addiction were studied, such as initiation, occasional short-term use, or relapse. Additionally, the endophenotypes must be clearly defined and different aspects of the behavior under consideration may react differently to treatment. For example, very different effects have been reported for novelty reactive vs. novelty-seeking rats depending on the specific measures of addiction (Belin et al. 2011). If baclofen or progesterone, for instance, were more effective in reducing drug intake in sweet-preferring vs. -non-preferring human cocaine abusers, as was found with the clinical effects of naltrexone on alcohol abstinence (Garbutt et al. 2009; Laaksonen et al. 2011), more experiments would be needed to understand the translational limitations of our model.

Behavioral phenotype alone may not be sensitive enough a measure for predicting response to all pharmacological treatment options, and may need to be refined and used in combination with other assessments, such as genotyping. However, searching for genetic variants that induce liability for complex behavioral diseases, such as addiction, may be substantially less effective than the application of environmental modifications (e.g., public health initiatives, behavioral interventions, etc.) due to the high number of genetic variants, each with small effect size (Merikangas and Risch 2003). Selectively bred drug-prone and -resistant models offer additional value, in that the experimentally delineated genomes in these animals may also help elucidate salient predictive variants that could guide clinical pharmacogenetic efforts. As suggested by Agrawal et al. (2012), the future of effective substance dependence therapy should involve personalized treatments shaped by the patients' endophenotypes (electroencephalogram activity, Stroop task performance, neuroimaging outcomes, etc.), biomarkers, addiction severity, and genotype.

Conclusion

This review has highlighted a number of behavioral endophenotypes that predict addictive behavior. In some cases rats have been selectively bred for specific behaviors, and in other circumstances, they are selected from the high and low ends of the range for the behavior of interest. While not many differences between selection and selective breeding have surfaced in terms of predictability of drug abuse-related behaviors, there is a wide range of behaviors that predict drug seeking; such as, novelty-seeking, novelty preference, sweet intake, impulsivity, emotional reactivity, physical activity, and incentive salience attribution (sign-tracking *vs.* goal-tracking). While many of these factors partially overlap with one or two others, there is not complete overlap, suggesting multiple factors that predict drug abuse rather than a fundamental underlying trait (Carroll et al. 2008, 2009). Additionally, some predictors (e.g., female, adolescent) are additive with others to enhance the risk for drug abuse. Identifying the major drug-prone and -resistant traits and the high-risk combinations of traits will be useful for designing and customizing treatment strategies for optimum benefit.

References

Acewicz, A., Mierzejewski, P., Dyr, W., Jastrzebska, A., Korkosz, I., Wyszogrodzka, E., et al. (2012). Cocaine self-administration in Warsaw alcohol high-preferring (WHP) and Warsaw alcohol low-preferring (WLP) rats. *European Journal of Pharmacology, 674*(2-3), 275–279.

Agrawal, A., & Lynskey, M. T. (2008). Are there genetic influences on addiction: Evidence from family, adoption and twin studies. *Addiction, 103*(7), 1069–1081.

Agrawal, A., Verweij, K. J., Gillespie, N. A., Heath, A. C., Lessov-Schlaggar, C. N., Martin, N. G., et al. (2012). The genetics of addiction-a translational perspective. *Translational Psychiatry, 2*, e140.

Aldao, A., Nolen-Hoeksema, S., & Schweizer, S. (2010). Emotion-regulation strategies across psychopathology: A meta-analytic review. *Clinical Psychology Review, 30*(2), 217–237.

Alsop, B. (2007). Problems with spontaneously hypertensive rats (SHR) as a model of attention-deficit/hyperactivity disorder (AD/HD). *Journal of Neuroscience Methods, 162*(1–2), 42–48.

Ambrosio, E., Goldberg, S. R., & Elmer, G. I. (1995). Behavior genetic investigation of the relationship between spontaneous locomotor activity and the acquisition of morphine self-administration behavior. *Behavioural Pharmacology, 6*(3), 229–237.

American Psychiatric Association, & Task Force on DSM-IV. (1994). *Diagnostic and statistical manual of mental disorders: DSM-IV* (4th ed.). Washington, DC: American Psychiatric Association.

Anderson, K. G., & Diller, J. W. (2010). Effects of acute and repeated nicotine administration on delay discounting in Lewis and Fischer 344 rats. *Behavioural Pharmacology, 21*(8), 754–764.

Anker, J. J., & Carroll, M. E. (2011). Females are more vulnerable to drug abuse than males: Evidence from preclinical studies and the role of ovarian hormones. *Current Topics in Behavioral Neurosciences, 8*, 73–96.

Anker, J. J., Gliddon, L. A., & Carroll, M. E. (2008). Impulsivity on a Go/No-go task for intravenous cocaine or food in male and female rats selectively bred for high and low saccharin intake. *Behavioural Pharmacology, 19*(5–6), 615–629.

Anker, J. J., Holtz, N. A., & Carroll, M. E. (2012). Effects of progesterone on escalation of intravenous cocaine self-administration in rats selectively bred for high or low saccharin intake. *Behavioural Pharmacology, 23*(2), 205–210.

Anker, J. J., Perry, J. L., Gliddon, L. A., & Carroll, M. E. (2009a). Impulsivity predicts the escalation of cocaine self-administration in rats. *Pharmacology, Biochemistry, and Behavior, 93*(3), 343–348.

Anker, J. J., Zlebnik, N. E., Gliddon, L. A., & Carroll, M. E. (2009b). Performance under a Go/No-go task in rats selected for high and low impulsivity with a delay-discounting procedure. *Behavioural Pharmacology, 20*(5–6), 406–414.

Antoniou, K., Papathanasiou, G., Papalexi, E., Hyphantis, T., Nomikos, G. G., Spyraki, C., et al. (2008). Individual responses to novelty are associated with differences in behavioral and neurochemical profiles. *Behavioural Brain Research, 187*(2), 462–472.

Armario, A., Gavalda, A., & Marti, J. (1995). Comparison of the behavioural and endocrine response to forced swimming stress in five inbred strains of rats. *Psychoneuroendocrinology, 20*(8), 879–890.

Atkins, A. L., Helms, M. L., O'Toole, L. A., & Belknap, J. K. (2001). Stereotypic behaviors in mice selectively bred for high and low methamphetamine-induced stereotypic chewing. *Psychopharmacology, 157*(1), 96–104.

Badishtov, B. A., Overstreet, D. H., Kashevskaya, O. P., Viglinskaya, I. V., Kampov-Polevoy, A. B., Seredenin, S. B., et al. (1995). To drink or not to drink: Open field behavior in alcohol-preferring and nonpreferring rat strains. *Physiology and Behavior, 57*(3), 585–589.

Baekeland, F., & Lundwall, L. (1975). Dropping out of treatment: A critical review. *Psychological Bulletin, 82*(5), 738–783.

Bauman, A. E., Reis, R. S., Sallis, J. F., Wells, J. C., Loos, R. J., & Martin, B. W. (2012). Correlates of physical activity: Why are some people physically active and others not? *Lancet, 380*(9838), 258–271.

Belin, D., Berson, N., Balado, E., Piazza, P. V., & Deroche-Gamonet, V. (2011). High-novelty-preference rats are predisposed to compulsive cocaine self-administration. *Neuropsychopharmacology, 36*(3), 569–579.

Belin, D., Mar, A. C., Dalley, J. W., Robbins, T. W., & Everitt, B. J. (2008). High impulsivity predicts the switch to compulsive cocaine-taking. *Science, 320*(5881), 1352–1355.

Belknap, J. K., Haltli, N. R., Goebel, D. M., & Lame, M. (1983). Selective breeding for high and low levels of opiate-induced analgesia in mice. *Behavior Genetics, 13*(4), 383–396.

Bell, R. L., Rodd, Z. A., Lumeng, L., Murphy, J. M., & McBride, W. J. (2006). The alcohol-preferring P rat and animal models of excessive alcohol drinking. *Addiction Biology, 11*(3–4), 270–288.

Berridge, K. C. (2012). From prediction error to incentive salience: Mesolimbic computation of reward motivation. *European Journal of Neuroscience, 35*(7), 1124–1143.

Bjornebekk, A., Mathe, A. A., & Brene, S. (2005). The antidepressant effect of running is associated with increased hippocampal cell proliferation. *International Journal of Neuropsychopharmacology, 8*(3), 357–368.

Blanchard, M. M., Mendelsohn, D., & Stamp, J. A. (2009). The HR/LR model: Further evidence as an animal model of sensation seeking. *Neuroscience and Biobehavioral Reviews, 33*(7), 1145–1154.

Bock, G., & Whelan, J. (1992). *Cocaine—Scientific and social dimensions*. Chichester: J. Wiley.

Breland, K., & Breland, M. (1961). The misbehavior of organisms. *American Psychologist, 16*, 681–683.

Broos, N., Schmaal, L., Wiskerke, J., Kostelijk, L., Lam, T., Stoop, N., et al. (2012). The relationship between impulsive choice and impulsive action: A cross-species translational study. *PloS One, 7*(5), e36781.

Camp, D. M., Browman, K. E., & Robinson, T. E. (1994). The effects of methamphetamine and cocaine on motor behavior and extracellular dopamine in the ventral striatum of Lewis versus Fischer 344 rats. *Brain Research, 668*(1–2), 180–193.

Carlson, K. R., & Perez, L. (1997). Ethanol and cocaine intake by rats selectively bred for oral opioid acceptance. *Pharmacology, Biochemistry, and Behavior, 57*(1–2), 309–313.

Carmody, T. P., Vieten, C., & Astin, J. A. (2007). Negative affect, emotional acceptance, and smoking cessation. *Journal of Psychoactive Drugs, 39*(4), 499–508.

Carroll, M. E., Anderson, M. M., & Morgan, A. D. (2007). Higher locomotor response to cocaine in female (vs. male) rats selectively bred for high (HiS) and low (LoS) saccharin intake. *Pharmacology, Biochemistry, and Behavior, 88*(1), 94–104.

Carroll, M. E., & Anker, J. J. (2010). Sex differences and ovarian hormones in animal models of drug dependence. *Hormones and Behavior, 58*(1), 44–56.

Carroll, M. E., Anker, J. J., & Perry, J. L. (2009). Modeling risk factors for nicotine and other drug abuse in the preclinical laboratory. *Drug and Alcohol Dependence, 104*(Suppl 1), S70–S78.

Carroll, M. E., Morgan, A. D., Anker, J. J., Perry, J. L., & Dess, N. K. (2008). Selective breeding for differential saccharin intake as an animal model of drug abuse. *Behavioural Pharmacology, 19*(5–6), 435–460.

Chaouloff, F., Kulikov, A., Sarrieau, A., Castanon, N., & Mormede, P. (1995). Male Fischer 344 and Lewis rats display differences in locomotor reactivity, but not in anxiety-related behaviours: Relationship with the hippocampal serotonergic system. *Brain Research, 693*(1–2), 169–178.

Chen, F., Rezvani, A., Jarrott, B., & Lawrence, A. J. (1998). Distribution of GABAA receptors in the limbic system of alcohol-preferring and non-preferring rats: In situ hybridisation histochemistry and receptor autoradiography. *Neurochemistry International, 32*(2), 143–151.

Christensen, C. J., Kohut, S. J., Handler, S., Silberberg, A., & Riley, A. L. (2009). Demand for food and cocaine in Fischer and Lewis rats. *Behavioral Neuroscience, 123*(1), 165–171.

Cools, A. R., & Gingras, M. A. (1998). Nijmegen high and low responders to novelty: A new tool in the search after the neurobiology of drug abuse liability. *Pharmacology, Biochemistry, and Behavior, 60*(1), 151–159.

Crabbe, J. C. (2002). Genetic contributions to addiction. *Annual Review of Psychology, 53*, 435–462.

Crabbe, J. C. (2008). Review. Neurogenetic studies of alcohol addiction. *Philosophical Transactions of the Royal Society of London. Series B: Biological Sciences, 363*(1507), 3201–3211.

Cummings, J. A., Gowl, B. A., Westenbroek, C., Clinton, S. M., Akil, H., & Becker, J. B. (2011). Effects of a selectively bred novelty-seeking phenotype on the motivation to take cocaine in male and female rats. *Biology of Sex Differences, 2*, 3.

Darwin, C. (1874). *The descent of man, and selection in relation to sex* (2nd ed.). New York: A. L. Burt.

Davis, B. A., Clinton, S. M., Akil, H., & Becker, J. B. (2008). Behavioural characterisation of high impulsivity on the 5-choice serial reaction time task: Specific deficits in 'waiting' versus 'stopping'. *Pharmacology Biochemistry and Behavior, 90*(3), 331–338.

de Fiebre, N. C., Dawson, R., Jr., & de Fiebre, C. M. (2002). The selectively bred high alcohol sensitivity (HAS) and low alcohol sensitivity (LAS) rats differ in sensitivity to nicotine. *Alcoholism: Clinical and Experimental Research, 26*(6), 765–772.

Deitrich, R. A. (1993). Selective breeding for initial sensitivity to ethanol. *Behavior Genetics, 23*(2), 153–162.

Dellu, F., Piazza, P. V., Mayo, W., Le Moal, M., & Simon, H. (1996). Novelty-seeking in rats—Biobehavioral characteristics and possible relationship with the sensation-seeking trait in man. *Neuropsychobiology, 34*(3), 136–145.

Deminiere, J. M., Piazza, P. V., Le Moal, M., & Simon, H. (1989). Experimental approach to individual vulnerability to psychostimulant addiction. *Neuroscience and Biobehavioral Reviews, 13*(2–3), 141–147.

Deroche, V., Piazza, P. V., Le Moal, M., & Simon, H. (1993). Individual differences in the psychomotor effects of morphine are predicted by reactivity to novelty and influenced by corticosterone secretion. *Brain Research, 623*(2), 341–344.

Deroche-Gamonet, V., Belin, D., & Piazza, P. V. (2004). Evidence for addiction-like behavior in the rat. *Science, 305*(5686), 1014–1017.

Desor, J. A., & Beauchamp, G. K. (1987). Longitudinal changes in sweet preferences in humans. *Physiology and Behavior, 39*(5), 639–641.

DeSousa, N. J., Bush, D. E., & Vaccarino, F. J. (2000). Self-administration of intravenous amphetamine is predicted by individual differences in sucrose feeding in rats. *Psychopharmacology, 148*(1), 52–58.

Dess, N. K., Arnal, J., Chapman, C. D., Sidebel, S., VanderWeele, D. A., & Green, K. F. (2000). Exploring adaptations to famine: Rats selectively bred for differential intake of saccharin differ on deprivation-induced hyperactivity and emotionality. *International Journal of Comparative Psychology, 13*(1), 34–52.

Dess, N. K., Badia-Elder, N. E., Thiele, T. E., Kiefer, S. W., & Blizard, D. A. (1998). Ethanol consumption in rats selectively bred for differential saccharin intake. *Alcohol, 16*(4), 275–278.

Dess, N. K., & Minor, T. R. (1996). Taste and emotionality in rats selectively bred for high versus low saccharin intake. *Learning and Behavior, 24*, 105–115.

Diaz-Moran, S., Palencia, M., Mont-Cardona, C., Canete, T., Blazquez, G., Martinez-Membrives, E., et al. (2012). Coping style and stress hormone responses in genetically heterogeneous rats: Comparison with the Roman rat strains. *Behavioural Brain Research, 228*(1), 203–210.

Diergaarde, L., Pattij, T., Nawijn, L., Schoffelmeer, A. N., & De Vries, T. J. (2009). Trait impulsivity predicts escalation of sucrose seeking and hypersensitivity to sucrose-associated stimuli. *Behavioral Neuroscience, 123*(4), 794–803.

Dommett, E. J., & Rostron, C. L. (2013). Appetitive and consummative responding for liquid sucrose in the spontaneously hypertensive rat model of attention deficit hyperactivity disorder. *Behavioural Brain Research, 238*, 232–242.

Eagle, D. M., Bari, A., & Robbins, T. W. (2008). The neuropsychopharmacology of action inhibition: cross-species translation of the stop-signal and go/no-go tasks. *Psychopharmacology, 199*(3), 439–456.

Ersche, K. D., Jones, P. S., Williams, G. B., Smith, D. G., Bullmore, E. T., & Robbins, T. W. (2012). Distinctive personality traits and neural correlates associated with stimulant drug use versus familial risk of stimulant dependence. *Biological Psychiatry, 74*(2), 137–144.

Fagergren, P., Overstreet, D. H., Goiny, M., & Hurd, Y. L. (2005). Blunted response to cocaine in the Flinders hypercholinergic animal model of depression. *Neuroscience, 132*(4), 1159–1171.

Fattore, L., Piras, G., Corda, M. G., & Giorgi, O. (2009). The Roman high- and low-avoidance rat lines differ in the acquisition, maintenance, extinction, and reinstatement of intravenous cocaine self-administration. *Neuropsychopharmacology, 34*(5), 1091–1101.

Fernandez-Teruel, A., Driscoll, P., Gil, L., Aguilar, R., Tobena, A., & Escorihuela, R. M. (2002). Enduring effects of environmental enrichment on novelty seeking, saccharin and ethanol intake in two rat lines (RHA/Verh and RLA/Verh) differing in incentive-seeking behavior. *Pharmacology, Biochemistry, and Behavior, 73*(1), 225–231.

Ferreira, A., Lamarque, S., Boyer, P., Perez-Diaz, F., Jouvent, R., & Cohen-Salmon, C. (2006). Spontaneous appetence for wheel-running: A model of dependency on physical activity in rat. *European Psychiatry, 21*(8), 580–588.

Flagel, S. B., Clark, J. J., Robinson, T. E., Mayo, L., Czuj, A., Willuhn, I., et al. (2011). A selective role for dopamine in stimulus-reward learning. *Nature, 469*(7328), 53–57.

Flagel, S. B., Robinson, T. E., Clark, J. J., Clinton, S. M., Watson, S. J., Seeman, P., et al. (2010). An animal model of genetic vulnerability to behavioral disinhibition and responsiveness to reward-related cues: implications for addiction. *Neuropsychopharmacology, 35*(2), 388–400.

Flagel, S. B., Watson, S. J., Akil, H., & Robinson, T. E. (2008). Individual differences in the attribution of incentive salience to a reward-related cue: Influence on cocaine sensitization. *Behavioural Brain Research, 186*(1), 48–56.

Foroud, T., Ritchotte, A., Spence, J., Liu, L., Lumeng, L., Li, T. K., et al. (2003). Confirmation of alcohol preference quantitative trait loci in the replicate high alcohol drinking and low alcohol drinking rat lines. *Psychiatric Genetics, 13*(3), 155–161.

Fox, A. T., Hand, D. J., & Reilly, M. P. (2008). Impulsive choice in a rodent model of attention-deficit/hyperactivity disorder. *Behavioural Brain Research, 187*(1), 146–152.

Freeman, K. B., Kearns, D. N., Kohut, S. J., & Riley, A. L. (2009). Strain differences in patterns of drug-intake during prolonged access to cocaine self-administration. *Behavioral Neuroscience, 123*(1), 156–164.

Froehlich, J. C. (2010). What aspects of human alcohol use disorders can be modeled using selectively bred rat lines? *Substance Use and Misuse, 45*(11), 1727–1741.

Gahtan, E., Labounty, L. P., Wyvell, C., & Carroll, M. E. (1996). The relationships among saccharin consumption, oral ethanol, and i.v. cocaine self-administration. *Pharmacology, Biochemistry, and Behavior, 53*(4), 919–925.

Gallaher, E. J., Hollister, L. E., Gionet, S. E., & Crabbe, J. C. (1987). Mouse lines selected for genetic differences in diazepam sensitivity. *Psychopharmacology, 93*(1), 25–30.

Garbutt, J. C., Osborne, M., Gallop, R., Barkenbus, J., Grace, K., Cody, M., et al. (2009). Sweet liking phenotype, alcohol craving and response to naltrexone treatment in alcohol dependence. *Alcohol and Alcoholism (Oxford, Oxfordshire), 44*(3), 293–300.

Garcia, A., & Kirkpatrick, K. (2012). Impulsive choice behavior in four strains of rats: Evaluation of possible models of Attention-Deficit/Hyperactivity Disorder. *Behavioural Brain Research, 238*, 10–22.

Garcia-Lecumberri, C., Torres, I., Martin, S., Crespo, J. A., Miguens, M., Nicanor, C., et al. (2011). Strain differences in the dose-response relationship for morphine self-administration and impulsive choice between Lewis and Fischer 344 rats. *Journal of Psychopharmacology, 25*(6), 783–791.

Gingras, M. A., & Cools, A. R. (1996). Analysis of the biphasic locomotor response to ethanol in high and low responders to novelty: A study in Nijmegen Wistar rats. *Psychopharmacology, 125*(3), 258–264.

Giorgi, O., Corda, M. G., Carboni, G., Frau, V., Valentini, V., & Di Chiara, G. (1997). Effects of cocaine and morphine in rats from two psychogenetically selected lines: A behavioral and brain dialysis study. *Behavior Genetics, 27*(6), 537–546.

Giorgi, O., Piras, G., Lecca, D., & Corda, M. G. (2005). Behavioural effects of acute and repeated cocaine treatments: A comparative study in sensitisation-prone RHA rats and their sensitisation-resistant RLA counterparts. *Psychopharmacology, 180*(3), 530–538.

Godfrey, C. D., Froehlich, J. C., Stewart, R. B., Li, T. K., & Murphy, J. M. (1997). Comparison of rats selectively bred for high and low ethanol intake in a forced-swim-test model of depression: Effects of desipramine. *Physiology and Behavior, 62*(4), 729–733.

Gordon, T. L., Meehan, S. M., & Schechter, M. D. (1993). Differential effects of nicotine but not cathinone on motor activity of P and NP rats. *Pharmacology, Biochemistry, and Behavior, 44*(3), 657–659.

Gosnell, B. A. (2000). Sucrose intake predicts rate of acquisition of cocaine self-administration. *Psychopharmacology, 149*(3), 286–292.

Gosnell, B. A., Krahn, D. D., Yracheta, J. M., & Harasha, B. J. (1998). The relationship between intravenous cocaine self-administration and avidity for saccharin. *Pharmacology, Biochemistry, and Behavior, 60*(1), 229–236.

Gosnell, B. A., Lane, K. E., Bell, S. M., & Krahn, D. D. (1995). Intravenous morphine self-administration by rats with low versus high saccharin preferences. *Psychopharmacology, 117*(2), 248–252.

Gottesman, I. I., & Gould, T. D. (2003). The endophenotype concept in psychiatry: Etymology and strategic intentions. *The American Journal of Psychiatry, 160*(4), 636–645.

Guitart-Masip, M., Gimenez-Llort, L., Fernandez-Teruel, A., Canete, T., Tobena, A., Ogren, S. O., et al. (2006). Reduced ethanol response in the alcohol-preferring RHA rats and neuropeptide mRNAs in relevant structures. *European Journal of Neuroscience, 23*(2), 531–540.

Hall, F. S., Fong, G. W., Ghaed, S., & Pert, A. (2001). Locomotor-stimulating effects of indirect dopamine agonists are attenuated in Fawn hooded rats independent of postweaning social experience. *Pharmacology, Biochemistry, and Behavior, 69*(3–4), 519–526.

Hall, F. S., Markou, A., Levin, E. D., & Uhl, G. R. (2012). Mouse models for studying genetic influences on factors determining smoking cessation success in humans. *Annals of the New York Academy of Sciences, 1248*, 39–70.

Haney, M., Castanon, N., Cador, M., Le Moal, M., & Mormede, P. (1994). Cocaine sensitivity in Roman High and Low Avoidance rats is modulated by sex and gonadal hormone status. *Brain Research, 645*(1–2), 179–185.

He, S., Yang, Y., Mathur, D., & Grasing, K. (2008). Selective breeding for intravenous drug self-administration in rats: A pilot study. *Behavioural Pharmacology, 19*(8), 751–764.

Higgins, S. T. (2003). An historical note on Darwin and nonhuman drug self-administration. *Experimental and Clinical Psychopharmacology, 11*(4), 317.

Hinojosa, F. R., Spricigo, L., Jr., Izidio, G. S., Bruske, G. R., Lopes, D. M., & Ramos, A. (2006). Evaluation of two genetic animal models in behavioral tests of anxiety and depression. *Behavioural Brain Research, 168*(1), 127–136.

Holtz, N. A., & Carroll, M. E. (2011). Baclofen has opposite effects on escalation of cocaine self-administration: increased intake in rats selectively bred for high (HiS) saccharin intake and decreased intake in those selected for low (LoS) saccharin intake. *Pharmacology, Biochemistry, and Behavior, 100*(2), 275–283.

Holtz, N. A., & Carroll, M. E. (2013). Escalation of i.v. cocaine intake in peri-adolescent vs. adult rats selectively bred for high (HiS) vs. low (LoS) saccharin intake. *Psychopharmacology, 227*(2), 243–250.

Holtz, N. A., Radke, A. C., Gewirtz, J. C., & Carroll, M. E. (In preparation). Morphine-induced potentiation of the acoustic startle reflex and conditioned place aversion in Lewis (LEW) and Fischer (F344) rats.

Hooks, M. S., Jones, G. H., Smith, A. D., Neill, D. B., & Justice, J. B., Jr. (1991). Response to novelty predicts the locomotor and nucleus accumbens dopamine response to cocaine. *Synapse, 9*(2), 121–128.

Howard, M. O., Kivlahan, D., & Walker, R. D. (1997). Cloninger's tridimensional theory of personality and psychopathology: Applications to substance use disorders. *Journal of Studies on Alcohol, 58*(1), 48–66.

Huskinson, S. L., Krebs, C. A., & Anderson, K. G. (2010). Strain differences in delay discounting between Lewis and Fischer 344 rats at baseline and following acute and chronic administration of d-amphetamine. *Pharmacology, Biochemistry, and Behavior, 101*(3), 403–416.

Izidio, G. S., & Ramos, A. (2007). Positive association between ethanol consumption and anxiety-related behaviors in two selected rat lines. *Alcohol, 41*(7), 517–524.

Jama, A., Cecchi, M., Calvo, N., Watson, S. J., & Akil, H. (2008). Inter-individual differences in novelty-seeking behavior in rats predict differential responses to desipramine in the forced swim test. *Psychopharmacology, 198*(3), 333–340.

Janowsky, D. S., Pucilowski, O., & Buyinza, M. (2003). Preference for higher sucrose concentrations in cocaine abusing-dependent patients. *Journal of Psychiatric Research, 37*(1), 35–41.

Jupp, B., Caprioli, D., & Dalley, J. W. (2013). Highly impulsive rats: Modelling an endophenotype to determine the neurobiological, genetic and environmental mechanisms of addiction. *Disease Models & Mechanisms, 6*(2), 302–311.

Kabbaj, M. (2006). Individual differences in vulnerability to drug abuse: The high responders/low responders model. *CNS & Neurological Disorders Drug Targets, 5*(5), 513–520.

Kampov-Polevoy, A. B., Garbutt, J. C., Davis, C. E., & Janowsky, D. S. (1998). Preference for higher sugar concentrations and Tridimensional Personality Questionnaire scores in alcoholic and nonalcoholic men. *Alcoholism: Clinical and Experimental Research, 22*(3), 610–614.

Kampov-Polevoy, A., Garbutt, J. C., & Janowsky, D. (1997). Evidence of preference for a high-concentration sucrose solution in alcoholic men. *The American Journal of Psychiatry, 154*(2), 269–270.

Kampov-Polevoy, A. B., Tsoi, M. V., Zvartau, E. E., Neznanov, N. G., & Khalitov, E. (2001). Sweet liking and family history of alcoholism in hospitalized alcoholic and non-alcoholic patients. *Alcohol and Alcoholism (Oxford, Oxfordshire), 36*(2), 165–170.

Kearns, D. N., Gomez-Serrano, M. A., Weiss, S. J., & Riley, A. L. (2006). A comparison of Lewis and Fischer rat strains on autoshaping (sign-tracking), discrimination reversal learning and negative auto-maintenance. *Behavioural Brain Research, 169*(2), 193–200.

Keskitalo, K., Knaapila, A., Kallela, M., Palotie, A., Wessman, M., Sammalisto, S., et al. (2007a). Sweet taste preferences are partly genetically determined: Identification of a trait locus on chromosome 16. *American Journal of Clinical Nutrition, 86*(1), 55–63.

Keskitalo, K., Tuorila, H., Spector, T. D., Cherkas, L. F., Knaapila, A., Silventoinen, K., et al. (2007b). Same genetic components underlie different measures of sweet taste preference. *American Journal of Clinical Nutrition, 86*(6), 1663–1669.

Keskitalo, K., Tuorila, H., Spector, T. D., Cherkas, L. F., Knaapila, A., Kaprio, J., et al. (2008). The three-factor eating questionnaire, body mass index, and responses to sweet and salty fatty foods: A twin study of genetic and environmental associations. *American Journal of Clinical Nutrition, 88*(2), 263–271.

Khanna, J. M., Kalant, H., Chau, A. K., & Sharma, H. (1990). Initial sensitivity, acute tolerance and alcohol consumption in four inbred strains of rats. *Psychopharmacology, 101*(3), 390–395.

Kiianmaa, K., Stenius, K., & Sinclair, J. D. (1991). Determinants of alcohol preference in the AA and ANA rat lines selected for differential ethanol intake. *Alcohol and Alcoholism. Supplement, 1*, 115–120.

Kingwell, B. A., Arnold, P. J., Jennings, G. L., & Dart, A. M. (1998). The effects of voluntary running on cardiac mass and aortic compliance in Wistar-Kyoto and spontaneously hypertensive rats. *Journal of Hypertension, 16*(2), 181–185.

Klebaur, J. E., Bevins, R. A., Segar, T. M., & Bardo, M. T. (2001). Individual differences in behavioral responses to novelty and amphetamine self-administration in male and female rats. *Behavioural Pharmacology, 12*(4), 267–275.

Kosten, T. A., & Ambrosio, E. (2002). HPA axis function and drug addictive behaviors: Insights from studies with Lewis and Fischer 344 inbred rats. *Psychoneuroendocrinology, 27*(1–2), 35–69.

Kosten, T. A., Zhang, X. Y., & Haile, C. N. (2007). Strain differences in maintenance of cocaine self-administration and their relationship to novelty activity responses. *Behavioral Neuroscience, 121*(2), 380–388.

Kruzich, P. J., & Xi, J. (2006a). Differences in extinction responding and reinstatement of methamphetamine-seeking behavior between Fischer 344 and Lewis rats. *Pharmacology, Biochemistry, and Behavior, 83*(3), 391–395.

Kruzich, P. J., & Xi, J. (2006b). Different patterns of pharmacological reinstatement of cocaine-seeking behavior between Fischer 344 and Lewis rats. *Psychopharmacology, 187*(1), 22–29.

Laaksonen, E., Lahti, J., Sinclair, J. D., Heinala, P., & Alho, H. (2011). Predictors for the efficacy of naltrexone treatment in alcohol dependence: Sweet preference. *Alcohol and Alcoholism (Oxford, Oxfordshire), 46*(3), 308–311.

Lange, L. A., Kampov-Polevoy, A. B., & Garbutt, J. C. (2010). Sweet liking and high novelty seeking: Independent phenotypes associated with alcohol-related problems. *Alcohol and Alcoholism (Oxford, Oxfordshire), 45*(5), 431–436.

Larson, E. B., & Carroll, M. E. (2005). Wheel running as a predictor of cocaine self-administration and reinstatement in female rats. *Pharmacology, Biochemistry, and Behavior, 82*(3), 590–600.

Laudet, A. B., Stanick, V., & Sands, B. (2009). What could the program have done differently? A qualitative examination of reasons for leaving outpatient treatment. *Journal of Substance Abuse Treatment, 37*(2), 182–190.

Le Foll, B., Gallo, A., Le Strat, Y., Lu, L., & Gorwood, P. (2009). Genetics of dopamine receptors and drug addiction: A comprehensive review. *Behavioural Pharmacology, 20*(1), 1–17.

Le, A. D., Li, Z., Funk, D., Shram, M., Li, T. K., & Shaham, Y. (2006). Increased vulnerability to nicotine self-administration and relapse in alcohol-naive offspring of rats selectively bred for high alcohol intake. *Journal of Neuroscience, 26*(6), 1872–1879.

Lecca, D., Piras, G., Driscoll, P., Giorgi, O., & Corda, M. G. (2004). A differential activation of dopamine output in the shell and core of the nucleus accumbens is associated with the motor responses to addictive drugs: A brain dialysis study in Roman high- and low-avoidance rats. *Neuropharmacology, 46*(5), 688–699.

Li, T. K., Lumeng, L., McBride, W. J., & Murphy, J. M. (1994). Genetic and neurobiological basis of alcohol-seeking behavior. *Alcohol and Alcoholism (Oxford, Oxfordshire), 29*(6), 697–700.

Li, T. K., Lumeng, L., & Doolittle, D. P. (1993). Selective breeding for alcohol preference and associated responses. *Behavior Genetics, 23*(2), 163–170.

Lin, S. J., Epps, S. A., West, C. H., Boss-Williams, K. A., Weiss, J. M., & Weinshenker, D. (2012). Operant psychostimulant self-administration in a rat model of depression. *Pharmacology, Biochemistry, and Behavior, 103*(2), 380–385.

Lodge, D. J., & Lawrence, A. J. (2003). Comparative analysis of hepatic ethanol metabolism in Fawn-Hooded and Wistar-Kyoto rats. *Alcohol, 30*(1), 75–79.

Lovic, V., Saunders, B. T., Yager, L. M., & Robinson, T. E. (2011). Rats prone to attribute incentive salience to reward cues are also prone to impulsive action. *Behavioural Brain Research, 223*(2), 255–261.

MacKillop, J. (2013). Integrating behavioral economics and behavioral genetics: Delayed reward discounting as an endophenotype for addictive disorders. *Journal of the Experimental Analysis of Behavior, 99*(1), 14–31.

MacKillop, J., & Kahler, C. W. (2009). Delayed reward discounting predicts treatment response for heavy drinkers receiving smoking cessation treatment. *Drug and Alcohol Dependence, 104*(3), 197–203.

Madden, G. J., Smith, N. G., Brewer, A. T., Pinkston, J. W., & Johnson, P. S. (2008). Steady-state assessment of impulsive choice in Lewis and Fischer 344 rats: Between-condition delay manipulations. *Journal of the Experimental Analysis of Behavior, 90*(3), 333–344.

Manzo, L., Gomez, M. J., Callejas-Aguilera, J. E., Fernandez-Teruel, A., Papini, M. R., & Torres, C. (2012). Oral ethanol self-administration in inbred Roman high- and low-avoidance rats: Gradual versus abrupt ethanol presentation. *Physiology and Behavior, 108*, 1–5.

Marti, J., & Armario, A. (1996). Forced swimming behavior is not related to the corticosterone levels achieved in the test: A study with four inbred rat strains. *Physiology and Behavior, 59*(2), 369–373.

Marusich, J. A., Darna, M., Charnigo, R. J., Dwoskin, L. P., & Bardo, M. T. (2011a). A multivariate assessment of individual differences in sensation seeking and impulsivity as predictors of amphetamine self-administration and prefrontal dopamine function in rats. *Experimental and Clinical Psychopharmacology, 19*(4), 275–284.

Marusich, J. A., McCuddy, W. T., Beckmann, J. S., Gipson, C. D., & Bardo, M. T. (2011b). Strain differences in self-administration of methylphenidate and sucrose pellets in a rat model of attention-deficit hyperactivity disorder. *Behavioural Pharmacology, 22*(8), 794–804.

McBride, W. J., & Li, T. K. (1998). Animal models of alcoholism: Neurobiology of high alcohol-drinking behavior in rodents. *Critical Reviews in Neurobiology, 12*(4), 339–369.
Meneses, A., Perez-Garcia, G., Ponce-Lopez, T., Tellez, R., Gallegos-Cari, A., & Castillo, C. (2011). Spontaneously hypertensive rat (SHR) as an animal model for ADHD: A short overview. *Reviews in the Neurosciences, 22*(3), 365–371.
Merikangas, K. R., & Risch, N. (2003). Genomic priorities and public health. *Science, 302*(5645), 599–601.
Miguens, M., Botreau, F., Olias, O., Del Olmo, N., Coria, S. M., Higuera-Matas, A., et al. (2011). Genetic differences in the modulation of accumbal glutamate and gamma-amino butyric acid levels after cocaine-induced reinstatement. *Addiction Biology, 18*(4), 623–632.
Mikrouli, E., Wortwein, G., Soylu, R., Mathe, A. A., & Petersen, A. (2011). Increased numbers of orexin/hypocretin neurons in a genetic rat depression model. *Neuropeptides, 45*(6), 401–406.
Miserendino, M. J., Haile, C. N., & Kosten, T. A. (2003). Strain differences in response to escapable and inescapable novel environments and their ability to predict amphetamine-induced locomotor activity. *Psychopharmacology, 167*(3), 281–290.
Molander, A. C., Mar, A., Norbury, A., Steventon, S., Moreno, M., Caprioli, D., et al. (2011). High impulsivity predicting vulnerability to cocaine addiction in rats: Some relationship with novelty preference but not novelty reactivity, anxiety or stress. *Psychopharmacology, 215*(4), 721–731.
Moreno, M., Cardona, D., Gomez, M. J., Sanchez-Santed, F., Tobena, A., Fernandez-Teruel, A., et al. (2010). Impulsivity characterization in the Roman high- and low-avoidance rat strains: Behavioral and neurochemical differences. *Neuropsychopharmacology, 35*(5), 1198–1208.
Murphy, S. B., Reinarman, C., & Waldorf, D. (1989). An 11-year follow-up of a network of cocaine users. *British Journal of Addiction, 84*(4), 427–436.
Nichols, J. R., & Hsiao, S. (1967). Addiction liability of albino rats: Breeding for quantitative differences in morphine drinking. *Science, 157*(3788), 561–563.
Olsen, C. M. (2011). Natural rewards, neuroplasticity, and non-drug addictions. *Neuropharmacology, 61*(7), 1109–1122.
Overstreet, D. H., & Djuric, V. (2001). A genetic rat model of cholinergic hypersensitivity: Implications for chemical intolerance, chronic fatigue, and asthma. *Annals of the New York Academy of Sciences, 933*, 92–102.
Overstreet, D. H., Friedman, E., Mathe, A. A., & Yadid, G. (2005). The Flinders Sensitive Line rat: A selectively bred putative animal model of depression. *Neuroscience and Biobehavioral Reviews, 29*(4–5), 739–759.
Overstreet, D. H., Kampov-Polevoy, A. B., Rezvani, A. H., Murrelle, L., Halikas, J. A., & Janowsky, D. S. (1993). Saccharin intake predicts ethanol intake in genetically heterogeneous rats as well as different rat strains. *Alcoholism: Clinical and Experimental Research, 17*(2), 366–369.
Overstreet, D. H., & Wegener, G. (2013). The flinders sensitive line rat model of depression—25 years and still producing. *Pharmacological Reviews, 65*(1), 143–155.
Pepino, M. Y., & Mennella, J. A. (2007). Effects of cigarette smoking and family history of alcoholism on sweet taste perception and food cravings in women. *Alcoholism: Clinical and Experimental Research, 31*(11), 1891–1899.
Perry, J. L., Anderson, M. M., Nelson, S. E., & Carroll, M. E. (2007a). Acquisition of i.v. cocaine self-administration in adolescent and adult male rats selectively bred for high and low saccharin intake. *Physiology and Behavior, 91*(1), 126–133.
Perry, J. L., Nelson, S. E., Anderson, M. M., Morgan, A. D., & Carroll, M. E. (2007b). Impulsivity (delay discounting) for food and cocaine in male and female rats selectively bred for high and low saccharin intake. *Pharmacology, Biochemistry, and Behavior, 86*(4), 822–837.
Perry, J. L., & Carroll, M. E. (2008). The role of impulsive behavior in drug abuse. *Psychopharmacology, 200*(1), 1–26.
Perry, J. L., Morgan, A. D., Anker, J. J., Dess, N. K., & Carroll, M. E. (2006). Escalation of i.v. cocaine self-administration and reinstatement of cocaine-seeking behavior in rats bred for high and low saccharin intake. *Psychopharmacology, 186*(2), 235–245.

Perry, J. L., Nelson, S. E., & Carroll, M. E. (2008). Impulsive choice as a predictor of acquisition of IV cocaine self- administration and reinstatement of cocaine-seeking behavior in male and female rats. *Experimental and Clinical Psychopharmacology, 16*(2), 165–177.

Pettinati, H. M., O'Brien, C. P., & Dundon, W. D. (2013). Current status of co-occurring mood and substance use disorders: A new therapeutic target. *The American Journal of Psychiatry, 170*(1), 23–30.

Piazza, P. V., Deminiere, J. M., Le Moal, M., & Simon, H. (1989). Factors that predict individual vulnerability to amphetamine self-administration. *Science, 245*(4925), 1511–1513.

Picetti, R., Caccavo, J. A., Ho, A., & Kreek, M. J. (2012). Dose escalation and dose preference in extended-access heroin self-administration in Lewis and Fischer rats. *Psychopharmacology, 220*(1), 163–172.

Picetti, R., Ho, A., Butelman, E. R., & Kreek, M. J. (2010). Dose preference and dose escalation in extended-access cocaine self-administration in Fischer and Lewis rats. *Psychopharmacology, 211*(3), 313–323.

Pike, E., Stoops, W. W., Fillmore, M. T., & Rush, C. R. (2011). Drug-related stimuli impair inhibitory control in cocaine abusers. *Drug and Alcohol Dependence, 133*(2), 768–771.

Piras, G., Giorgi, O., & Corda, M. G. (2010). Effects of antidepressants on the performance in the forced swim test of two psychogenetically selected lines of rats that differ in coping strategies to aversive conditions. *Psychopharmacology, 211*(4), 403–414.

Piras, G., Lecca, D., Corda, M. G., & Giorgi, O. (2003). Repeated morphine injections induce behavioural sensitization in Roman high- but not in Roman low-avoidance rats. *Neuroreport, 14*(18), 2433–2438.

Pisula, W. (2003). The Roman high- and low-avoidance rats respond differently to novelty in a familiarized environment. *Behavioural Processes, 63*(2), 63–72.

Pomerleau, O. F. (1995). Individual differences in sensitivity to nicotine: Implications for genetic research on nicotine dependence. *Behavior Genetics, 25*(2), 161–177.

Pomerleau, C. S., Garcia, A. W., Drewnowski, A., & Pomerleau, O. F. (1991). Sweet taste preference in women smokers: Comparison with nonsmokers and effects of menstrual phase and nicotine abstinence. *Pharmacology, Biochemistry, and Behavior, 40*(4), 995–999.

Pucilowski, O., Overstreet, D. H., Rezvani, A. H., & Janowsky, D. S. (1993). Chronic mild stress-induced anhedonia: Greater effect in a genetic rat model of depression. *Physiology and Behavior, 54*(6), 1215–1220.

Reed, D. R., Bachmanov, A. A., Beauchamp, G. K., Tordoff, M. G., & Price, R. A. (1997). Heritable variation in food preferences and their contribution to obesity. *Behavior Genetics, 27*(4), 373–387.

Rezvani, A. H., Parsian, A., & Overstreet, D. H. (2002). The Fawn-Hooded (FH/Wjd) rat: A genetic animal model of comorbid depression and alcoholism. *Psychiatric Genetics, 12*(1), 1–16.

Riley, E. P., Worsham, E. D., Lester, D., & Freed, E. X. (1977). Selective breeding of rats for differences in reactivity to alcohol. An approach to an animal model of alcoholism. II. Behavioral measures. *Journal of Studies on Alcohol, 38*(9), 1705–1717.

Robinson, E. S., Eagle, D. M., Economidou, D., Theobald, D. E., Mar, A. C., Murphy, E. R., et al. (2009). Behavioural characterisation of high impulsivity on the 5-choice serial reaction time task: Specific deficits in 'waiting' versus 'stopping'. *Behavioural Brain Research, 196*(2), 310–316.

Roman, E., Stewart, R. B., Bertholomey, M. L., Jensen, M. L., Colombo, G., Hyytia, P., et al. (2012). Behavioral profiling of multiple pairs of rats selectively bred for high and low alcohol intake using the MCSF test. *Addiction Biology, 17*(1), 33–46.

Roth-Deri, I., Friedman, A., Abraham, L., Lax, E., Flaumenhaft, Y., Dikshtein, Y., et al. (2009). Antidepressant treatment facilitates dopamine release and drug seeking behavior in a genetic animal model of depression. *European Journal of Neuroscience, 30*(3), 485–492.

Sagvolden, T., Johansen, E. B., Woien, G., Walaas, S. I., Storm-Mathisen, J., Bergersen, L. H., et al. (2009). The spontaneously hypertensive rat model of ADHD—The importance of selecting the appropriate reference strain. *Neuropharmacology, 57*(7–8), 619–626.

Saunders, B. T., & Robinson, T. E. (2011). Individual variation in the motivational properties of cocaine. *Neuropsychopharmacology, 36*(8), 1668–1676.
Saunders, B. T., Yager, L. M., & Robinson, T. E. (2013). Cue-evoked cocaine "craving": Role of dopamine in the accumbens core. *Journal of Neuroscience, 33*(35), 13989–14000.
Shiffman, S. (1989). Tobacco "chippers"—Individual differences in tobacco dependence. *Psychopharmacology, 97*(4), 539–547.
Sinclair, J. D., Kampov-Polevoy, A., Stewart, R., & Li, T. K. (1992). Taste preferences in rat lines selected for low and high alcohol consumption. *Alcohol, 9*(2), 155–160.
Sinha, R. (2011). New findings on biological factors predicting addiction relapse vulnerability. *Current Psychiatry Reports, 13*(5), 398–405.
Smolen, A., Marks, M. J., DeFries, J. C., & Henderson, N. D. (1994). Individual differences in sensitivity to nicotine in mice: Response to six generations of selective breeding. *Pharmacology, Biochemistry, and Behavior, 49*(3), 531–540.
Stark, M. J. (1992). Dropping out of substance-abuse treatment—A clinically oriented review. *Clinical Psychology Review, 12*(1), 93–116.
Stewart, R. B., Russell, R. N., Lumeng, L., Li, T. K., & Murphy, J. M. (1994). Consumption of sweet, salty, sour, and bitter solutions by selectively bred alcohol-preferring and alcohol-nonpreferring lines of rats. *Alcoholism: Clinical and Experimental Research, 18*(2), 375–381.
Suzuki, T., George, F. R., & Meisch, R. A. (1988). Differential establishment and maintenance of oral ethanol reinforced behavior in Lewis and Fischer 344 inbred rat strains. *Journal of Pharmacology and Experimental Therapeutics, 245*(1), 164–170.
Swendsen, J., & Le Moal, M. (2011). Individual vulnerability to addiction. *Annals of the New York Academy of Sciences, 1216*, 73–85.
Tomie, A., Aguado, A. S., Pohorecky, L. A., & Benjamin, D. (1998). Ethanol induces impulsive-like responding in a delay-of-reward operant choice procedure: Impulsivity predicts autoshaping. *Psychopharmacology, 139*(4), 376–382.
Tomie, A., Brooks, W., & Zito, B. (1989). Sign-tracking: The search for reward. In S. B. Klein & R. R. Mowrer (Eds.), *Pavlovian conditioning and the status of traditional learning theory* (pp. 191–223). Hillsdale, NJ: Erlbaum.
Tomie, A., Grimes, K. L., & Pohorecky, L. A. (2008). Behavioral characteristics and neurobiological substrates shared by Pavlovian sign-tracking and drug abuse. *Brain Research Reviews, 58*(1), 121–135.
Tordoff, M. G., Alarcon, L. K., & Lawler, M. P. (2008). Preferences of 14 rat strains for 17 taste compounds. *Physiology and Behavior, 95*(3), 308–332.
Tzilos, G. K., Rhodes, G. L., Ledgerwood, D. M., & Greenwald, M. K. (2009). Predicting cocaine group treatment outcome in cocaine-abusing methadone patients. *Experimental and Clinical Psychopharmacology, 17*(5), 320–325.
Uhl, G. R., Drgon, T., Johnson, C., & Liu, Q. R. (2009). Addiction genetics and pleiotropic effects of common haplotypes that make polygenic contributions to vulnerability to substance dependence. *Journal of Neurogenetics, 23*(3), 272–282.
van den Bergh, F. S., Bloemarts, E., Chan, J. S., Groenink, L., Olivier, B., & Oosting, R. S. (2006). Spontaneously hypertensive rats do not predict symptoms of attention-deficit hyperactivity disorder. *Pharmacology, Biochemistry, and Behavior, 83*(3), 380–390.
Vendruscolo, L. F., Izidio, G. S., & Takahashi, R. N. (2009). Drug reinforcement in a rat model of attention deficit/hyperactivity disorder—The Spontaneously Hypertensive Rat (SHR). *Current Drug Abuse Reviews, 2*(2), 177–183.
Weiss, G. (1982). Food fantasies of incarcerated drug users. *International Journal of the Addictions, 17*(5), 905–912.
Weiss, J. M., Cierpial, M. A., & West, C. H. (1998). Selective breeding of rats for high and low motor activity in a swim test: Toward a new animal model of depression. *Pharmacology, Biochemistry, and Behavior, 61*(1), 49–66.
Weiss, J. M., West, C. H., Emery, M. S., Bonsall, R. W., Moore, J. P., & Boss-Williams, K. A. (2008). Rats selectively-bred for behavior related to affective disorders: Proclivity for intake of alcohol and drugs of abuse, and measures of brain monoamines. *Biochemical Pharmacology, 75*(1), 134–159.

Werme, M., Thoren, P., Olson, L., & Brene, S. (2000). Running and cocaine both upregulate dynorphin mRNA in medial caudate putamen. *European Journal of Neuroscience, 12*(8), 2967–2974.
West, C. H., & Weiss, J. M. (2006). Intake of ethanol and reinforcing fluids in rats bred for susceptibility to stress. *Alcohol, 38*(1), 13–27.
Wheeler, J. M., Reed, C., Burkhart-Kasch, S., Li, N., Cunningham, C. L., Janowsky, A., et al. (2009). Genetically correlated effects of selective breeding for high and low methamphetamine consumption. *Genes, Brain and Behavior, 8*(8), 758–771.
White, D. A., Kalinichev, M., & Holtzman, S. G. (2007). Locomotor response to novelty as a predictor of reactivity to aversive stimuli in the rat. *Brain Research, 1149*, 141–148.
Wilhelm, C. J., & Mitchell, S. H. (2008). Rats bred for high alcohol drinking are more sensitive to delayed and probabilistic outcomes. *Genes, Brain and Behavior, 7*(7), 705–713.
Wronski, M., Skrok-Wolska, D., Samochowiec, J., Ziolkowski, M., Swiecicki, L., Bienkowski, P., et al. (2007). Perceived intensity and pleasantness of sucrose taste in male alcoholics. *Alcohol and Alcoholism (Oxford, Oxfordshire), 42*(2), 75–79.
Yerkes, R. M., & Dodson, J. D. (1908). The relation of strength of stimulus to rapidity of habit-formation. *Journal of Comparative Neurology and Psychology, 18*, 459–482.

Chapter 11
Potentiation of the Startle Reflex as a Behavioral Measure of Anxiety

Jonathan C. Gewirtz and Anna K. Radke

Introduction

Anxiety disorders are widespread in human populations. In the United States, for example, the 12-month prevalence for anxiety disorders has been estimated at 18.1 % (Kessler et al. 2005). Despite the availability of some anxiolytic drugs, the need for reliable, translational models of fear and anxiety is therefore still present. Potentiation of the acoustic startle reflex has been widely employed as such in both rat and human studies, but so far little work has been done to establish this as a reliable measure in mice. This is unfortunate, as mouse behavioral models offer rich opportunities to investigate the genetic basis of both healthy and pathological fear states. The purpose of this review is therefore to examine the use of potentiated startle as a reliable and standardized behavioral measure of fear and exaggerated fear in mice. We will start by reviewing the literature on startle-based models of fear and anxiety in rodents and humans and conclude with a number of recommendations for implementation of procedures to obtain robust potentiated startle in mice, based on published studies and experiences from our laboratory.

J.C. Gewirtz (✉)
Department of Psychology, University of Minnesota, 75 East River Road, Minneapolis, MN 55455, USA
e-mail: jgewirtz@umn.edu

A.K. Radke
Department of Psychology, Miami University, 90 North Patterson Avenue, Oxford, OH 45056, USA
e-mail: radkeak@miamioh.edu

J.C. Gewirtz, Y.-K. Kim (eds.), *Animal Models of Behavior Genetics*, Advances in Behavior Genetics, DOI 10.1007/978-1-4939-3777-6_11

Startle Reflex

Startle vs. Potentiated Startle

The acoustic startle reflex is a rapid, complex, and coordinated set of movements in response to loud and sudden acoustic stimuli (Landis and Hunt 1939). It is mediated in rodents by a simple, three-synapse neural circuit. Auditory signals are relayed via axon collaterals of cochlear nerve fibers to a small population of large neurons, scattered along the cochlear nerve root. The principle targets of these cochlear root neurons are cells in the caudal pontine reticular nucleus (PnC), which in turn project to motor neurons in the spinal cord and facial motor nucleus (Koch & Schnitzler 1997; Lee et al. 1996; Lopez et al. 1999). Because the startle reflex is conserved across mammalian, and even nonmammalian species (Landis and Hunt 1939; Eaton 1984), it is a useful tool for research in human and animal models.

In rodents, nonhuman primates, and humans, conditioned and unconditioned aversive or threatening cues increase startle, and this increase (potentiated startle) is used as an objective measure of fear or anxiety (Davis 2006). Conversely, the startle reflex may be attenuated to cues associated with reward in rodents (Schmid et al. 1995) and in humans (Bradley et al. 2001), although this phenomenon may not be as robust as that of startle potentiation to threat (e.g., Grillon and Baas 2003; Engelmann et al. 2011). These alterations in startle magnitude are transient and are reversed shortly after termination of the emotive cues. Exposure of rats or mice to certain forms of stress—and in particular to predator odor—can bring about longer term, and perhaps permanent changes in startle magnitude (Adamec et al. 1999, 2006; Hebb et al. 2003; Sink et al. 2011). Such changes in behavior, reflective of alterations in "trait" rather than "state", will not be reviewed in the current chapter. Finally, startle is also modulated by weak cues (prepulses) that occur very shortly before the startle-eliciting stimulus. Reduction of startle in this manner (prepulse inhibition) is commonly used as a measure of sensorimotor gating.

In rodents, startle is measured as the magnitude of the full-body reflex via displacement of a test cage, generally located in a sound-attenuating chamber. All experimental cues, including lights, tones, and shocks can be administered within this chamber, allowing training and testing to be conducted with the same equipment. If necessary, contextual cues such as ambient lighting or scent can also be easily modified. Startle in nonhuman primates has also been measured as amplitude of the whole-body reflex (Davis et al. 2008). In human subjects, as occasionally in rodents (Canli and Brown 1996; Choi et al. 2001), startle is assessed as the amplitude of the eyeblink reflex (Greenwald et al. 1998). Human studies also differ from rodent studies in the use of tactile (i.e., air-puff), in addition to acoustic stimuli, to elicit startle (Lissek et al. 2005a). These studies can be conducted in a variety of contexts and more recently investigators have employed virtual reality technology to modulate contextual variables (Baas et al. 2004).

Interspecies Translation

Perhaps, the primary strength of potentiated startle as a behavioral measure is its "translational" value in modeling fear and anxiety in nonhuman species. Because potentiated startle can be measured relatively reliably in humans, where it is closely associated with other measures of fear and anxiety, the insights gained from studying nonhuman species can reasonably be assumed to apply to our understanding of fear and anxiety in humans. This is particularly useful in studying neural circuitry or in screening potential anxiolytic compounds. Attempts to delineate the neural substrates underlying fear and anxiety states are also aided by the fact that the startle circuit is well characterized. Finally, startle is a useful model for the study of anxiety disorders because it is increased in the presence of anxiogenic, but not other, aversive stimuli (Balaban and Taussig 1994).

From a methodological standpoint, the measurement of startle in nonhuman species also holds a number of advantages. For example, some tests of anxiety in rodents, such as the elevated-plus maze and shock-probe defensive burying task, are most commonly performed only once on a single subject because of changes in the potency of the anxiogenic stimulus across repeated exposures (e.g., Treit et al. 1990, 1993). Startle, on the other hand, can be measured repeatedly within and across days so that within-subject comparisons and long-term changes in fear can be determined. Startle potentiation paradigms also offer flexibility in the choice of anxiogenic stimuli. This means that investigators can assess fear to explicit cues (tones, lights, etc.), contexts, and inherently anxiogenic stimuli such as bright lights and predator odors. Because the intensity of the startle stimulus can be varied and the reflex shows a wide dynamic range, ceiling and floor issues can generally be avoided. Fourth, measuring startle in the presence and absence of conditioned stimuli allows for any nonspecific effects of experimental manipulations to be assessed. And, on purely practical grounds, the fact that another startle-based behavioral index (prepulse inhibition) is already used routinely in models of psychosis and schizophrenia makes potentiated startle measurement a sound choice for experimenters wishing to make the most of their behavioral equipment.

Current Topics

Startle as a Measure of Fear and Anxiety

Evidence

Support for the contention that increases in startle reflect fearful states can be derived from a variety of sources. For example, startle potentiation in the presence of anxiogenic cues is attenuated by drugs that are anxiolytic in humans. A range of drugs—including benzodiazepines, monoamine reuptake inhibitors, adrenergic compounds (e.g., clonidine and propranolol), opiates, and buspirone—reduce potentiated startle in the presence of aversively conditioned cues and contexts at doses that do not significantly affect startle measured in the absence of those cues (Davis 1979a, b; Davis et al. 1979, 1988; Hijzen et al. 1995; Rothwell et al. 2009; Smith et al. 2011). Lesions or pharmacological inactivation of the amygdala, well-established as the "hub" of fear in a number of paradigms in rodents and humans (LeDoux 1998), also reduce startle potentiation induced by fear and anxiety (Hitchcock and Davis 1986; Walker and Davis 1997; Walker et al. 2005; Harris et al. 2006). Further, the fact that increased startle is seen only in the presence of aversive stimuli and can be attenuated by rewarding stimuli indicates that it is valence-specific and more than just a measure of arousal (Schmid et al. 1995; Bradley et al. 2001). The occurrence of startle potentiation is also closely associated with other behavioral models of fear, such as freezing and active avoidance (Leaton and Borszcz 1985; Miller et al. 1999). In human populations, the magnitude of the startle reflex while viewing aversive pictures also correlates strongly with measures of trait fear (Grillon et al. 1993; Vaidyanathan et al. 2009), further supporting the idea that fearful states underlie increased startle.

Fear-Potentiated Startle

One of the widest applications of the startle reflex in behavioral research is as a measure of conditioned fear. Pavlovian fear conditioning procedures produce an association between a neutral, conditioned stimulus (CS), such as a light or a tone, and an aversive, unconditioned stimulus (US), such as a shock. Following conditioning, a number of behavioral indices of fear can be observed during presentation of the CS, including an increase in the magnitude of the startle reflex. "Fear-potentiated startle" has been demonstrated in rats, mice, monkeys, and humans. In humans, fear-potentiated startle can also be induced through instruction (i.e., telling subjects that a particular stimulus will bring on a shock) or presentation of fear-inducing pictures (Bradley et al. 2001; Funayama et al. 2001; Baas et al. 2002).

Reliable measurement of fear-potentiated startle, like other behavioral expressions of fear (e.g., freezing) in rodents, has been central to our understanding of how fear memories are formed and fear- and anxiety-induced responses are generated in the brain. These studies have contributed to our understanding of the means by which inputs from different sensory modalities converge on neurons in the basolateral complex of the amygdala to produce cellular plasticity necessary for the formation and retrieval of long-term fear memories (Davis 2006; Johansen et al. 2011).

While rodent models have aided in delineating the neural mechanisms of fear, human fear-potentiated startle studies have been used to study patients with clinical anxiety. More complex conditioning methods have helped pinpoint-specific deficits in fear processing such as impairments in the ability to discriminate between safe and neutral cues and inhibit fear in the presence of safety signals. These types of findings are reviewed in greater detail in the section, "Modeling Mechanisms Underlying Anxiety".

Startle Potentiation to Contextual Cues

The fear-potentiated startle paradigms described above rely on conditioning to short-duration, discrete cues, but fear can also be conditioned to contextual cues. Context conditioning in rodents involves delivery of shock in a unique location, often distinguished by visual, tactile, and olfactory stimuli. Following conditioning, the magnitude of the startle reflex is higher in the shock-paired context than in a neutral one (McNish et al. 1997; Walker et al. 2005). In humans, subjects can be tested in distinct physical locations or virtual reality technology can be used to simulate a variety of test locations (Baas et al. 2004; Grillon et al. 2006; Alvarez et al. 2007). This type of conditioning produces a sustained state of fear in response to nonspecific cues (Grillon 2008a), as opposed to the phasic state of fear generated to explicit cues (Davis et al. 1989; Burman and Gewirtz 2004).

A rapid increase in startle is also observed immediately after a series of foot shocks (Davis 1989). There is some evidence that this phenomenon, formerly referred to simply as "sensitization" of the startle reflex (implying a nonassociative form of learning), is in fact a rapid form of context conditioning (Richardson and Elsayed 1998). On the other hand, the fact that tail shock delivered in a separate context sensitizes the startle reflex suggests that there is likely a nonassociative component to this process as well (Garrick et al. 1997, 2001; Gewirtz et al. 1998).

Startle Potentiation to Non-associative Stimuli

In addition to its utility as a model of conditioned fear, the startle reflex is increased in the presence of a number of unconditioned, anxiogenic stimuli. Startle potentiation to these nonassociative stimuli is found humans and rodents, making these paradigms potentially useful in the study of anxiety disorders.

Startle Potentiation to High or Low Levels of Illumination

Rodents are nocturnal animals that avoid brightly lit environments. This emotionally driven response can be captured in the "light-enhanced startle" paradigm, in which startle is elevated in a brightly illuminated, compared to dark, testing chamber (Walker and Davis 1997, 2002; Veening et al. 2009). In humans, this relationship is reversed, leading to what has been termed "darkness-facilitation" of startle (Grillon et al. 1997, 1999). Human startle potentiation in the dark is attenuated by benzodiazepines (Baas et al. 2002), increased in PTSD patients (Grillon et al. 1998b), and increased by prior stress exposure, a significant risk factor for the development of anxiety disorders (Grillon et al. 2007b).

Drug Withdrawal

Withdrawal from drugs of abuse is anxiogenic in both rodent models and humans, indicated, for instance, by potentiation of the startle reflex as the effects of some addictive drugs begin to wear off. This phenomenon, which we have termed "withdrawal-potentiated startle," is observed during withdrawal from morphine (Kalinichev and Holtzman 2003; Harris and Gewirtz 2004; Cabral et al. 2009; Rothwell et al. 2009; Radke et al. 2011), heroin (Park et al. 2013), nicotine (Engelmann et al. 2009), and ethanol (Rassnick et al. 1992) in rats. In human subjects, startle potentiation by unpredictable threats (see below) is exaggerated during nicotine withdrawal (Grillon et al. 2007a; Hogle et al. 2010) and yohimbine-potentiated startle (see below) is exaggerated during withdrawal from methadone (Stine et al. 2001). Startle amplitude has also been shown to correlate with number of previous ethanol detoxifications (Krystal et al. 1997). Considering the substantial comorbidity of addictive and anxiety disorders increases in startle during drug withdrawal could be a useful tool for studying the development of anxiety in humans.

Yohimbine

The noradrenergic α-2 receptor antagonist yohimbine elicits anxiety-like behavior in rodents (Davis et al. 1979; Venault et al. 1993; Shimada et al. 1995), healthy human subjects, and patients with anxiety disorders (Charney et al. 1984, 1989). Yohimbine increases startle in rats (Kehne and Davis 1985; Schulz-Klaus et al. 2005) and, at a dose that is too low to affect startle per se, enhances fear-potentiated startle (Davis et al. 1979). The more highly selective noradrenergic α-2 receptor antagonist atipamezole potentiates startle in mice (Gresack and Risbrough 2011). Human subjects also show yohimbine-potentiated startle (Morgan et al. 1993), an effect that is more pronounced in patients diagnosed with PTSD (Morgan et al. 1995).

The effects of yohimbine form an interesting bridge between the anxiety induced by drug withdrawal and the anxiety induced by stressors, described above. In addition to potentiating startle, administration of yohimbine is a potent cue for both inducing and potentiating reinstatement of drug-seeking behavior in rodents (Shepard et al. 2004; Buffalari and See 2011). And, as described above, yohimbine-potentiated startle is exacerbated during methadone withdrawal (Stine et al. 2001). These findings suggest a commonality between the anxiety-like state induced by associative or non-associative cues, and the internal state induced by stressors that is partially responsible for the reinstatement and persistence of drug-taking behaviors.

Modeling Mechanisms Underlying Anxiety

Thus far, we have described a variety of circumstances that induce a change in emotional state that can be quantified as potentiation of the startle reflex. Because potentiated startle can be measured in humans as well as rodents, this set of paradigms offers a test bed for pinpointing maladaptive processes underlying anxiety disorders and, at the same time, for testing the effects of targeting these processes to treat anxiety. As will be described below, several potential sources of dysfunction have been identified using this approach. Development of similar paradigms in mice will be essential in characterizing the contribution of specific genes to the etiology of anxiety disorders.

Exaggerated Acquisition or Expression of Fear

It is reasonable, and perhaps most straightforward, to postulate that a tendency to acquire greater fear, or for fear-eliciting cues to induce more intense states of fear, constitutes a maladaptive process producing chronic anxiety. In Pavlovian fear-conditioning paradigms, such a dysfunction would be manifested as exaggerated conditioned fear to a wide range of CSs. In anatomical terms, differences in functioning of the amygdala complex would be most likely implicated in this pattern of behavior, since this set of structures (and in particular its basolateral complex and central nucleus) is virtually a *sine qua non* for Pavlovian fear conditioning. In rats, prior exposure to stress increases acquisition of fear, measured using freezing, to both explicit, auditory and contextual cues (Conrad et al. 1999; Rau et al. 2005; Ponomarev et al. 2010). Exposure to stress after training does not have the same effect, suggesting an effect on acquisition, rather than expression (Rau et al. 2005). A number of studies on anxiety disorder patients have examined this possibility, often utilizing fear-potentiated startle. Although the results have been mixed, depending on the diagnosed group studied or the conditioning paradigms used, a meta-analysis indicated that fear conditioned to a CS in a single-cue (i.e., CS+) conditioning paradigm does indeed tend to be exaggerated in anxiety disorder patients (Lissek et al. 2005b).

Overgeneralization of Fear

A second possibility is that anxiety arises, or at least is exacerbated, by a tendency to express fear to a broader range of cues that share some features with specific cues that reliably signal threat. In both monkey and rat, such broadening of the "generalization gradient" of predictive cues has been observed following exposure to inescapable shock (Sidman et al. 1957; Hearst 1962). An analogous broadening of generalization gradients has been observed in human subjects (Dunsmoor et al. 2009). Moreover, fear conditioning, measured using fear-potentiated startle, produces broader generalization in panic and generalized anxiety disorder patients than in controls (Lissek et al. 2010, 2013b). A similar generalization effect has been reported in PTSD patients, where fear has been measured using either fear-potentiated startle or autonomic arousal (skin conductance and heart rate) (Grillon and Morgan 1999; Peri et al. 2000). However, other studies have reported exaggerated fear to CS+, but intact discrimination between CS+ and CS– (Orr et al. 2000; Norrholm et al. 2011).

The alterations in perceptual processing that underlie excessive generalization presumably involve modulation by stress-related circuitry (e.g., the basolateral complex of the amygdala) of cortical regions mediating perceptual processing. Consistent with this, generalization to faces in a fear-conditioning paradigm is correlated with the degree of functional connectivity between the amygdala and the fusiform gyrus (i.e., the "face" area) (Dunsmoor et al. 2011). A recent study that measured fMRI BOLD responses during a visual discrimination fear conditioning task found generalization gradients associated with activity in anterior insula cortex, the supplementary motor area, and the inferior parietal lobule (all of which were maximally activated by CS+) and a reverse gradient in the ventral hippocampus (which was maximally activated by CS−). The latter finding is consistent with the proposed role of the hippocampus in learning to inhibit fear (see below). It should be noted that regions of interest did not include visual cortex in this study (Lissek et al. 2013a).

Exaggerated Fear to Contextual Cues

The two theories discussed above share the notion that anxiety is produced by an expansion in the range of cues that generate states of fear. An alternative scenario is that pathological fear and anxiety arise from excessive responses to a single class, or specific classes, of cues. Conditioning studies of fear-potentiated startle in patients with anxiety disorders suggest a specific vulnerability toward the expression of excessive fear to the context in which fear conditioning takes place. Excessive fear to contextual cues has been demonstrated in PTSD patients (Grillon and Morgan 1999). Similarly, basal startle (i.e., startle measured prior to conditioning) is higher in subjects with PTSD and panic disorder during exposure to a room in which shock delivery is anticipated (Grillon et al. 1998a). Increased context conditioning is also associated with risk factors for anxiety, such as being female (Grillon 2008b). Finally, exaggerated contextual fear-potentiated startle is itself a predictor of vulnerability to develop PTSD: the risk of developing PTSD in police cadets was predicted by the degree of startle potentiation that occurred when electrical contacts were placed on the subject's finger, even though they were explicitly told they would not yet be shocked (Pole et al. 2009).

There is, of course, a distinction between discovering that excessive contextual fear contributes to the manifestation of anxiety and ascertaining the extent to which it does so. Nonetheless, these findings are a useful starting point for identifying neural mechanisms of anxiety, since a number of neural substrates subserving contextual fear conditioning have been identified. Rodent models have revealed that context conditioning preferentially involves at least two structures: the rostral portion of the hippocampus and the lateral portion of the bed nucleus of the stria terminalis (BNST) (Hammack et al. 2004; Sullivan et al. 2004; Maren 2008; Duvarci et al. 2009; Arruda-Carvalho et al. 2011; Hott et al. 2012). The latter is part of the "extended" amygdala, a macrostructure in the forebrain, and projects to many of the same brainstem targets as the CeA (de Olmos and Heimer 1999). Human studies of contextual fear conditioning corroborate the role of the hippocampus (Alvarez et al. 2008; Marschner et al. 2008). In part as a function of the challenge of isolating BNST activity through fMRI measures, BNST-specific activation has not yet been reported in a standard contextual fear-conditioning task. However, BNST function has recently been implicated in a related task (see section "Exaggerated Fear to Cues That Are Imprecise or Sustained Predictors of Threat" below). The relevance of contextual fear conditioning to anxiety is also supported by findings that knock-out or knockdown of the genes coding for three neuropeptides implicated in anxiety—brain-derived neurotrophic factor (BDNF), corticotropin-releasing factor (CRF), and neuronal PAS domain protein (NPAS4) (Le-Niculescu et al. 2011; Autry and Monteggia 2012; Koob and Zorrilla 2012)—produces pronounced deficits in fear to contextual, and not explicit cues (Liu et al. 2004; Risbrough et al. 2009; Ramamoorthi et al. 2011).

Exaggerated Fear to Cues That Are Imprecise or Sustained Predictors of Threat

A key difference between the characteristics of contextual and explicit cues in Pavlovian fear conditioning lies in their duration. Thus, contextual cues constitute less precise predictors of the timing of aversive events. This uncertainty may in turn produce a more sustained state of fear. Consistent with this idea, startle in healthy individuals is increased during the intertrial interval when the predictive value of the CS is degraded (Grillon et al. 2004).

Excessive susceptibility to sustained fear generated under these conditions is an additional possible factor underlying the symptomatology of anxiety disorders (Grillon 2008a). This hypothesis has received direct support, in that panic disorder patients show excessive fear-potentiated startle to cues of limited temporal predictive value (Grillon et al. 2008). In animals, in addition to contextual cues, potentiation of startle by nonassociative (i.e., innate, or unlearned) threat cues, such as bright lights or predator odors, and intracerebroventricular (ICV) infusion of the peptide CRF, would all fall into this category. The striking fact that all these phenomena also require activity of the BNST (Lee and Davis 1997; Walker and Davis 1997, 2002; Fendt et al. 2003) supports the hypothesis that the BNST plays a core role in mounting sustained responses to threatening cues that are temporally diffuse (Walker and Davis 2008; Davis et al. 2010). This hypothesis was tested directly by training animals with a CS that varied in duration across trials (3 s–8 min). This form of training produces a sustained state of fear, as indicated by fear-potentiated startle probed at various time points after CS onset that is blocked by inactivation of the BNST (Walker et al. 2009). A vigilance task, which produced a similarly sustained state of preparedness for imminent threat, results in activation of the BNST, as well as the insular cortex, in human subjects (Somerville et al. 2010). Moreover, intensity of activation of these structures correlated heavily with a composite measure of anxiety self-report indices. This study is the first to provide a bridge between BNST function, sustained fear potentiation of the startle reflex, and anxiety.

Impaired Inhibition of Fear

Just as cues can acquire the power to excite a state of fear, they can acquire the power to inhibit expression of one as well. Through Pavlovian "conditioned inhibition" paradigms a cue comes to signal safety or that a given aversive US will not occur (Falls and Davis 1997). A second form of inhibitory Pavlovian conditioning procedure is "extinction," whereby a stimulus that was formerly paired with a US is now presented on its own. The reduction of fear that occurs over repeated presentations does not indicate forgetting of fear, but rather an active inhibition that is similar in many respects to the inhibition exerted by a safety signal (Bouton 2002).

The final mechanism that has been proposed to contribute to the psychopathology of anxiety is that of inadequate inhibition of fear (Davis et al. 2000; Quirk and Mueller 2008). This hypothesis comports with clinical evidence that exposure therapy—a procedure that procedurally resembles Pavlovian extinction—is the most effective treatment for some forms of anxiety disorder (McLean and Foa 2011; Barlow et al. 2013). A number of studies now indicate that exposure therapy can be accelerated when administered in conjunction with the NMDA glutamate receptor co-agonist D-cycloserine (Norberg et al. 2008; Bontempo et al. 2012; Ressler et al. 2004). The rationale for applying this treatment in human patients was derived entirely from experiments indicating that the same drug accelerated extinction of conditioned fear (Walker et al. 2002; Ressler et al. 2004; Norberg et al. 2008). Finally, the most direct evidence that anxiety disorders are linked to an impairment in the ability to inhibit fear comes from a study of fear-potentiated startle in which individuals with PTSD showed levels of fear expression comparable to those of control subjects, but impaired inhibition of fear when the CS was signaled by a conditioned inhibitor (Jovanovic et al. 2009).

As in the case of the contextual fear hypothesis, one of the strengths of the fear-inhibition hypothesis is that a considerable amount is now known about the neural substrates of the inhibition of fear. Extinction of fear recruits a network of limbic structures, specifically, the amygdala, hippocampus, and medial prefrontal cortex (mPFC). Though difficult to do justice to an extensive corpus of work in a brief summary, fear extinction in both amygdala and mPFC involves NMDA receptor-dependent plasticity, stimulation of cannabinoid receptors, modulation of $GABA_A$ receptor-mediated neurotransmission, and stimulation of the BDNF trkB receptor (for reviews, see Myers and Davis 2007; Milad and Quirk 2012).

The latter finding should be highlighted, since BDNF appears to be both necessary and sufficient in this limbic network for the development of extinction. Direct, local infusion of the peptide into the mPFC actually induces extinction of fear (Peters et al. 2010), and female mice, which show less fear extinction than males, also exhibit reduced BDNF mRNA expression in the mPFC (Baker-Andresen et al. 2013). Reduction of BDNF mRNA transcription in the dorsal hippocampus also disrupts fear extinction (Heldt et al. 2007a), an effect that occurs due to reduced release of BDNF in the mPFC from neurons originating in the hippocampus (Peters et al. 2010). Finally, a more recent study has pointed to the clinical relevance of BDNF's role in extinction: PTSD patients possessing a copy of the Met allele of the

Val66Met BDNF polymorphism showed poorer responses to exposure therapy (Felmingham et al. 2013). This relatively rare variant of the single nucleotide polymorphism at codon 66 is thought to reduce the efficiency of BDNF trafficking (Chen et al. 2008). A systems-level model that integrates the roles of BDNF in the amygdala, hippocampus, and mPFC, as well as those of the other neurotransmitter systems implicated in extinction, has yet to be developed.

In sum, there are putative malfunctions of several processes underlying Pavlovian fear conditioning that may contribute to manifestations of anxiety. Whether one or more of these mechanisms is of paramount importance is an open question, but it is most likely that anxiety, a heterogeneous category of mental disorder, is related to disruptions of interacting emotional processes.

Use of Startle as a Behavioral Model of Anxiety in Mice

The adoption of mouse models for behavioral analysis, with the range of tools to alter genetic signaling they provide, has been important in the acceleration of progress in the discovery of the neural bases of fear and anxiety. Surprisingly few such studies have made use of potentiated startle, however, since this measure was first introduced in 1997.

Most studies of potentiated startle in mice have utilized the fear-potentiated startle paradigm (Table 11.1). Initial characterizations on this phenomenon in mice determined that fear responses are generally stronger in C57BL/6J than in the DBA/2J strain (Falls et al. 1997; McCaughran et al. 2000; Waddell et al. 2004). Deficits in fear-potentiated startle have been demonstrated following posttraining lesions of the amygdala (Heldt et al. 2000) and after administration of anxiolytic drugs (Risbrough et al. 2003a; Smith et al. 2011). cAMP response element binding protein (CREB), the ERK/MAPK pathway, and the dopaminergic system have been found necessary for the acquisition of fear-potentiated startle (Falls et al. 2000; Di Benedetto et al. 2008; Fadok et al. 2009, 2010), whereas BDNF is important in both its consolidation (Choi et al. 2010) and extinction (see above, Heldt et al. 2007a). The importance of CRF in contextual fear memories has also been confirmed in a mouse model of fear-potentiated startle (Risbrough et al. 2009).

Recommended Startle Parameters

The most likely, and entirely practical, reason that potentiated startle studies have not been more commonplace in mice is its exquisite sensitivity to parametric variations, exacerbated by pronounced differences between different inbred lines. In contrast, judging by the range of parameters variously adopted in the literature, freezing—the most common measure of fear in mice—appears to have more relaxed tolerance limits. Nevertheless, as first laid out systematically by Falls (2002), reliable measures of fear can be obtained, provided appropriate procedures are

Table 11.1 Potentiated startle studies in mice

Reference	Paradigm	Strain	CS	Notes
Barrenha and Chester (2007)	FPS	High and low alcohol preferring	Light	Mice bred for high alcohol preference have greater FPS
Barrenha et al. (2011)	FPS	High and low alcohol preferring	Light	Alcohol impairs expression of FPS in high alcohol preferring mice; diazepam impairs expression in both strains
Di Benedetto et al. (2008)	FPS	C57BL/6J	Tone	ERK/MAPK in LA is necessary for acquisition of FPS
Choi et al. (2010)	FPS	BDNF ^{tm3Jae}/J, cortex-BDNF KO	Tone	Cortical BDNF deletion impairs consolidation of fear memories
Fadok et al. (2009)	FPS	$Th^{fs/fs}$; $Dbh^{Th/+}$, D_1R-KO, D_2R-KO (on C57BL/6 background)	Light	Dopamine depletion and D_1R knockout impairs cued fear conditioning
Fadok et al. (2010)	FPS	$Th^{fs/fs}$; $Dbh^{Th/+}$ (on C57BL/6 background)	Light	Dopamine is necessary in the BLA and NAc for fear learning
Falls et al. (1997)	FPS	C57BL/6J, DBA/2J	Tone	First demonstration of mouse FPS, C57 > DBA
Falls et al. (2000)	FPS	$CREB^{\alpha\delta-/-}$	Tone	CREB knockout mice do not show FPS
Falls (2002)	FPS	Various strains	Tone	Procedural variables in FPS in mice
Gresack and Risbrough (2011)	CRF and atipamezole	C57BL/6J	N/A	Role of norepinephrine and CRF signaling in CRF- and atipamezole-enhanced startle
Heldt et al. (2000)	FPS	C57BL/6J	Light, tone	Posttraining amygdala lesions disrupt FPS
Heldt et al. (2007a)	FPS	BDNF ^{tm3Jae}/J (on a mixed B6, 129S4, BALB/c background)	Tone	Hippocampal BDNF deletion reduces extinction
Jones et al. (2005)	FPS	C57BL/6J	Odor	Demonstration of olfactory fear conditioning in mice
McCaughran et al. (2000)	FPS	C57BL/6J, DBA/2J	Light	Response after 5 trials, max response at 20 trials; C57 > DBA
Powers et al. (2010)	FPS	High and low alcohol preferring	Light	Inhibition of endocannabinoid uptake impairs FPS in high alcohol preferring mice
Risbrough et al. (2003a)	FPS	C57BL/6J, DBA/1J	Light	Response after 8 trials; diazepam, chlordiazepoxide, and buspirone decreased FPS
Risbrough et al. (2003b)	CRF	C57BL/6J, 129S6/SvEvTac	N/A	CRF-R1 and CRF-R2 involved in CRF-enhanced startle
Risbrough and Geyer (2005)	FPS	DBA/1J	Light	FPS was not potentiated by anxiogenic compounds

Risbrough et al. (2009)	FPS	CRF_1 and CRF_2 KO (on C57BL/6 background)	Light	CRF-R1 and CRF-R2 knockout mice impaired in conditioning to contextual but not discrete cues
Smith et al. (2011)	FPS	C57BL/6J	Tone	Modified parameters to strengthen paradigm in C57 mice, reduce unconditioned effects of tone; benzodiazepines reduced FPS
Smith et al. (2012)	FPS	$GABA_A$ receptor subtype knockouts (on C57/BL/6J background)	Tone	Benzodiazepines ineffective in $GABA_A$ receptor knockout mice
Toth et al. (2013)	CRF	C57BL/6	N/A	Protein kinase C (PKC) inhibitor blocks CRF-enhanced startle
Waddell et al. (2004)	FPS	C57BL/6J, DBA/2J	Light	DBA extinguish more quickly than C57 and have deficit in context specificity of extinction

followed. Except where otherwise stated, procedures are broadly similar to those used in rats. Further, the parameters described are recommended for use with C57/BL6 mice, the most commonly used strain in these studies. Variations to a number of parameters are almost certainly required for other strains (see below).

Phases of Studies

The three phases generally applied in experiments to measure fear-potentiated startle in rodents are (1) a preconditioning startle test, (2) fear conditioning, and (3) a post-conditioning startle test. Sessions typically commence with a "preperiod" after the animals are placed in their cages, typically of 5-min duration. The initial test of startle is used to establish the magnitude of preconditioning startle (i.e., basal startle). It also serves to reduce the novelty of the startle stimulus and produce some long-term habituation of the startle reflex, which is at its largest when first elicited (Davis 1970).

Startle Stimulus Parameters

The startle stimulus is a brief (e.g., 20–40 ms) burst of white noise. Because there are wide variations in the input–output function of the startle reflex (i.e., relationship between startle stimulus intensity and response intensity), it is recommended that subjects are tested across a range of startle stimulus intensities so as to reduce ceiling or floor effects (Falls 2002). The typical range of startle stimuli used has been 85–100 dB, somewhat lower than the range conventionally used in rats. This may be because mouse fear-potentiated startle paradigms (in contrast to rats) have tended not to present the startle stimulus against a white background noise, which raises the level of noise burst intensity required to elicit startle (Davis 1974b). The interstimulus interval (ISI) between noise bursts is routinely set at 30 s, consistent with rat studies.

The CS

Based on exploratory studies, William Falls at the University of Vermont showed that CS onset has behavioral activating effects (personal communication). We have recently made similar observations (Gewirtz 2014). Hence, tone duration has typically been set at 30 s, compared to a typical CS duration of around 4 s in rats. To test fear-potentiated startle, the startle stimulus is presented in the absence versus presence of the CS. In the latter trials, it is presented at the same point in time at which the US is presented in training (see below), that is, toward the end of the period of CS presentation. Because the startle-eliciting noise burst occurs on a background of the auditory CS (Davis 1974a) the unconditioned effects of the tone on reflex magnitude can be substantial (Falls 2002). It is therefore important to select an acoustic CS at a frequency and intensity that produce relatively little unconditioned potentiation or inhibition of startle. This selection process becomes something of a balancing act because lowering stimulus intensity may reduce CS salience and hence the strength of conditioning acquired (Scavio and Gormezano 1974). It is also recommended to ramp the intensity of CS onset (Smith et al. 2011) so as to avoid startle reflex elicitation by the CS itself. This measure has not proven necessary in rats, where startle is only elicited by stimuli substantially louder than those used as the CS. Smith et al. (2011) found that a 12-KHz, 70-dB tone, with a 3-s rise-time has little unconditioned effect on startle amplitude while at the same time supporting robust fear conditioning. We have confirmed these findings, but found unconditioned effects to be at a minimum at 10 KHz (unpublished data).

It is worth noting that the use of a pure tone of 30-s duration (as opposed to a band pass-filtered noise of 4-s duration in rats) brings fear-potentiated startle into line with standard protocols for assessing freezing behavior. This means that the same animals can be tested sequentially in equipment specialized for measuring freezing and fear-potentiated startle, respectively (Heldt et al. 2007b). The ability to test animals sequentially with two different, and well-established, behavioral measures of fear, is a distinct advantage of this arrangement.

A light has also been successfully used as a CS in some studies. A 7- or 8-W light of 10–30-s duration is typically employed. While use of a light CS may prevent the behavioral activating effects of a tone, extra training trials may be necessary to obtain the same level of conditioning (Heldt et al. 2000).

CS Pretest

Because an auditory CS can itself have such significant effects on startle magnitude, it has been recommended to include a test for fear-potentiated startle prior to training (Falls 2002). This may not be absolutely necessary for C57Bl/6 mice if the guidelines for CS characteristics described above are observed. Furthermore, CS preexposure may produce the Pavlovian phenomenon of latent inhibition, which retards conditioning (Lubow and Gewirtz 1995). The strength of latent inhibition is partly a function of the number of preexposures to the CS (Lubow 1973). Thus, if a CS pretest is included, it should contain no more than 15 presentations of the CS (Lubow 1973).

Contextual Fear

Contextual fear is measured as the increase in startle amplitude after training when no CS is present and when tested in the training environment. If robust, contextual fear may interfere with measurement of explicit cue fear. That is why freezing to a tone CS is typically tested in a context distinct from that used in training. In rats, context-specific startle potentiation is seen only in initial test trials (McNish et al. 1997). Hence, provided a series of startle stimuli are presented prior to the CS in test (see above), contextual fear does not interfere with measurement of fear to the

CS. In C57BL/6 male mice, however, contextual potentiation of startle is robust, often substantial, and sustained across extensive repetitions of the startle stimulus. This effect can be mitigated either by testing in a novel context or by extinguishing contextual fear (through one or more days of startle testing, without the CS or US present) prior to testing fear to the CS. The advantages of the latter procedure are (a) that one does not need to maintain an alternative, distinctive context and (b) that the extinction procedure also serves as a test for contextual fear. Interestingly, female C57BL/6 mice show relatively little contextual fear-potentiated startle so that it is possible to test for fear to an explicit cue without the need for such additional contextual manipulations.

Fear Decay

In rats, fear decays quickly after offset of the CS (Davis et al. 1989; Burman and Gewirtz 2004), thus permitting fear-potentiated startle to be measured as the difference between responses on CS versus no-CS startle trials, presented in an intermixed fashion. Nevertheless, in conditions under which startle does not fully return to its baseline rapidly after CS offset, using intermixed no-CS startle responses as a baseline will cause one to underestimate the strength of fear-potentiated startle. This is rarely an issue in rats, but in C57BL/6 mice we observe robust residual fear, even at longer intervals. This may be because fear is slower to decay after CS offset in mice than in rats, or because the CS acts as a retrieval cue and induces reinstatement of extinguished fear to the context. If this issue arises, we recommend using the mean startle response to the initial startle stimuli, prior to first presentation of the CS, as the baseline above which potentiation is assessed.

Variation Between Mouse Strains

Falls (2002) has reported considerable variation between different mouse strains in (1) the amplitude of the startle reflex as a function of noise-burst intensity, (2) the magnitude of the nonassociative effects of a tone CS on startle responses, and (3) the relationship between footshock intensity and magnitude of fear conditioning. Hence, thorough parametric analysis of these variables is recommended if one is testing mice that are not maintained solely on a C57BL/6 background.

Conclusions and Future Directions

To date, the use of fear conditioning paradigms in animals and humans has greatly advanced our understanding of the etiology of fear and anxiety-related disorders. Studies employing startle are of particular value as translational tools because the reflex is highly conserved across mammalian species. Unfortunately, the potential of mouse startle studies to elucidate genetic contributions to anxiety disorders has remained largely untapped. It is our hope that, by following the recommendations outlined here and in Falls (2002), investigators will find these studies more feasible and the literature in this area will begin to flourish.

Because only a few mouse startle studies have been conducted so far, early studies will need to focus on extending this groundwork. One particularly important area of research will be to test fear-potentiated startle across mouse strains, as preliminary work indicates there are considerable variations among them (Falls 2002) and to further characterize sex differences. It will also be important to develop a range of CSs, since these are necessary components of more complex fear conditioning paradigms. Finally, many of the fear paradigms that have been established in rats have yet to be established in mice. Phenomena such as CS generalization and conditioned inhibition, which are affected in human anxiety patients, would all add utility to the experimenter's armamentarium. Concurrent with laying this groundwork, researchers can begin to exploit the full potential of startle potentiation to explore the genetic underpinnings of anxiety in genetically modified mice.

References

Adamec, R. E., Burton, P., Shallow, T., & Budgell, J. (1999). Unilateral block of NMDA receptors in the amygdala prevents predator stress-induced lasting increases in anxiety-like behavior and unconditioned startle—Effective hemisphere depends on the behavior. *Physiology and Behavior, 65*, 739–751.

Adamec, R., Head, D., Blundell, J., Burton, P., & Berton, O. (2006). Lasting anxiogenic effects of feline predator stress in mice: Sex differences in vulnerability to stress and predicting severity of anxiogenic response from the stress experience. *Physiology and Behavior, 88*, 12–29.

Alvarez, R. P., Biggs, A., Chen, G., Pine, D. S., & Grillon, C. (2008). Contextual fear conditioning in humans: Cortical-hippocampal and amygdala contributions. *Journal of Neuroscience, 28*, 6211–6219.

Alvarez, R. P., Johnson, L., & Grillon, C. (2007). Contextual-specificity of short-delay extinction in humans: Renewal of fear-potentiated startle in a virtual environment. *Learning and Memory, 14*, 247–253.

Arruda-Carvalho, M., Sakaguchi, M., Akers, K. G., Josselyn, S. A., & Frankland, P. W. (2011). Posttraining ablation of adult-generated neurons degrades previously acquired memories. *Journal of Neuroscience, 31*, 15113–15127.

Autry, A. E., & Monteggia, L. M. (2012). Brain-derived neurotrophic factor and neuropsychiatric disorders. *Pharmacological Reviews, 64*, 238–258.

Baas, J. M., Grillon, C., Bocker, K. B., Brack, A. A., Morgan, C. A., 3rd, Kenemans, J. L., et al. (2002). Benzodiazepines have no effect on fear-potentiated startle in humans. *Psychopharmacology, 161*, 233–247.

Baas, J. M., Nugent, M., Lissek, S., Pine, D. S., & Grillon, C. (2004). Fear conditioning in virtual reality contexts: A new tool for the study of anxiety. *Biological Psychiatry, 55*, 1056–1060.

Baker-Andresen, D., Flavell, C. R., Li, X., & Bredy, T. W. (2013). Activation of BDNF signaling prevents the return of fear in female mice. *Learning and Memory, 20*, 237–240.

Balaban, M. T., & Taussig, H. N. (1994). Salience of fear/threat in the affective modulation of the human startle blink. *Biological Psychology, 38*, 117–131.

Barlow, D. H., Bullis, J. R., Comer, J. S., & Ametaj, A. A. (2013). Evidence-based psychological treatments: An update and a way forward. *Annual Review of Clinical Psychology, 9*, 1–27.

Barrenha, G. D., & Chester, J. A. (2007). Genetic correlation between innate alcohol preference and fear-potentiated startle in selected mouse lines. *Alcoholism, Clinical and Experimental Research, 31*, 1081–1088.
Barrenha, G. D., Coon, L. E., & Chester, J. A. (2011). Effects of alcohol on the acquisition and expression of fear-potentiated startle in mouse lines selectively bred for high and low alcohol preference. *Psychopharmacology, 218*, 191–201.
Bontempo, A., Panza, K. E., & Bloch, M. H. (2012). D-cycloserine augmentation of behavioral therapy for the treatment of anxiety disorders: A meta-analysis. *The Journal of Clinical Psychiatry, 73*, 533–537.
Bouton, M. E. (2002). Context, ambiguity, and unlearning: Sources of relapse after behavioral extinction. *Biological Psychiatry, 52*, 976–986.
Bradley, M. M., Codispoti, M., Cuthbert, B. N., & Lang, P. J. (2001). Emotion and motivation I: Defensive and appetitive reactions in picture processing. *Emotion, 1*, 276–298.
Buffalari, D. M., & See, R. E. (2011). Inactivation of the bed nucleus of the stria terminalis in an animal model of relapse: Effects on conditioned cue-induced reinstatement and its enhancement by yohimbine. *Psychopharmacology, 213*, 19–27.
Burman, M. A., & Gewirtz, J. C. (2004). Timing of fear expression in trace and delay conditioning measured by fear-potentiated startle in rats. *Learning and Memory, 11*, 205–212.
Cabral, A., Ruggiero, R. N., Nobre, M. J., Brandao, M. L., & Castilho, V. M. (2009). GABA and opioid mechanisms of the central amygdala underlie the withdrawal-potentiated startle from acute morphine. *Progress in Neuropsychopharmacology and Biological Psychiatry, 33*, 334–344.
Canli, T., & Brown, T. H. (1996). Amygdala stimulation enhances the rat eyeblink reflex through a short-latency mechanism. *Behavioral Neuroscience, 110*, 51–59.
Charney, D. S., Heninger, G. R., & Breier, A. (1984). Noradrenergic function in panic anxiety: Effects of yohimbine in healthy subjects and patients with agoraphobia and panic disorder. *Archives of General Psychiatry, 41*, 751.
Charney, D. S., Woods, S. W., & Heninger, G. R. (1989). Noradrenergic function in generalized anxiety disorder: Effects of yohimbine in healthy subjects and patients with generalized anxiety disorder. *Psychiatry Research, 27*, 173–182.
Chen, Z. Y., Bath, K., McEwen, B., Hempstead, B., & Lee, F. (2008). Impact of genetic variant BDNF (Val66Met) on brain structure and function. *Novartis Foundation Symposium, 289*, 180–188; discussion 188–195.
Choi, J. S., Lindquist, D. H., & Brown, T. H. (2001). Amygdala lesions block conditioned enhancement of the early component of the rat eyeblink reflex. *Behavioral Neuroscience, 115*, 764–775.
Choi, D. C., Maguschak, K. A., Ye, K., Jang, S.-W., Myers, K. M., & Ressler, K. J. (2010). Prelimbic cortical BDNF is required for memory of learned fear but not extinction or innate fear. *Proceedings of the National Academy of Sciences, 107*, 2675–2680.
Conrad, C. D., LeDoux, J. E., Magarinos, A. M., & McEwen, B. S. (1999). Repeated restraint stress facilitates fear conditioning independently of causing hippocampal CA3 dendritic atrophy. *Behavioral Neuroscience, 113*, 902–913.
Davis, M. (1970). Effects of interstimulus interval length and variability on startle-response habituation in the rat. *Journal of Comparative and Physiological Psychology, 72*(2), 177.
Davis, M. (1974a). Sensitization of the rat startle response by noise. *Journal of Comparative and Physiological Psychology, 87*, 571–581.
Davis, M. (1974b). Signal-to-noise ratio as a predictor of startle amplitude and habituation in the rat. *Journal of Comparative and Physiological Psychology, 86*, 812–825.
Davis, M. (1979a). Diazepam and flurazepam: Effects on conditioned fear as measured with the potentiated startle paradigm. *Psychopharmacology, 62*, 1–7.
Davis, M. (1979b). Morphine and naloxone: Effects on conditioned fear as measured with the potentiated startle paradigm. *European Journal of Pharmacology, 54*, 341–347.
Davis, M. (1989). Sensitization of the acoustic startle reflex by footshock. *Behavioral Neuroscience, 103*, 495.

Davis, M. (2006). Neural systems involved in fear and anxiety measured with fear-potentiated startle. *The American Psychologist, 61*, 741–756.
Davis, M., Antoniadis, E. A., Amaral, D. G., & Winslow, J. T. (2008). Acoustic startle reflex in rhesus monkeys: A review. *Reviews in the Neurosciences, 19*, 171–186.
Davis, M., Cassella, J. V., & Kehne, J. H. (1988). Serotonin does not mediate anxiolytic effects of buspirone in the fear-potentiated startle paradigm: Comparison with 8-OH-DPAT and ipsapirone. *Psychopharmacology, 94*, 14–20.
Davis, M., Falls, W. A., Gewirtz, J. (2000). Neural systems involved in fear inhibition: Extinction and conditioned inhibition. In M. S. Myslobodsky, & I. Weiner (Eds.) *Contemporary issues in modeling psychopathology* (pp 113–141). Boston: Kluwer.
Davis, M., Redmond, D. E., Jr., & Baraban, J. M. (1979). Noradrenergic agonists and antagonists: Effects on conditioned fear as measured by the potentiated startle paradigm. *Psychopharmacology, 65*, 111–118.
Davis, M., Schlesinger, L. S., & Sorenson, C. A. (1989). Temporal specificity of fear conditioning: Effects of different conditioned stimulus-unconditioned stimulus intervals on the fear-potentiated startle effect. *Journal of Experimental Psychology: Animal Behavior Processes, 15*, 295–310.
Davis, M., Walker, D. L., Miles, L., & Grillon, C. (2010). Phasic vs sustained fear in rats and humans: Role of the extended amygdala in fear vs anxiety. *Neuropsychopharmacology, 35*, 105–135.
de Olmos, J. S., & Heimer, L. (1999). The concepts of the ventral striatopallidal system and extended amygdala. *Annals of the New York Academy of Sciences, 877*, 1–32.
Di Benedetto, B., Kallnik, M., Weisenhorn, D. M. V., Falls, W. A., Wurst, W., & Hölter, S. M. (2008). Activation of ERK/MAPK in the lateral amygdala of the mouse is required for acquisition of a fear-potentiated startle response. *Neuropsychopharmacology, 34*, 356–366.
Dunsmoor, J. E., Mitroff, S. R., & LaBar, K. S. (2009). Generalization of conditioned fear along a dimension of increasing fear intensity. *Learning and Memory, 16*, 460–469.
Dunsmoor, J. E., Prince, S. E., Murty, V. P., Kragel, P. A., & LaBar, K. S. (2011). Neurobehavioral mechanisms of human fear generalization. *NeuroImage, 55*, 1878–1888.
Duvarci, S., Bauer, E. P., & Pare, D. (2009). The bed nucleus of the stria terminalis mediates inter-individual variations in anxiety and fear. *Journal of Neuroscience, 29*, 10357–10361.
Eaton, R. C. (1984). *Neural mechanisms of startle behavior*. New York: Plenum Publishing Corporation.
Engelmann, J. M., Gewirtz, J. C., & Cuthbert, B. N. (2011). Emotional reactivity to emotional and smoking cues during smoking abstinence: Potentiated startle and P300 suppression. *Psychophysiology, 48*, 1656–1668.
Engelmann, J. M., Radke, A. K., & Gewirtz, J. C. (2009). Potentiated startle as a measure of the negative affective consequences of repeated exposure to nicotine in rats. *Psychopharmacology, 207*, 13–25.
Fadok, J. P., Darvas, M., Dickerson, T. M., & Palmiter, R. D. (2010). Long-term memory for pavlovian fear conditioning requires dopamine in the nucleus accumbens and basolateral amygdala. *PloS One, 5*, e12751.
Fadok, J. P., Dickerson, T. M., & Palmiter, R. D. (2009). Dopamine is necessary for cue-dependent fear conditioning. *Journal of Neuroscience, 29*, 11089–11097.
Falls WA (2002) Fear potentiated startle in mice. *Current protocols in neuroscience*:8.11 B. 11-18.11 B. 16.
Falls, W. A., Carlson, S., Turner, J. G., & Willott, J. F. (1997). Fear-potentiated startle in two strains of inbred mice. *Behavioral Neuroscience, 111*, 855.
Falls, W. A., & Davis, M. (1997). Inhibition of fear-potentiated startle can be detected after the offset of a feature trained in a serial feature-negative discrimination. *Journal of Experimental Psychology: Animal Behavior Processes, 23*, 3–14.
Falls, W. A., Kogan, J. H., Silva, A. J., Willott, J. F., Carlson, S., & Turner, J. G. (2000). Fear-potentiated startle, but not prepulse inhibition of startle, is impaired in CREBalphadelta–/– mutant mice. *Behavioral Neuroscience, 114*, 998–1004.

Felmingham, K. L., Dobson-Stone, C., Schofield, P. R., Quirk, G. J., & Bryant, R. A. (2013). The brain-derived neurotrophic factor Val66Met polymorphism predicts response to exposure therapy in posttraumatic stress disorder. *Biological Psychiatry, 73*, 1059–1063.

Fendt, M., Endres, T., & Apfelbach, R. (2003). Temporary inactivation of the bed nucleus of the stria terminalis but not of the amygdala blocks freezing induced by trimethylthiazoline, a component of fox feces. *Journal of Neuroscience, 23*, 23–28.

Funayama, E. S., Grillon, C., Davis, M., & Phelps, E. A. (2001). A double dissociation in the affective modulation of startle in humans: Effects of unilateral temporal lobectomy. *Journal of Cognitive Neuroscience, 13*, 721–729.

Garrick, T., Morrow, N., Eth, S., Marciano, D., & Shalev, A. (1997). Psychophysiologic parameters of traumatic stress disorder in rats. *Annals of the New York Academy of Sciences, 821*, 533–537.

Garrick, T., Morrow, N., Shalev, A. Y., & Eth, S. (2001). Stress-induced enhancement of auditory startle: An animal model of posttraumatic stress disorder. *Psychiatry, 64*, 346–354.

Gewirtz, J. C. (2014). Deficits in social interaction and social cognition in mice in relation to neurodevelopmental psychiatric disorders. In *Proceedings of the International Summit Forum on Development and Education of Children*. Liaoning Normal University, Dalian, China.

Gewirtz, J. C., McNish, K. A., & Davis, M. (1998). Lesions of the bed nucleus of the stria terminalis block sensitization of the acoustic startle reflex produced by repeated stress, but not fear-potentiated startle. *Progress in Neuropsychopharmacology and Biological Psychiatry, 22*, 625–648.

Greenwald, M. K., Bradley, M. M., Cuthbert, B. N., & Lang, P. J. (1998). Startle potentiation: Shock sensitization, aversive learning, and affective picture modulation. *Behavioral Neuroscience, 112*, 1069–1079.

Gresack, J. E., & Risbrough, V. M. (2011). Corticotropin-releasing factor and noradrenergic signaling exert reciprocal control over startle reactivity. *International Journal of Neuropharmacology, 2011*(4), 1179–1194.

Grillon, C. (2008a). Models and mechanisms of anxiety: Evidence from startle studies. *Psychopharmacology, 199*, 421–437.

Grillon, C. (2008b). Greater sustained anxiety but not phasic fear in women compared to men. *Emotion, 8*, 410–413.

Grillon, C., Ameli, R., Foot, M., & Davis, M. (1993). Fear-potentiated startle: Relationship to the level of state/trait anxiety in healthy subjects. *Biological Psychiatry, 33*, 566–574.

Grillon, C., Avenevoli, S., Daurignac, E., & Merikangas, K. R. (2007a). Fear-potentiated startle to threat, and prepulse inhibition among young adult nonsmokers, abstinent smokers, and nonabstinent smokers. *Biological Psychiatry, 62*, 1155–1161.

Grillon, C., Duncko, R., Covington, M. F., Kopperman, L., & Kling, M. A. (2007b). Acute stress potentiates anxiety in humans. *Biological Psychiatry, 62*, 1183–1186.

Grillon, C., & Baas, J. (2003). A review of the modulation of the startle reflex by affective states and its application in psychiatry. *Clinical Neurophysiology, 114*, 1557–1579.

Grillon, C., Baas, J. M., Cornwell, B., & Johnson, L. (2006). Context conditioning and behavioral avoidance in a virtual reality environment: Effect of predictability. *Biological Psychiatry, 60*, 752–759.

Grillon, C., Baas, J. P., Lissek, S., Smith, K., & Milstein, J. (2004). Anxious responses to predictable and unpredictable aversive events. *Behavioral Neuroscience, 118*, 916–924.

Grillon, C., Lissek, S., Rabin, S., McDowell, D., Dvir, S., & Pine, D. S. (2008). Increased anxiety during anticipation of unpredictable but not predictable aversive stimuli as a psychophysiologic marker of panic disorder. *The American Journal of Psychiatry, 165*, 898–904.

Grillon, C., Merikangas, K. R., Dierker, L., Snidman, N., Arriaga, R. I., Kagan, J., et al. (1999). Startle potentiation by threat of aversive stimuli and darkness in adolescents: A multi-site study. *International journal of psychophysiology, 32*, 63–73.

Grillon, C., & Morgan, C. A., 3rd. (1999). Fear-potentiated startle conditioning to explicit and contextual cues in Gulf War veterans with posttraumatic stress disorder. *Journal of Abnormal Psychology, 108*, 134–142.

Grillon, C., Morgan, C. A., 3rd, Davis, M., & Southwick, S. M. (1998a). Effects of experimental context and explicit threat cues on acoustic startle in Vietnam veterans with posttraumatic stress disorder. *Biological Psychiatry, 44*, 1027–1036.

Grillon, C., Morgan, C. A., Davis, M., & Southwick, S. M. (1998b). Effect of darkness on acoustic startle in Vietnam veterans with PTSD. *American Journal of Psychiatry, 155*, 812–817.

Grillon, C., Pellowski, M., Merikangas, K. R., & Davis, M. (1997). Darkness facilitates the acoustic startle reflex in humans. *Biological Psychiatry, 42*, 453–460.

Hammack, S. E., Richey, K. J., Watkins, L. R., & Maier, S. F. (2004). Chemical lesion of the bed nucleus of the stria terminalis blocks the behavioral consequences of uncontrollable stress. *Behavioral Neuroscience, 118*, 443–448.

Harris, A. C., Atkinson, D. M., Aase, D. M., & Gewirtz, J. C. (2006). Double dissociation in the neural substrates of acute opiate dependence as measured by withdrawal-potentiated startle. *Neuroscience, 139*, 1201–1210.

Harris, A. C., & Gewirtz, J. C. (2004). Elevated startle during withdrawal from acute morphine: A model of opiate withdrawal and anxiety. *Psychopharmacology, 171*, 140–147.

Hearst, E. (1962). Concurrent generalization gradients for food-controlled and shock-controlled behavior. *Journal of the Experimental Analysis of Behavior, 5*, 19–31.

Hebb, A. L., Zacharko, R. M., Gauthier, M., & Drolet, G. (2003). Exposure of mice to a predator odor increases acoustic startle but does not disrupt the rewarding properties of VTA intracranial self-stimulation. *Brain Research, 982*, 195–210.

Heldt, S., Stanek, L., Chhatwal, J., & Ressler, K. (2007a). Hippocampus-specific deletion of BDNF in adult mice impairs spatial memory and extinction of aversive memories. *Molecular Psychiatry, 12*, 656–670.

Heldt, S. A., Stanek, L., Chhatwal, J. P., & Ressler, K. J. (2007b). Hippocampus-specific deletion of BDNF in adult mice impairs spatial memory and extinction of aversive memories. *Molecular Psychiatry, 12*, 656–670.

Heldt, S., Sundin, V., Willott, J. F., & Falls, W. A. (2000). Posttraining lesions of the amygdala interfere with fear-potentiated startle to both visual and auditory conditioned stimuli in C57BL/6J mice. *Behavioral Neuroscience, 114*, 749.

Hijzen, T. H., Houtzager, S. W., Joordens, R. J., Olivier, B., & Slangen, J. L. (1995). Predictive validity of the potentiated startle response as a behavioral model for anxiolytic drugs. *Psychopharmacology, 118*, 150–154.

Hitchcock, J., & Davis, M. (1986). Lesions of the amygdala, but not of the cerebellum or red nucleus, block conditioned fear as measured with the potentiated startle paradigm. *Behavioral Neuroscience, 100*, 11–22.

Hogle, J. M., Kaye, J. T., & Curtin, J. J. (2010). Nicotine withdrawal increases threat-induced anxiety but not fear: Neuroadaptation in human addiction. *Biological Psychiatry, 68*, 719–725.

Hott, S. C., Gomes, F. V., Fabri, D. R., Reis, D. G., Crestani, C. C., Correa, F. M., et al. (2012). Both alpha1- and beta1-adrenoceptors in the bed nucleus of the stria terminalis are involved in the expression of conditioned contextual fear. *British Journal of Pharmacology, 167*, 207–221.

Johansen, J. P., Cain, C. K., Ostroff, L. E., & LeDoux, J. E. (2011). Molecular mechanisms of fear learning and memory. *Cell, 147*, 509–524.

Jones, S. V., Heldt, S. A., Davis, M., & Ressler, K. J. (2005). Olfactory-mediated fear conditioning in mice: Simultaneous measurements of fear-potentiated startle and freezing. *Behavioral Neuroscience, 119*, 329–335.

Jovanovic, T., Norrholm, S. D., Fennell, J. E., Keyes, M., Fiallos, A. M., Myers, K. M., et al. (2009). Posttraumatic stress disorder may be associated with impaired fear inhibition: Relation to symptom severity. *Psychiatry Research, 167*, 151–160.

Kalinichev, M., & Holtzman, S. G. (2003). Changes in urination/defecation, auditory startle response, and startle-induced ultrasonic vocalizations in rats undergoing morphine withdrawal: Similarities and differences between acute and chronic dependence. *The Journal of Pharmacology and Experimental Therapeutics, 304*, 603–609.

Kehne, J. H., & Davis, M. (1985). Central noradrenergic involvement in yohimbine excitation of acoustic startle: Effects of DSP4 and 6-OHDA. *Brain Research, 330*, 31–41.

Kessler, R. C., Chiu, W. T., Demler, O., Merikangas, K. R., & Walters, E. E. (2005). Prevalence, severity, and comorbidity of 12-month DSM-IV disorders in the National Comorbidity Survey Replication. *Archives of General Psychiatry, 62*, 617–627.

Koch, M., & Schnitzler, H.-U. (1997). The acoustic startle response in rats—circuits mediating evocation, inhibition and potentiation. *Behavioural Brain Research, 89*(1), 35–49.

Koob, G. F., & Zorrilla, E. P. (2012). Update on corticotropin-releasing factor pharmacotherapy for psychiatric disorders: A revisionist view. *Neuropsychopharmacology, 37*, 308–309.

Krystal, J. H., Webb, E., Grillon, C., Cooney, N., Casal, L., Morgan, C. A., 3rd, et al. (1997). Evidence of acoustic startle hyperreflexia in recently detoxified early onset male alcoholics: Modulation by yohimbine and m-chlorophenylpiperazine (mCPP). *Psychopharmacology, 131*, 207–215.

Landis, C., & Hunt, W. (1939). *The startle pattern*. Oxford, England: Farrar & Rinehart.

Leaton, R. N., & Borszcz, G. S. (1985). Potentiated startle: Its relation to freezing and shock intensity in rats. *Journal of Experimental Psychology: Animal Behavior Processes, 11*, 421–428.

LeDoux, J. (1998). *The emotional brain: The mysterious underpinnings of emotional life*. New York: Simon and Schuster.

Lee, Y., & Davis, M. (1997). Role of the hippocampus, the bed nucleus of the stria terminalis, and the amygdala in the excitatory effect of corticotropin-releasing hormone on the acoustic startle reflex. *Journal of Neuroscience, 17*, 6434–6446.

Lee, Y., Lopez, D. E., Meloni, E. G., & Davis, M. (1996). A primary acoustic startle pathway: Obligatory role of cochlear root neurons and the nucleus reticularis pontis caudalis. *Journal of Neuroscience, 16*, 3775–3789.

Le-Niculescu, H., Balaraman, Y., Patel, S. D., Ayalew, M., Gupta, J., Kuczenski, R., et al. (2011). Convergent functional genomics of anxiety disorders: Translational identification of genes, biomarkers, pathways and mechanisms. *Translational psychiatry, 1*, e9.

Lissek, S., Baas, J. M., Pine, D. S., Orme, K., Dvir, S., Nugent, M., et al. (2005a). Airpuff startle probes: An efficacious and less aversive alternative to white-noise. *Biological Psychology, 68*, 283–297.

Lissek, S., Powers, A. S., McClure, E. B., Phelps, E. A., Woldehawariat, G., Grillon, C., et al. (2005b). Classical fear conditioning in the anxiety disorders: A meta-analysis. *Behaviour Research and Therapy, 43*, 1391–1424.

Lissek, S., Bradford, D. E., Alvarez, R. P., Burton, P., Espensen-Sturges, T., Reynolds, R. C., et al. (2013a). Neural substrates of classically conditioned fear-generalization in humans: A parametric fMRI study. *Social Cognitive and Affective Neuroscience, 9*, 1134–1142.

Lissek, S., Kaczkurkin, A. N., Rabin, S., Geraci, M., Pine, D. S., & Grillon, C. (2013b). Generalized anxiety disorder is associated with overgeneralization of classically conditioned fear. *Biological Psychiatry, 75*(11), 909–915.

Lissek, S., Rabin, S., Heller, R. E., Lukenbaugh, D., Geraci, M., Pine, D. S., et al. (2010). Overgeneralization of conditioned fear as a pathogenic marker of panic disorder. *The American Journal of Psychiatry, 167*, 47–55.

Liu, I. Y., Lyons, W. E., Mamounas, L. A., & Thompson, R. F. (2004). Brain-derived neurotrophic factor plays a critical role in contextual fear conditioning. *Journal of Neuroscience, 24*, 7958–7963.

Lopez, D. E., Saldana, E., Nodal, F. R., Merchan, M. A., & Warr, W. B. (1999). Projections of cochlear root neurons, sentinels of the rat auditory pathway. *The Journal of Comparative Neurology, 415*, 160–174.

Lubow, R. E. (1973). Latent inhibition. *Psychological Bulletin, 79*, 398–407.

Lubow, R. E., & Gewirtz, J. C. (1995). Latent inhibition in humans: Data, theory, and implications for schizophrenia. *Psychological Bulletin, 117*, 87–103.

Maren, S. (2008). Pavlovian fear conditioning as a behavioral assay for hippocampus and amygdala function: Cautions and caveats. *European Journal of Neurology, 28*, 1661–1666.

Marschner, A., Kalisch, R., Vervliet, B., Vansteenwegen, D., & Buchel, C. (2008). Dissociable roles for the hippocampus and the amygdala in human cued versus context fear conditioning. *Journal of Neuroscience, 28*, 9030–9036.

McCaughran, J. A., Jr., Bell, J., III, & Hitzemann, R. J. (2000). Fear-potentiated startle response in mice: Genetic analysis of the C57BL/6J and DBA/2J intercross. *Pharmacology Biochemistry and Behavior, 65*, 301–312.

McLean, C. P., & Foa, E. B. (2011). Prolonged exposure therapy for post-traumatic stress disorder: A review of evidence and dissemination. *Expert Review of Neurotherapeutics, 11*, 1151–1163.

McNish, K. A., Gewirtz, J. C., & Davis, M. (1997). Evidence of contextual fear after lesions of the hippocampus: A disruption of freezing but not fear-potentiated startle. *Journal of Neuroscience, 17*, 9353–9360.

Milad, M. R., & Quirk, G. J. (2012). Fear extinction as a model for translational neuroscience: Ten years of progress. *Annual Review of Psychology, 63*, 129–151.

Miller, M. W., Curtin, J. J., & Patrick, C. J. (1999). A startle-probe methodology for investigating the effects of active avoidance on negative emotional reactivity. *Biological Psychology, 50*, 235–257.

Morgan, C., III, Grillon, C., Southwick, S. M., Nagy, L. M., Davis, M., Krystal, J. H., et al. (1995). Yohimbine facilitated acoustic startle in combat veterans with post-traumatic stress disorder. *Psychopharmacology, 117*, 466–471.

Morgan, C., III, Southwick, S. M., Grillon, C., Davis, M., Krystal, J., & Charney, D. (1993). Yohimbine—Facilitated acoustic startle reflex in humans. *Psychopharmacology, 110*, 342–346.

Myers, K. M., & Davis, M. (2007). Mechanisms of fear extinction. *Molecular Psychiatry, 12*, 120–150.

Norberg, M. M., Krystal, J. H., & Tolin, D. F. (2008). A meta-analysis of D-cycloserine and the facilitation of fear extinction and exposure therapy. *Biological Psychiatry, 63*, 1118–1126.

Norrholm, S. D., Jovanovic, T., Olin, I. W., Sands, L. A., Karapanou, I., Bradley, B., et al. (2011). Fear extinction in traumatized civilians with posttraumatic stress disorder: Relation to symptom severity. *Biological Psychiatry, 69*, 556–563.

Orr, S. P., Metzger, L. J., Lasko, N. B., Macklin, M. L., Peri, T., & Pitman, R. K. (2000). De novo conditioning in trauma-exposed individuals with and without posttraumatic stress disorder. *Journal of Abnormal Psychology, 109*, 290–298.

Park, P. E., Vendruscolo, L. F., Schlosburg, J. E., Edwards, S., Schulteis, G., & Koob, G. F. (2013). Corticotropin-releasing factor (CRF) and alpha 2 adrenergic receptors mediate heroin withdrawal-potentiated startle in rats. *The international journal of neuropsychopharmacology/ official scientific journal of the Collegium Internationale Neuropsychopharmacologicum, 16*(8), 1867–1875.

Peri, T., Ben-Shakhar, G., Orr, S. P., & Shalev, A. Y. (2000). Psychophysiologic assessment of aversive conditioning in posttraumatic stress disorder. *Biological Psychiatry, 47*, 512–519.

Peters, J., Dieppa-Perea, L. M., Melendez, L. M., & Quirk, G. J. (2010). Induction of fear extinction with hippocampal-infralimbic BDNF. *Science, 328*, 1288–1290.

Pole, N., Neylan, T. C., Otte, C., Henn-Hasse, C., Metzler, T. J., & Marmar, C. R. (2009). Prospective prediction of posttraumatic stress disorder symptoms using fear potentiated auditory startle responses. *Biological Psychiatry, 65*, 235–240.

Ponomarev, I., Rau, V., Eger, E. I., Harris, R. A., & Fanselow, M. S. (2010). Amygdala transcriptome and cellular mechanisms underlying stress-enhanced fear learning in a rat model of posttraumatic stress disorder. *Neuropsychopharmacology, 35*, 1402–1411.

Powers, M. S., Barrenha, G. D., Mlinac, N. S., Barker, E. L., & Chester, J. A. (2010). Effects of the novel endocannabinoid uptake inhibitor, LY2183240, on fear-potentiated startle and alcohol-seeking behaviors in mice selectively bred for high alcohol preference. *Psychopharmacology, 212*, 571–583.

Quirk, G. J., & Mueller, D. (2008). Neural mechanisms of extinction learning and retrieval. *Neuropsychopharmacology, 33*, 56–72.

Radke, A. K., Rothwell, P. E., & Gewirtz, J. C. (2011). An anatomical basis for opponent process mechanisms of opiate withdrawal. *Journal of Neuroscience, 31*, 7533–7539.

Ramamoorthi, K., Fropf, R., Belfort, G. M., Fitzmaurice, H. L., McKinney, R. M., Neve, R. L., et al. (2011). Npas4 regulates a transcriptional program in CA3 required for contextual memory formation. *Science, 334*, 1669–1675.

Rassnick, S., Koob, G. F., & Geyer, M. A. (1992). Responding to acoustic startle during chronic ethanol intoxication and withdrawal. *Psychopharmacology, 106*, 351–358.

Rau, V., DeCola, J. P., & Fanselow, M. S. (2005). Stress-induced enhancement of fear learning: An animal model of posttraumatic stress disorder. *Neuroscience and Biobehavioral Reviews, 29*, 1207–1223.

Ressler, K. J., Rothbaum, B. O., Tannenbaum, L., Anderson, P., Graap, K., Zimand, E., et al. (2004). Cognitive enhancers as adjuncts to psychotherapy: Use of D-cycloserine in phobic individuals to facilitate extinction of fear. *Archives of General Psychiatry, 61*, 1136–1144.

Richardson, R., & Elsayed, H. (1998). Shock sensitization of startle in rats: The role of contextual conditioning. *Behavioral Neuroscience, 112*, 1136–1141.

Risbrough, V. B., Brodkin, J. D., & Geyer, M. A. (2003a). GABA-A and 5-HT1A receptor agonists block expression of fear-potentiated startle in mice. *Neuropsychopharmacology, 28*, 654–663.

Risbrough, V. B., Hauger, R. L., Pelleymounter, M. A., & Geyer, M. A. (2003b). Role of corticotropin releasing factor (CRF) receptors 1 and 2 in CRF-potentiated acoustic startle in mice. *Psychopharmacology, 170*, 178–187.

Risbrough, V. B., & Geyer, M. A. (2005). Anxiogenic treatments do not increase fear-potentiated startle in mice. *Biological Psychiatry, 57*, 33–43.

Risbrough, V. B., Geyer, M. A., Hauger, R. L., Coste, S., Stenzel-Poore, M., Wurst, W., et al. (2009). CRF1 and CRF2 receptors are required for potentiated startle to contextual but not discrete cues. *Neuropsychopharmacology, 34*, 1494–1503.

Rothwell, P. E., Thomas, M. J., & Gewirtz, J. C. (2009). Distinct profiles of anxiety and dysphoria during spontaneous withdrawal from acute morphine exposure. *Neuropsychopharmacology, 34*, 2285–2295.

Scavio, M. J., Jr., & Gormezano, I. (1974). CS intensity effects on rabbit nictitating membrane conditioning, extinction and generalization. *The Pavlovian Journal of Biological Science, 9*, 25–34.

Schmid, A., Koch, M., & Schnitzler, H. U. (1995). Conditioned pleasure attenuates the startle response in rats. *Neurobiology of Learning and Memory, 64*, 1–3.

Schulz-Klaus, B., Fendt, M., & Schnitzler, H. U. (2005). Temporary inactivation of the rostral perirhinal cortex induces an anxiolytic-like effect on the elevated plus-maze and on the yohimbine-enhanced startle response. *Behavioural Brain Research, 163*, 168–173.

Shepard, J. D., Bossert, J. M., Liu, S. Y., & Shaham, Y. (2004). The anxiogenic drug yohimbine reinstates methamphetamine seeking in a rat model of drug relapse. *Biological Psychiatry, 55*, 1082–1089.

Shimada, T., Matsumoto, K., Osanai, M., Matsuda, H., Terasawa, K., & Watanabe, H. (1995). The modified light/dark transition test in mice: Evaluation of classic and putative anxiolytic and anxiogenic drugs. *General Pharmacology, 26*, 205–210.

Sidman, M., Herrnstein, R. J., & Conrad, D. G. (1957). Maintenance of avoidance behavior by unavoidable shocks. *Journal of Comparative and Physiological Psychology, 50*, 553–557.

Sink, K. S., Walker, D. L., Yang, Y., & Davis, M. (2011). Calcitonin gene-related peptide in the bed nucleus of the stria terminalis produces an anxiety-like pattern of behavior and increases neural activation in anxiety-related structures. *Journal of Neuroscience, 31*, 1802–1810.

Smith, K. S., Engin, E., Meloni, E. G., & Rudolph, U. (2012). Benzodiazepine-induced anxiolysis and reduction of conditioned fear are mediated by distinct $GABA_A$ receptor subtypes in mice. *Neuropharmacology, 63*, 250–258.

Smith, K. S., Meloni, E. G., Myers, K. M., Van't Veer, A., Carlezon, W. A., Jr., & Rudolph, U. (2011). Reduction of fear-potentiated startle by benzodiazepines in C57BL/6J mice. *Psychopharmacology, 213*, 697–706.

Somerville, L. H., Whalen, P. J., & Kelley, W. M. (2010). Human bed nucleus of the stria terminalis indexes hypervigilant threat monitoring. *Biological Psychiatry, 68*, 416–424.

Stine, S. M., Grillon, C. G., Morgan, C. A., 3rd, Kosten, T. R., Charney, D. S., & Krystal, J. H. (2001). Methadone patients exhibit increased startle and cortisol response after intravenous yohimbine. *Psychopharmacology, 154*, 274–281.

Sullivan, G. M., Apergis, J., Bush, D. E., Johnson, L. R., Hou, M., & Ledoux, J. E. (2004). Lesions in the bed nucleus of the stria terminalis disrupt corticosterone and freezing responses elicited by a contextual but not by a specific cue-conditioned fear stimulus. *Neuroscience, 128*, 7–14.

Toth, G. J. E., Hauger, R. L., Halberstadt, A. L., & Risbrough, V. B. (2013). The role of PKC signaling in CRF-induced modulation of startle. *Psychopharmacology, 229*, 579–589.

Treit, D. (1990). A comparison of anxiolytic and nonanxiolytic agents in the shock-probe/burying test for anxiolytics. *Pharmacology Biochemistry and Behavior, 36*(1), 203–205.

Treit, D., Menard, J., & Royan, C. (1993). Anxiogenic stimuli in the elevated plus-maze. *Pharmacology Biochemistry and Behavior, 44*(2), 463–469.

Vaidyanathan, U., Patrick, C. J., & Bernat, E. M. (2009). Startle reflex potentiation during aversive picture viewing as an indicator of trait fear. *Psychophysiology, 46*, 75–85.

Veening, J. G., Bocker, K. B., Verdouw, P. M., Olivier, B., De Jongh, R., & Groenink, L. (2009). Activation of the septohippocampal system differentiates anxiety from fear in startle paradigms. *Neuroscience, 163*, 1046–1060.

Venault, P., Jacquot, F., Save, E., Sara, S., & Chapouthier, G. (1993). Anxiogenic-like effects of yohimbine and idazoxan in two behavioral situations in mice. *Life Sciences, 52*, 639–645.

Waddell, J., Dunnett, C., & Falls, W. A. (2004). C57BL/6J and DBA/2J mice differ in extinction and renewal of extinguished conditioned fear. *Behavioural Brain Research, 154*, 567–576.

Walker, D. L., & Davis, M. (1997). Double dissociation between the involvement of the bed nucleus of the stria terminalis and the central nucleus of the amygdala in startle increases produced by conditioned versus unconditioned fear. *Journal of Neuroscience, 17*, 9375–9383.

Walker, D. L., & Davis, M. (2002). Light-enhanced startle: Further pharmacological and behavioral characterization. *Psychopharmacology, 159*, 304–310.

Walker, D. L., & Davis, M. (2008). Role of the extended amygdala in short-duration versus sustained fear: A tribute to Dr. Lennart Heimer. *Brain Structure and Function, 213*, 29–42.

Walker, D. L., Miles, L. A., & Davis, M. (2009). Selective participation of the bed nucleus of the stria terminalis and CRF in sustained anxiety-like versus phasic fear-like responses. *Progress in Neuropsychopharmacology and Biological Psychiatry, 33*, 1291–1308.

Walker, D. L., Paschall, G. Y., & Davis, M. (2005). Glutamate receptor antagonist infusions into the basolateral and medial amygdala reveal differential contributions to olfactory vs. context fear conditioning and expression. *Learning and Memory, 12*, 120–129.

Walker, D. L., Ressler, K. J., Lu, K. T., & Davis, M. (2002). Facilitation of conditioned fear extinction by systemic administration or intra-amygdala infusions of D-cycloserine as assessed with fear-potentiated startle in rats. *Journal of Neuroscience, 22*, 2343–2351.

Part IV
Future Directions for Animal Models in Behavior Genetics

Chapter 12
Future Directions for Animal Models in Behavior Genetics

Patrick E. Rothwell and Marc V. Fuccillo

Introduction

Animal model systems have been an essential tool in attempts to understand how differences in genetic architecture alter whole organism behavioral output. These models have numerous experimental advantages over human subjects—(1) the freedom to precisely alter genomic structure, (2) the ability to directly assess the physiological function of gene manipulations both in vitro and in vivo, and (3) the feasibility of propagating mutations over several generations or combining multiple alleles with altered function. In addition, the marked sequence similarity of coding regions across species increases the likelihood of conserved function at the molecular level, and raises hope that discoveries in animal models will translate to humans. Going forward, progress using animal models for behavior genetic research relies both on increasing sophistication in the design, analysis, and interpretation of behavioral protocols as well as novel molecular technologies that allow us to both document and mechanistically explore genetic diversity.

P.E. Rothwell (✉)
Department of Neuroscience, University of Minnesota,
Wallin Medical Biosciences Building, Room 4-142, 2101 6th St SE, Minneapolis, MN 55455, USA
e-mail: rothwell@umn.edu

M.V. Fuccillo
Department of Neuroscience, Perelman School of Medicine, University of Pennsylvania,
415 Curie Blvd, Clinical Research Building, Room 211, Philadelphia, PA 19104, USA
e-mail: fuccillo@mail.med.upenn.edu

J.C. Gewirtz, Y.-K. Kim (eds.), *Animal Models of Behavior Genetics*,
Advances in Behavior Genetics, DOI 10.1007/978-1-4939-3777-6_12

Current Topics: Technologies for Uncovering and Exploring Genetic Diversity

Genetic diversity, together with environmental experience, is thought to be an essential ingredient from which behavioral variability is generated. Demonstrations of this long-held belief depend heavily on two methodologies—(1) the ability to survey, catalogue, and correlate genetic differences with alterations in behavior and (2) techniques that allow subtle, precise changes in gene sequence to test how genetic variation relates to altered physiology and ultimately behavior. The former methodology has largely been the domain of human geneticists while the latter, the realm of molecular biologists. Despite these historical divisions, both fields have experienced significant technical advances in the past 10 years that open up a larger range of synergistic experimental pursuits and increase the chance for a more integrated understanding of genetic and behavioral diversity.

Identifying Genetic Variation Associated with Alterations in Behavior

The exploration of genetic diversity and its relationship to behavior has received sustained attention through studies of the genetic etiology of human diseases, including neuropsychiatric disorders. The completion of the human genome-sequencing project was met with great hope regarding advances in our understanding and treatment of human disease. Nowhere was this more true than in the field of psychiatry, where the extensive heritability of many neuropsychiatric disorders, as observed through high twin monozygotic concordance rates, was taken to mean that the genesis of these poorly understood disorders could be easily uncovered through discovery of sequence abnormalities. However, markedly reduced dizygotic concordance rates painted a more complex picture that included potential roles of multiple genetic abnormalities, background genetic modifiers, and environmental contributions. In the following sections, we review progress in understanding the genetic architecture of neuropsychiatric disorders, focusing on advancements in genetic sequencing that have enabled increasingly sophisticated analysis of this complex issue.

Linkage Analysis

Work from the early 1980s to mid-2000s focused heavily on gene linkage and candidate-gene association studies. Success with linkage analysis was largely predicated on the idea that single genes with large effect were causing neuropsychiatric disorders, as observed in previously modeled Mendelian diseases. If true, studying familial cohorts with many affected individuals would allow genetic evidence to coalesce around single candidate genes. This approach turned out to be disappointing for elucidating the underlying genetics of neuropsychiatric disorders, as proposed candidate molecules were rarely validated in secondary studies (Burmeister et al. 2008). In retrospect, given the polygenic nature of neuropsychiatric disorders and the low penetrance of a given genetic perturbation (International Schizophrenia Consortium 2009; Purcell et al. 2014; Weiss et al. 2009), linkage methodologies were poorly suited to produce major gains, as linkage evidence would be spread across multiple genomic loci thereby requiring massive cohorts to reach adequate statistical power. In parallel, candidate-gene association studies analyzed the relationship of genetic variation to brain disorders for specific genes selected by their hypothesized function in disease pathophysiology. This approach generated an extensive literature for several major molecular players that did not withstand the scrutiny of subsequent, larger unbiased studies (Drago et al. 2007; Sullivan 2007).

Single-Nucleotide Polymorphisms and Copy Number Variation

Earlier attempts at linking genetic and behavioral diversity in humans were hampered by the polygenic nature of neuropsychiatric disorders, as well as our poor conceptual framework for how genetic networks potentially contribute to the pathophysiology of disease. Technological and conceptual advances in the past 10 years regarding the acquisition and interpretation of human genetic data have dramatically changed our thinking about the relationship between genetics and behavior (Malhotra and Sebat 2012; McCarroll et al. 2014). Since the herculean effort to sequence the human genome, the price of sequencing technology has dramatically dropped, making the detection of single-nucleotide polymorphisms and copy number variations broadly accessible to the genetics community. Single-nucleotide polymorphisms (SNPs) are small, common sequence variations that are frequently linked to neighboring sequence polymorphisms to form transmissible genetic haplotypes that can be routinely assessed through affordable SNP arrays (McCarroll et al. 2014). Interrogating a large number of SNP haplotypes across the genome demands substantial statistical power to control for multiple hypothesis testing, and the polygenic nature of behavioral disorders further increases the number of subjects required to uncover candidate genes. However the affordability of SNP analysis and its implementation with growing test cohorts provides a scalability that can eventually achieve the large numbers required for making statistically significant predictions. Thus far, SNP analyses have identified candidate genes for schizophrenia, bipolar disorder, and Alzheimer's disease that have reached statistical significance and been replicated across distinct experimental cohorts (Lambert et al. 2013).

In addition to describing small sequence differences scattered across the genome, SNP-based technology has afforded a glimpse into more substantial alterations in genome architecture that have been tied to behavioral abnormalities. Copy number variations (CNVs) are submicroscopic structural variations, typically less than 500 kb in size, that occur throughout the human genome (Iafrate et al. 2004; Sebat et al. 2004). These structural variants include duplication, insertion, deletion, and inversion of genomic sequences and account for equivalent amounts of sequence diversity as SNPs (Malhotra and Sebat 2012). CNVs have been discovered in populations with autism and schizophrenia (Sebat et al. 2007; Stefansson et al. 2008; Walsh et al. 2008), and have quickly provided essential information about rare, intermediate effect genetic loci as well as key insights into the role of *de novo* mutations in the pathogenesis of these disorders (Levy et al. 2011; Sanders et al. 2011).

Exome and Whole-Genome-Sequencing

The next frontier in comprehensive exploration of genetic variation is complete sequencing of all expressed regions within the genome (i.e., the "exome"), as well as all intervening intergenic regions where much of the SNP variation is found. This type of large-scale genome-sequencing has long been cost-prohibitive, but continual improvements in sequencing efficiency are leading to increased availability and implementation of these approaches, which are being applied in the realm of psychiatric disorders (e.g., Fromer et al. 2014; Purcell et al. 2014) and other diseases (e.g., Waddell et al. 2015). The high resolution of these comprehensive sequencing technologies will continue to expand and improve our understanding of genetic variation in the coming years.

The Power of Genome Manipulation in Assessing Causality

It seems clear that establishing a causal link between an organism's genetic code and particular patterns of behavioral output will require direct manipulations of DNA sequence followed by physiological and behavioral testing to assess the resulting changes. Animal models from worm to mouse have been powerful tools to test genetic and molecular causality, owing largely to the pliability of their genomes and the ease of behavioral testing. Forward-genetic screens have been extensively performed in drosophila and zebrafish and have provided a powerful, non-biased approach for relating behavioral phenotypes with underlying molecular genetic mechanisms (e.g., Lorent et al. 2001; Yapici et al. 2008). Recent efforts in rodents have also focused on targeting specific genes via homologous recombination, allowing for a more directed genetic manipulation as well as the introduction of subtle genetic variations (e.g., point mutations in coding sequences) that mimic mutations found in human populations (Capecchi 2005).

Gene Targeting in Mice

Since the major technological breakthroughs of the mid-1980s, which demonstrated the existence of homologous recombination in mice and the ability to implement this technique in mouse embryonic stem cells (Joyner et al. 1989; Thomas and Capecchi 1987), nearly all known mouse genes have been targeted to generate loss-of-function alleles (Bradley et al. 2012). For two and a half decades, these lines have served as the backbone of research aimed at understanding gene function through physiological, molecular, and behavioral analyses. In recent years, gene targeting technology has expanded to permit conditional genetic deletion, using targeting constructs in which key exons are flanked by loxP sites ("floxed"), which are recognized and deleted by Cre recombinase (Orban et al. 1992). Floxed alleles and conditional deletion have enabled gene manipulation in mice with an unprecedented degree of temporal, regional, and cell-type-specificity (to be discussed in Section "Mapping the Molecular Circuitry of Behavior").

Another important development in mouse genetics has been the expanding use of bacterial artificial chromosome (BAC) transgenes, which allow the insertion of large DNA sequences that can span hundreds of kilobases (Heintz 2001). BAC transgenes can thus include coding sequences together with large flanking regions of intergenic DNA, which often contain regulatory elements that contribute to specific patterns of expression. This has enabled expression of fluorescent proteins (Gong et al. 2003) and Cre recombinase (Gerfen et al. 2013) in genetically defined brain cell types with high precision. This approach also spares direct manipulation of the endogenous genomic locus, which can alter endogenous expression levels and yield hypomorphic alleles with significant physiological and behavioral phenotypes. However, it still remains important to thoroughly characterize new transgenic mouse lines to confirm the accurate localization of protein expression (Lammel et al. 2015) and assess the potential impact on behavior, which may depend on the background strain (Chan et al. 2012).

While gene targeting in mice remains a formidable tool in uniting genetics and behavior, the significant time investment required for animal generation may limit the overall utility of this technology when exploring the effects of multiple genetic polymorphisms or modifier alleles. In the face of growing genetic data implicating complex genetic architectures and numerous non-coding sequence polymorphisms in behavioral function, it has become necessary to develop higher-throughput approaches for genetic manipulation, as discussed in the following sections.

RNA Interference

One potential technique for analyzing the in vivo function of gene networks on behavior has been RNA interference (RNAi): the introduction of small interfering RNA (siRNA) molecules that bind mRNA transcripts through sequence homology and mediate their cleavage and degradation (Mohr et al. 2014). Due to the incomplete nature of mRNA removal, RNAi typically leads to a "knock-down" of gene function as a small population of transcript continues to be translated into protein (Krueger et al. 2007). Despite this potential drawback, libraries of RNAi constructs have been generated and used in cell culture preparations to permit rapid exploration of genetic pathways (Boutros and Ahringer 2008).

One of the biggest caveats limiting the expansion of RNAi techniques to in vivo systems that permit behavioral analysis is the widely documented off-target gene silencing that occurs when the 6–7 nucleotide "seed" region of the siRNA sequence matches with the 3′ untranslated region (UTR) of non-targeted transcripts throughout the genome (Birmingham et al. 2006; Jackson et al. 2003). While these potential off-target effects can be addressed by rescue experiments, wherein an RNAi-insensitive version of the gene being analyzed is expressed together with the siRNA molecule (e.g., Soler-Llavina et al. 2011), the feasibility of this overall approach rapidly diminishes as the number of genes being targeted by RNAi expands. Furthermore, RNAi techniques operate exclusively at the level of mRNA degradation and as such are incapable of yielding direct insights into the function of multiple small sequence polymorphisms on whole organism behavior.

Programmable Nucleases

Given the extensive genetic heterogeneity being implicated in behavioral function, the ideal genome editing technique would allow for multiple sequence alterations, each at precisely pre-defined genetic loci. The advent of programmable nucleases, which use either DNA- or RNA-mediated sequence binding to cause double-strand breaks within the genome, may represent the best solution to the challenges of modeling such complex genetic structures (Kim and Kim 2014). Zinc-finger nucleases (ZFNs) use two separate arrays of repeated zinc-finger DNA-binding motifs to bring together the FokI nuclease, which requires dimerization to cleave DNA (Urnov et al. 2010). Similarly, transcription activator-like effector nucleases (TALENs) recruit the FokI nuclease domain to the DNA (Christian et al. 2010). In contrast, the type-II clustered regularly interspaced short palindromic regions (CRISPR) system uses a set of RNA guide molecules to target and activate the endonuclease properties of the Crispr-associated protein (Cas9). Following the double-strand cleavage elicited by these three methods, the DNA can be repaired either by non-homologous end joining, which typically introduces insertions or deletions, or through homology-directed repair, whereby donor sequences surrounded by homology arms can be incorporated at the break site (Kim and Kim 2014).

The ability for precise genome editing at multiple target sites makes these three approaches enormously attractive for attempts to recreate complex genomic variations and study their functional effects in vivo (Singh et al. 2015). A sustained technological push, particularly for the CRISPR/Cas9 technology, is currently generating site-specific nucleases that can be implemented in a range of animal models as well as human cells (Konermann et al. 2015; Liang et al. 2015). Recent studies support the feasibility of producing large genetic perturbations, such as copy number variations and chromosomal rearrangements, by the targeting of nucleases at broadly spaced intervals (Fujii et al. 2013; Kraft et al. 2015). In the presence of sufficient non-coding sequence homology, it should also be possible to recreate precise SNP haplotypes so that their effects on gene expression, interactions with other genetic modifiers, and resulting behavioral changes can be explored.

Viral Vectors for Genetic Manipulation of the Nervous System

The genome targeting and editing strategies reviewed in preceding sections are often applied at embryonic stages of development, in an attempt to constitutively target all cells in the body and enable germline transmission of genetic variation to progeny. However, it is often desirable to perform more targeted genetic manipulations that are confined to the nervous system—for example, if a constitutive genetic mutation causes lethal cardiovascular dysfunction that precludes behavioral analysis. In the past decade, viral vectors have become an increasingly popular and versatile vehicle for genetic manipulation of the nervous system (for recent reviews, see Kantor et al. 2014; Nassi et al. 2015). Many different neurotropic viruses can be used to efficiently transduce brain cells with Cre recombinase, siRNA, or programmable nucleases. These viral vectors can be injected into specific parts of the brain using stereotaxic surgical procedures, leading to localized infection of cells at the site of injection.

Certain types/strains of virus are also transported by neurons in either the retrograde or anterograde direction, and in some cases can cross synapses and thereby spread to infect synaptically connected cells (reviewed by Nassi et al. 2015). For example, both rabies virus and canine adenovirus vectors can infect neurons at axon terminals, and are then transported in the retrograde direction to the cell bodies from which these axons originate. Further spread of these viruses can be restricted by using mutant strains in which key genes have been deleted from the viral genome (Soudais et al. 2004; Wickersham et al. 2007). Exogenous proteins like Cre recombinase can also be expressed from the genome of these viruses, allowing for genetic manipulations based on anatomical connectivity (e.g., Dolen et al. 2013; Hnasko et al. 2006). One advantage of the rabies virus system is the potential to genetically restrict the population of initially infected "starter cells" (reviewed by Callaway and Luo 2015). However, a limitation of rabies virus vectors is toxicity that develops after weeks of infection (Wickersham et al. 2007), in contrast to canine adenovirus, which is well-tolerated over long periods of infection (Soudais et al. 2004).

Mapping the Molecular Circuitry of Behavior

A major challenge in behavioral genetics is identifying the alterations in brain function that result from genetic variation and lead to changes in behavior. This is an endeavor that will require contributions from behavioral genetics as well as many other areas of neuroscience (Akil et al. 2010). This venture begins with a simple but vexing question: where in the brain does genetic variation act to influence behavior? Constitutive genetic mutations affect gene expression throughout the brain as well as the rest of the body, making it a daunting challenge to narrow down the specific locations where genetic variation influences physiology to alter behavior. In the following sections, we outline technological advancements that are enabling new approaches to address this important question.

Identifying Candidates

One approach to winnowing down the plethora of locations where a gene may be critical is to map expression levels of a given gene in multiple locations within a candidate circuit. Many genes are not expressed ubiquitously throughout the nervous system, but instead exhibit enriched expression in particular brain regions and cell types. The Allen Brain Atlas (http://www.brain-map.org) provides an enormously useful public resource for examining the topography of gene expression across development in the brains of mice, humans, and non-human primates. The ability to isolate and quantify gene expression in single brain cells has greatly advanced through recent developments in microfluidic technology and RNA sequencing (e.g., Portmann et al. 2014; Zeisel et al. 2015). These approaches have been used in conjunction with transgenic mouse lines that label genetically defined neuronal cell types (Sugino et al. 2006), and can also be coupled with anatomical techniques for tracing neural circuits based on synaptic inputs and outputs, including the virus-based approaches discussed in Section "Viral Vectors for Genetic Manipulation of the Nervous System" (for a recent example, see Fuccillo et al. 2015). Implementation of these approaches may lead to the identification of one or more subpopulations of brain cells with enriched expression of a gene of interest, suggesting a key site of action for that gene.

Casting a Wide Net

An alternative yet complimentary approach to narrowing down potential sites of gene action is to perform conditional genetic deletion (with Cre-loxP technology) using manipulations that affect large yet specific families of brain cells. For example, several elegant papers have compared the effects of conditional gene deletion in excitatory versus inhibitory brain cells, using mouse lines with Cre expression driven by canonical markers for these types of cells (Chao et al. 2010; Cui et al. 2008; Monory et al. 2006). In each example, the behavioral phenotypes caused by constitutive gene knockout were recapitulated by conditional genetic deletion from either excitatory or inhibitory brain cells. These types of "broad stroke" genetic manipulations provide another way to pinpoint the population of brain cells in which a particular genetic deletion causes behavioral effects.

Narrowing Down the Possibilities

Results obtained using broad stroke manipulations can be followed up with more targeted genetic deletions. For example, subpopulations of inhibitory interneurons are located throughout the brain, and can be distinguished using differential expression of markers like parvalbumin, cholecystokinin, and somatostatin (Fishell and Rudy 2011). These different molecular markers correspond to interneuron subtypes that exhibit unique physiological and functional properties, and can be selectively targeted using Cre expression driven by the appropriate promoter. Similarly, the inhibitory projection neurons of the striatum comprise two distinct subtypes of medium spiny neuron (MSN) that make discrete projections to downstream structures within the basal ganglia. These MSN subtypes differ in expression of several molecular markers, including expression of either the *Drd1a* or *Drd2* dopamine receptor. Each MSN subtype can be uniquely targeted using Cre expression driven by the promoter of the appropriate dopamine receptor (Gerfen et al. 2013).

Genetic approaches provide a powerful means to access defined populations of cells that may be distributed in a widespread fashion throughout the brain. However, in some cases this widespread distribution can be a disadvantage for precise localization of behaviorally relevant genetic circuits. For example, parvalbumin is expressed by a subpopulation of fast-spiking interneurons found in a number of brain regions, including cortex, hippocampus, striatum, and cerebellum (Fishell and Rudy 2011). Transgenic mouse lines that express Cre from the parvalbumin locus (e.g., Hippenmeyer et al. 2005) provide a powerful means of genetic access to these neurons, but conditional gene knockout using parvalbumin-Cre lines will impact these cells in all brain regions.

A complementary approach that enables better control over the spatial extent of genetic manipulation is the intracranial injection of viral vectors (see section "Viral Vectors for Genetic Manipulation of the Nervous System"). For example, a viral vector expressing Cre recombinase can be stereotaxically injected into the brain of a conditional knockout mouse, deleting a gene and protein of interest only in infected neurons at the site of injection. While this viral approach provides more precise spatial manipulation of gene expression, it lacks the cell-type-specificity afforded by genetic approaches described in the preceding paragraph. Convergent evidence from both genetic and virus-based manipulations thus provides very strong evidence for the function of a gene in a particular cell type and brain region (e.g., Rothwell et al. 2014).

Functional Consequences of Genetic Variation

The ability to localize behaviorally relevant gene function to a specific subpopulation of brain cells is a major accomplishment in and of itself. However, this raises obvious questions regarding the specific biological function of a given gene in this cell type: how does genetic deletion affect cellular structure or function in a way that ultimately influences behavioral output? While such questions are difficult to answer and have traditionally fallen beyond the scope of behavioral genetics, a number of exciting developments in modern neuroscience are making pursuit of such questions increasingly tractable.

Monitoring Brain Activity and Structure

One of the most obvious ways for a genetic perturbation to affect brain function is through alterations in overall patterns of neural activity, including changes in the frequency or timing of action potentials fired during behavior. Advancements in electrophysiological recordings now permit the recording of these firing patterns during complex behavioral tasks—a powerful tool for determining the impact of genetic variation on the physiological regulation of behavior (e.g., Burguiere et al. 2013). Measurement of single unit spiking can be complemented by recording of local field potentials, which represent synchronized activity of large populations of neurons and can reveal oscillatory patterns that are linked to behavioral state (Ray 2015). An alternative approach to monitor the individual activity of large populations of neurons is the use of genetically encoded fluorescent calcium indicators, which can be imaged in superficial brain structures through cranial windows (Yang et al. 2014), or in deeper brain structures using miniaturized microscopes (Ziv et al. 2013), allowing for more global correlations of activity with behavior.

Genetic variation may also impact brain development and influence the morphology of individual neurons as well as the pattern of synaptic connections between neurons. The visualization of brain structure in three dimensions has long been hampered by light scattering caused by the inhomogeneity of biological tissue (reviewed by Richardson and Lichtman 2015). Recent innovations like CLARITY (Chung et al. 2013) and iDISCO (Renier et al. 2014) have helped obviate this problem by "clearing" tissue so that it becomes transparent, allowing for three-dimensional imaging of intact biological systems. The resolution of imaging can also be effectively enhanced by physical expansion of the tissue specimen, a process termed "expansion microscopy" (Chen et al. 2015). Synaptic connections can be mapped with high specificity by combining retrograde viral vectors (discussed in Section "Viral Vectors for Genetic Manipulation of the Nervous System") to trace the relationship between synaptic input and output, a method known as TRIO (Schwarz et al. 2015)

Measuring Cellular Properties

Electrophysiological experiments in acute brain slice preparations permit more detailed analysis of the cellular factors that may underlie changes in activity patterns. In a brain slice preparation, whole-cell patch-clamp recordings can be obtained from individual brain cells, and can be targeted to identified subtypes of neurons using genetically encoded fluorescent reporters (Gong et al. 2003), or cells infected with a virus that expresses a fluorescent marker as well as a molecular manipulation (e.g., siRNA or Cre recombinase). These recordings can be used to measure ionic conductances that determine intrinsic excitability (e.g., Tsai et al. 2012), as well as currents generated by activation of various excitatory and inhibitory synaptic inputs (e.g., Rothwell et al. 2014). Changes in the properties or plasticity of synaptic inputs are particularly important to assess in light of the numerous genetic mutations in synaptic signaling proteins that have been associated with neuropsychiatric disorders (Fromer et al. 2014; Grant 2012; Sudhof 2008; Zoghbi and Bear 2012).

Manipulating Cellular Activity

Functional readouts of the impact of different genetic mutations at the cellular level are important, but still correlative: a given genetic mutation may produce a myriad of changes in cellular function, making it difficult to parse the critical changes that ultimately influence behavior. It thus becomes important to manipulate patterns of activity in genetically defined populations of neurons, to directly test the relationship between cellular activity and behavioral output. Recent technical breakthroughs are making it possible to accomplish this lofty goal, using the combination of transgenic mice expressing Cre recombinase in defined populations of brain cells, coupled with Cre-dependent expression constructs encoded by viral vectors (see section "Viral Vectors for Genetic Manipulation of the Nervous System"). These constructs, referred to as flip excision (FLEX; Atasoy et al. 2008; Schnutgen et al. 2003) or double inverse open reading frame (DIO; Sohal et al. 2009; Tsai et al. 2009), are initially oriented in an inverse direction and therefore minimally expressed (Fig. 12.1a, top row). However, in the presence of Cre recombinase, the coding sequence is flipped and expressed at high levels when driven by neighboring promoter sequences (Fig. 12.1a, bottom row). More recent and sophisticated variations on this approach involve coding sequences that are only expressed in the simultaneous presence of multiple recombinases, permitting "intersectional" control of gene expression (e.g., Fenno et al. 2014; Madisen et al. 2015; Fig. 12.1b). The combination of genetic recombinase expression and recombinase-dependent viral constructs can thus be applied in several ways to manipulate brain activity with unprecedented specificity.

Optogenetics

One increasingly popular approach is optogenetics: the expression of light-sensitive proteins that can be activated with high temporal precision by specific wavelengths of light (for review, see Fenno et al. 2011). This includes ion channels like channelrhodopsin-2 (Nagel et al. 2003), which open in response to light and depolarize brain cells to increase activity, as well as ion pumps like halorhodopsin (Zhang et al. 2007) and archaerhodopsin (Chow et al. 2010), which are activated by light and inhibit neural activity. Viral expression of these proteins is followed by intracranial implantation of a small optic fiber above the site of injection, allowing specific patterns of light stimulation that trigger corresponding changes in neural activity. Channelrhodopsin-2 and other opsins have undergone extensive engineering since their discovery, leading to identification of new variants with improved performance characteristics, including the capacity to follow light stimulation at high frequencies (Gunaydin et al. 2010; Lin et al. 2009). This high degree of temporal precision represents a major advantage of optogenetic approaches to manipulating brain activity.

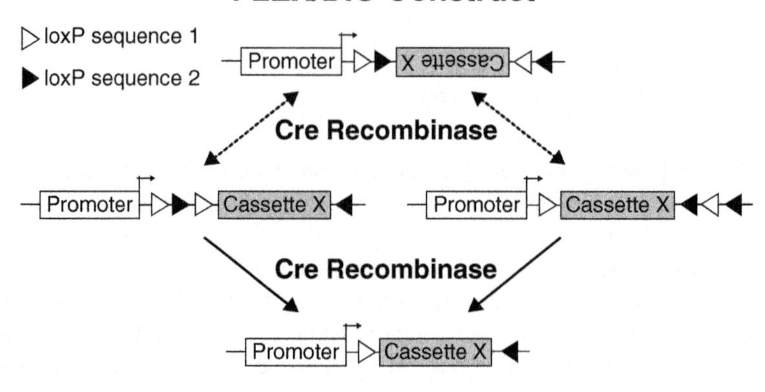

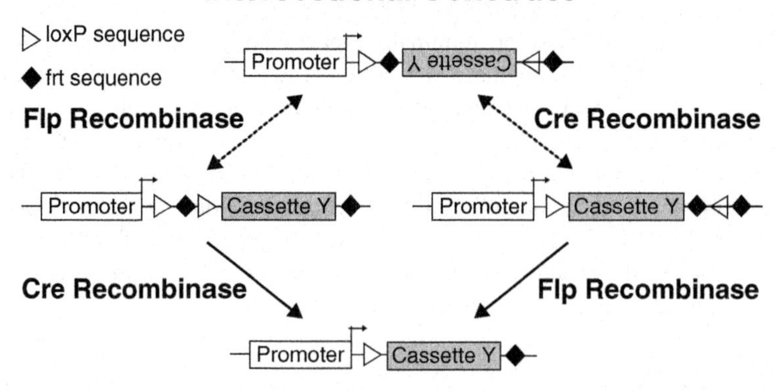

Fig. 12.1 Illustration of viral constructs that are initially oriented in an inverse direction, but are conditionally expressed in the presence of specific recombinase enzymes. (**a**) A flip excision (FLEX) or double inverse open reading frame (DIO) construct, with an inverted expression cassette surrounded by two sets of non-complementary loxP sequences (1 and 2) oriented in opposite directions (*top row*). When either set of loxP sequences is recognized by Cre recombinase, the intervening DNA is flipped into the forward orientation (*middle row*), and may flip back to the reverse orientation (*dotted arrows*). However, Cre may delete the region between parallel loxP sites (*solid arrows*), locking the construct in the forward direction and driving expression of the cassette from the promoter (*bottom row*). (**b**) Intersectional constructs involve an expression cassette surrounded by two sets of DNA sequences recognized by different recombinase enzymes. In this example, the expression cassette is flanked by loxP sequences (recognized by Cre) and frt sequences (recognized by Flp), each oriented in opposite directions (*top row*). Initial recognition by either Cre or Flp will reversibly flip the intervening DNA into the forward orientation (*middle row*). In this state, the second recombinase can delete the region between parallel loxP or frt sequences, locking the construct in the forward direction (*bottom row*)

Chemogenetics

An alternative approach to modulating brain activity in a genetically precise fashion involves expression of Designed Receptors Exclusively Activated by Designer Drugs (DREADDs; Sternson and Roth 2014). DREADDs are G-protein-coupled receptors that have been engineered through a process of directed molecular evolution, so that they are no longer activated by endogenous ligands. Instead, DREADDs are activated by synthetic compounds like clozapine N-oxide (CNO), which do not bind to endogenous targets in the brain. Thus, DREADDs are only activated in the presence of CNO, which can be administered systemically as it crosses the blood–brain barrier. Upon activation, different types of DREADDs engage in different signaling pathways, which can lead to either increased or decreased cellular excitability. The time course of these cellular changes are largely determined by the pharmacokinetics of CNO, which allows for prolonged activation, but does not permit the same degree of high-frequency temporal precision afforded by optogenetics. However, a significant advantage of this "chemogenetic" or "pharmacogenetic" approach is that DREADDs can be activated by systemic administration of CNO, making the method relatively noninvasive.

Testing Necessity and Sufficiency

If a particular genetic mutation is found to alter the activity of genetically defined populations of brain cells, optogenetic and chemogenetic approaches open the door to highly stringent tests of sufficiency and necessity. For example, if the activity of a specific brain cell type is increased by a given genetic mutation, one can ask whether stimulating this cell type in normal (wild-type) control groups recapitulates a behavioral phenotype of interest (i.e., sufficiency). Conversely, one can inhibit the cell type in the presence of the genetic mutation, and ask whether this reverses or "rescues" a behavioral phenotype (i.e., necessity). These are challenging experiments that depend on first localizing the behavioral impact of a genetic mutation to a specific cell type, and there are likely to be instances where genetic mutations affect more than one node in a given circuit, or multiple circuits in parallel. However, these approaches permit the exploration of the relationship between neural circuit function and behavior that were not previously possible, providing exciting new avenues to pursue the relationship between genetics, brain function, and behavior.

Connecting Genetic Variation to Behavior through Endophenotypes

A critical final question for the future pertains to the behavioral assays that are used to read out the consequences of genetic variation in animal models. In the context of animal models of human disease, an important challenge is selecting behavioral assays that are likely to inform our understanding of disease pathophysiology. However, the validity of specific models and behavioral readouts has become a complex and controversial issue (Nestler and Hyman 2010). One issue is construct validity—whether behavioral changes in the animal model are generated by a process similar to the human condition. An obvious advantage of genetic manipulation is that the same genetic variants associated with human disease can be introduced in the genome of animal models, providing a solid basis for construct validity. A second issue is predictive validity—whether behavioral changes in the animal model are ameliorated by treatments known to have therapeutic efficacy in humans. Predictive validity can be addressed for some disorders with well-established therapeutic interventions—for instance, chronic treatment with specific serotonin reuptake inhibitors in animal models of depression. Predictive validity is more difficult to address for disorders like autism, where effective treatments for human patients are limited, though some progress is being made in this direction (e.g., Penagarikano et al. 2011).

The Challenge of Face Validity

A third issue is face validity—the correspondence between behavioral phenotypes in animal models and human symptoms of disease. There is general consensus that it is unrealistic to expect any given animal model to fully recapitulate all features and symptoms of a human disease (Fernando and Robbins 2011; Gould and Gottesman 2006; Nestler and Hyman 2010). In fact, the behavioral consequences of a given genetic mutation may present much differently across species. This is true even for relatively simple behavioral states, like anxiety. For example, because rodents are nocturnal, they prefer to spend time in dark versus brightly lit locations, and this forms the basis for assays of anxiety-like behavior like the light–dark box. This preference for dark over light is not a behavioral phenotype with direct face validity for human anxiety—if anything, humans would be expected to prefer brightly lit locations over dark locations. In addition to species-specific presentation of behavioral phenotypes, many features of complex mental disorders like schizophrenia and autism simply cannot be directly assessed in animals.

The Endophenotype Approach in Animal Models

These concerns regarding strict face validity have led to a shift in focus away from modeling symptoms of disease and towards modeling intermediate phenotypes, simpler behavioral constructs with well-characterized neurobiological underpinnings. One driving force for this shift is the endophenotype approach that emerged in schizophrenia research over 40 years ago (Gottesman and Shields 1973), and has had a major impact on research in psychiatry. At its core, the endophenotype concept refers to intermediate features that are directly associated with a specific disease, and caused by the same underlying genetic variation as the disease itself. These features may be behavioral, but are not necessarily equivalent to (and may in fact be independent of) clinical symptoms.

A major advantage of focusing on endophenotypes is that they are often mediated by brain circuits that are well-characterized and conserved across species. In the case of anxiety described above, it is well-established that fear and anxiety-related behavior in rodents depends on a threat-detection system that includes the amygdala and related structures (Davis et al. 2010), and these same neural circuits are also implicated in human anxiety disorders (Shin and Liberzon 2010). Activation of this threat system may produce some behavioral manifestations that are similar in humans and rodents (e.g., elevation of heart rate and blood pressure), but other manifestations will be species-specific, like the aversion for bright light exhibited by rodents. Yet it is the underlying activation of well-characterized and conserved neural circuits that leads to behavioral phenotypes across species, and which matters more than the exact manner in which activation of these circuits is expressed. A focus on endophenotypes that are less wedded to clinical symptoms and more closely aligned with underlying neurobiology should facilitate mechanistic studies of the relationship between genetic diversity and behavior, particularly in the case of complex mental disorders like autism and schizophrenia.

Specificity

A final issue that merits discussion is the specificity with which both genetic variability and behavioral phenotypes map onto particular classes of disorders. In terms of genetics, single genes or mutations are often linked with more than one type of disorder. For example, mutations in NRXN1 are commonly associated with both autism and schizophrenia (for review, see Sudhof 2008). This is perhaps not surprising, as comorbidity between mental disorders is quite common, and it is widely recognized that boundaries between different diagnostic categories in psychiatry are fuzzy rather than firm. However, this lack of specificity makes it difficult to align an animal model carrying one of these mutations with a particular disorder.

In terms of behavior, different genetic mutations that cause a similar behavioral phenotype in animal models can be associated with distinct clinical syndromes. For instance, repetitive self-grooming is an interesting and robust endophenotype in several mouse models of neuropsychiatric disorders, including knockout of *Sapap3* (Welch et al. 2007) and *Nrxn1* (Etherton et al. 2009). However, these genetic mutations are associated with distinct disorders in humans: SAPAP with obsessive-compulsive disorder (Zuchner et al. 2009), and neurexin-1 with autism and schizophrenia (for review, see Sudhof 2008).

How should researchers deal with this variability as we continue to develop animal models in behavioral genetics? Although it may seem counterintuitive, we would argue that focusing on correspondence between a particular animal model and a specific human disorder is a distraction that impedes progress in the field. The direct action of genetic mutations is not on clinical symptoms per se, but instead on brain function and neural circuits that underlie these symptoms. Altered neural circuit function in animal models can be read out as behavioral endophenotypes that may be relevant to more than one human disorder, but this lack of specificity does not undermine the mechanistic insight gleaned from this approach. The same line of thought has recently led the US National Institute of Mental Health to launch the Research Domain Criteria project (Insel et al. 2010), which shifts the focus of clinical studies towards well-defined behavioral and neurobiological constructs that cut across classic diagnostic boundaries in psychiatry. Thus, even at the clinical level, strict correspondence to one particular psychiatric diagnosis should not impede ongoing investigation of well-defined behavioral processes and their underlying neurobiology—a philosophy that should also inspire future work in animal models.

Conclusions and Future Directions

In this concluding chapter, we have highlighted a number of pertinent issues related to animal models of behavioral genetics, as well as technical advances that will make some of these issues more tractable in coming years. We believe the best way forward is to introduce genetic variability in animal models using a variety of different techniques, and focus on behavioral constructs with well-defined neural substrates that can be investigated using a range of different assays. The precise manifestation of a behavioral phenotype in animals versus humans is not critical—instead, it is critical that animal behavior reflects a species-appropriate change in neural circuitry that mediates a particular behavior, and thus provides an opportunity to explore cellular and molecular mechanisms. This approach grants validity to behavioral endophenotypes in animals that present differently than in humans, but still reflect a change in a conserved neural pathway that is likely relevant to the human condition. This should open the door to more mechanistic studies that can forge links between genes and behavior across multiple levels of brain function, from molecules to cell to circuits.

References

Akil, H., Brenner, S., Kandel, E., Kendler, K. S., King, M. C., Scolnick, E., et al. (2010). Medicine. The future of psychiatric research: Genomes and neural circuits. *Science, 327*, 1580–1581.

Atasoy, D., Aponte, Y., Su, H. H., & Sternson, S. M. (2008). A FLEX switch targets Channelrhodopsin-2 to multiple cell types for imaging and long-range circuit mapping. *The Journal of Neuroscience, 28*, 7025–7030.

Birmingham, A., Anderson, E. M., Reynolds, A., Ilsley-Tyree, D., Leake, D., Fedorov, Y., et al. (2006). 3′ UTR seed matches, but not overall identity, are associated with RNAi off-targets. *Nature Methods, 3*, 199–204.

Boutros, M., & Ahringer, J. (2008). The art and design of genetic screens: RNA interference. *Nature Reviews Genetics, 9*, 554–566.

Bradley, A., Anastassiadis, K., Ayadi, A., Battey, J. F., Bell, C., Birling, M. C., et al. (2012). The mammalian gene function resource: The International Knockout Mouse Consortium. *Mammalian Genome, 23*, 580–586.

Burguiere, E., Monteiro, P., Feng, G., & Graybiel, A. M. (2013). Optogenetic stimulation of lateral orbitofronto-striatal pathway suppresses compulsive behaviors. *Science, 340*, 1243–1246.

Burmeister, M., McInnis, M. G., & Zollner, S. (2008). Psychiatric genetics: Progress amid controversy. *Nature Reviews Genetics, 9*, 527–540.

Callaway, E. M., & Luo, L. (2015). Monosynaptic circuit tracing with glycoprotein-deleted rabies viruses. *The Journal of Neuroscience, 35*, 8979–8985.

Capecchi, M. R. (2005). Gene targeting in mice: Functional analysis of the mammalian genome for the twenty-first century. *Nature Reviews Genetics, 6*, 507–512.

Chan, C. S., Peterson, J. D., Gertler, T. S., Glajch, K. E., Quintana, R. E., Cui, Q., et al. (2012). Strain-specific regulation of striatal phenotype in Drd2-eGFP BAC transgenic mice. *The Journal of Neuroscience, 32*, 9124–9132.

Chao, H. T., Chen, H., Samaco, R. C., Xue, M., Chahrour, M., Yoo, J., et al. (2010). Dysfunction in GABA signalling mediates autism-like stereotypies and Rett syndrome phenotypes. *Nature, 468*, 263–269.

Chen, F., Tillberg, P. W., & Boyden, E. S. (2015). Optical imaging. Expansion microscopy. *Science, 347*, 543–548.

Chow, B. Y., Han, X., Dobry, A. S., Qian, X., Chuong, A. S., Li, M., et al. (2010). High-performance genetically targetable optical neural silencing by light-driven proton pumps. *Nature, 463*, 98–102.

Christian, M., Cermak, T., Doyle, E. L., Schmidt, C., Zhang, F., Hummel, A., et al. (2010). Targeting DNA double-strand breaks with TAL effector nucleases. *Genetics, 186*, 757–761.

Chung, K., Wallace, J., Kim, S. Y., Kalyanasundaram, S., Andalman, A. S., Davidson, T. J., Wallace, J., Kim, S. Y., Kalyanasundaram, S., Andalman, A. S., Davidson, T. J., et al. (2013). Structural and molecular interrogation of intact biological systems. *Nature, 497*, 332–337.

Cui, Y., Costa, R. M., Murphy, G. G., Elgersma, Y., Zhu, Y., Gutmann, D. H., et al. (2008). Neurofibromin regulation of ERK signaling modulates GABA release and learning. *Cell, 135*, 549–560.

Davis, M., Walker, D. L., Miles, L., & Grillon, C. (2010). Phasic vs sustained fear in rats and humans: Role of the extended amygdala in fear vs anxiety. *Neuropsychopharmacology, 35*, 105–135.

Dolen, G., Darvishzadeh, A., Huang, K. W., & Malenka, R. C. (2013). Social reward requires coordinated activity of nucleus accumbens oxytocin and serotonin. *Nature, 501*, 179–184.

Drago, A., De Ronchi, D., & Serretti, A. (2007). Incomplete coverage of candidate genes: A poorly considered bias. *Current Genomics, 8*, 476–483.

Etherton, M. R., Blaiss, C. A., Powell, C. M., & Sudhof, T. C. (2009). Mouse neurexin-1alpha deletion causes correlated electrophysiological and behavioral changes consistent with cognitive impairments. *Proceedings of the National Academy of Sciences of the United States of America, 106*, 17998–18003.

Fenno, L., Yizhar, O., & Deisseroth, K. (2011). The development and application of optogenetics. *Annual Review of Neuroscience, 34*, 389–412.

Fenno, L. E., Mattis, J., Ramakrishnan, C., Hyun, M., Lee, S. Y., He, M., Mattis, J., Ramakrishnan, C., Hyun, M., Lee, S. Y., He, M., et al. (2014). Targeting cells with single vectors using multiple-feature Boolean logic. *Nature Methods, 11*, 763–772.

Fernando, A. B., & Robbins, T. W. (2011). Animal models of neuropsychiatric disorders. *Annual Review of Clinical Psychology, 7*, 39–61.

Fishell, G., & Rudy, B. (2011). Mechanisms of inhibition within the telencephalon: "Where the wild things are". *Annual Review of Neuroscience, 34*, 535–567.

Fromer, M., Pocklington, A. J., Kavanagh, D. H., Williams, H. J., Dwyer, S., Gormley, P., et al. (2014). De novo mutations in schizophrenia implicate synaptic networks. *Nature, 506*, 179–184.

Fuccillo, M. V., Foldy, C., Gokce, O., Rothwell, P. E., Sun, G. L., Malenka, R. C., et al. (2015). Single-cell mRNA profiling reveals cell-type-specific expression of neurexin isoforms. *Neuron, 87*, 326–340.

Fujii, W., Kawasaki, K., Sugiura, K., & Naito, K. (2013). Efficient generation of large-scale genome-modified mice using gRNA and CAS9 endonuclease. *Nucleic Acids Research, 41*, e187.

Gerfen, C. R., Paletzki, R., & Heintz, N. (2013). GENSAT BAC cre-recombinase driver lines to study the functional organization of cerebral cortical and basal ganglia circuits. *Neuron, 80*, 1368–1383.

Gong, S., Zheng, C., Doughty, M. L., Losos, K., Didkovsky, N., Schambra, U. B., et al. (2003). A gene expression atlas of the central nervous system based on bacterial artificial chromosomes. *Nature, 425*, 917–925.

Gottesman, I. I., & Shields, J. (1973). Genetic theorizing and schizophrenia. *The British Journal of Psychiatry, 122*, 15–30.

Gould, T. D., & Gottesman, I. I. (2006). Psychiatric endophenotypes and the development of valid animal models. *Genes, Brain, and Behavior, 5*, 113–119.

Grant, S. G. (2012). Synaptopathies: Diseases of the synaptome. *Current Opinion in Neurobiology, 22*, 522–529.

Gunaydin, L. A., Yizhar, O., Berndt, A., Sohal, V. S., Deisseroth, K., & Hegemann, P. (2010). Ultrafast optogenetic control. *Nature Neuroscience, 13*, 387–392.

Heintz, N. (2001). BAC to the future: The use of bac transgenic mice for neuroscience research. *Nature Reviews Neuroscience, 2*, 861–870.

Hippenmeyer, S., Vrieseling, E., Sigrist, M., Portmann, T., Laengle, C., Ladle, D. R., et al. (2005). A developmental switch in the response of DRG neurons to ETS transcription factor signaling. *PLoS Biology, 3*, e159.

Hnasko, T. S., Perez, F. A., Scouras, A. D., Stoll, E. A., Gale, S. D., Luquet, S., et al. (2006). Cre recombinase-mediated restoration of nigrostriatal dopamine in dopamine-deficient mice reverses hypophagia and bradykinesia. *Proceedings of the National Academy of Sciences of the United States of America, 103*, 8858–8863.

Iafrate, A. J., Feuk, L., Rivera, M. N., Listewnik, M. L., Donahoe, P. K., Qi, Y., et al. (2004). Detection of large-scale variation in the human genome. *Nature Genetics, 36*, 949–951.

Insel, T., Cuthbert, B., Garvey, M., Heinssen, R., Pine, D. S., Quinn, K., et al. (2010). Research domain criteria (RDoC): Toward a new classification framework for research on mental disorders. *The American Journal of Psychiatry, 167*, 748–751.

International Schizophrenia Consortium. (2009). Common polygenic variation contributes to risk of schizophrenia and bipolar disorder. *Nature, 460*, 748–752.

Jackson, A. L., Bartz, S. R., Schelter, J., Kobayashi, S. V., Burchard, J., Mao, M., et al. (2003). Expression profiling reveals off-target gene regulation by RNAi. *Nature Biotechnology, 21*, 635–637.

Joyner, A. L., Skarnes, W. C., & Rossant, J. (1989). Production of a mutation in mouse En-2 gene by homologous recombination in embryonic stem cells. *Nature, 338*, 153–156.

Kantor, B., Bailey, R. M., Wimberly, K., Kalburgi, S. N., & Gray, S. J. (2014). Methods for gene transfer to the central nervous system. *Advances in Genetics, 87*, 125–197.

Kim, H., & Kim, J. S. (2014). A guide to genome engineering with programmable nucleases. *Nature Reviews Genetics, 15*, 321–334.

Konermann, S., Brigham, M. D., Trevino, A. E., Joung, J., Abudayyeh, O. O., Barcena, C., et al. (2015). Genome-scale transcriptional activation by an engineered CRISPR-Cas9 complex. *Nature, 517*, 583–588.

Kraft, K., Geuer, S., Will, A. J., Chan, W. L., Paliou, C., Borschiwer, M., et al. (2015). Deletions, inversions, duplications: Engineering of structural variants using CRISPR/Cas in Mice. *Cell Reports, 10*(5), 833–839.

Krueger, U., Bergauer, T., Kaufmann, B., Wolter, I., Pilk, S., Heider-Fabian, M., et al. (2007). Insights into effective RNAi gained from large-scale siRNA validation screening. *Oligonucleotides, 17*, 237–250.

Lambert, J. C., Ibrahim-Verbaas, C. A., Harold, D., Naj, A. C., Sims, R., Bellenguez, C., et al. (2013). Meta-analysis of 74,046 individuals identifies 11 new susceptibility loci for Alzheimer's disease. *Nature Genetics, 45*, 1452–1458.

Lammel, S., Steinberg, E. E., Foldy, C., Wall, N. R., Beier, K., Luo, L., et al. (2015). Diversity of transgenic mouse models for selective targeting of midbrain dopamine neurons. *Neuron, 85*, 429–438.

Levy, D., Ronemus, M., Yamrom, B., Lee, Y. H., Leotta, A., Kendall, J., et al. (2011). Rare de novo and transmitted copy-number variation in autistic spectrum disorders. *Neuron, 70*, 886–897.

Liang, P., Xu, Y., Zhang, X., Ding, C., Huang, R., Zhang, Z., et al. (2015). CRISPR/Cas9-mediated gene editing in human tripronuclear zygotes. *Protein & Cell, 6*, 363–372.

Lin, J. Y., Lin, M. Z., Steinbach, P., & Tsien, R. Y. (2009). Characterization of engineered channelrhodopsin variants with improved properties and kinetics. *Biophysical Journal, 96*, 1803–1814.

Lorent, K., Liu, K. S., Fetcho, J. R., & Granato, M. (2001). The zebrafish space cadet gene controls axonal pathfinding of neurons that modulate fast turning movements. *Development, 128*, 2131–2142.

Madisen, L., Garner, A. R., Shimaoka, D., Chuong, A. S., Klapoetke, N. C., Li, L., et al. (2015). Transgenic mice for intersectional targeting of neural sensors and effectors with high specificity and performance. *Neuron, 85*, 942–958.

Malhotra, D., & Sebat, J. (2012). CNVs: Harbingers of a rare variant revolution in psychiatric genetics. *Cell, 148*, 1223–1241.

McCarroll, S. A., Feng, G., & Hyman, S. E. (2014). Genome-scale neurogenetics: Methodology and meaning. *Nature Neuroscience, 17*, 756–763.

Mohr, S. E., Smith, J. A., Shamu, C. E., Neumuller, R. A., & Perrimon, N. (2014). RNAi screening comes of age: Improved techniques and complementary approaches. *Nature Reviews Molecular Cell Biology, 15*, 591–600.

Monory, K., Massa, F., Egertová, M., Eder, M., Blaudzun, H., Westenbroek, R., et al. (2006). The endocannabinoid system controls key epileptogenic circuits in the hippocampus. *Neuron, 51*, 455–466.

Nagel, G., Szellas, T., Huhn, W., Kateriya, S., Adeishvili, N., Berthold, P., et al. (2003). Channelrhodopsin-2, a directly light-gated cation-selective membrane channel. *Proceedings of the National Academy of Sciences of the United States of America, 100*, 13940–13945.

Nassi, J. J., Cepko, C. L., Born, R. T., & Beier, K. T. (2015). Neuroanatomy goes viral! *Frontiers in Neuroanatomy, 9*, 80.

Nestler, E. J., & Hyman, S. E. (2010). Animal models of neuropsychiatric disorders. *Nature Neuroscience, 13*, 1161–1169.

Orban, P. C., Chui, D., & Marth, J. D. (1992). Tissue- and site-specific DNA recombination in transgenic mice. *Proceedings of the National Academy of Sciences of the United States of America, 89*, 6861–6865.

Penagarikano, O., Abrahams, B. S., Herman, E. I., Winden, K. D., Gdalyahu, A., Dong, H., Abrahams, B. S., Herman, E. I., Winden, K. D., Gdalyahu, A., Dong, H., et al. (2011). Absence of CNTNAP2 leads to epilepsy, neuronal migration abnormalities, and core autism-related deficits. *Cell, 147*, 235–246.

Portmann, T., Yang, M., Mao, R., Panagiotakos, G., Ellegood, J., Dolen, G., et al. (2014). Behavioral abnormalities and circuit defects in the basal ganglia of a mouse model of 16p11.2 deletion syndrome. *Cell Reports, 7*, 1077–1092.

Purcell, S. M., Moran, J. L., Fromer, M., Ruderfer, D., Solovieff, N., Roussos, P., et al. (2014). A polygenic burden of rare disruptive mutations in schizophrenia. *Nature, 506*, 185–190.

Ray, S. (2015). Challenges in the quantification and interpretation of spike-LFP relationships. *Current Opinion in Neurobiology, 31*, 111–118.

Renier, N., Wu, Z., Simon, D. J., Yang, J., Ariel, P., & Tessier-Lavigne, M. (2014). iDISCO: A simple, rapid method to immunolabel large tissue samples for volume imaging. *Cell, 159*, 896–910.

Richardson, D. S., & Lichtman, J. W. (2015). Clarifying tissue clearing. *Cell, 162*, 246–257.
Rothwell, P. E., Fuccillo, M. V., Maxeiner, S., Hayton, S. J., Gokce, O., Lim, B. K., et al. (2014). Autism-associated neuroligin-3 mutations commonly impair striatal circuits to boost repetitive behaviors. *Cell, 158*, 198–212.
Sanders, S. J., Ercan-Sencicek, A. G., Hus, V., Luo, R., Murtha, M. T., Moreno-De-Luca, D., et al. (2011). Multiple recurrent de novo CNVs, including duplications of the 7q11.23 Williams syndrome region, are strongly associated with autism. *Neuron, 70*, 863–885.
Schnutgen, F., Doerflinger, N., Calleja, C., Wendling, O., Chambon, P., & Ghyselinck, N. B. (2003). A directional strategy for monitoring Cre-mediated recombination at the cellular level in the mouse. *Nature Biotechnology, 21*, 562–565.
Schwarz, L. A., Miyamichi, K., Gao, X. J., Beier, K. T., Weissbourd, B., DeLoach, K. E., et al. (2015). Viral-genetic tracing of the input-output organization of a central noradrenaline circuit. *Nature, 524*, 88–92.
Sebat, J., Lakshmi, B., Troge, J., Alexander, J., Young, J., Lundin, P., et al. (2004). Large-scale copy number polymorphism in the human genome. *Science, 305*, 525–528.
Sebat, J., Lakshmi, B., Malhotra, D., Troge, J., Lese-Martin, C., Walsh, T., et al. (2007). Strong association of de novo copy number mutations with autism. *Science, 316*, 445–449.
Shin, L. M., & Liberzon, I. (2010). The neurocircuitry of fear, stress, and anxiety disorders. *Neuropsychopharmacology, 35*, 169–191.
Singh, P., Schimenti, J. C., & Bolcun-Filas, E. (2015). A mouse geneticist's practical guide to CRISPR applications. *Genetics, 199*, 1–15.
Sohal, V. S., Zhang, F., Yizhar, O., & Deisseroth, K. (2009). Parvalbumin neurons and gamma rhythms enhance cortical circuit performance. *Nature, 459*, 698–702.
Soler-Llavina, G. J., Fuccillo, M. V., Ko, J., Sudhof, T. C., & Malenka, R. C. (2011). The neurexin ligands, neuroligins and leucine-rich repeat transmembrane proteins, perform convergent and divergent synaptic functions in vivo. *Proceedings of the National Academy of Sciences of the United States of America, 108*, 16502–16509.
Soudais, C., Skander, N., & Kremer, E. J. (2004). Long-term in vivo transduction of neurons throughout the rat CNS using novel helper-dependent CAV-2 vectors. *FASEB Journal, 18*, 391–393.
Stefansson, H., Rujescu, D., Cichon, S., Pietiläinen, O. P., Ingason, A., Steinberg, S., et al. (2008). Large recurrent microdeletions associated with schizophrenia. *Nature, 455*, 232–236.
Sternson, S. M., & Roth, B. L. (2014). Chemogenetic tools to interrogate brain functions. *Annual Review of Neuroscience, 37*, 387–407.
Sudhof, T. C. (2008). Neuroligins and neurexins link synaptic function to cognitive disease. *Nature, 455*, 903–911.
Sugino, K., Hempel, C. M., Miller, M. N., Hattox, A. M., Shapiro, P., Wu, C., et al. (2006). Molecular taxonomy of major neuronal classes in the adult mouse forebrain. *Nature Neuroscience, 9*, 99–107.
Sullivan, P. F. (2007). Spurious genetic associations. *Biological Psychiatry, 61*, 1121–1126.
Thomas, K. R., & Capecchi, M. R. (1987). Site-directed mutagenesis by gene targeting in mouse embryo-derived stem cells. *Cell, 51*, 503–512.
Tsai, H. C., Zhang, F., Adamantidis, A., Stuber, G. D., Bonci, A., de Lecea, L., et al. (2009). Phasic firing in dopaminergic neurons is sufficient for behavioral conditioning. *Science, 324*, 1080–1084.
Tsai, P. T., Hull, C., Chu, Y., Greene-Colozzi, E., Sadowski, A. R., Leech, J. M., et al. (2012). Autistic-like behaviour and cerebellar dysfunction in Purkinje cell Tsc1 mutant mice. *Nature, 488*, 647–651.
Urnov, F. D., Rebar, E. J., Holmes, M. C., Zhang, H. S., & Gregory, P. D. (2010). Genome editing with engineered zinc finger nucleases. *Nature Reviews Genetics, 11*, 636–646.
Waddell, N., Pajic, M., Patch, A. M., Chang, D. K., Kassahn, K. S., Bailey, P., et al. (2015). Whole genomes redefine the mutational landscape of pancreatic cancer. *Nature, 518*, 495–501.
Walsh, T., et al. (2008). Rare structural variants disrupt multiple genes in neurodevelopmental pathways in schizophrenia. *Science, 320*, 539–543.

Weiss, L. A., Arking, D. E., Gene Discovery Project of Johns Hopkins & the Autism Consortium, Daly, M. J., & Chakravarti, A. (2009). A genome-wide linkage and association scan reveals novel loci for autism. *Nature, 461*, 802–808.

Welch, J. M., Lu, J., Rodriguiz, R. M., Trotta, N. C., Peca, J., Ding, J. D., et al. (2007). Cortico-striatal synaptic defects and OCD-like behaviours in Sapap3-mutant mice. *Nature, 448*, 894–900.

Wickersham, I. R., Finke, S., Conzelmann, K. K., & Callaway, E. M. (2007). Retrograde neuronal tracing with a deletion-mutant rabies virus. *Nature Methods, 4*, 47–49.

Yang, G., Lai, C. S., Cichon, J., Ma, L., Li, W., & Gan, W. B. (2014). Sleep promotes branch-specific formation of dendritic spines after learning. *Science, 344*, 1173–1178.

Yapici, N., Kim, Y. J., Ribeiro, C., & Dickson, B. J. (2008). A receptor that mediates the post-mating switch in Drosophila reproductive behaviour. *Nature, 451*, 33–37.

Zeisel, A., Muñoz-Manchado, A. B., Codeluppi, S., Lönnerberg, P., La Manno, G., Juréus, A., et al. (2015). Cell types in the mouse cortex and hippocampus revealed by single-cell RNA-seq. *Science, 347*(6226), 1138–1142.

Zhang, F., Wang, L. P., Brauner, M., Liewald, J. F., Kay, K., Watzke, N., et al. (2007). Multimodal fast optical interrogation of neural circuitry. *Nature, 446*, 633–639.

Ziv, Y., Burns, L. D., Cocker, E. D., Hamel, E. O., Ghosh, K. K., Kitch, L. J., et al. (2013). Long-term dynamics of CA1 hippocampal place codes. *Nature Neuroscience, 16*, 264–266.

Zoghbi, H. Y., & Bear, M. F. (2012). Synaptic dysfunction in neurodevelopmental disorders associated with autism and intellectual disabilities. *Cold Spring Harbor Perspectives in Biology, 4*, a009886.

Zuchner, S., et al. (2009). Multiple rare SAPAP3 missense variants in trichotillomania and OCD. *Molecular Psychiatry, 14*, 6–9.

Index

J.C. Gewirtz, Y.-K. Kim (eds.), *Animal Models of Behavior Genetics*,
Advances in Behavior Genetics, DOI 10.1007/978-1-4939-3777-6

V

W

Y

Z

CPSIA information can be obtained
at www.ICGtesting.com
Printed in the USA
BVOW07*1051140816
458990BV00001B/2/P